ALLE ZEIT WACH 1842

G. Rozenberg A. Salomaa (Eds.)

Lindenmayer Systems

Impacts on Theoretical Computer Science, Computer Graphics, and Developmental Biology

With 116 Figures

Springer-Verlag
Berlin Heidelberg New York
London Paris Tokyo
Hong Kong Barcelona
Budapest

Grzegorz Rozenberg
Department of Computer Science
Leiden University
P. O. Box 9512
2300 RA Leiden, The Netherlands

Arto Salomaa
The Academy of Finland
Department of Mathematics
University of Turku
SF-20 500 Turku, Finland

ISBN 3-540-55320-7 Springer-Verlag Berlin Heidelberg New York
ISBN 0-387-55320-7 Springer-Verlag New York Berlin Heidelberg

Library of Congress Cataloging-in-Publibation Data. Lindenmayer systems : impacts on theoretical computer science, computer graphics, and developmental biology / Grzegorz Rozenberg, Arto Salomaa [editors]. p. cm. Includes bibliographical references and index. ISBN 3-540-55320-7 (Berlin : acid-free paper). – ISBN 0-387-55320-7 (New York : acid-free paper) 1. Computer graphics. 2. Developmental biology. 3. L systems. I. Rozenberg, Grzegorz. II. Salomaa, Arto.
T385.L558 1992 511.3–dc20 92-14822

Printed in the United States of America

Typesetting: Camera ready by authors
45/3140 - 5 4 3 2 1 0 - Printed on acid-free paper

Preface

This is a collection of papers, mostly reporting original research, dealing with or related to L systems, which is short for Lindenmayer systems. Just as in our previous volume *The Book of L* (Springer-Verlag 1986), the range of contributions is large and shows the wide applicability of L systems.

We would like to point out two very recent developments as examples of the strength of L systems both in theoretical and practical considerations. (i) The methods and techniques invented originally in connection with D0L systems have turned out to be applicable in various areas of automata theory and combinatorics of words not at all associated with L systems. (ii) It suffices to study the book *The Algorithmic Beauty of Plants* by P. Prusinkiewicz and A. Lindenmayer (Springer-Verlag 1990) to become convinced of the significance of L systems in computer graphics.

We believe that this volume will be useful for quite a variegated audience. In addition to students and researchers working in L systems and in related areas of automata and language theory, as well as in combinatorics of words, the volume should be of definite interest also to people working in computer graphics and in developmental biology. Moreover, after reading the introductory contribution by Jürgensen and Lindenmayer that was originally intended to be published in the *Scientific American*, even a non-initiated reader should find much of the material accessible. The metatitle of this first part, In the Footsteps of L, was the working title and still is our "secret code" for the book.

Aristid Lindenmayer is known to us as the father of L systems. He was convinced that for many biological phenomena, relational and set-theoretical concepts were more suitable than those from mathematical analysis. The discrete and combinatorial nature of biological structures, everywhere from the molecular to the cellular and population levels, was obvious to him. The same seems to apply to physical structures as well and, therefore, it is apparent that new tools and models are needed for practicing scientists.

L systems were first visualized as everywhere expanding arrays of finite automata. Such arrays provided realistic simulations of growing cellular filaments as found in algae, mosses and fungi. The model was suitable for describing the global development of multicellular structures on subunits present where the growth increments are applied simultaneously in each derivation step. The first of these requirements is needed because there are natural subunits, each carrying the same genetic material, in developing organisms. The second requirement is necessary because one cannot apply local productions here and there in an asynchronous manner: time runs in the same way in every cell. Instead of arrays of finite automata, models from formal language theory turned out to be more economical, especially for computer implementations. We all know the success of L systems in computer graphics, as well as the fact that by now several books and about a thousand papers exist dealing with the mathematical aspects of the theory alone. Lindenmayer's invention has turned out to be one of the most beautiful examples of interdisciplinary science: work in one area has induced most fruitful ideas and results in other areas.

It is a great pleasure for us that almost without exception the leading authorities in the field have been able to contribute to this volume. We are pleased to have also many of the most promising young researchers in the area among the contributors.

We express our gratitude to all contributors. Special thanks are due to Juha Honkala for suggesting the working title of the volume, from which the part metatitles are derived, as well as to Springer-Verlag, in particular Dr. Hans Wössner and Mrs. Ingeborg Mayer, for valuable suggestions and continuing good cooperation.

It is very difficult, if not impossible, to find such a combination of a great scientist and warm, modest personality as Aristid Lindenmayer. He was always most helpful and ready to discuss also with beginners all aspects, in particular, biological implications.

We dedicate this book to the memory of Aristid Lindenmayer (1925–1989). In doing so we are thinking also of his wife Jane and of her importance in creating the atmosphere for Aristid's unique work.

February 1992 Grzegorz Rozenberg Arto Salomaa

Table of Contents

1. In the Footsteps of L:

Exposition of L Systems

2. The Main Path Continues Further:

Studies of Basic L Systems

5. Enjoying Flora Along the Path:

Biological Aspects and Models

6. Crossroads and Branching Path:

Variations and Generalizations of L Systems, Related Models

1. In the Footsteps of L:
Exposition of L Systems

Grammars of Development: Discrete-State Models for Growth, Differentiation, and Gene Expression in Modular Organisms[1]

A. Lindenmayer and H. Jürgensen[2]

[2] Department of Computer Science, The University of Western Ontario, London, Ontario, Canada, N6A 5B7.

Abstract: L systems were introduced to model biological development. We discuss some of the related biological and general scientific issues and survey some of the results and applications.

Keywords: L system, plant modelling, gene control, developmental biology.

Preface by H. Jürgensen: This paper is a survey of some ideas and results concerning L systems and their applications, which Aristid and I were preparing, originally with a non-specialized journal in mind. The first idea to write this paper resulted from a discussion which we had with A. K. Dewdney, a colleague of mine who writes the Computer Recreations articles for the Scientific American, during one of Aristid's visits at The University of Western Ontario about 7 years ago. Since then and until the fall of 1988, Aristid and I met on several occasions in London and Utrecht to complete this paper, which in many ways turned out to be a greater challenge than a normal research paper for a readership of specialists. The final revisions were planned for the first half of 1989, but never got done.

When the call for contributions to this volume was issued I suggested to include this paper, which Aristid had been very fond of. Indeed, the editors as well as Jane Lindenmayer, whom I asked for an opinion and permission, found this was a good idea.

Thus, in the spring of 1991 I revised our draft. Since I was on my own in decisions now and without the help of those stimulating and exciting discussions in which Aristid and I had sometimes struggled even with single words, I resisted from making any really major changes except updating or eliminating some obsolete information. Hence the paper as revised for this publication is probably closer to the draft on which it is based than originally planned. Nevertheless, I hope that the present text essentially reflects what both Aristid and I had in mind when we last talked about it.

1. Gene Control in Development

The development of multicellular organisms presents formidable theoretical difficulties. Even if we think only of the combinations of active genes which can be present in different cells of the same organism, we realize that we are dealing with immense numbers. And then we have not considered yet the possible cytoplasmic and membrane configurations for all its cells. Clearly, for any kind of attempt to describe development in terms of gene expression and molecular mechanisms of differentiation, we have to accept severe abstractions. One of the abstractions which is widely discussed is the concept of a

[1] This work was supported by the Natural Science and Engineering Research Council of Canada, Grant OGP0000243.

'program' or 'grammar' for development. Developmental programs or grammars would provide a framework within which all the facets of observations could be organized.

In favour of such concepts, Sydney Brenner [16] puts the problem in sharp focus as follows: "The total explanation of all organisms resides within them [the genes], and you feel there has to be a grammar in it somewhere. Ultimately, the organism must be explicable in terms of its genes, simply because evolution has come about through alterations in DNA. But the representation will not be explicit. We need to understand the grammar of development to make sense of it." But he adds cautiously: "We tend to talk loosely about genetic programs and we should be careful about the implications of this language, even when used metaphorically."

A sample of an opposing view, that of Gunther Stent [20], is: "An essential ideological component of the genetic approach to development is the view that development follows a 'program,' and that this program is embodied in the genes. But this view is rooted in a semantic confusion about the concept of 'program.' Once that confusion is cleared up, it becomes evident that development from egg to adult is unlikely to be a programmatic phenomenon." Stent bases his opinion on a definition of programmatic phenomena which requires "that in addition to the phenomenon itself, there exists a second thing, the 'program,' whose structure is isomorphic with, that is, can be brought into one-to-one correspondence with, the phenomenon." Considering development, "it is most unlikely that the overall sequence of its events is isomorphic with the genome."

It appears that talking about developmental programs or grammars gets one immediately embroiled in the controversy of how much control genes have over development, as well as in possible misunderstandings about the meaning of those terms. The controversy about gene control in development is fought on many fronts, from purely technical to general philosophical levels—the latter, since it has direct bearing on reductionist and anti-reductionist standpoints in biological research. We do not wish to go into the substance of this question here (one of us, A. L., has published on the reduction problem before). It is sufficient to say that we think that genes as well as cytoplasmic and membrane components participate in developmental control, and it is rather meaningless to try to assign primary responsibility to one or the other set of elements. The fact that all cellular components must eventually be coded in DNA does not mean that all cellular processes are under its direct control. For instance membrane properties like polarity are propagated from cell to cell without DNA involvement. In general spatial aspects of development often cannot be directed by on-off states of genes; they have to be specified by pre-existing cytoplasmic or wall (membrane) structures. On the other hand, it is difficult to underestimate the rôle which genes play in development. We take into account both the 'genetic' and 'structural' control elements in our models.

The main question we wish to address is whether a 'program' or 'grammar' can be defined and used profitably in the study of multicellular development "from egg to adult organism." It is quite clear that if we speak of a 'program'—a set of rules to be executed—then these rules must be applied to many different cells or groups of cells (clones) simultaneously in the developing organism. The various parts of the 'program' would be executed in various parts of the organism simultaneously and we certainly would not expect any simple one-to-one correspondence between the 'program' and the different developmental processes taking place. It is because of this distributed character of the multicellular development that the term 'grammar' may be better applicable. A 'grammar' is also a set of rules, but it specifies substitutions of subunits of a structure by other subunits or groups of subunits. In formal language theory, structures are 'words' and the subunits are 'letters;' a grammar specifies possible transitions from a word to some other words by replacement of one or more of the letters by single letters, by groups of letters, or by erasing a letter. Grammars have been defined for structures other than just strings of letters as well, for instance for arrays, graphs, and maps.

Grammar-like constructs have been defined and used by us to describe development. They have come to be called *L systems* and have been the object of a considerable amount of mathematical work. They are distinguished from grammars (in the sense of Chomsky grammars) in two ways: all subunits are transformed simultaneously in a structure in every derivation step; there is no distinction between terminal and non-terminal subunits. These two differences from grammars in the usual sense are quite important mathematically. Thus the introduction and subsequent study of L systems has led to new questions in formal language theory and to deeper insights into language generating mechanisms such as grammars.

L systems are discrete mathematical constructs. They are discrete in three senses: the state transformations are defined on discrete subunits (cells); each subunit may be present in one of a finite set of states; and the transformations are performed in discrete time steps. The discreteness of subunits need not be defended in the case of biological models: cells and some multicellular structures (like internodes or apices in plants, segments in insects) are naturally discrete units. The fact that cellular states are chosen from a finite set is also defensible on the grounds that they represent partly discrete combinations of active genes and partially cytoplasmic states which are usually equilibrium states of metabolic components. The least obvious is the requirement that there should be discrete time steps in these models. However, the alternative would be to deal with an unmanageable number of differential equations for which no tractable mathematical manipulation is available. The choice of a suitable length for the time steps presents usually no problem as most developmental processes have a natural time unit (for instance the cell cycle length or the plastochron interval).

An important component of all developmental control is by gene activity. As cells divide, active genes in the daughter cells may be the same as those in the mother cell or different between the two daughter cells and/or the mother cell. Change in gene expression may be brought about by the activity of other genes in the same cell or by inducers or inhibitors produced in other cells. This distinction gives rise to a basic dichotomy of developmental control: on the one hand, control of gene expression by cell lineages, and, on the other hand, control by cell interactions. This dichotomy plays an important rôle in our definitions and conclusions.

2. L Systems as 'Grammars' for Development

We start defining the simplest kind of L systems, those for development of essentially one-dimensional (filamentous) organisms where control is entirely by cell lineages. These are called *string 0L systems*, 'string' referring to sequences of cellular states and '0L system' being an abbreviation of an 'L system with zero-sided interactions.' The cellular states are at first not separated into states of active genes, of cytoplasm and of cell walls; we shall consider these later. There is a tacit assumption of overall polarity when we denote filaments by strings of symbols. This polarity could be expressed by placing arrows between the symbols as in: $\ldots \rightarrow b \rightarrow a \rightarrow c \rightarrow a \rightarrow \ldots$ But we write simply: $\ldots baca \ldots$

A string 0L system consists of a finite set Σ of symbols (the *alphabet*) standing for cellular states, a finite set P of *state transition rules* (productions), and a *starting string* α. The rules in P are of the form $a \rightarrow w$ where $a \in \Sigma$ and $w \in \Sigma^*$; a rule $a \rightarrow w$ represents the transformation of the cell state a into the string w of cell states.[3] State transition rules allow a cell to remain in its state, to change its state, to divide into several cells with specified states in a given order, or to disappear from the string. Figure 1 shows some of these transitions with the corresponding rules.

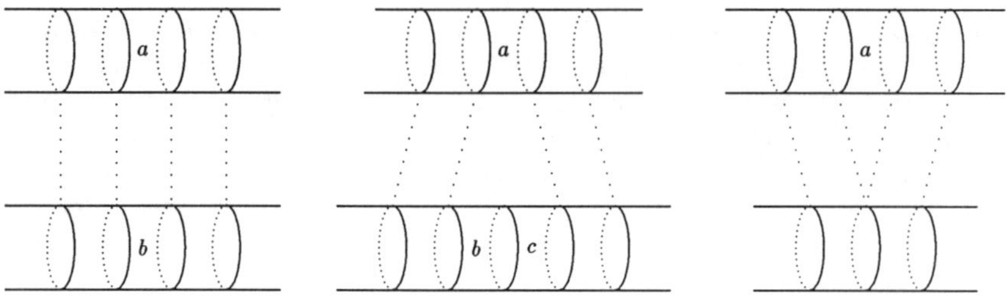

$a \rightarrow b$ $a \rightarrow bc$ $a \rightarrow \Lambda$

Figure 1. Three state transitions and their rules.

Development starts with a single cell or several cells in states given by the string α. It takes place in steps called *derivation steps*. A derivation step from string w consists of replacing each occurrence of a symbol in w by a string according to a transition rule and of concatenating the resulting strings in the proper order. Different selections of transition rules for the same symbol may lead to different possibilities for a string to result from w in a derivation step. Thus, derivation steps may be non-deterministic.

A derivation consists of any number of consecutive derivation steps. If there is a single transition rule for each state symbol then the string produced in each derivation step is unique, and the system is called a *deterministic 0L system* (*D0L system* for short). Such a system produces a single infinite derivation sequence which represents the *global* behaviour of the system resulting from the *local* transition rules. Non-deterministic 0L systems produce many derivation sequences. For these systems the set of all strings which could be produced is the only meaningful expression of the global behaviour; such a set is called the *language* of the given system. This terminology is motivated by analogous constructions in the theory of formal languages.

3. Structural Properties of D0L Systems and Sequences

We consider now a simple example of a D0L system. Let us assume that cells of a certain filamentous organism are either in a dividing state a or in a waiting state b one

[3] Σ^* is the set of all strings that can be formed from symbols in Σ, including the string with no symbols, the *empty string* Λ.

time step away from division. We further assume that each cell has a wall state and an internal polarity indicated by an arrow. The transition rules are

$$\overrightarrow{a} \rightarrow \overleftarrow{a}\,\overrightarrow{b}\,, \quad \overrightarrow{b} \rightarrow \overrightarrow{a}\,, \quad \overleftarrow{a} \rightarrow \overleftarrow{b}\,\overrightarrow{a}\,, \quad \overleftarrow{b} \rightarrow \overleftarrow{a}\,,$$

and the starting string is $\overrightarrow{b}$.

These rules are to be interpreted as follows: if a cell with a given polarity divides it produces two daughter cells in such a way that they occur in the order a–b in the direction of the mother cell's polarity, and the sister cell in state a has the opposite polarity while the sister cell in state b has the same polarity as the mother cell. This can be seen in Figure 2 where development starts with a cell in state b with polarity directed to the right. In the next step, this cell is transformed to one in state a with the same polarity according to the second rule. In the following step division takes place with states and polarities of the daughter cells assigned by the first rule. This procedure can be continued indefinitely.

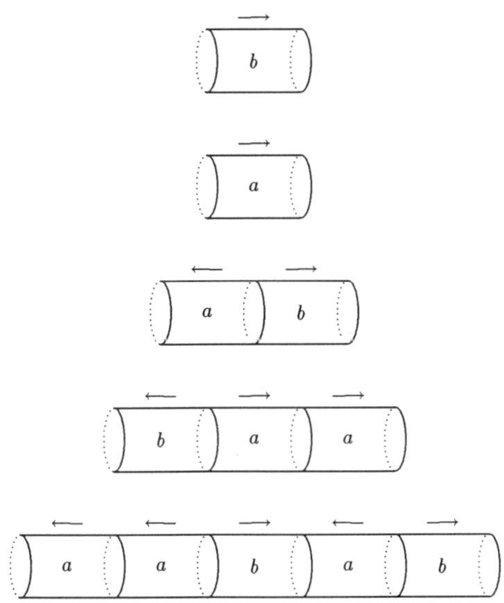

Figure 2. Example of a D0L system.

It is interesting to note that in this case the number of cells increases according to the main Fibonacci sequence, defined as the sequence $f(n) = f(n-1) + f(n-2)$ and named after Leonardo Pisano Fibonacci.[4] The polarity changing rules of this D0L system are

[4] This sequence is obtained as the solution of an exercise in Fibonacci's *Liber abbaci* [9], p. 283, of 1202 and 1228 concerning the offspring of a single pair of rabbits in one year.

the same as the rules found with high regularity in the blue-green bacterium *Anabaena catenula* [18]. The timing of cell divisions in the filaments is different, however. The ratio of cell cycle time between a and b cells in *Anabaena* is 4:5, while in the above system we assume a ratio of 1:2 for the sake of simplicity. Other polarity transforming rules, for instance, the set

$$\overrightarrow{a} \to \overrightarrow{a}\,\overleftarrow{b}, \quad \overrightarrow{b} \to \overrightarrow{a}, \quad \overleftarrow{a} \to \overrightarrow{b}\,\overleftarrow{a}, \quad \overleftarrow{b} \to \overleftarrow{a}$$

have been found to hold for the green algae *Chaetomorpha linum* and *Ulothrix aequalis* [17].

One can consider branching filaments as well and show that similar transition rules can be used. Branching is indicated by special branching symbols—brackets, for instance—which have to be included in the alphabet.

If the transition rules are

$$\overrightarrow{a} \to \overrightarrow{c}\,[\,\overrightarrow{b}\,]\,\overrightarrow{a}, \quad \overrightarrow{b} \to \overrightarrow{a}, \quad \overrightarrow{c} \to \overrightarrow{c}, \quad [\,\to [, \quad]\to],$$

and the starting string consists of b, then a developmental sequence as shown in Figure 3 results. Cells in states a and b are apical cells, and those in state c are basal cells. In this case, polarity does not change directions in any of the cells. We can assume, for instance, that all cells have polarity from base to apex and the corresponding arrows are placed accordingly. The brackets in the rules determine attachment of each branch to its basal cell, but not its geometric position according to the subtended angle, or the angle along the circumference (we placed the branches arbitrarily right and left to their axes in the drawings). Omitting the arrows, the last structure shown in Figure 3 can be written in string form as $c[c[b]a]c[a]c[b]a$. This string is transformed in one derivation step according to the given D0L system into the string $c[c[a]c[b]a]c[c[b]c]c[a]c[b]a$.

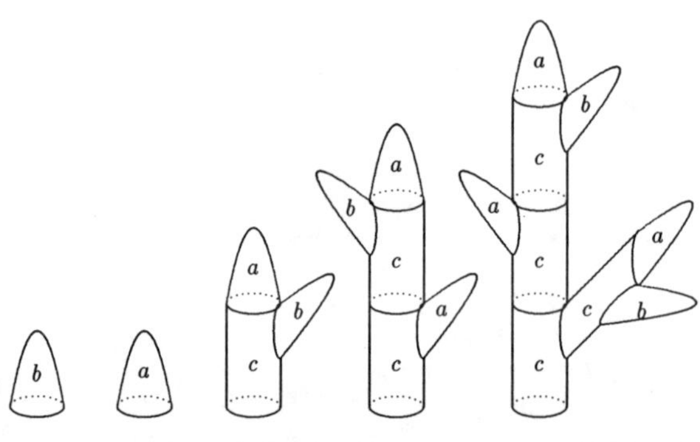

Figure 3. Developmental sequence of a branching filament.

Many branching filaments with such structures can be found among lower plants, for instance in the red alga *Callithamnion roseum* [15]. The precise shape of a plant depends on delays in the development of branch filaments with respect to the axis on which they were borne and possibly on stochastic factors influencing individual cells or branches. But the main principle to be noted is that all branch apices repeat essentially the same development.

To capture the generation of repeating structures in development, a feature which is more common in plants than in animals, one studies the recursive behaviour of D0L systems. Recursion can be found in the strings of the previous example by observing that in each string which is generated the entire previous string is repeated in the right-hand portion, and the string of two steps in the past is repeated between brackets. This can be expressed by the recursive formula

$$S_n = c\,[\,S_{n-2}\,]\,S_{n-1}$$

which holds for every derived structure at step n, $n > 1$, (the starting structure is at step 0). In this formula, three state symbols appear besides the recursive terms. The state symbols c, [, and] are constants in this recursion (they are states generating only themselves).

In general, every D0L system can be written as a set of recursive formulae. This is trivially true since every transition rule can be considered to be a recursive equation. For instance, the D0L system of the example in Figure 2 can be directly re-written in the form of four recursive equations:

$$\overrightarrow{A}_n = \overleftarrow{A}_{n-1}\overrightarrow{B}_{n-1}, \quad \overrightarrow{B}_n = \overrightarrow{A}_{n-1}, \quad \overleftarrow{A}_n = \overleftarrow{B}_{n-1}\overrightarrow{A}_{n-1}, \quad \overleftarrow{B}_n = \overleftarrow{A}_{n-1}.$$

These equations, together with the initial conditions

$$\overrightarrow{A}_0 = \overrightarrow{a}\,, \quad \overrightarrow{B}_0 = \overrightarrow{b}\,, \quad \overleftarrow{A}_0 = \overleftarrow{a}\,, \quad \overleftarrow{B}_0 = \overleftarrow{b}$$

define four string sequences of which the one starting with $\overrightarrow{B}_0$ corresponds to the original D0L-sequence. Let us note that the same D0L-sequence can be described by reducing the number of recursive equations to two

$$\overrightarrow{B}_n = \overleftarrow{B}_{n-1}\overrightarrow{B}_{n-2}, \quad \overleftarrow{B}_n = \overleftarrow{B}_{n-2}\overrightarrow{B}_{n-1}$$

with the initial conditions

$$\overrightarrow{B}_0 = \overrightarrow{b}\,, \quad \overrightarrow{B}_1 = \overrightarrow{a}\,, \quad \overleftarrow{B}_0 = \overleftarrow{b}\,, \quad \overleftarrow{B}_1 = \overleftarrow{a}\,.$$

If a set of recursive equations can be reduced to a single one, then the corresponding D0L system is called *locally catenative*. In such a case, every string is entirely composed of strings which occurred previously in the derivation. This means that the system of Figure 3 is not locally catenative because of the constants present in the recursive formula.

It is a long-standing open problem whether there is an algorithm which, for a given D0L system, decides the question: is the system locally catenative or not? This problem appears to be at least as difficult as the D0L-equivalence problem which is discussed below. A simple result in this context states that a D0L system with a state-graph such that all loops in the graph pass through a single symbol is locally catenative.

The existence of a single or a few recursive formulae for a string generating system means that each additional structure can be generated by combining previously obtained structures in a specific order. Thus we have two different ways in which a sequence of growing structures could be specified: by local state-transition rules or by the global composition rules. In observing living organisms, one sees the global repeating structure more readily as this does not require any knowledge of development in time. From these larger and larger repeating units one can infer that the underlying developmental scheme is most likely controlled by cell lineages and not by cell interactions.

It is easy to see that different D0L systems can generate the same sequence of structures. For example, take the following systems G_1 and G_2 with the same alphabet $\{a, b, c\}$:

$$G_1 : a \to a, \quad b \to bc, \quad c \to c; \qquad G_2 : a \to ca, \quad b \to b, \quad c \to c.$$

One can easily verify that these systems derive the same infinite sequence of strings from ba as the starting string, but that the derivations are quite different.

We call two D0L systems with this property *sequence-equivalent.* The general task to decide the sequence-equivalence of two different systems is rather difficult. In fact, after this question has been first proposed in 1972, in the 'L system problem book' of Århus, it took five years and many attempts until the decidability of this question was proven by K. Culik II and I. Fris [5]. This means that an algorithm exists which, after a guaranteed finite number of steps will produce a 'yes' or 'no' answer to the equivalence problem for two arbitrary D0L systems. The upper bound of the number of derivation steps which have to be tested, may be very large in terms of the number of symbols in the alphabet. To illustrate the problem, we show two D0L systems which generate the same strings for two initial steps, and their results diverge afterwards:

$$G_1 : a \to abb, \ b \to aabba; \qquad G_2 : a \to abbaabb, \ b \to a.$$

The starting string is ab. From just looking at these systems one could not predict their equivalence or non-equivalence.

4. Ehrenfeucht's Conjecture and the HD0L Equivalence Problem

The solution of the D0L equivalence problem has stimulated a great deal of further mathematical work on the combinatorial behaviour of strings under homomorphisms—the transition rules of a D0L system define a homomorphism of the set of all strings over the alphabet into itself.

For instance, a D0L sequence can be transformed into a new sequence by applying a homomorphism to each of its members. The resulting sequence is called an HD0L sequence. Clearly, the class of all HD0L sequences contains the class of D0L sequences—in fact, it is far larger. Sequences obtained in this way by the application of a homomorphism can often be interpreted as corresponding to incomplete biological observations, that is, to observations which cannot distinguish certain cellular states.

The HD0L sequence equivalence problem is to decide whether two given D0L systems G_1 and G_2 and two homomorphisms h_1 and h_2 define the same sequence—as illustrated in Figure 4. A few years ago a well-known mathematical problem called the *Ehrenfeucht conjecture* was settled. As a consequence of this result we know that the HD0L sequence equivalencce problem is decidable.

$$\begin{array}{ccccc}
G_1\text{-sequence} & \text{homomorphic images} & & & G_2\text{-sequence} \\
w_1 & \longrightarrow h_1(w_1) \stackrel{?}{=} & h_2(v_1) & \longleftarrow & v_1 \\
\Downarrow & & & & \Downarrow \\
w_2 & \longrightarrow h_1(w_2) \stackrel{?}{=} & h_2(v_2) & \longleftarrow & v_2 \\
\Downarrow & & & & \Downarrow \\
\vdots & \vdots \quad \vdots & \vdots & & \vdots
\end{array}$$

Figure 4. The HD0L sequence equivalence problem.

The Ehrenfeucht conjecture made a rather surprising claim: every set S—possibly infinite and of arbitrary complexity—of strings over a finite alphabet possesses a finite subset T which can be used as a "test set" for S; that is, if g and h are any two homomorphisms then, in order to decide whether $g(x) = h(x)$ holds true for all $x \in S$, it is already sufficient to check this fact on T.

This conjecture was proved to hold by M. H. Albert and J. Lawrence [1] and indepenently also by V. S. Guba [11]. An earlier result [6] said that the existence of finite test sets for the sets of strings generated by D0L systems would imply that these test sets could be constructed effectively and, thus, that the HD0L sequence equivalence problem would be decidable. The proof of the Ehrenfeucht conjecture has generated several exciting mathematical consequences. A brief survey of some of these was given by K. Culik II and J. Karhumäki [7].

5. Adult Structures of 0L Systems

Development with continued growth is easier to program than one with terminating growth. The latter leads to "adult" organs or "adult" organisms. Animals as whole organisms often have an adult phase while plants usually don't. On the other hand, in plants there are many adult organs—like leaves or flowers—which arise during their development.

An adult structure is a structure which remains globally the same while its parts may be regenerated. Abstractly it is defined as a structure which generates only itself. The *adult set* of a string 0L system is the set of all strings which derive only themselves and which are derived from the starting string. Adult strings may be composed of cell states which give rise to themselves only, but they may also be composed of cells which are displaced and die while others are produced—as in the case of stem cells. The simple case shown in Figure 5 illustrates the latter possibility.

Let us now take an example of leaf development. We consider the margin of a leaf which is bilaterally symmetric with respect to the midrib. The row of cells on the margin is to be generated by a non-deterministic 0L system with the rules:

$$a \rightarrow kbk, \quad b \rightarrow jbj, \quad b \rightarrow jcj, \quad c \rightarrow c, \quad j \rightarrow j, \quad k \rightarrow k.$$

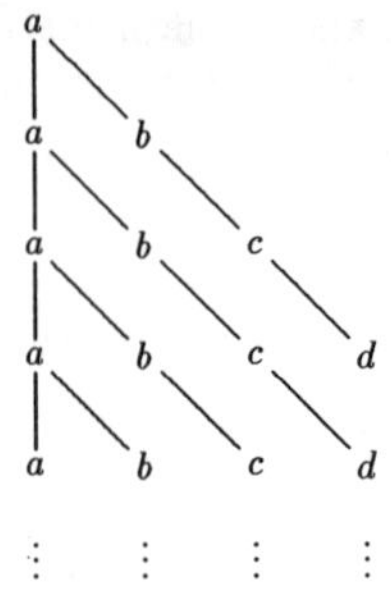

Figure 5. Adult strings with dying cells produced by the rules $a \to ab$, $b \to c$, $c \to d$, and $d \to \Lambda$.

The first few steps of possible derivations starting with a are shown in Figure 6. The set of these strings, the adult set of this system, is thus $\{kj^ncj^nk \mid n \geq 1\}$. The cells in states b or c represent the tip of the leaf, the ones in state j are the margin cells and those in state k are on the petiole (see Figure 7). Note that the same number of j cells is present on both sides of every leaf whose growth has terminated—indicated by the exponents n in the set expression. As long as a cell in state b is present at the leaf tip it generates a new pair of j cells in each step, and as soon as its further growth is terminated by its transition to state c the number of j cells and the size of the leaf are fixed.

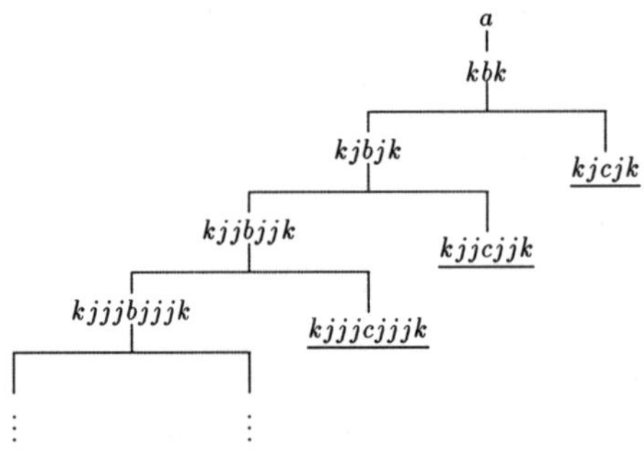

Figure 6. Leaf development, the first few derivation steps. The strings which generate only themselves are underlined.

This very simple case serves as an example for a very large class of problems concerning interactionless development. The general principle is that in strings unbounded and synchronously terminating growth can take place at most pairwise, and not in triples, quadruples, and so on. This general result—known in formal language theory as the "pumping lemma for context-free languages"—is a consequence of the fact that a growth centre somewhere along the string can produce similar structures and stop growing both of them at the same time at most in two directions. In other words, a growth centre can "pump" specific terminal sequences only singularly or in pairs along a string. In order to produce more than two synchronized sequences one would need to have several growth centres which exchange signals. The reason that we can apply the pumping lemma to 0L systems is because of a theorem proved by G. T. Herman and A. Walker [13] which states that the class of adult sets of 0L systems is precisely the same as the class of context-free languages—generated by context-free Chomsky grammars.

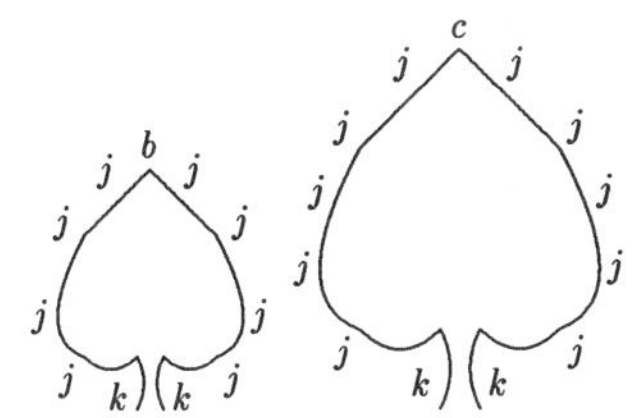

Figure 7. Leaf margin generated by a 0L system.

To illustrate the meaning of the pumping lemma we mention a result quite remote from biological applications: it can be used to show that context-free—that is, interactionless—grammars are not sufficient as tools for defining high-level programming languages such as Pascal, ALGOL, ADA, and so on. One of the features which cannot be modelled in a context-free manner is the requirement that variables be declared and and that they be used consistently with their declared types. This points to an inherent complexity in programming languages and their compilers.

We should note that the context-free restriction of unbounded and synchronized growth to pairs is also a consequence of the one-dimensional nature of strings. If one were dealing with a branching system in which a growth centre is located at the base of three or more branches, it could produce similar elements synchronously in all the branches. Such a development could not be expressed in string notation; one would need a graph rewriting system instead. This kind of development is actually hardly ever observed in plant branching where growth centres are mostly located at the apices of branches rather than at their bases.

The theorem of Herman and Walker and the pumping lemma imply that a set of the form $\{a^n b^n c^n \mid n \geq 1\}$ cannot be the adult set of any 0L system. Such an adult set would consist of all strings composed of three equally long segments. Its generation would correspond to the development of filamentous organisms with terminal stages of this kind but with various lengths. This is the kind of development that is meant by the 'French flag pattern' [22]. Originally this pattern generation problem was proposed by Wolpert with regard to the development of *Hydra*, that is, to the mechanism by which the proportions of the head, body, and foot portions of *Hydra* are formed and maintained. Besides asking how segments of constant proportions are formed, independently of the size, the French flag problem also asks for a mechanism which would provide for the regeneration of segments with the same proportions after one of the original structures has been subdivided into smaller ones.

By the above result it is clear that no cellular array with finitely many states and without interactions grown from a few cells could produce such 3-partite strings of arbitrary lengths. It is also clear that the regeneration of such strings cannot be accomplished without cell interactions. In fact, using cell interaction Wolpert proposed a mechanism of producing and maintaining gradients along the cell arrays as a solution to the French flag problem. There are also other mechanism available to achieve this, such as signal propagation along the strings.

An adult set of bi-partite strings such as $\{a^n b^n \mid n \geq 1\}$ can of course easily be produced by a 0L system. The same is true for adult sets with two equal parts and any number of additional parts with independent lengths such as $\{a^n b^m c^n \mid m, n \geq 1\}$. We should point out that sets with 3 equal parts such as $\{xa^n yb^n zc^n \mid n \geq 1\}$ can be produced by a 0L system, in fact a deterministic one. However, such a set cannot be the adult set of a 0L system.

The above mentioned mathematical result is just one example of 'characterization' theorems which are available for deterministic as well as for adult 0L sets and 0L sequences. They provide guidance in the choice between different types of developmental mechanisms for various organisms. Such results provide insights into the possibility or impossibility of an underlying cell lineage or cell interaction mechanism, as well as into the need for deterministic or stochastic developmental models, or for taking into account environmental control factors.

6. Cell Lineages

A few years ago a monumental effort has been made to describe cellular descent and movement in the embryonic development of a nematode, *Caenorhabditis elegans*, identifying the lineages of all 671 cells which are present at the end of embryogenesis. These cells are programmed to differentiate terminally, or die, or—for post-embryonic blast cells—to give rise to specific tissues [21]. This is an addition to many cell lineage studies of which outstanding examples are known in the case of molluscs and insects among vertebrates, and of mosses and ferns among lower plants. In which way can the theoretical approach outlined above be applied in order to obtain new insights from such lineage studies?

Of course the main question concerns the underlying control factors. The nematode study has been undertaken primarily to search for the genetic factors involved in cell determination and differentiation. "... it took some time to appreciate that cells are the units of development and that we must find out how the genes get hold of the cells"—S. Brenner, in his description of this project [3].

The problems involved are manifold. First of all, having found a repeatably uniform cell lineage pattern in a certain organism points toward but does not necessarily mean that, therefore, genetic programming of development takes place without cell interactions. As we remarked before one can imagine two fundamentally different ways in which the genomic state—the combination of active genes—of a cell can change during the life of a cell or during cell division: either genes are repressed or activated by products of other genes in the same cell, or by products of genes in other cells which are then transported to the first cell. One usually assumes the first alternative in the case of regularly occurring cell lineages, that is, that changing cell states and unequal cell divisions are due to internal factors. Of course, this need not always be the case: they could possibly be caused by signals from neighbouring cells. Evidence to distinguish between these possibilities, such as from killing or removing the neighbour cells, is sometimes difficult to obtain.

Finding the control genes is the next problem. By 'control genes' we mean those genes which channel a cell or the entire progeny of a cell into a differentiation pathway. Good examples of such developmental control genes are the segmentation genes and the homeotic genes like those in the *Bothorax* and *Antennapedia* complexes of *Drosophila*— the latter containing the "homeo box" sequences as common units (see [10]). The protein transcripts of one of these genes have been detected in just those sections of the wild-type embryos (in seven evenly spaced belts of blastoderm cells) which are missing from embryos with a mutant genotype (such as the *fushi tarazu* mutant (see [12]).

Evidence of this kind is hard to come by for organisms with known lineage histories. It is possible that the number of genes involved is rather small. With a few control genes in active or in inactive form a large number of differentiation states can be determined— for instance, with only 10 genes one could obtain $2^{10} = 1024$ different combinations. Repetitions or other regularities in the lineage diagrams can point to cells in which the same gene has become activated. In Figures 8 and 9, as an example of such patterns we show a small portion of the lineages of the wild-type and mutant *Caenorhabditis elegans* individuals (from V. Ambros and H. R. Horwitz [2]). Lineages are shown which begin with a tail seam cell (T) in hermaphrodite (⚥) and male (♂) embryos of the nematode *Caenorhabditis elegans*. They are shown for the wild-type as well as for embryos with lin-14 or lin-28 heterochronic mutant genes. For each lineage a D0L system is provided indicating the cell states which are necessary to generate the sublineages designated by P, Q, U, V, and W. We see that with relatively few states (some of which occur in different positions) one can derive the given patterns. As pointed out by the authors, a comparison of these lineages indicates that the mutant gene (lin-14) has the effect of switching on the same sublineage block again and again which in the wild-type occurs only once.

The study and comparison of complex lineages can be efficiently carried out by assigning state transitions to the cells with or without interactions, as described above, and by applying some of the powerful characterization theorems for 0L systems and their derivatives.

7. Growth Functions

For each deterministic (or stochastic) 0L system a *growth function* can be defined. The growth function $f(n)$ of a D0L system is simply the (expected) number of subunits present in the n-th string generated. The growth function $f(n)$ can be directly obtained by finding the *growth matrix* M of the system and raising this matrix to the n-th power.

The growth matrix contains the number of occurrences of each symbol in the rules as illustrated in the example of Figure 9. In addition, the row vector π contains the symbol frequencies in the starting string, and the column vector η contains 1 in every position. Then the growth function $f(n)$ can be expressed as $f(n) = \pi M^n \eta$. From such a matrix expression one can always obtain an analytical formula of the form $f(n) = \sum_i p_i(n) c_i^n$ where each $p_i(n)$ is a polynomial and each c_i is a constant. In fact, these constants are the eigenvalues of the matrix M. This implies that D0L systems can be classified as having exponential, polynomial, or bounded growth. The example of Figure 9 has

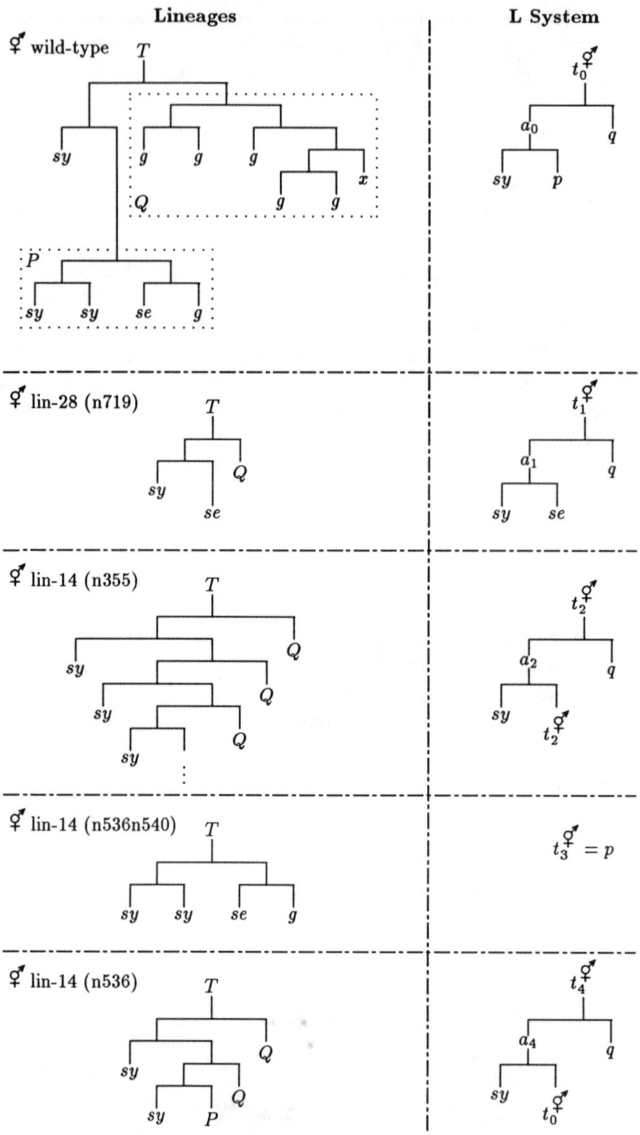

Figure 8. Lineages and D0L systems for the tail seam cell (T) in hermaphrodite ⚥ embryos of the nematode *Caenorhabditis elegans*.

polynomial growth with a single eigenvalue which is equal to 1 and a formula $f(n) = (n^2 + 2n + 2) \cdot 1^n$. The growth matrix for the branching system in Figure 3—omitting the branching indicators—is

$$\begin{pmatrix} 1 & 1 & 1 \\ 1 & 0 & 0 \\ 0 & 0 & 1 \end{pmatrix}.$$

This matrix has three eigenvalues (the c_i):

$$\frac{1+\sqrt{5}}{2}, \quad 1, \quad \text{and} \quad \frac{1-\sqrt{5}}{2}.$$

The first three values of the growth function are $f(0) = 1$, $f(1) = 1$, and $f(2) = 3$. From these the p_i can be computed, these being constants in this case. The full growth function is

$$f(n) = \left(1 + \frac{1}{\sqrt{5}}\right) \cdot \left(\frac{1+\sqrt{5}}{2}\right)^n - 1^n + \left(1 - \frac{1}{\sqrt{5}}\right) \cdot \left(\frac{1-\sqrt{5}}{2}\right)^n.$$

The above classification result also implies that any observed cellular growth process with a growth function which is neither exponential nor polynomial nor bounded cannot be programmed by a D0L system. Growth functions such as logarithmic or certain subexponential functions can, however, be programmed by deterministic L systems with cell interactions.

If a growth matrix can be obtained for an extensively branching plant then its global growth characteristics and its productivity in terms of amount of leaf area, fruits, etc. can be calculated according to the above method. This approach is used in studies on "plant demography," where the modular components of plants, like apices, internodes, leaves, flowers, are the subunits counted. Growth functions can be obtained not only for individual plants in this way but also for entire stands or crops. In this connection *Leslie-matrices* are used frequently; they are related to our growth matrices but may have also non-integer elements, and they have a special shape. It should be pointed out that this "demographic" approach can only be used if there is no interaction among the subunits.

General non-negative and possibly non-integer matrices arise in connection with stochastic 0L systems. Such growth functions are also considered in the theory of (mathematical) branching processes.

8. Advanced L Systems

L systems are applied extensively in certain areas of computer graphics, modelling, and animation. For a discussion of some aspects of this application see [4]. The recent book [19] contains many details and many beautiful images generated using L systems.

Complex branching structures like that of the flowering development of *Mycelis muralis*—the wild lettuce [14]—require interaction between parts. To model these one uses interactive L systems. In the case of *Mycelis muralis,* flowering signals and apical inhibitors are transmitted along the branches.

Multidimensional development like that of cell layers can also be modelled using D0L production rules. While the modelling of the cellular filaments can be achieved

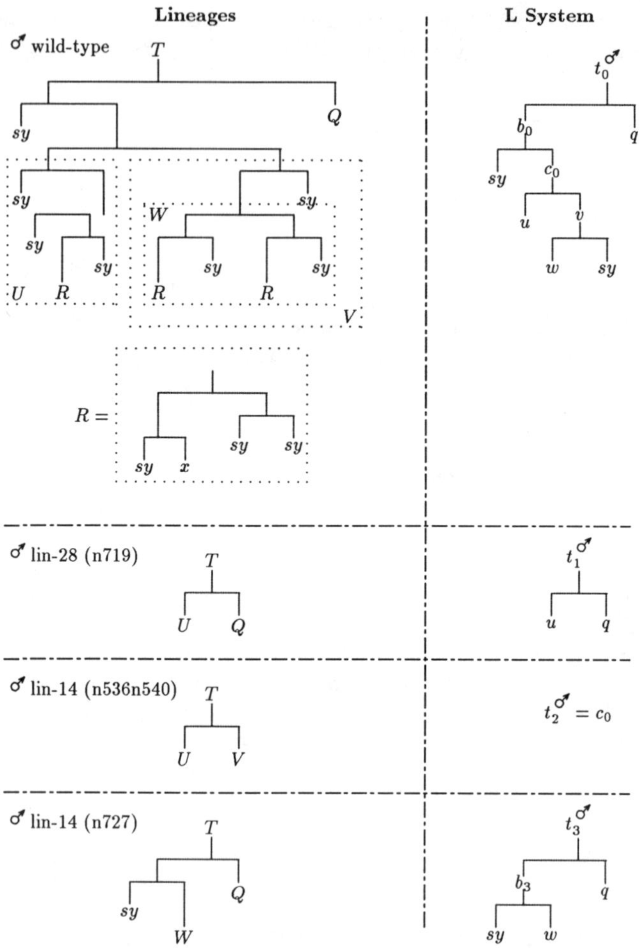

Figure 9. Lineages and D0L systems for the tail seam cell (T) in male ♂ embryos of the nematode *Caenorhabditis elegans.*

by transforming strings of symbols, and the modelling of branching structures can be accomplished by considering string expressions with brackets, a 2- or 3-dimensional cellular structure cannot be usefully modelled by operations on string expressions. Instead, one can represent a cellular layer by a "map," a structure consisting of line segments and enclosed surfaces. The former are considered to form the boundaries—walls or membranes—of cells, and the latter to represent the cells themselves. Cell expansion and cell division can then be programmed by assigning state symbols to the line seg-

$$\text{rules}: \begin{cases} a \rightarrow a\,b\,c\,b \\ b \rightarrow b\,c \\ c \rightarrow c \end{cases} \qquad M = \begin{pmatrix} 1 & 2 & 1 \\ 0 & 1 & 1 \\ 0 & 0 & 1 \end{pmatrix}$$

$$\text{starting string}: \quad a\,c \qquad \pi = (1 \quad 0 \quad 1) \qquad \eta = \begin{pmatrix} 1 \\ 1 \\ 1 \end{pmatrix}$$

$$\begin{array}{ll} ac & f(0) = 2 \\ \Downarrow & \\ abcbc & f(1) = 5 \\ \Downarrow & \\ abcbbccbcc & f(2) = 10 \\ \Downarrow & \\ abcbbccbcbcccbccc & f(3) = 17 \\ \vdots & \vdots \\ & f(n) = \pi M^n \eta = (n+1)^2 + 1 \end{array}$$

Figure 10. The growth function of a D0L system.

ments and to the cells. The rules for transforming the line segments can be just like D0L rules, while the rules for cell divisions are such that they specify the insertion of new line segments into existing cells. Similar formalisms can be devised for 3-dimensional groups of cells. Such a map generating system to simulate the leaf development of the moss *Phascum cuspidatum* is given in [8].

9. Modelling by L Systems—an Inference Problem

L systems like all mathematical models in science are useful as simulation tools. In addition, since they are specifically introduced on the basis of assumptions current in developmental biology they are able to provide insights into the underlying mechanisms of development and into their complexity. The simulation procedures and the theoretical results for L systems together provide a context within which specific developmental studies can be conducted.

The first problem one faces when attempting to apply L systems to a set of developmental observations is to decide which type of system to use: deterministic or stochastic, interactive or non-interactive, with a single or with more than one table—tables correspond to changes of sets of rules according to environmental changes, for instance. Normally when trying to make such a decision, one would start with models of lower complexity and progress to the more complex ones if no acceptable simpler ones can be found. Thus one would start with D0L systems unless other considerations exclude them right from the start.

The problem of finding an L system of a given type for a set of observations is called the 'inference problem' for that type. It is related to the grammatical inference problem in formal language theory: there a grammar is to be inferred algorithmically from examples of correctly and incorrectly formed words. Related questions arise in artificial intelligence in the context of natural language understanding by computer.

The first step of an inference procedure would consist of agreeing on a general pattern for the developmental rules. In the case of D0L inference the pattern would be expressed as a 0L system. It would incorporate accepted biological principles of the developmental process studied—for instance, excluding the death of subunits, or assumptions about the place of insertions of subunits as in subapical branching.

The second step of the inference procedure consists of finding a specific set of transition rules which are consistent with the observations and with the pattern. If no such system can be found then we have to revise either our underlying pattern or the type of system we are looking for. For instance, if the search for a D0L system fails in this way, one might then consider stochastic or other types of systems.

However, even if a system is found which meets the original criteria it need not be an acceptable one. For instance a D0L system which needs approximately as many states as there are subunits in all observations is hardly acceptable. Further reliability or acceptability criteria can be formulated rigorously. The rejection of a solution under the additional criteria would again force one to consider other types of systems.

There may be more than one acceptable solution. Computer programs have been prepared which carry out the inference procedure for deterministic 0L systems. The D0L systems shown in Figures 8 and 9 could have been obtained with these programs.

Whereas the inference procedures for deterministic or stochastic 0L systems with or without tables seem to be computationally feasible, the inference procedures for L systems with interactions—to the extent that they exist at all—are of a far greater complexity. There seems to be no immediate hope that inference algorithms for these more complicated systems with acceptable run times can be found.

Grammars obtained in this way for the development of organisms can be used to study developmental mechanisms in general. Their comparison can help to detect evolutionary relationships between organisms and to determine developmental patterns. The mathematical theory of grammars, formal languages, and automata provides some of the necessary tools.

10. References

1. M. H. Albert, J. Lawrence: A Proof of Ehrenfeucht's Conjecture. *Theoret. Comput. Sci.* **41** (1985), 121–123.
2. V. Ambros, H. R. Horwitz: Heterochronic Mutants of the Nematode *Caenorhabditis elegans*. *Science* **226** (1984), 409–416.
3. S. Brenner: Nematode Research. *Trends in Biochemical Sciences* **9** (1984), 172.
4. T. W. Chien, H. Jürgensen: Parameterized L Systems for Modelling: Potential and Limitations. This volume.
5. K. Culik II, I. Fris: The Decidability of the Equivalence Problem for D0L Systems. *Information and Control* **35** (1977), 20–39.
6. K. Culik II, J. Karhumäki: On the Ehrenfeucht Conjecture for D0L Languages. *RAIRO, Informatique Théorique* **17** (1983), 205–230.
7. K. Culik II, J. Karhumäki: Decision Problems Solved with the Help of the Ehrenfeucht Conjecture. *Bulletin of the European Ass. for Theoret. Comp. Sci.* **27** (1985), 30–35.

8. M. de Does, A. Lindenmayer: Algorithms for the Generation and Drawing of Maps Representing Cell Clones. In: H. Ehrig, M. Nagl, G. Rozenbergs (eds.), *Graph Grammars and Their Application to Computer Science. Lecture Notes in Computer Science* **153** (1983), 39–57.
9. Leonardo Pisano Fibonacci: *Liber abbaci. (Il Liber Abbaci di Leonardo Pisano, edited by B. Boncompagni)*, Rome, 1857.
10. W. J. Gehring: The Homeo Box: A Key to the Understanding of Development. *Cell* **40** (1985), 3–5.
11. V. S. Guba: (Compactness for algebraic equations), unpublished.
12. E. Hafen, A. Kuroiwa, W. J. Gehring: Spatial Distribution of Transcripts from the Segmentation Gene *fushi tarazu* during Drosophila Embryonic Development. *Cell* **37** (1984), 833–841.
13. G. T. Herman, A. Walker: Context Free Languages in Biological Systems. *Internat. J. Comp. Math.* **4** (1975), 369–391.
14. J. M. Janssen, A. Lindenmayer: Models of Branching Positions and Flowering Sequences of Capitula in *Mycelis muralis* (L.) Dumont (Compositae). *New Phytologist* **105** (1987), 191–220.
15. E. Konrad-Hawkins: Developmental Studies on Regenerates of *Callithamnion roseum* Harvey. Parts I and II. *Protoplasma* **58** (1964), 42–74.
16. R. Lewin: Why Is Development So Illogical. (Interview with S. Brenner) *Science* **224** (1984), 1327–1329.
17. H. B. Lück, J. Lück: Cell Number and Cell Size in Filamentous Organisms in Relation to Ancestrally and Positionally Dependent Generation Times. In: A. Lindenmayer, G. Rozenberg (eds.), *Automata, Languages, Development*, North-Holland, Amsterdam, 1976, 109–124.
18. G. J. Mitchinson, M. Wilcox: Rule Governing Cell Division in *Anabaena*. *Nature* **239** (1972), 110–111.
19. P. Prusinkiewicz, A. Lindenmayer: *The Algorithmic Beauty of Plants.* Springer-Verlag, New York, 1990.
20. G. Stent quoted in J. C. Gerhart's Report on The Cellular Basis of Morphogenetic Change: In: J. T. Bonner (ed.), *Report of the Dahlem Workshop on Evolution and Development*, Springer-Verlag, Berlin, 1982, 87–114.
21. J. E. Sulston, E. Schierenberg, J. G. White, J. N. Thomson: The Embryonic Cell Lineage of the Nematode *Caenorhabditis elegans*. *Developmental Biology* **100** (1983), 64–119.
22. L. Wolpert: Pattern Formation in Biological Development. *Scientific American* **239**(4) (1978), 154–164.

2. The Main Path Continues Further:

Studies of Basic L Systems

REGULARITY PROPERTIES OF L AMBIGUITIES OF MORPHISMS

Juha Honkala

Department of Mathematics, University of Turku, 20500 Turku 50, Finland

Abstract: We study the L ambiguity of morphisms of the free monoid. We define four basic ambiguity sets and establish their effective regularity in many cases. Decidability results concerning L codes are obtained as consequences.

Keywords: Code, L code, ambiguity, regularity, elementary morphism, number system.

1. INTRODUCTION

In [11] Maurer, Salomaa and Wood generalized the notion of a code by defining an L code. By definition, the morphism $h: X^* \longrightarrow X^*$ is an *L code* if the *L associate* $\bar{h}$ of h is an injective mapping. The word $\bar{h}(w)$ is obtained by applying h to the first letter of w, h^2 to the second letter of w, h^3 to the third letter of w, and so on, and catenating the results. The mapping $\bar{h}$ is a morphism only in very special cases. In [11] Maurer, Salomaa and Wood showed that every code is an L code but not vice versa. They also established the interconnection with number systems. This line of studies was continued in [3,6-8]. For closely related work, see [1,5,10].

The theory of L codes is an extension of the theory of codes, [2]. A notion in the theory of codes which is very important in the theory of L systems is bounded delay. This notion for L codes was defined and various resulting families studied in [12]. Many classical cryptosystems can be viewed as codes, [14]. The cryptographic properties of L codes are discussed in [15].

It has turned out to be difficult to decide whether or not a given morphism is an L code. Clearly, the morphism $h: X^* \longrightarrow X^*$ is an L code if and only if there does not exist an *L ambiguity* of the form

$$h(a_1)h^2(a_2)\dots h^m(a_m) = h(b_1)h^2(b_2)\dots h^n(b_n),$$
$$a_1a_2\dots a_m \neq b_1b_2\dots b_n \quad (a_i, b_j \in X).$$

The interconnection with number systems makes it possible to decide the existence of such ambiguities if h is a unary morphism. A decision method is known also if the morphism h does not permute any nonempty subset of X or, equivalently, if each letter of X is growing, [9]. However, the general case is still open. In this paper we give a decision method provided that no letter of X is semibounded. By definition, the letter $x \in X$ is semibounded if there is a positive integer k and words $u,v \in X^*$ containing only bounded letters such that $h^k(x) = uxv$ and the word uv is nonempty.

In this paper the main goal, however, is to study L ambiguities in a slightly more general framework to be explained now.

By definition, a DOL *system with immigration* (shortly, ImDOL system) is a triple (X,h,B). Here X is an alphabet, $h: X^* \longrightarrow X^*$ is a morphism and B is a finite subset of X^*. If $G = (X,h,B)$ is an ImDOL system, its language is defined by

$$L(G) = \{b_0h(b_1)\dots h^n(b_n) \mid n \geq 0,\ b_i \text{ in } B\}.$$

Intuitively, the words of L(G) describe the various developmental stages of the immigrants. To ask whether or not a morphism $h: X^* \longrightarrow X^*$ is an L code amounts to asking whether or not the population development described by the ImDOL system (X,h,X) is backwards deterministic. In this paper we are going to study the L ambiguities of an arbitrary ImDOL system (X,h,B).

If $h: X^* \longrightarrow X^*$ is a morphism and $w_1,\dots,w_n$ are words of X^*, we denote

$$\bar{h}(w_1;w_2;\dots;w_n) = h(w_1)h^2(w_2)\dots h^n(w_n).$$

Suppose $B \subseteq X^*$ is a finite set not containing the empty word. The L *ambiguity sets* $L_1(h,B)$ and $Amb_1(h,B)$ are now defined as follows:

$$L_1(h,B) = \{(b_1,b_1')\ldots(b_m,b_m')(b_{m+1},\lambda)\ldots(b_{m+n},\lambda) \mid b_i,b_j' \in B,\ \bar{h}(b_1;\ldots;b_{m+n}) = \bar{h}(b_1';\ldots;b_m'),\ m \geq 0,\ n \geq 0\}$$

and

$$Amb_1(h,B) = \{b_1\ldots b_n \mid n \geq 1,\ b_i \in B \text{ and, for some } b_1',\ldots,b_m' \in B \text{ with } (b_1,\ldots,b_n) \neq (b_1',\ldots,b_m'),\ \bar{h}(b_1;\ldots;b_n) = \bar{h}(b_1';\ldots;b_m')\}.$$

We show that $L_1(h,B)$ and $Amb_1(h,B)$ are effectively regular, if every word of B contains a growing letter and h is nonerasing. If some word of B does not contain a growing letter, the languages $L_1(h,B)$ and $Amb_1(h,B)$ are not, in general, regular. This is partly due to the possible presence of arbitrarily long overflows consisting of bounded letters only. By ruling out such overflows we define smaller L ambiguity sets $L_2(h,B)$ and $Amb_2(h,B)$. We show that these sets are effectively regular if h is nonerasing and no letter of X is semibounded.

A brief outline of the contents of this paper follows. Section 2 contains the basic definitions and examples of nonregularity. Section 3 contains the constructions of the automata recognizing the ambiguity sets of an elementary morphism. In Section 4 the finiteness of the automata is proved. In Section 5 a reduction to elementary morphisms is given and the proofs of the main results are completed. The paper is largely self-contained, only the basic properties of elementary morphisms and regular languages are used. If the need arises, [13] may be consulted.

2. DEFINITIONS

If X is a finite alphabet, the *free monoid* (resp. *free semigroup*) generated by X is denoted by X^* (resp. X^+). The *length* of a word $w \in X^*$ is denoted by $|w|$. The length of the *empty word* λ is zero. If $u,v \in X^*$ we denote the longest common suffix of u and v by $u \wedge v$. The last letter of a nonempty word $u \in X^+$ is denoted by last(u).

If u,v belong to X^*, $u^{-1}v$ is an element of the *free group* $X^{(*)}$ generated by X. It belongs to X^* if and only if u is a prefix of v.

A *morphism* h: $X^* \longrightarrow Y^*$ is specified by the images of the letters of X. The morphism h is called *nonerasing* if the empty word is not the image of any letter. If h: $X^* \longrightarrow X^*$ is a morphism, a letter $x \in X$ is termed *h-growing* (resp. *h-bounded*) if the length sequence $|h^n(x)|$ is unbounded (resp. bounded). The letter $x \in X$ is called *h-permuted* if there exists a positive integer n such that $h^n(x) = x$. The letter $x \in X$ is called *h-semibounded* (or *h-pumping*) if there exist a positive integer m and words $u,v \in X^*$ containing only h-bounded letters such that $h^m(x) = uxv$ and the word uv is nonempty. If the morphism h is understood, we speak simply of growing letters etc. It is very important to notice that a semibounded letter is growing and a permuted letter is bounded. Clearly, a bounded letter is not necessarily permuted. Neither is a growing letter necessarily semibounded. The image of a bounded letter contains bounded letters only. If $x \in X$ is semibounded, each word of the sequence $(h^n(x))$ contains exactly one growing letter.

If h: $X^* \longrightarrow X^*$ is an injective morphism each h-bounded letter is h-permuted. Consequently, an injective morphism permutes the bounded letters.

By definition, a morphism h: $X^* \longrightarrow Y^*$ has *bounded delay from right to left* if there exists a positive integer d such that if $h(a_1a_2...a_d)$ is a suffix of $h(b_1b_2...b_r)$ then $a_d = b_r$. (Here a's and b's are letters of X.) The minimal d is called the *delay* of h. It is well-known that if the morphisms h: $X^* \longrightarrow Y^*$ and g: $Y^* \longrightarrow Z^*$ have bounded delay from right to left so does gh. Clearly, if h: $X^* \longrightarrow Y^*$ has bounded delay, h is an injective morphism. An important class of morphisms of bounded delay consists of the elementary morphisms. By definition, a morphism h: $X^* \longrightarrow Y^*$ is *elementary* if there do not exist an alphabet X_1 and morphisms h_1: $X^* \longrightarrow X_1^*$ and h_2: $X_1^* \longrightarrow Y^*$ such that $h = h_2h_1$ and $card(X_1) < card(X)$. (The possibility that X_1 is the empty set is not excluded here.)

Let Σ be a finite alphabet. A (*nondeterministic*) *automaton* $\mathcal{A} = (Q,\delta,I,F)$ over Σ is specified by the set Q of states of $\mathcal{A}$, the subset I of initial states, the subset F of final states and a subset δ of $Q \times \Sigma \times Q$. The elements of δ are called the *transitions* of $\mathcal{A}$. If $(q,a,q') \in \delta$, we denote $q' \in \delta(q,a)$ or $qa \Longrightarrow q'$. If $(q_0,a_1,q_1),(q_1,a_2,q_2),\ldots,(q_{n-1},a_n,q_n) \in \delta$ ($n \geq 0$) we denote

$$q_0a_1a_2\ldots a_n \Longrightarrow^* q_n.$$

The sequence $(q_0,a_1,q_1)(q_1,a_2,q_2)\ldots(q_{n-1},a_n,q_n)$ is called a *path in* $\mathcal{A}$ *from* q_0 *to* q_n *through the states in* $\{q_1,q_2,\ldots,q_{n-1}\}$.

We now define the languages we are going to study in this paper. Suppose $h: X^* \longrightarrow X^*$ is a morphism and $w_1,w_2,\ldots,w_n \in X^*$. Following Maurer, Salomaa and Wood, [11], we denote

$$\bar{h}(w_1;w_2;\ldots;w_n) = h(w_1)h^2(w_2)\ldots h^n(w_n).$$

If $B \subseteq X^+$ is a finite nonempty set we denote $\check{B} = \{(b) \mid b \in B\}$ and regard $\check{B}$ as an alphabet with card(B) letters. As a notational simplification we identify (b) and b. Hence, the word $(b_1)(b_2)\ldots(b_m)$ is written simply as $b_1b_2\ldots b_m$. This should not lead into confusion.

Definition 1. Suppose $h: X^* \longrightarrow X^*$ is a morphism and $B \subseteq X^+$. Denote $\Sigma = B \times B \cup B \times \{\lambda\}$. The language $L_1(h,B)$ over the alphabet Σ is defined by

$$L_1(h,B) = \{(b_1,b_1')(b_2,b_2')\ldots(b_m,b_m')(b_{m+1},\lambda)\ldots(b_{m+n},\lambda) \mid \bar{h}(b_1;b_2;\ldots;b_m;b_{m+1};\ldots;b_{m+n}) = \bar{h}(b_1';b_2';\ldots;b_m'); m \geq 0,\ n \geq 0,\ b_i,b_j' \in B\}.$$

The language $Amb_1(h,B)$ over the alphabet $\check{B}$ is defined by

$$Amb_1(h,B) = \{b_1b_2\ldots b_n \mid \text{there exist } m \geq 1 \text{ and } b_1',b_2',\ldots,b_m' \in B \text{ such that } \bar{h}(b_1;b_2;\ldots;b_n) = \bar{h}(b_1';b_2';\ldots;b_m') \text{ and } (b_1,b_2,\ldots,b_n) \neq (b_1',b_2',\ldots,b_m');\ b_1,b_2,\ldots,b_n \in B,\ n \geq 1\}.$$

Example 1. Suppose X is an alphabet, $a \in X$ is a letter and $h: X^* \longrightarrow X^*$ is a morphism such that $h(a) = a$. Furthermore, suppose $B \subseteq X^+$ is a finite set containing the words a and a^2. Denote the letters (a,a^2) and (a^2,a) of $B \times B$ by x and y, respectively. Then

$$L_1(h,B) \cap \{x,y\}^+ = \{w \mid x \text{ has as many occurrences in } w \text{ as } y\}.$$

Hence $L_1(h,B)$ is not a regular language.

Even if $L_1(h,B)$ and $Amb_1(h,B)$ are nonregular, regularity results can be obtained for the smaller languages $L_2(h,B)$ and $Amb_2(h,B)$ defined below.

Definition 2. Suppose $h: X^* \longrightarrow X^*$ is a morphism and $B \subseteq X^+$. Denote $\Sigma = B \times B \cup B \times \{\lambda\}$. The language $L_2(h,B)$ over the alphabet Σ is defined by

$$L_2(h,B) = \{(b_1,b_1')(b_2,b_2')\ldots(b_m,b_m')(b_{m+1},\lambda)\ldots(b_{m+n},\lambda) \in L_1(h,B) \mid \text{ if } \bar{h}(b_1;b_2;\ldots;b_j) = \bar{h}(b_1';b_2';\ldots;b_j')w \text{ or } \bar{h}(b_1;b_2;\ldots;b_j)w = \bar{h}(b_1';b_2';\ldots;b_j') \text{ where } b_k \neq b_k' \text{ for some } k\ (1 \le k \le j) \text{ and the word } w \in X^* \text{ contains only h-permuted letters, then } j = m \text{ and } w = h^{m+1}(b_{m+1})\ldots h^{m+n}(b_{m+n})\}.$$

The language $Amb_2(h,B)$ over the alphabet $\check{B}$ is defined by

$$Amb_2(h,B) = \{b_1b_2\ldots b_n \mid \text{ there exist } m \ge 1,\ b_1',\ldots,b_m' \in B \text{ such that } \bar{h}(b_1;b_2;\ldots;b_n) = \bar{h}(b_1';b_2';\ldots;b_m'), (b_1,b_2,\ldots,b_n) \neq (b_1',b_2',\ldots,b_m') \text{ and, furthermore, if } j \le n,\ j \le m \text{ and } \bar{h}(b_1;b_2;\ldots;b_j) = \bar{h}(b_1';b_2';\ldots;b_j')w \text{ or } \bar{h}(b_1;b_2;\ldots;b_j)w = \bar{h}(b_1';b_2';\ldots;b_j'), \text{ where } b_k \neq b_k' \text{ holds for some } k\ (1 \le k \le j) \text{ and } w \text{ contains only h-permuted letters, then either } j = m \text{ and } w = h^{m+1}(b_{m+1})\ldots h^n(b_n) \text{ or } j = n \text{ and } w = h^{n+1}(b_{n+1}')\ldots h^m(b_m');\ b_1,b_2,\ldots,b_n \in B,\ n \ge 1\}.$$

Notice that $L_2(h,B) \subseteq L_1(h,B)$ and $Amb_2(h,B) \subseteq Amb_1(h,B)$. Furthermore, if each h-permuted letter belongs to B, the language $Amb_1(h,B)$ is empty if and only if $Amb_2(h,B)$ is empty.

The definition due to Maurer, Salomaa and Wood, [11], reads as follows in our terminology.

Definition 3. The morphism $h: X^* \longrightarrow X^*$ is an *L code* if and only if $Amb_1(h,X)$ is empty.

This definition gives the original motivation for the study of L ambiguity.

Example 2. Denote $X = \{a,b,c,d\}$ and define $h: X^* \longrightarrow X^*$ by $h(a) = c^2$, $h(b) = cbc$, $h(c) = c$, $h(d) = \lambda$. To simplify the notation we denote $\bar{h}(a_1a_2\ldots a_n)$ instead of $\bar{h}(a_1;a_2;\ldots;a_n)$ if $a_1,a_2,\ldots,a_n$ are letters of X. It is not difficult to verify that for any $n \geq 1$,

$$\bar{h}(aba^3ba^5b\ldots a^{2n-1}ba^n) = \bar{h}(d^{2+1}bd^{2+3}bd^{2+5}b\ldots d^{2+2n-1}b).$$

Notice that

$$|d^{2+1}bd^{2+3}bd^{2+5}b\ldots d^{2+2n-1}b| - |aba^3ba^5b\ldots a^{2n-1}ba^n| = n.$$

This implies that the language $L_1(h,X)$ is not regular. However

$$\begin{aligned} L_2(h,X) = \{(\alpha,\alpha)\,|\,\alpha \in X\}^*[&(c,a)(d,\lambda)^*(c,\lambda)(d,\lambda)^* \\ &\cup (d,c)(d,\lambda)^*(c,\lambda)(d,\lambda)^* \cup (d,a)(d,\lambda)^*(a,\lambda)(d,\lambda)^* \\ &\cup (d,a)(d,\lambda)^*(c,\lambda)(d,\lambda)^*(c,\lambda)(d,\lambda)^*]. \end{aligned}$$

3. CONSTRUCTIONS OF THE AUTOMATA

Suppose $p: Y^* \longrightarrow Y^*$ is an injective morphism and $B \subseteq Y^+$ is a finite set. Our first goal in this section is to construct a nondeterministic automaton $\mathcal{A}_1(p,B)$ which recognizes $L_1(p,B)$. In general, $\mathcal{A}_1(p,B)$ is not a finite automaton.

Define

$$S_0 = \{(\lambda,\lambda)\},$$

$$S_{n+1} = \{(u,v) \in Y^* \times Y^* \,|\, \text{there exist } (u_n,v_n) \in S_n \text{ and } b,b' \in B \text{ such that } u_n^{-1}bp(u) = v_n^{-1}b'p(v) \text{ and, if } u \neq \lambda \text{ and } v \neq \lambda \text{ then } \mathrm{last}(u) \neq \mathrm{last}(v)\},\quad n \geq 0$$

and

$$S = \bigcup_{n\geq 0} S_n.$$

Furthermore, define

$$T = \{((b_j,b_{j+1},\ldots,b_m),\lambda,\lambda) \,|\, 2 \leq j \leq m, \text{ there exist } b_1,\ldots,b_{j-1} \in B \text{ such that } (b_1\bar{p}(b_2;\ldots;b_m),\lambda) \in S,\ b_j,b_{j+1},\ldots,b_m \in B\} \cup \{(\lambda,\lambda,\lambda)\}.$$

Now define the state set of $\mathcal{A}_1(p,B)$ by $Q_1 = S \cup T$. The only initial state of $\mathcal{A}_1(p,B)$ is $s_0 = (\lambda,\lambda)$ and the final states of

$\mathcal{A}_1(p,B)$ are (λ,λ) and $(\lambda,\lambda,\lambda)$. The set δ_1 of transitions of $\mathcal{A}_1(p,B)$ is defined in the natural way. Hence

$$\delta_1((u,v),(b,b')) = \{(u',v') \in S \mid u^{-1}bp(u') = v^{-1}b'p(v')\},$$
$$(u,v) \in S,\ b,b' \in B,$$
$$(\lambda,\lambda,\lambda) \in \delta_1((b,\lambda),(b,\lambda)),\ b \in B,\ (b,\lambda) \in S$$

and

$$((b_2,\dots,b_m),\lambda,\lambda) \in \delta_1((b_1\bar{p}(b_2;\dots;b_m),\lambda),(b_1,\lambda)),$$
$$m \geq 2,\ b_1,\dots,b_m \in B,\ (b_1\bar{p}(b_2;\dots;b_m),\lambda) \in S.$$

Furthermore,

$$\delta_1(((b_1,\dots,b_m),\lambda,\lambda),(b_1,\lambda)) = \{((b_2,\dots,b_m),\lambda,\lambda)\},$$
$$m \geq 1,\ b_1,\dots,b_m \in B,\ ((b_1,\dots,b_m),\lambda,\lambda) \in T.$$

(If $m = 1$, it is understood that $(b_2,\dots,b_m) = \lambda$.)

Lemma 1. Suppose $p\colon Y^* \longrightarrow Y^*$ is an injective morphism and $u,v,w_1,w_2 \in Y^*$ are words such that u is a prefix of w_1 and v is a prefix of w_2. If $p(u)^{-1}p(w_1) = p(v)^{-1}p(w_2)$ then $u^{-1}w_1 = v^{-1}w_2$.

Proof. Suppose $w_1 = uw_1'$ and $w_2 = vw_2'$ where w_1', w_2' are in Y^*. Then $p(w_1) = p(u)p(w_1')$ and $p(w_2) = p(v)p(w_2')$. By the assumption, $p(w_1') = p(w_2')$ which implies that $w_1' = w_2'$. Hence $u^{-1}w_1 = v^{-1}w_2$. □

Lemma 1 does not hold without the prefix conditions. For example, if $g\colon \{a,b,c\}^* \longrightarrow \{a,b,c\}^*$ is defined by $g(a) = a$, $g(b) = ab$, $g(c) = ab^2$, we have $g(a)^{-1}g(b) = g(b)^{-1}g(c)$ but $a^{-1}b \neq b^{-1}c$.

Lemma 2. $\mathcal{A}_1(p,B)$ accepts $L_1(p,B)$.

Proof. Suppose $\bar{p}(b_1;\dots;b_m;\dots;b_{m+n}) = \bar{p}(b_1';\dots;b_m')$ where $b_i, b_i' \in B$. Because p is injective, we have $b_1\bar{p}(b_2;\dots;b_{m+n}) = b_1'\bar{p}(b_2';\dots;b_m')$. Now suppose inductively that

$$s_0(b_1,b_1')\dots(b_j,b_j') \Longrightarrow^* (u_j,v_j)$$

in $\mathcal{A}_1(p,B)$ where $(u_j,v_j) \in S_j$ and

$$u_j^{-1}b_{j+1}\bar{p}(b_{j+2};\dots;b_{m+n}) = v_j^{-1}b_{j+1}'\bar{p}(b_{j+2}';\dots;b_m'),$$

$j \leq m-2$. If $j = 0$, we denote $u_0 = v_0 = \lambda$. Choose a prefix u_{j+1} of $b_{j+2}\bar{p}(b_{j+3};\dots;b_{m+n})$ and a prefix v_{j+1} of $b_{j+2}'\bar{p}(b_{j+3}';\dots;b_m')$

such that $u_j^{-1}b_{j+1}p(u_{j+1}) = v_j^{-1}b'_{j+1}p(v_{j+1})$ and the words u_{j+1} and v_{j+1} do not have a common nonempty suffix. This is possible because $u_j = \lambda$ or $v_j = \lambda$ or $\mathrm{last}(u_j) \neq \mathrm{last}(v_j)$. Hence $(u_{j+1}, v_{j+1}) \in S_{j+1}$ and

$$s_0(b_1,b'_1)\dots(b_{j+1},b'_{j+1}) \Longrightarrow^* (u_{j+1},v_{j+1})$$

in $\mathcal{A}_1(p,B)$. Furthermore,

$$p(u_{j+1})^{-1}\bar{p}(b_{j+2};\dots;b_{m+n}) = p(v_{j+1})^{-1}\bar{p}(b'_{j+2};\dots;b'_m)$$

which implies by Lemma 1 that

$$u_{j+1}^{-1}b_{j+2}\bar{p}(b_{j+3};\dots;b_{m+n}) = v_{j+1}^{-1}b'_{j+2}\bar{p}(b'_{j+3};\dots;b'_m).$$

In this way we find a state $(u_{m-1},v_{m-1}) \in S_{m-1}$ such that

$$s_0(b_1,b'_1)\dots(b_{m-1},b'_{m-1}) \Longrightarrow^* (u_{m-1},v_{m-1})$$

in $\mathcal{A}_1(p,B)$ and

$$u_{m-1}^{-1}b_m\bar{p}(b_{m+1};\dots;b_{m+n}) = v_{m-1}^{-1}b'_m .$$

Now the automaton reaches the state $(b_{m+1}\bar{p}(b_{m+2};\dots;b_{m+n}),\lambda)$ when the letter (b_m,b'_m) is read. If $n = 0$, the automaton is in a final state. If $n = 1$, the final state $(\lambda,\lambda,\lambda)$ is reached when (b_{m+1},λ) is read. If $n > 1$, the final state $(\lambda,\lambda,\lambda)$ is reached through the states $((b_{m+2},\dots,b_{m+n}),\lambda,\lambda),\dots,$ $((b_{m+n}),\lambda,\lambda)$. Hence $\mathcal{A}_1(p,B)$ accepts the word

$$(b_1,b'_1)(b_2,b'_2)\dots(b_m,b'_m)(b_{m+1},\lambda)\dots(b_{m+n},\lambda).$$

Conversely, suppose that $\mathcal{A}_1(p,B)$ accepts the word

$$(b_1,b'_1)(b_2,b'_2)\dots(b_m,b'_m)(b_{m+1},\lambda)\dots(b_{m+n},\lambda).$$

Suppose that in $\mathcal{A}_1(p,B)$

$$(\lambda,\lambda)(b_1,b'_1) \Longrightarrow (u_1,v_1)$$
$$(u_1,v_1)(b_2,b'_2) \Longrightarrow (u_2,v_2)$$
$$\dots$$
$$(u_{m-1},v_{m-1})(b_m,b'_m) \Longrightarrow (u_m,v_m) .$$

Therefore

$$b_1p(u_1) = b'_1p(v_1)$$
$$u_1^{-1}b_2p(u_2) = v_1^{-1}b'_2p(v_2)$$
$$\dots$$
$$u_{m-1}^{-1}b_mp(u_m) = v_{m-1}^{-1}b'_mp(v_m) .$$

This implies that

$$p(b_1)p^2(u_1) = p(b'_1)p^2(v_1)$$
$$p^2(u_1)^{-1}p^2(b_2)p^3(u_2) = p^2(v_1)^{-1}p^2(b'_2)p^3(v_2)$$
$$\dots$$
$$p^m(u_{m-1})^{-1}p^m(b_m)p^{m+1}(u_m) = p^m(v_{m-1})^{-1}p^m(b'_m)p^{m+1}(v_m) .$$

Hence

$$p(b_1)p^2(b_2)\ldots p^m(b_m)p^{m+1}(u_m) = p(b_1')p^2(b_2')\ldots p^m(b_m')p^{m+1}(v_m).$$

Because $\mathcal{A}_1(p,B)$ reaches a final state when $(b_{m+1},\lambda)\ldots(b_{m+n},\lambda)$ is read in state (u_m,v_m) we have $u_m = b_{m+1}\bar{p}(b_{m+2};\ldots;b_{m+n})$ and $v_m = \lambda$. This implies $\bar{p}(b_1;\ldots;b_m;\ldots b_{m+n}) = \bar{p}(b_1';\ldots;b_m')$. □

Our second goal is to construct an automaton $\mathcal{A}_2(p,B)$ which accepts the language $L_2(p,B)$. For that purpose define

$S^{(1)} = \{(u,v) \in S \mid$ neither u nor v contains a growing letter$\}$,

$S^{(2)} = S - S^{(1)}$,

$S^{(3)} = \{(u,v) \in S \mid$ there exists a path from (λ,λ) to (u,v) in $\mathcal{A}_1(p,B)$ through the states of $S^{(2)}\}$

and

$T^{(3)} = \{t \in T \mid$ there exists a path from (λ,λ) to t in $\mathcal{A}_1(p,B)$ through the states of $S^{(3)} \cup T\}$.

Suppose now that $\bar{p}(b_1;\ldots;b_n)w = \bar{p}(b_1';\ldots;b_n')$ where $w \in Y^+$ does not contain a growing letter. If

$$(u,v) \in \delta_1((\lambda,\lambda),(b_1,b_1')\ldots(b_n,b_n'))$$

we have

$$p^{n+1}(u) = wp^{n+1}(v).$$

If $v \neq \lambda$, clearly $u \neq \lambda$. Because p permutes the bounded letters, there is a word $w' \in Y^+$ such that $p^{n+1}(w') = w$. The injectivity of p implies that $u = w'v$. This is not possible because $\mathrm{last}(u) \neq \mathrm{last}(v)$. Therefore $v = \lambda$ and $(u,v) \in S^{(1)}$. Hence $\mathcal{A}_2(p,B)$ can be constructed as follows.

The state set of $\mathcal{A}_2(p,B)$ is $\{i\} \cup S^{(3)} \cup T^{(3)}$. Here i is a new state which is the only initial state of $\mathcal{A}_2(p,B)$. The final states of $\mathcal{A}_2(p,B)$ are those among i, (λ,λ) and $(\lambda,\lambda,\lambda)$ which are states of $\mathcal{A}_2(p,B)$. The set δ_2 of transitions of $\mathcal{A}_2(p,B)$ is defined as follows:

$\delta_2(i,(b,b)) = i$, if $b \in B$,

$\delta_2(i,(b,b')) = \delta_1((\lambda,\lambda),(b,b'))$, if $b,b' \in B$ and $b \neq b'$,

$\delta_2(s,\sigma) = \delta_1(s,\sigma)$, if $s \in S^{(2)} \cup T^{(3)}$, $\sigma \in \Sigma$,

$\delta_2((u,\lambda),\sigma) = \delta_1((u,\lambda),\sigma) \cap T^{(3)}$, if $(u,\lambda) \in S^{(1)}$ and $u \neq \lambda$, $\sigma \in \Sigma$.

(Here s and (u,λ) are states of $\mathcal{A}_2(p,B)$.)

The following lemma is now immediate.

Lemma 3. $\mathcal{A}_2(p,B)$ accepts $L_2(p,B)$.

4. FINITENESS OF THE AUTOMATA

In this section we suppose that $p: Y^* \longrightarrow Y^*$ is a morphism which has bounded delay from right to left and $B \subseteq Y^+$ is a finite set. First we show that if each word of B contains at least one growing letter then $\mathcal{A}_1(p,B)$ is a finite automaton.

Lemma 4. If $(u,v) \in S$ then last(u) or last(v) is not a bounded letter if $u \neq \lambda$ and $v \neq \lambda$.

Proof. The claim holds if $(u,v) = (\lambda,\lambda)$. Suppose $(u,v) \in S_{n+1}$ where $n \geq 0$. By the definition of S there exist $(u_n,v_n) \in S_n$ and $b,b' \in B$ such that $u_n^{-1}bp(u) = v_n^{-1}b'p(v)$. Clearly $u_n^{-1}bp(u) = v_n^{-1}b'p(v)$ belongs to Y^*. If $u_n^{-1}bp(u) = v_n^{-1}b'p(v) = \lambda$ we have $bp(u) = u_n$ and $b'p(v) = v_n$ and the claim follows inductively. If $u_n^{-1}bp(u) = v_n^{-1}b'p(v) \in Y^+$ and $u \neq \lambda$ and $v \neq \lambda$, the letters last(u) and last(v) are not both bounded because p permutes the bounded letters. □

Lemma 5. Suppose that each word of $B \subseteq Y^+$ contains a growing letter. Then there is a positive integer M_1 such that, if $(u,v) \in S$, neither u nor v has a suffix longer than M_1 consisting solely of bounded letters.

Proof. If $(u,v) = (\lambda,\lambda)$ the claim trivially holds. Otherwise there exist $b_1,\ldots,b_n,b_1',\ldots,b_n' \in B$ such that

$$p(b_1)\ldots p^n(b_n)p^{n+1}(u) = p(b_1')\ldots p^n(b_n')p^{n+1}(v).$$

Without loss of generality we suppose $n > \text{card}(Y)$. Denote $w_1 = b_np(u)$ and $w_2 = b_n'p(v)$. By the assumption concerning B both w_1 and w_2 contain a growing letter. Suppose that the last growing letters of w_i and $p^n(w_i)$ are y_i and x, respectively. (Notice that $p^n(w_1)$ is a suffix of $p^n(w_2)$, or vice versa.) Clearly, there exist nonnegative integers $n_i \le \mathrm{card}(Y)$ such that the last growing letter of $p^{n_i}(w_i)$ is x. We may assume $n_1 = n_2$. Suppose next that w_i and $p^{n_i}(w_i)$ have suffixes of e_i and f_i bounded letters, respectively. It is easy to see that $e_i \le f_i$ and $f_1 = f_2$.

We now consider three cases separately. If $u = \lambda$, the numbers e_1 and f_1 are small. Consequently, e_2 is small. The case $v = \lambda$ is similar. Suppose $u \ne \lambda$ and $v \ne \lambda$. By Lemma 4 last(u) or last(v), say last(u), is a growing letter. Therefore e_1 and f_1 are small. Hence e_2 is small. This implies the existence of M_1. An upper bound for M_1 is easily obtained. □

Denote by M_2 the length of the longest common suffix of p(u) and p(v), if u and v are words of Y^+ and last(u) ≠ last(v). Denote $M_3 = M_1 + M_2 + \max\{|b| \mid b \in B\} + 2$, where M_1 is as in Lemma 5.

Lemma 6. Suppose that each word of $B \subseteq Y^+$ contains a growing letter. If $(u,v) \in S$ then neither u nor v contains M_3 consecutive bounded letters.

Proof. Because $S = \bigcup_{n\ge 0} S_n$ we again use induction on n. Suppose $(u,v) \in S_{n+1}$ and $u_n^{-1}bp(u) = v_n^{-1}b'p(v)$ where $b,b' \in B$ and $(u_n,v_n) \in S_n$ and, furthermore, that the claim holds for (u_n,v_n). We prove the claim concerning u. Because $u_n^{-1}bp(u) = v_n^{-1}b'p(v)$ belongs to Y^*, the word u_n is a prefix of bp(u). Suppose first that u_n is a prefix of b. If b′ is a prefix of v_n, the length of p(u) is at most M_2. If v_n is a prefix of b′, the length of p(u) is at most $|b'| + M_2$. Hence the claim holds in both cases. Suppose then that b is a proper prefix of u_n. If $bp(u) = u_n$ the claim follows immediately by the inductive hypothesis. Otherwise, choose the longest prefix u′ of u such that p(u′) is a prefix of $b^{-1}u_n$ and denote u = u′yu" where y is a letter. Now u′ does not contain M_3 consecutive bounded letters. Otherwise u_n would contain M_3 consecutive bounded letters. The length of the word yu" is at most $1 + M_2 + |b'|$.

The claim follows if y is a growing letter. Suppose that y is a bounded letter. Then $bp(u') = u_n$ and at least one of the last $M_1 + 1$ letters of u′ is a growing letter by Lemma 5, or $|u'|$ is at most M_1. Because $M_3 \geq M_1 + M_2 + |b'| + 2$ the word u does not contain M_3 consecutive bounded letters. By symmetry, the same holds for v. □

Lemma 7. If each word of B contains a growing letter, the state set of $\mathcal{A}_1(p,B)$ is finite.

Proof. It is enough to show that the set S is a finite set.

Choose an integer n_0 such that the length of $p^{n_0}(x)$ is greater than one if x is a growing letter. Consider the set

$$\{p^{n_0}(u) \wedge p^{n_0}(v) \mid u,v \in Y^+,\ \mathrm{last}(u) \neq \mathrm{last}(v)\}.$$

Because p^{n_0} has a finite delay from right to left, this is a finite set. Denote by M_4 the length of its longest word. Denote by M_5 the length of the longest word in the set $\{u,v \mid (u,v) \in S_n$ for some n with $0 \leq n \leq n_0\}$. Furthermore, denote by M_6 the length of the longest word in the set $\{c_1p(c_2)\ldots p^{n_0-1}(c_{n_0}) \mid c_i \in B\}$. Choose an integer M greater than M_5 and $M_3(M_6 + M_4 + 1)$. We are going to show inductively that if $(u,v) \in S$ then the length of u (resp. v) is at most M.

Suppose therefore that $(u,v) \in S_n$, where $n > n_0$. By the definition of S there exist $(u',v') \in S_{n-n_0}$ and $b_1,\ldots,b_{n_0}$, $b'_1,\ldots,b'_{n_0} \in B$ such that

$$u'^{-1}b_1p(b_2)p^2(b_3)\ldots p^{n_0-1}(b_{n_0})p^{n_0}(u) =$$
$$v'^{-1}b'_1p(b'_2)p^2(b'_3)\ldots p^{n_0-1}(b'_{n_0})p^{n_0}(v).$$

If $p^{n_0}(u)$ is a suffix of $p^{n_0}(v)$, the length of u is at most M_4. Otherwise,

$$b_1p(b_2)\ldots p^{n_0-1}(b_{n_0})p^{n_0}(u) = u'(p^{n_0}(u) \wedge p^{n_0}(v)) \qquad (1)$$

or

$$b_1p(b_2)\ldots p^{n_0-1}(b_{n_0})p^{n_0}(u) =$$

$$u'[v'^{-1}b_1'p(b_2')\dots p^{n_0-1}(b_{n_0}')]p^{n_0}(v) \ . \qquad (2)$$

Furthermore, in (2), $v'^{-1}b_1'p(b_2')\dots p^{n_0-1}(b_{n_0}')$ is a word in Y^*. By Lemma 6 and the choice of n_0, the length of the left hand side of (1) (resp. (2)) is at least

$$2\lfloor \frac{|u|}{M_3} \rfloor + |u| - \lfloor \frac{|u|}{M_3} \rfloor \ .$$

The length of the right hand side of (1) (resp. (2)) is at most $M + M_4$ (resp. $M + M_6 + M_4$). Therefore

$$\frac{|u|}{M_3} - 1 + |u| \le M + M_6 + M_4$$

which implies $|u| \le M$. □

In the second part of this section we again suppose that $p: Y^* \longrightarrow Y^*$ is a morphism which has bounded delay from right to left. Now we do not make any assumption concerning the finite set $B \subseteq Y^+$ but assume that no letter of Y is semibounded. We are going to show that this implies the finiteness of $\mathcal{A}_2(p,B)$.

Choose a positive integer k such that if $x \in Y$ is a growing letter then $p^k(x)$ contains at least two growing letters. Denote $N_1 = \max\{|p^k(u) \wedge p^k(v)| \mid u,v \in Y^+, \text{last}(u) \ne \text{last}(v)\}$ and $N_2 = \max\{|u|,|v| \mid$ there exists $i \le k$ such that $(u,v) \in S_i\}$. Furthermore, denote $N_3 = \max\{|c_1p(c_2)\dots p^{k-1}(c_k)| \mid c_i \in B\}$ and $N_4 = \max\{N_1+N_3, N_2\}$. If $u \in Y^*$ we denote by $|u|_g$ the number of growing letters in u.

Lemma 8. Suppose that no letter of Y is semibounded. If $(u,v) \in S^{(3)}$, then neither u nor v contains more than N_4 growing letters.

Proof. If $(u,v) \in S_i$ for some i less than or equal to k, the claim holds. Otherwise, there exist $b_1,b_2,\dots,b_k$, $b_1',b_2',\dots,b_k' \in B$ and $(u',v') \in S$ such that

$$u'^{-1}b_1p(b_2)\dots p^{k-1}(b_k)p^k(u) = v'^{-1}b_1'p(b_2')\dots p^{k-1}(b_k')p^k(v) \qquad (3)$$

and neither u' nor v' contains more than N_4 growing letters.

We prove the claim concerning u. Suppose first that $p^k(u) \wedge$

$p^k(v) = p^k(u)$. If $u = \lambda$ the claim is trivial. Otherwise also $v \neq \lambda$ and therefore $\mathrm{last}(u) \neq \mathrm{last}(v)$. Consequently, the length of u is at most N_1. Suppose then that $p^k(u) \wedge p^k(v) = p^k(v)$. By (3) we have

$$b_1p(b_2)\ldots p^{k-1}(b_k)p^k(u) = u'[v'^{-1}b_1'p(b_2')\ldots p^{k-1}(b_k')]p^k(v)$$

where $v'^{-1}b_1'p(b_2')\ldots p^{k-1}(b_k')$ belongs to Y^*. By the choice of k we have

$$2|u|_g \leq N_4 + N_3 + N_1.$$

Hence $|u|_g \leq N_4$.

Suppose finally that $p^k(u) \wedge p^k(v)$ is neither $p^k(u)$ nor $p^k(v)$. Then we have by (3)

$$b_1p(b_2)\ldots p^{k-1}(b_k)p^k(u) = u'(p^k(u) \wedge p^k(v)).$$

As before, this implies that $|u|_g \leq N_4$. The claim concerning v follows by symmetry. □

For the next lemma denote by N_5 the length of the longest common suffix of the words $p^{N_4k}(u)$ and $p^{N_4k}(v)$ if u and v are in Y^+ and $\mathrm{last}(u) \neq \mathrm{last}(v)$. Denote by N_6 the length of the longest word in the set $\{c_1p(c_2)\ldots p^{N_4k-1}(c_{N_4k}) \mid c_i \in B\}$. Furthermore, denote by N_7 the length of the longest word in the set $\{u,v \mid$ there is $i \leq N_4k$ such that $(u,v) \in S_i\}$ and $N_8 = \max\{N_5+N_6, N_7\}$.

Lemma 9. Suppose that no letter of Y is semibounded. If $(u,v) \in S^{(3)}$ then every growing letter in u (resp. v) is contained in the suffix of length N_8 of u (resp. v).

Proof. If $(u,v) \in S_i$ for some i less than or equal to N_4k the claim holds. Otherwise, there are $b_1,b_2,\ldots,b_{N_4k},b_1',b_2',\ldots,b_{N_4k}' \in B$ and $(u',v') \in S^{(3)}$ such that

$$u'^{-1}b_1p(b_2)\ldots p^{N_4k-1}(b_{N_4k})p^{N_4k}(u) =$$
$$v'^{-1}b_1'p(b_2')\ldots p^{N_4k-1}(b_{N_4k}')p^{N_4k}(v) .$$

We prove the claim concerning u. If $p^{N_4k}(u) \wedge p^{N_4k}(v) = p^{N_4k}(u)$ the length of u is at most N_5 and the claim trivially holds.

Suppose then that $p^{N_4k}(u) \wedge p^{N_4k}(v) = p^{N_4k}(v)$. Then

$$b_1p(b_2)\ldots p^{N_4k-1}(b_{N_4k})p^{N_4k}(u) =$$

$$u'[v'^{-1}b'_1p(b'_2)\ldots p^{N_4k-1}(b'_{N_4k})]p^{N_4k}(v)$$

where in the brackets we have a word of Y^*. If we had $u = u_1xu_2$ where x is a growing letter and the length of u_2 is at least N_8, the word u' would contain at least 2^{N_4} growing letters. This is impossible by Lemma 8 because $2^{N_4} > N_4$. Therefore the claim is true. The claim follows similarly in the remaining case. □

Lemma 10. Suppose that no letter of Y is semibounded. There exists a positive integer N_9 such that if $(u,v) \in S^{(3)} \cap S^{(2)}$ and v (resp. u) contains only bounded letters then the length of v (resp. u) is at most N_9.

Proof. We only sketch the argument because no new idea is used. Suppose that $(u,v) \in S^{(3)} \cap S^{(2)}$ and v contains only bounded letters. There exist $b_1,\ldots,b_n,b'_1,\ldots,b'_n \in B$ such that

$$p(b_1)p^2(b_2)\ldots p^n(b_n)p^{n+1}(u) =$$
$$p(b'_1)p^2(b'_2)\ldots p^n(b'_n)p^{n+1}(v) .$$

Choose s such that b'_s contains a growing letter and $b'_{s+1},\ldots,b'_n$ contain only bounded letters. If $s = n$, an argument similar to the proof of Lemma 5 shows that v is not very long. If $s \neq n$, there exists $(u',v') \in S^{(3)} \cap S^{(2)}$ such that

$$u'^{-1}b_{s+1}p(b_{s+2})\ldots p^{n-s-1}(b_n)p^{n-s}(u) =$$
$$v'^{-1}b'_{s+1}p(b'_{s+2})\ldots p^{n-s-1}(b'_n)p^{n-s}(v) . \quad (4)$$

Because v' and the right hand side of (4) contain only bounded letters, the integer n-s is smaller than kN_4. Otherwise u' would contain more than N_4 growing letters. Because

$$p(b_1)p^2(b_2)\ldots p^s(b_s)p^{s+1}(u') =$$
$$p(b'_1)p^2(b'_2)\ldots p^s(b'_s)p^{s+1}(v'),$$

an argument similar to the proof of Lemma 5 shows that v' is short if it is nonempty. Because n-s is small and last(u) is a growing letter this implies by (4) that v is not very long. □

Choose now a positive integer N greater than N_8 and N_9. Suppose also that N is greater than the length of the longest word in the sets B and $\{p(u) \wedge p(v) \mid u,v \in Y^+$ and $last(u) \neq last(v)\}$.

Lemma 11. Suppose that no letter of Y is semibounded. If $(u,v) \in S^{(3)}$ then u (resp. v) does not contain 3N consecutive bounded letters.

Proof. We prove the claim concerning u. The claim holds if $(u,v) = (\lambda,\lambda)$. Otherwise there exist $b,b' \in B$ and $(u',v') \in S^{(3)}$ such that $u'^{-1}bp(u) = v'^{-1}b'p(v)$. The claim holds if $(u',v') = (\lambda,\lambda)$. Suppose therefore that $(u',v') \in S^{(2)}$. If $p(u) \wedge p(v) = p(u)$, the length of u is less than N and the claim is trivial. Otherwise, we have

$$bp(u) = u'(p(u) \wedge p(v))$$

or

$$bp(u) = u'[v'^{-1}b']p(v)$$

according to whether $p(u) \wedge p(v) \neq p(v)$ or $p(u) \wedge p(v) = p(v)$. Here $v'^{-1}b'$ belongs to Y^*. Suppose now that the length of u is at least 3N. Then the length of u' is greater than N. By Lemmas 10 and 9 the suffix of u' of length N contains a growing letter. Hence, if u contains 3N consecutive bounded letters, so does u'. Therefore the claim follows inductively. □

Lemma 12. If no letter of Y is semibounded, the state set of $\mathcal{A}_2(p,B)$ is finite.

Proof. By Lemmas 9 and 11 the set $S^{(3)}$ is finite. This implies the finiteness of $T^{(3)}$. □

5. THE MAIN RESULTS

In this section we complete the proofs of the regularity properties of the basic ambiguity sets.

The method used in the proof of the following lemma is due to Ehrenfeucht and Rozenberg, [4].

Lemma 13. Suppose $h: X^* \longrightarrow X^*$ is a nonerasing morphism and $n = \mathrm{card}(X)$. Then there exist an alphabet Y, an elementary morphism $p: Y^* \longrightarrow Y^*$, a morphism $g: Y^* \longrightarrow X^*$ which has bounded delay from right to left and a nonerasing morphism $f: X^* \longrightarrow Y^*$ such that $h^{m+n} = gp^m f$ for every $m \geq 0$. Furthermore, $p^n = fg$.

Proof. If h is elementary, the claim follows immediately. Otherwise, there is an alphabet X_1 and morphisms $f_1: X^* \longrightarrow X_1^*$ and $g_1: X_1^* \longrightarrow X^*$ such that $h = g_1 f_1$, g_1 has bounded delay from right to left and $\mathrm{card}(X_1) < \mathrm{card}(X)$. Denote $h_1 = f_1 g_1$. If h_1 is not elementary then there exist an alphabet X_2 and morphisms $f_2: X_1^* \longrightarrow X_2^*$ and $g_2: X_2^* \longrightarrow X_1^*$ such that $h_1 = g_2 f_2$, g_2 has bounded delay from right to left and $\mathrm{card}(X_2) < \mathrm{card}(X_1)$. Denote $h_2 = f_2 g_2$. Define in the same way $X_3, \ldots, X_k$, $f_3, \ldots, f_k$, $g_3, \ldots, g_k$, $h_3, \ldots, h_k$ until h_k is elementary. If $m \geq 0$ then

$$h^{m+k} = (g_1 f_1)^{m+k} = g_1 (f_1 g_1)^{m+k-1} f_1 = g_1 (g_2 f_2)^{m+k-1} f_1$$
$$= g_1 g_2 (f_2 g_2)^{m+k-2} f_2 f_1 = \ldots = g_1 g_2 \ldots g_k h_k^m f_k \ldots f_2 f_1.$$

Furthermore,

$$f_k \ldots f_2 (f_1 g_1) g_2 \ldots g_k = f_k \ldots f_3 (f_2 g_2)^2 g_3 \ldots g_k$$
$$= f_k \ldots f_4 (f_3 g_3)^3 g_4 \ldots g_k = \ldots = (f_k g_k)^k = h_k^k.$$

Hence we can choose $Y = X_k$, $p = h_k$, $f = f_k \ldots f_2 f_1$ and $g = g_1 g_2 \ldots g_k h_k^{n-k}$. □

Lemma 14. Suppose X, Y, h, p, g, f and n are as in Lemma 13. If $a \in Y$ is p-bounded then g(a) is an h-permuted letter of X. If $a \in X$ is h-bounded then f(a) contains only p-bounded letters of Y. If $a \in X$ is h-growing then the word f(a) contains a p-growing letter. If Y contains a semibounded letter then so does X.

Proof. Suppose first that $a \in Y$ is p-bounded. Then $p^n(a) = fg(a)$ is a letter of Y. Because f is nonerasing, g(a) is a letter of X. Because p permutes the p-bounded letters there is a positive integer m_1 such that $p^{m_1+n}(a) = a$. Consequently, $h^{m_1+n}(g(a)) = gp^{m_1}fg(a) = gp^{m_1+n}(a) = g(a)$. Hence $g(a)$ is h-permuted. The proof of the second claim is immediate by the equality $h^{m+n} = gp^m f$. Suppose then that $a \in X$ is h-growing. If $f(a)$ contains only p-bounded letters, the word $h^{m+n}(a) = gp^m f(a)$ is short for every m. Because this is not true, $f(a)$ contains a p-growing letter. For the proof of the last claim suppose that $y \in Y$ is semibounded. This implies the existence of words $u,v \in Y^*$ which contain only p-bounded letters and a semibounded letter $y' \in Y$ such that $p^n(y) = uy'v = f(g(y))$. By the earlier claims the word $g(y)$ contains exactly one h-growing letter, say z. Clearly, $f(z) = u'y'v'$, where $u',v' \in Y^*$ contain only p-bounded letters. Now $h^n(z) = gf(z) = g(u'y'v')$. The word $g(u'y'v')$ contains exactly one h-growing letter. Otherwise $fg(u'y'v') = p^n(u'y'v')$ would not contain exactly one p-growing letter. Hence $h^n(z)$ contains exactly one h-growing letter. This implies that X contains a semibounded letter. □

Theorem 1. Suppose $h: X^* \longrightarrow X^*$ is a nonerasing morphism and $B \subseteq X^+$ is a finite set. If each word of B contains a growing letter, then $L_1(h,B)$ and $Amb_1(h,B)$ are (effectively obtainable) regular languages.

Proof. If h is elementary, the first claim follows by Lemmas 2 and 7. Suppose h is not elementary. Let n, Y, p, g and f be as in Lemma 13. Denote the delay of g from right to left by d. Suppose now that $(b_1,b_1')(b_2,b_2')\dots(b_{n+d},b_{n+d}')$ is an arbitrary word of length n+d over $B \times B$ which is a prefix of a word in $L_1(h,B)$. To prove the first claim it is enough to show that the language

$$L' = [(b_1,b_1')\dots(b_{n+d},b_{n+d}')]^{-1}L_1(h,B)$$

is regular. Suppose without loss of generality that

$$\bar{h}(b_1;\dots;b_{n-1}) = \bar{h}(b_1';\dots;b_{n-1}')w,$$

where $w \in X^*$. By the choice of d there exists a word $w' \in Y^*$ such that

$$wgf(b_n)gpf(b_{n+1})\dots gp^d f(b_{n+d}) = g(w').$$

Suppose $c_1,\dots,c_s,c_{s+1},\dots,c_{s+t},c_1',\dots,c_s' \in B$. Now

$$(c_1,c_1')\ldots(c_s,c_s')(c_{s+1},\lambda)\ldots(c_{s+t},\lambda) \in L' \tag{5}$$

if and only if

$$\bar{h}(b_1;\ldots;b_{n+d};c_1;\ldots;c_{s+t}) = \bar{h}(b_1';\ldots;b_{n+d}';c_1';\ldots;c_s')$$

which is equivalent to

$$wgf(b_n)gpf(b_{n+1})\ldots gp^d f(b_{n+d})gp^d\bar{p}(f(c_1);\ldots;f(c_{s+t}))$$
$$= gf(b_n')gpf(b_{n+1}')\ldots gp^d f(b_{n+d}')gp^d\bar{p}(f(c_1');\ldots;f(c_s'))$$

or, equivalently,

$$w'p^d\bar{p}(f(c_1);\ldots;f(c_{s+t})) = f(b_n')pf(b_{n+1}')\ldots p^d f(b_{n+d}')$$
$$p^d\bar{p}(f(c_1');\ldots;f(c_s')). \tag{6}$$

Denote $w'' = f(b_n')pf(b_{n+1}')\ldots p^d f(b_{n+d}')$. Then (6) holds if and only if

$$\bar{p}(w';p^d f(c_1);\ldots;p^d f(c_{s+t})) = \bar{p}(w'';p^d f(c_1');\ldots;p^d f(c_s')).$$

Denote $D = \{w',w''\} \cup \{p^d f(b) \mid b \in B\}$. By Lemma 14 each word of D contains a growing letter. Therefore, the language $L_1(p,D)$ is regular. The regularity of L' follows because we have seen that (5) holds if and only if

$$(p^d f(c_1),p^d f(c_1'))\ldots(p^d f(c_s),p^d f(c_s'))(p^d f(c_{s+1}),\lambda)\ldots$$
$$(p^d f(c_{s+t}),\lambda) \in (w',w'')^{-1}L_1(p,D).$$

This proves the first claim.

The regularity of $Amb_1(h,B)$ is a consequence of the regularity of $L_1(h,B)$. □

Theorem 2. Suppose $h\colon X^* \longrightarrow X^*$ is a nonerasing morphism and $B \subseteq X^+$ is a finite set. If no letter of X is semibounded then $L_2(h,B)$ and $Amb_2(h,B)$ are (effectively obtainable) regular languages.

Proof. If h is elementary, the first claim follows by Lemmas 3 and 12. Otherwise we proceed as in the proof of Theorem 1. Now the word $(b_1,b_1')\ldots(b_{n+d},b_{n+d}')$ is chosen such that for some (b,b'), the word $(b_1,b_1')\ldots(b_{n+d},b_{n+d}')(b,b')$ is a prefix of a word in $L_2(h,B)$. Furthermore,

$$L' = [(b_1,b_1')\ldots(b_{n+d},b_{n+d}')]^{-1}L_2(h,B)$$

and w', w'', p, f and D are as in the proof of Theorem 1. Again it suffices to prove the regularity of L'. By Lemma 14 the alphabet Y contains no semibounded letter. Hence $L_2(p,D)$ is regular. To complete the proof of the first claim we show that $(c_1,c_1')\ldots(c_s,c_s')(c_{s+1},\lambda)\ldots(c_{s+t},\lambda)$ belongs to L' or, equivalently,

$$(b_1,b_1')\dots(b_{n+d},b_{n+d}')(c_1,c_1')\dots(c_s,c_s')(c_{s+1},\lambda)\dots(c_{s+t},\lambda) \in L_2(h,B)$$

if and only if

$$(w',w'')(p^d f(c_1),p^d f(c_1'))\dots(p^d f(c_s),p^d f(c_s'))(p^d f(c_{s+1}),\lambda)\dots(p^d f(c_{s+t}),\lambda) \in L_2(p,D)$$

provided that, furthermore, if $b_1 = b_1',\dots,b_{n+d} = b_{n+d}'$, $c_1 = c_1',\dots,c_{k-1} = c_{k-1}'$, $c_k \neq c_k'$ and $f(c_k) = f(c_k')$ then $k = s$ and $t = 0$. This follows as in the proof of Theorem 1. However, the following additional fact has to be established.

Suppose that

$$\bar{h}(b_1;\dots;b_{n+d};c_1;\dots;c_u)v = \bar{h}(b_1';\dots;b_{n+d}';c_1';\dots;c_u')$$

and

$$\bar{p}(w';p^d f(c_1);\dots;p^d f(c_u))\bar{v} = \bar{p}(w'';p^d f(c_1');\dots;p^d f(c_u'))$$

where u is a nonnegative integer, $v \in X^*$ and $\bar{v} \in Y^*$. We show that v contains only h-permuted letters if and only if $\bar{v}$ contains only p-bounded letters. Clearly, v is empty if and only if $\bar{v}$ is empty. Suppose first that $v = v_1v_2\dots v_i$ where the letters $v_1,v_2,\dots,v_i \in X$ are h-permuted. Then there are h-permuted letters $v_1',v_2',\dots,v_i' \in X$ such that $v = h^{n+d+u+1}(v_1')\dots h^{n+d+u+i}(v_i')$. Now we have $\bar{v} = p^{u+2+d}f(v_1')\dots p^{u+i+1+d}f(v_i')$. By Lemma 14, $\bar{v}$ contains only p-bounded letters. Suppose then that $\bar{v} = \bar{v}_1\bar{v}_2\dots\bar{v}_i$, where $\bar{v}_1,\bar{v}_2,\dots,\bar{v}_i \in Y$ are p-bounded. For each j with $1 \le j \le i$ choose a p-bounded letter $\bar{v}_j'$ such that $\bar{v}_j = p^{u+1+j+d+n}(\bar{v}_j') = p^{u+1+j+d}f(g(\bar{v}_j'))$. Then $v = h^{n+d+u+1}g(\bar{v}_1')\dots h^{n+d+u+i}g(\bar{v}_i')$. By Lemma 14, $g(\bar{v}_1'),\dots,g(\bar{v}_i')$ are h-permuted letters. Hence v contains only h-permuted letters.

The second claim is again a consequence of the first claim. □

In the balance of this section we discuss shortly some generalizations and applications of Theorems 1 and 2.

Theorem 3. Suppose $h: X^* \longrightarrow X^*$ is a nonerasing morphism and $B \subseteq X^+$ is a finite set. Assume that R is a regular language over Σ satisfying the following condition. There is an $M \ge 1$ such that, if $(b_1,b_1')\dots(b_M,b_M')$ is a subword of a word in R, then at least one of the words $b_1,\dots,b_M$ (resp. $b_1',\dots,b_M'$) contains a growing letter ($b_i, b_i' \in B$). Then $L_1(h,B) \cap R$ is effectively regular.

Proof. It is not difficult to see that there is an $M_0 \geq 1$ such that if $(c_1,c_1')\dots(c_n,c_n')(c_{n+1},\lambda)\dots(c_{n+m},\lambda) \in L_1(h,B) \cap R$ then $m \leq M_0$. Therefore Theorem 3 is a corollary of Theorem 1. □

Theorem 4. Suppose $h: X^* \longrightarrow X^*$ is a morphism. If no letter of X is semibounded then it is decidable whether or not h is an L code.

Proof. If there exists a letter $x \in X$ such that $h(x) = \lambda$ the morphism h is not an L code. Otherwise, by Theorem 2 the emptiness of $Amb_2(h,X)$ is a decidable property. The claim follows because $Amb_1(h,X)$ is empty if and only if $Amb_2(h,X)$ is empty. □

Given a morphism $h: X^* \longrightarrow X^*$ and a finite set $B \subseteq X^+$, the *degree of L ambiguity of* h *with respect to* B can be defined similarly as the degree of a context-free grammar, see [9]. Various regularity and decidability results can be obtained as above. However, a detailed discussion of the degrees of L ambiguity lies outside the scope of this paper.

ACKNOWLEDGEMENT. The author would like to express his gratitude to Professor Arto Salomaa for many useful comments.

REFERENCES

[1] J. Berstel. Fibonacci words - a survey. In G. Rozenberg and A. Salomaa (eds.) *The Book of L.* Springer-Verlag, Berlin, Heidelberg, New York (1986) 13-27.

[2] J. Berstel and D. Perrin. *Theory of Codes*. Academic Press, New York (1985).

[3] K. Culik II and A. Salomaa. Ambiguity and decision problems concerning number systems. *Information and Control* 56 (1983) 139-153.

[4] A. Ehrenfeucht and G. Rozenberg. Simplifications of homomorphisms. *Information and Control* 38 (1978) 298-309.

[5] C. Frougny. Linear numeration systems of order two. *Information and Computation* 77 (1988) 233-259.
[6] J. Honkala. Unique representation in number systems and L codes. *Discrete Applied Mathematics* 4 (1982) 229-232.
[7] J. Honkala. Bases and ambiguity of number systems. *Theoretical Computer Science* 31 (1984) 61-71.
[8] J. Honkala. A decision method for the recognizability of sets defined by number systems. *RAIRO Theoretical Informatics* 20 (1986) 395-403.
[9] J. Honkala. It is decidable whether or not a permutation-free morphism is an L code. *International Journal of Computer Mathematics* 22 (1987) 1-11.
[10] A. de Luca and A. Restivo. Star-free sets of integers. *Theoretical Computer Science* 43 (1986) 265-275.
[11] H. Maurer, A. Salomaa and D. Wood. L codes and number systems. *Theoretical Computer Science* 22 (1983) 331-346.
[12] H. Maurer, A. Salomaa and D. Wood. Bounded delay L codes. *Theoretical Computer Science*, to appear.
[13] A. Salomaa. *Jewels of Formal Language Theory*. Computer Science Press, Rockville (1981).
[14] A. Salomaa. *Public-Key Cryptography*. Springer-Verlag, Berlin, Heidelberg, New York (1990).
[15] A. Salomaa. Cryptographic properties of L codes. In preparation.

Iterated Substitutions and Locally Catenative Systems: a Decidability Result in the Binary Case

Christian Choffrut

Laboratoire d'Informatique Théorique et de Programmation, Université Paris 7, Tour 55-56, 1er étage, 2 Pl. Jussieu, Paris 75 251 Cedex 05, email: cc@litp.ibp.fr

Abstract : Given a substitution h over a binary alphabet {a,b} and the infinite DOL-sequence a, h(a), $h^2(a)$,..., $h^n(a)$, . . . we prove that for some m>0 the m-th term $h^m(a)$ of the sequence is an arbitrary product of the previous terms, if and only if this is true for m=3. This settles in a particular case the decidability of the local catenativity of the DOL-sequence, a longstanding open problem posed some fifteen years ago.

Key words: substitutions in the free monoids, DOL-systems, DOL-sequences, linear recurrence equations, locally catenative systems, codes, finite deciphering delay, equations in words.

1. INTRODUCTION

In the literature infinite words such as the Thue-Morse or the Fibonacci words are usually defined as the limit of an increasing sequence of finite words. Such sequences can be generated in basically two different ways. The first one makes use of DOL systems, i.e., fixed substitutions that are iterated infinitely many times (e.g., the Thue-Morse word). The other one utilizes catenative systems where the n-th term of the sequence is inductively defined by a fixed recurrence equation involving the last k terms, for a fixed integer k, and can be viewed as the natural generalization to words of linear recurrence equations for numbers (e.g. the Fibonacci sequence).This paper investigates the relationship between these two methods by determining in the case of binary alphabets when they are equivalent.

Let Σ^* be the set of all words on a finite alphabet Σ. A substitution (or morphism) h over Σ^* is defined as a rule that assigns to each letter of the alphabet a fixed word. Then the image by h of every word in Σ^* is the word obtained by substituting for every occurrence of each letter the word assigned by the rule. E.g., if h(a)=aba, h(b)=acb and h(c)=bc then h(aacb)=(aba)(aba)(bc)(acb). Given a word $w \in \Sigma^*$ (an axiom) we are interested in the sequence: w, h(w), $h(h(w))=h^2(w)$,..$h(h^{m-1}(w))=h^m(w)$,...These sequences of words were introduced by Lindenmayer in the seventies in order to model the growth of living organisms. They gave rise to a rich field of research one of the most

This work was supported by the PRC Mathématiqes et Informatique and partially written during a visit at the Universidade de São Paulo in September 1990

profound result of which is the decidability of the equivalence of two such sequences. Here we deal with a different problem concerning these objects. Our main result is as follows:

Theorem

Given a morphism $h:\Sigma^*\to\Sigma^*$ *with* $\Sigma=\{a,b\}$ *such that* $h(ab)\neq h(ba)$ *the following conditions are equivalent*:

(i) *there exists some* $m>0$ *such that* $h^m(a)$ *is a product of elements in* $\{a, h(a), h^2(a),..., h^{m-1}(a)\}$

(ii) $h^3(a)$ *is a product of elements in* $\{a, h(a), h^2(a)\}$

Note that the corresponding problem for free commutative monoids has, to our knowledge, not yet been solved except again in the binary case (cf. [La]) which covers the case $h(ab)=h(ba)$. In fact, apart for two very restricted families of morphisms (cf. (3.2.6) and (3.3.9)), it suffices to verify that $h^2(a)$ is a product of elements in $\{a, h(a)\}$. This is an interesting analogy of the present situation with the vector spaces where every endomorphism satisfies an algebraic equation of degree equal to the dimension of the space. Here, morphisms may not satisfy any equation but if they do then the equation has degree 2, disregarding very particular morphisms.

More generally, the sequence $w, h(w), h^2(w),..,h^n(w),...$ is *locally catenative* if there exist $m>0, r>0, 0<i_0,..., i_r<m$ such that we have:

(1.1) $$h^m(w)=h^{m-i_0}(w)...h^{m-i_r}(w)$$

The problem as to whether or not local catenativity is decidable for morphisms over arbitrary alphabets was first stated in [LR] and remains a challenging open issue. So far only partial solutions, assuming completely different restrictions, are known. Define the *width* of the previous recurrence equation as the integer r and its *depth* as $\max\{i_0,..., i_r\}$. Then the problem can be shown to be decidable if the width or if the depth is bounded (cf. [RS] Theorem 2.10 and [Ru] Theorem 1.1). Since the problem is trivial when the morphism is assumed to be injective, both these papers show how to construct an injective morphism from the original one. Another approach consists in giving sufficient conditions (the so-called "dependent" systems of [LR]) but they are much too strong in the binary case since they reduce to $h(b)$ being a power of the letter a.

Local catenativity not only shows up in connection with generating infinite words but also in other frameworks such as numeration systems. In [Sh] it is proven that an infinite sequence is automatic (in some system of numeration) if and only if it can be generated by an "algebraic homomorphism" which is slightly more general than the condition for a morphism to be locally catenative.

In their handbook on "The Mathematical Theory of L Systems" the authors claim local catenativity and the DOL sequence equivalence are "comparable in importance" (cf. [RS] p. 14). However the latter was given two different solutions some twelve years ago (cf. [CF] and [ER1]) while the former remains open. In fact, when considering the number of iterations on which two non equivalent DOL-sequences agree there exists a huge gap between the double exponential of the so far best upper bound (cf. [ER2]) and the linear lower bound (cf. [Ni]). Yet, in the case of binary alphabets a sharp result can be stated since it is known that exactly 4 iterations are sufficient for deciding that two DOL-sequences are equivalent (cf. [Ka1]). It is interesting to notice that the example proving the sharpness of the result consists of a pair of morphisms that are locally catenative.

What concerns locally catenative systems no bound what so ever is known on the value of the integer m in (1.1), if such a value exists at all, but again the binary case results in a very neat and surprinsingly simple statement. This is certainly due to the strong properties of two-word subsets (see also [Se] where a sharp result on the periodicity of words generated by a binary alphabet is established). Not only do two word subsets define codes under mild conditions but these codes enjoy the property of finite deciphering delay which intuitively means that if a message is decoded in real time, only the last k factors for some fixed k, might be incorrect. Moreover, in general a word can be considered in at most one way as a subword of an arbitrary product of two given words (e.g., if the subset is {a,aba} then the words that are subwords of products of a's and aba's do not have bb as subword and they are subwords in just one way). Furthermore we know exactly how to characterize those pairs of words for which this fails to be true.

Unfortunetaly we strongly suspect that our technique can not be extended to a significantly more general case. The most that can be expected is that it extends to morphisms with finite decipherability delay but this is not too clear either.

2.PRELIMINARIES

2.1 Free monoids

We start by recalling the standard definitions on free monoids. Let $\Sigma=\{a, b, \dots\}$ be a finite set - or *alphabet* - consisting of *letters* and let Σ^* be the free monoid it generates, i.e., the set of all *words* - or finite sequences of elements in Σ – . The *length* of a word w, i.e., the number of its letters is denoted by $|w|$. The word of length 0 is the *empty* word and is denoted by 1. We denote by $\Sigma^+=\Sigma^*-\{1\}$ the set of all non empty words, i.e., the free semigroup generated by Σ.

Given four words w,p,u,s with w=pus we say that p (resp., s, u) is a *prefix* (resp. *suffix, subword*) of w. A prefix (resp. suffix, subword) is *proper* if it is different from the word itself. Given two words $u,v \in \Sigma^*$ we write $u\leq v$ (resp. $u<v$) whenever u is a prefix (resp. a proper prefix) of v. The following notation is useful for arbitrary subsets $U, V\subseteq\Sigma^*$:

$$U^{-1}V=\{x\in\Sigma^* \mid \exists\, u\in U, \exists\, v\in V \;\; v=ux\}$$
$$VU^{-1}=\{x\in\Sigma^* \mid \exists\, u\in U, \exists\, v\in V \;\; v=xu\}$$

We say that two words are *comparable* whenever $u\leq v$ or $v\leq u$, otherwise they are *incomparable*: e.g., ab≤aba but ab and ba are incomparable. An *occurrence* u in a word w is a subword of w along with its position in the word. Formally it is a pair (p,u) such that pu is a prefix of w. We also say that (p,u) is the occurrence of u at *position* p. E.g., bababa has two occurrences of aba at positions b and bab: (b,aba) and (bab,aba).

2.2 Sequences of words

Let us give precise definitions of the two notions on sequences $(u_n)_{n\geq 0}$ of words that are the main concepts of our work.

A sequence is *locally catenative* if there exist integers $m>0$, $r>0$ and $0<i_0,\dots, i_r<m$ such that the following recurrence equation holds for all $n>m$:

$$u_n=u_{n-i_0}\cdots u_{n-i_r}$$

For instance the famous Fibonacci sequence a, b, ba, bab, babba,babbaba ...is defined by : u_0=b, u_1=a and $u_n = u_{n-1}u_{n-2}$ for all n>1 and is thus locally catenative.

A sequence is *generated by a* DOL-*system* (or an *iterated substitution*) if there exist a word $w\in\Sigma^*$ and a morphism $h:\Sigma^*\to\Sigma^*$ such that: $u_0=w$, $u_1=h(w)$, $u_2=h(u_1)=h^2(w)$,..., $u_n=h(u_{n-1})=h^n(w)$,...

The question naturally arose as to whether or not a given sequence generated by a DOL-system is locally catenative. More precisely:

Question:

Given a word $w\in\Sigma^*$ *and a morphism* $h:\Sigma^*\to\Sigma^*$ *is it decidable whether or not there exist integers* m>0, r >0, $0<i_0,\ldots, i_r<m$ *such that the following holds*:

$$h^m(w)=h^{m-i_0}(w)\ldots h^{m-i_r}(w)$$

Notice that if this equality holds for some value m then it holds for all values greater than m. So far only partial solutions, assuming completely different restrictions on the sequence $i_0,\ldots, i_r$, are known. Here we solve the problem under the hypothesis that Σ is a binary alphabet {a,b} and that w is a letter:

Theorem

Given a morphism $h:\Sigma^*\to\Sigma^*$ *with* $\Sigma=\{a,b\}$ *such that* $h(ab)\neq h(ba)$ *the following conditions are equivalent*:

(i) *there exists some* m>0 *such that* $h^m(a)$ *is a product of elements in* $\{a, h(a), h^2(a),\ldots, h^{m-1}(a)\}$

(ii) $h^3(a)$ *is a product of elements in* $\{a, h(a), h^2(a)\}$

More precisely we can classify the morphisms that satisfy condition (i) according to their smallest exponent m:

m=1:

(2.1) *for some* i>0 *we have*
$$h(a)=a^i$$

m=2:

(2.2) *for some* i>0 *we have*
$$h(b)=a^i$$

(2.3) *for some* n>1, i,p>0, *and* $z\in\{a,(b^na)^pb^n\}^*$ *we have:*
$$h(a)=[(za^{2i})^{n-1}za^i]^p(za^{2i})^{n-1}z \quad \textit{and} \quad h(b)=a^iza^i$$

(2.4) *for some* $q,p,i>0$ *and* $z\in\{a,(ba)^pb\}^*$ *we have*
$u=(za^i)^pz$ *and* $v=a^i(za^i)^q$

(2.5) *for some* $q>0$, $p>$, $i+j>0$ *and* $z\in\{a,b^p\}^*$ *we have*
$u=(za^{i+j})^{p-1}z$ *and* $v=(a^iza^j)^q$

m=3:

(2.6) *for some* $i>0$, $k\geq 0$ *we have*
$u=b^{2k+1}a^ib^{2k+1}$ *and* ($v=a^ib^{2k+1}$ or $v=b^{2k+1}a^i$)

(2.7) *for some* $i>0$, $k,q\geq 0$ and ($z=(ab)^{2k+1}$ *or* $z=(ba)^{2k+1}$) *we have*
$u=za^iz$ *and* $v=a^i(za^i)^{2q+1}$

Observe that the converse, to wit whether or not a given locally catenative sequence can be generated by a DOL-system, is trivial because the morphism can only be one among finitely many candidates.

2.3 Equations Codes

Testing local catenativity of a morphism amounts to verifying that certain relations hold between words and we speak of *equations* satisfied by *unknowns* that are words. These relations involve overlappings of occurrences that result in strong restrictions on words. We illustrate the techniques constantly used in the proofs by quoting the so called *conjugacy equation* (cf. [LS] or [Lo], Proposition 1.3.4):

Proposition 2.1 *Given three words* $u,v,w\in\Sigma^*$ *the following conditions are equivalent*:

(i) $uv=vw$

(ii) *there exist two words* $x,y\in\Sigma^*$ *and an integer* $i\geq 0$ *such that*:
$u=xy$ $v=(xy)^ix$ and $w=yx$

A word v two occurrences of which overlap gives rise to a conjugacy equation. The Proposition asserts that v has a *period* equal to xy.

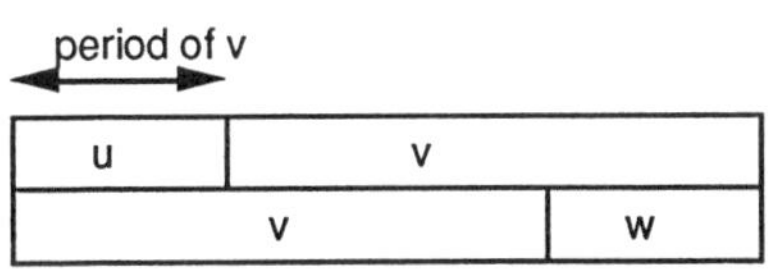

The equation $xty=yzx$ in four unknowns can not be solved in the sense that there is no finite description of the set of all its solutions. In this work the equation shows up in connection with overlappings of occurrences. It does not present itself in all its generality but under the hypothesis that $t=a^i$ for some integer $i\geq 0$, in which case the equation can be solved.

Lemma 2.2 *A word* $w \in \Sigma^*$ *admits a factorization of the form:*

(i) $w = xa^iy = yzx$

for some x,y,z *and* $i \geq 0$ *if and only if there exists* $t \in \Sigma^*$ *where*

(ii) $x, y \in (t\,a^i)^*t$ *and* $z = a^i$

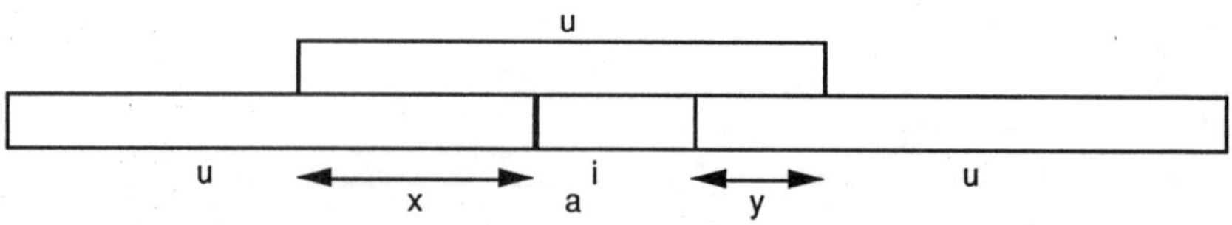

Proof : Clearly only the case $w \notin a^*$ need be considered. Since the condition is trivially sufficient we prove that it is necessary. By symmetry we may assume without loss of generality that $|y| \leq |x|$. Furthermore the condition (ii) obviously holds when y=x which allows us to suppose that y is a proper prefix of x.

Observe that the condition $|y|<|x|<|yz|$ may not occur since it would entail $x=yz_1$ and $z=z_1z_2$ for some $z_1, z_2 \in \Sigma^*$. After left cancellation of (i) by yz_1 we obtain: $a^iy=z_2x=z_2yz_1$. Because of $|z|=i$, we get $z_2 \in a^*$ and thus $a^jy=yz_1$ for some $0 \leq j \leq i$. This implies $z_1, y \in a^*$ and therefore $w \in a^*$ by Proposition 2.1.

We are left with the condition $|yz|<|x|$, i.e., $x=yzx_1$ for some $x_1 \in \Sigma^*$. After left simplification of (i) by yz we get: $x_1a^iy=yzx_1$. Since (ii) obviously holds when w has minimal length we may conclude by induction. □

On subsets $X \subseteq \Sigma^*$ of words we need two defintions. As usual X^* (resp. X^+) is the submonoid (resp. subsemigroup) generated by X, i.e., the set of all products of $n \geq 0$ (resp. $n>0$) words in X, where the product is the empty word when n=0.

We say X is a *code* whenever all words in the submonoid X^* have a unique factorization in words of X. Formally:

> if $x_1 \ldots x_n = y_1 \ldots y_m$ holds where $x_i, y_j \in X$ for i=1,...,n and j=1,...,m then n=m and $x_i=y_i$ for i=1,...,n

Example 2.1 : {a,ba} is a code but {a,ba,ab} is not since (a)(ba)=(ab)(a). □

A X-*prefix* (resp. X-*suffix*) of a word $w \in X^*$ is a word $u \in X^*$ such that $u^{-1}w \in X^*$ (resp. $wu^{-1} \in X^*$).

Example 2.2 : with X={a,ba}, $a \in X$ is not a X-suffix of w=ba. □

When X has two elements it is simple to determine when it is a code (cf., e.g.[LS]):

Proposition 2.3 *Given two words* $x, y \in \Sigma^*$ *the following conditions are equivalent:*
(i) $xy \neq yx$
(ii) $x^+ \cap y^+ = \emptyset$
(iii) $\{x,y\}$ *is a code*
(iv) *if* w *is a prefix that is common to the subsemigroups* $x\{x,y\}^*$ *and* $y\{x,y\}^*$ *then* $|w| \leq |x| + |y| - 1$

The fourth property is known as the *finite deciphering delay* -abbreviated *f.d.d.* - property (cf. [BP], p.128 or [RS], p. 130) . If further the two words satisfy $y\Sigma^* \cap x^* = \emptyset$ we have the stronger property:

(2.8) *if* $X=\{x,y\} \subseteq \Sigma^+$ *satisfy* $y\Sigma^* \cap x^* = \emptyset$ *and if* w *is a prefix of* xX^* *and* yX^* *then* $|w| < \max\{|x|, |y|\}$

Indeed, if $|w| \geq \max\{|x|, |y|\}$ then we have $|y| > |x|$ since otherwise y is a prefix of x. Furthermore, y is a prefix of xX^* and because of condition $y\Sigma^* \cap x^* = \emptyset$ we get $y = x^i y_1$ where $i>0$ and y_1 is a proper prefix of y. This implies $y_1\Sigma^* \cap x^* \neq \emptyset$ by Proposition 2.1, i.e., $x^i y_1 \Sigma^* \cap x^* \neq \emptyset$ and finally $y\Sigma^* \cap x^* = \emptyset$, a contradiction.□

Under the same assumption, by repeated applications of (2.8) we get:
(2.9) $(X^*y)^{-1}X^* \subseteq X^*$

We also make constant use of the following elementary properties on morphisms (we recall that a morphism h is *prefix*, if for all $a,b \in \Sigma$, the condition $h(a)$ is a prefix of $h(b)$ implies $a=b$):

(2.10) *for all* $m>0$ *the morphism* h *is injective if and only if so is* h^m
(2.11) *if the morphism* h *is prefix then for all words* $u,v \in \Sigma^*$ *we have:* u *is a prefix of* v *if and only if* $h(u)$ *is a prefix of* v

2.4 Coverings

Let $X \subseteq \Sigma^*$ be a fixed non empty code and let $w \in \Sigma^*$ be an arbitrary word. A *X-covering* of w (or for short a *covering* when X is understood) is a product $u = x_1 \ldots x_n$ of elements of X such that w is a prefix of $x_1 \ldots x_n$ and $x_1 \ldots x_{n-1}$ is a proper prefix of w. Also we say that X *covers* w and that w is *X-covered*by u. E.g., let $X=\{a,ab\}$ and $w=aaba$. Then (a)(ab)(a) and (a)(ab)(ab) are two coverings of w. A covering u of w *meets* a prefix $x \in X^*$ of w if $u=xx'$ for some $x' \in X^*$. Eg., if $X=\{a,ab,bb\}$ and $w=abbb$ then the covering (a)(bb)(bb) meets the prefix abb but the covering (ab)(bb) does not.

More generally, a *X-interpretation* of $w \in \Sigma^*$ (or for short an *interpretation* when X is understood) is a sequence $x_1', x_2, \ldots x_{n-1}, x_n'$ of words satifying the following conditions:
(i) $n>1$
(ii) $x_i \in X$ for $i=2\ldots,n-1$
(iii) x_1' is a proper suffix of a word in X and x_n' is a proper prefix of a word in X
(iv) $w = x_1' x_2 \ldots x_{n-1} x_n'$

Given a covering $x_1 \ldots x_n$ of w and a factorization $w=pus$ such that the occurrence (p,u) overlaps two distinct factors of the covering, we say that $x_i', x_{i+1}, \ldots, x_{j-1}, x_j'$ is the interpretation of the factor u induced by the covering $x_1 \ldots x_n$ if the following holds (cf. Figure):
$i<j$, $u= x_i' x_{i+1} \ldots x_{j-1} x_j'$, $x_i=x_i'' x_i'$, $x_j=x_j' x_j''$

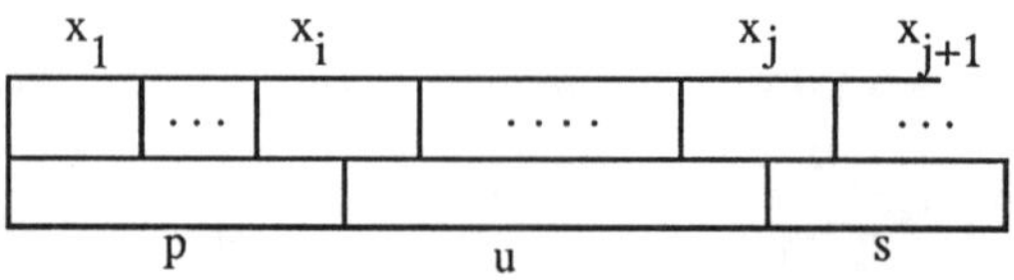

The following deals with what is known as the operation of composition of codes.

Lemma 2.4

Let $X=\{x,y\} \subseteq \Sigma^+$ *satisfy* $y\Sigma^* \cap x^* = \emptyset$. *Assume* $u \in xX^*$ *and* $v \in yX^*$ *and set* $U=\{u,v\}$. *If a word* w *in* $X^+ - U^* x^+$ *has a* U*-covering it is unique.*

Proof

Assume there exists a word w in $X^+ - U^* x^+$ that has two different U-coverings: $z_1 \ldots z_p$ and $z_1' \ldots z_q'$ for some $p,q>0$. Because of (2.8), without loss of generality we may assume that:

$$z_1=z_1', \ldots, z_{p-1}= z_{p-1}' , z_p=v, \text{ and } z_p'= \ldots = z_q' \in x^+$$

But this implies $w= z_1' \ldots z_q' \in U^* x^+$, a contradiction. ▫

Lemma 2.5

Assume $Z=\{z,t\} \subseteq \Sigma^+$ *satisfy* $t\Sigma^* \cap z^* = \emptyset$ *and* $u=(tz^i)^r t$, $v=z$. *If for* $0<s<r$, w *is a prefix of some word in* $(tz^i)^s t \{u,v\}^*$ *and some word in* $\{u,v\}^*$ *then* w *is a prefix of* $(tz^i)^*$.

Proof

Because of the condition $t\Sigma^* \cap z^* = \emptyset$ the first factor of w in $\{u,v\}^*$ is $(tz^i)^r t$. If w is shorter than $(tz^i)^r t$ then we are done. Otherwise, the equality $(tz^i)^s t z^i (tz^i)^{r-s-1} t = (tz^i)^r t$ shows that w is comparable with $(tz^i)^s t z^i u = (tz^i)^s t z^i (tz^i)^r t$. Consider the word $w' = [(tz^i)^s t z^i]^{-1} w$. If $|w'| \leq |(tz^i)^{r-s-1} t|$ then we are done. Otherwise it is a prefix of some word in $(tz^i)^{r-s-1} t \{u,v\}^*$ and some word in $\{u,v\}^*$ which completes the proof by induction. ▫

2.5 An outline of the proof

What follows is an informal presentation of the proof of the Theorem. Our ambition here is to give some insight of its structure which might be blurred by the technicality of the verifications. It must be taken as an attempt to facilitate the comprehension. A word of caution: we are aware that rigorously speaking, almost all our statements can be disproved by an appropriate counterexample. This also explains the length of our proof because "in general" everything goes well except for some "naughty" cases that have to be handled directly. Let us illustrate this idea. Without loss of generality we may assume that u has a prefix of the form a^ib and v a prefix of the form a^jb some some $i,j\geq 0$. The case where $j<i$, which under all reasonable measures can be considered as half of all the possible cases, can be treated in less than half a page (see case 3.4.1). The remaining cases are thus those that require our attention in this work.

Set $h(a)=u$, $h(b)=v$ and for all $m\geq 0$ define:

$$u_m = h^m(a) = h^{m-1}(u) \quad \text{and} \quad v_m = h^m(b) = h^{m-1}(v)$$

The idea of the proof is to verify by induction on $m>0$ that (with one exception cf. case 3.3.1), if u_2 is not a product of elements in $\{u_0,u_1\}=\{a,u\}$ then the sequence:

$$a,\ h(a),\ ...,\ h^{m-1}(a),..$$

is not locally catenative. More precisely, a stronger result holds in general, namely if u_2 is not $\{a,u_1\}$-covered then for all $m>0$, u_m is not $\{u_0,u_1,...,u_{m-1}\}$-covered. Set:

(2.12) $\qquad u\in tb\Sigma^*$

where t is the maximum prefix of u such that $h(t)\in\{a,u\}^*$. Then $u_2=h(u)\in h(t)v\Sigma^*$ and all $\{a,u\}$-coverings of u_2 start with $h(t)$. It is thus natural to consider the maximum factor x' of u starting at position t such that $h(x')$ is $\{a,u\}$-covered.

Example 2.3: if $u=abaabb$ and $v=aab$, then $t=a$ and $x'=b$ since $vaabb=au$ and vu is not $\{a,u\}$-covered.

Example 2.4: if $u=b^3a^2b^3$ and $v=ab^3a$ then $t=1$ and $x'=b^3$ since $v^3ab^3=auaau$ and v^3u is not $\{a,u\}$-covered.

More generally, we are led to investigate the elements of the semigroup $v\{u,v\}^*$ that are $\{a,u\}$-covered. It turns out that these words are in general all $\{u,v\}$-prefixes of a unique (possibly infinite) word s (v^2 and the infinite sequence $v^\omega=vv...$ respectively in the previous examples). This property may be compared with the result of [Ka3] on the intersection of two subsemigroups generated by two elements but it can not be reduced to it since it is stronger due to the specific situation we deal with. Furthermore the properties we use rely on the connection between the words u,v and the word s that controls the relations between these words: intuitively, the longer and more complicated the word s is, the more the occurrences u and v will overlap and therefore the more they will be restricted.

Set $x=x'c$ where c is the next letter of the occurrence (t,x') in u. The $\{a,u\}$-covering of $h(x')$ is in the subset $\{a,u\}^*u'$ where u' is a proper prefix of u.

Example 2.3 (continued): $x=ba$ and $u'=ab$ ▫

Example 2.4 (continued): $x=bbba$ and $u'=bbba$ ▫

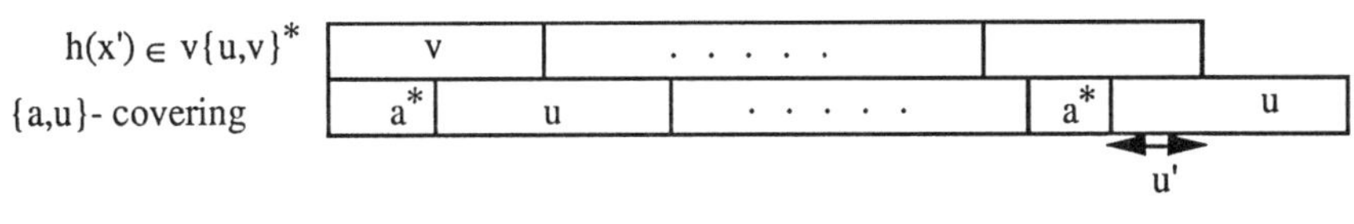

Assume $h^2(a) \notin \{a,h(a)\}^*$ and for all m>0 set $x_m = h^{m-1}(x)$, $t_m = h^{m-1}(t)$ and $x'_m = h^{m-1}(x')$. We prove the following predicate by induction on m>0 which clearly implies that the sequence is not locally catenative:

A_m: u_m is not covered by $\{u_0, ..,u_{m-1}\}$

Indeed, because of (2.12) we have $u_m \in t_m x_m \Sigma^*$ and u_m is not covered if x_m is not covered by $\{u_0, ..,u_{m-1}\}$ (this statement is not obvious but our presentation is informal and we're sweeping under the rug). We thus add the new predicate:

B_m: x_m is not covered by $\{u_0, ..,u_{m-1}\}$

This last predicate can be proven along the following lines. In general we have (as in Example 2.4):

(2.13) $\quad u'h(c) \in tx\Sigma^*$

and then it is not hard to show that all $\{u_0, ..,u_{m-1}\}$-coverings of x_m start with the $\{u_0, ..,u_{m-1}\}$-covering of x'_m. When this is not the case as in Example 1 it falls short of being true, meaning in the worst case the leftmost two letters are missing, and then we need to add new predicates. The word $h(x)=h(x'c)$ is not $\{a,u\}$-covered and it happens (once again due to the very specific solutions mentioned earlier) that is has a maximum prefix, say p, that is common with the word in $\{a,u\}^*$ that covers h(x'). Let w and z be corresponding suffixes, i.e., in particular h(x)=pw . Both w and z start with two different letters (in the Figure we assume as an illustration that these letters are a and b).

Example 2.4 (continued): $p=abbbaabbbaabbba$, $h(x)=v^3u=pbbba^2bbb$ and auaau=pabbb. We also have: h(x')=auaabbba=au(a)(a)(a)x, i.e., (2.13) is satisfied □

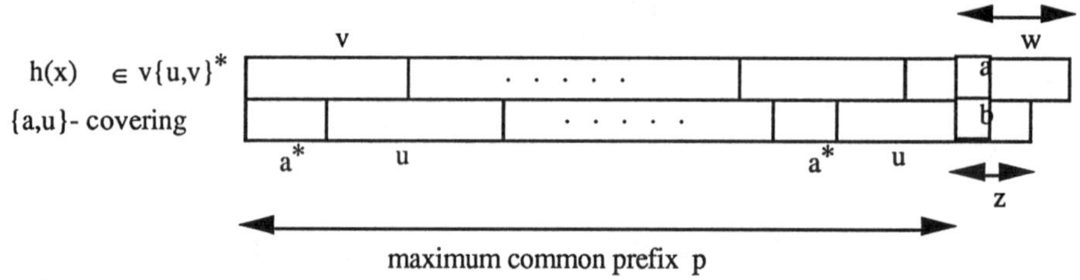

Assume $w \in a\Sigma^*$ and $z \in b\Sigma^*$ to fix ideas and take the m-th iterates $w_m \in u_{m-1}\{u_{m-1}, v_{m-1}\}^*$ and $z_m \in v_{m-1}\{u_{m-1}, v_{m-1}\}^*$ of w and z respectively. Under a mild hypothesis (such as w containing the letter b and z containing the letter a for example) the fact that w_m is a prefix of z_m (or vice versa) contradicts the finite deciphering delay property of the two element code $\{u_{m-1}, v_{m-1}\}$ cf. Proposition 2.3 (iv). Again if this hypothesis is not met then we must add new predicates.

3. PROOF OF THEOREM

Based on the classification of the pairs $\{u,v\}$ given at the end of the Appendix, we prove the theorem by a case study. In all instances, with the only exception of the subcase 3.3.1, we prove that if $h^2(a)$ is not $\{a,u\}$-covered then the sequence $a, h(a), ..., h^{m-1}(a),..$ is not locally catenative. We set $h(a)=u$ and $h(b)=v$ and for all $m>0$:

$$u_m=h^{m-1}(u) \quad \text{with} \quad u_0=a$$
$$v_m=h^{m-1}(v) \quad \text{with} \quad v_0=b$$
$$U_m=\{u_0,u_1,...,u_m\}$$

More generally, and unless otherwise stated, for all words $x\in\Sigma^*$ and all integers $m>0$ we use the notation:

$$x_m= h^{m-1}(x)$$

As we go through the proof, we reduce the number of cases by making use of the following elementary observation. To the morphism $h:\Sigma^*\rightarrow\Sigma^*$ assign the *reverse* morphism $\tilde{h}:\Sigma^*\rightarrow\Sigma^*$ defined by $\tilde{h}(a)=\tilde{u}$ and $\tilde{h}(b)=\tilde{v}$. For all $x\in\Sigma^*$ if $h(x)=w$ then $\tilde{h}(\tilde{x})=\tilde{w}$. Thus we may associate with the sequence $a, h(a),..., h^i(a),..$ the reverse sequence $a, \tilde{h}(a),..., \tilde{h}^i(a),...$ In particular for all $m\geq0$ $\tilde{h}^m(a)=\tilde{u}_m$ holds and the first sequence is locally catenative if and only if so is the second.

3.1 Preliminary results

The different cases share certain technical properties stated in the following Lemmas that can be skipped in a first reading.

Let us explain the first Lemma intuitively. Assume a word $w\in\Sigma^*$ has a U_1-covering: $x_1x_2...x_p$ where $x_i\in\{a,u\}$ for $i=1,2,...,p$. Clearly, $w_m=h^{m-1}(w)$ admits the U_m-covering: $h^{m-1}(x_1)h^{m-1}(x_2)...h^{m-1}(x_p)$. The Lemma says that under certain assumptions this is its only possible U_m-covering.

Lemma 3.1.1 *Assume the following two conditions are satisfied*:

(i) u_p *is not* U_{p-1}*-covered for* $p=m-1,m$

(ii) u_{m-1} *is not a prefix of some word in* $U_{m-2}^+u_m$.

Then:

(1) u_{m-1} *can only be* U_m*-covered by* u_{m-1} *or by* u_m

(2) *all words in* $\{u_{m-1},u_m\}^*u_m$ *have a unique* U_m*-covering.*

Proof

(1) Indeed, if this were not the case we would have $u_{m-1}x=yu_{m-1}$ or $u_{m-1}x=yu_m$ with $x\in\Sigma^*$ and $y\in U_{m-2}^+$. The latter contradicts (ii) while the former leads via Proposition 2.1 to u_{m-1} being U_{m-1}-covered, contradicting (i).

(2) As in (1) u_m can only be U_m-covered by u_m itself. Now all elements in $\{u_{m-1}, u_m\}^*u_m$ can be uniquely factored into maximal products of the form $u_{m-1}^{\ell}u_m$ $(\ell\geq0)$. Such

a factor has a unique U_m-covering. Indeed, assume for some $0 \le j < \ell$, the j-th occurrence of u_{m-1} is covered by u_m. Then $u_{m-1}^{\ell-j} u_m \in u_m \Sigma^*$ implying via Proposition 2.1:

$$u_m \Sigma^* \cap u_{m-1}^{+} \neq \emptyset$$

a contradiction to (i). □

Lemma 3.1.2 *Assume* $u \in a^i x \Sigma^*$ *where* $x \in b\Sigma^*$. *If further the conditions* (i) *and* (ii) *of* Lemma 3.1.1 *are satisfied and if* x_m *is not* U_{m-1}*-covered then a word* $w = t u_{m-1}^j x_m$ *with* $t \in [\{u_{m-1}, u_m\}^* u_m + 1]$ *has a* U_m*-covering if and only if* $j \ge i$. *In this case the covering is unique and equals* $t u_{m-1}^{j-i} u_m$.

Proof

By Lemma 3.1.1 we may assume the word w is of the form $u_{m-1}^j x_m$. Observe the occurrence (u_{m-1}^j, x_m) can only be U_m-covered when $i=0$ and in this case it is covered by u_m. Indeed, in all other cases x_m is a prefix of $y u_m \in y u_{m-1}^i x_m \Sigma^*$ for some $y \in U_{m-1}^*$, i.e., by Proposition 2.1 $x_m \Sigma^* \cap (y u_{m-1}^i)^+ \neq \emptyset$, contradiction.

Therefore for some $0 \le k < j$ the occurrence (u_{m-1}^k, u_{m-1}) is covered by u_m. If $j < i$ then $u_{m-1}^{j-k} x_m$ is a prefix of $u_{m-1}^i x_m$, violating the f.d.d. property, cf. Proposition 2.3 (iv). Now if $j \ge i$ and for some integer $k \neq j-i$ the occurrence (u_{m-1}^{k-1}, u_{m-1}) is covered by u_m then $u_{m-1}^{j-k} x_m$ and $u_{m-1}^i x_m$ are comparable, violating the same property. □

Lemma 3.1.3 *Assume condition* (i) *of Lemma 3.1.1 is satisfied and* u, v *are incomparable. Set* $u \in a^i x c \Sigma^*$ *with* $c \in \Sigma$ *and* $x \notin a^*$ *and* $w = xd$ *where* $d \in \Sigma$, $c \neq d$. *Then for every* $j \ge 0$, *the* U_m*-covering. of the words* $u_{m-1}^j w_m$ *meet* u_{m-1}^j.

Proof

If for $0 \le k < j$, an occurrence (u_{m-1}^k, u_{m-1}) is covered by u_m then u_{m-1} and v_{m-1} are comparable contradiction. □

From now on, using the classification presented at the end of the Appendix, we divide the possible pairs $\{u,v\}$ into the following 7 different classes. For each of them (with the exceptions where we may have $h^2(a) \notin \{a, h(a)\}^*$ and yet $h^3(a) \in \{a, h(a), h^2(a)\}^*$, cf. (3.2.6) and (3.3.9)), we assume that $h^2(a) \notin \{a, h(a)\}^*$ and then we prove that the system is not locally catenative. All cases are examined in the following order. When studying a given case the conditions of the previous cases are assumed not to hold. Also since $v \in \{a,u\}^*$ clearly defines a locally catenative morphism we assume that $v \notin \{a,u\}^*$ holds.

1) $v^+ \cap \{a,u\}^* \neq \emptyset$
2) $v\{u,v\}^* u \{u,v\}^* v \cap \{a,u\}^* \neq \emptyset$
3) v is not $\{a,u\}$-covered
4) $v^n u$ is covered for some $n>0$
5) v^n is $\{a,u\}$-covered by $a^i u a^k u$ for some $i,k \ge 0$, $n>1$ and v has no prefix in $a^* u$

6) v has a prefix in $a^* u$.
7) v^n is a prefix of $a^i u$ for some $n>0$, $i\geq 0$ and v^{n+1} is not $\{a,u\}$-covered

3.2 Assumption: $v^+ \cap \{a,u\}^* \neq \emptyset$

By Lemma A.1 we have:

$$u=(za^{i+j})^{p-1}z \text{ and } v=(a^i za^j)^q \text{ with } q>0,\ p>1\ \text{g.c.d.}(p,q)=1 \text{ and } i+j>0$$

Furthermore,

(3.2.1) $v^p=(a^i ua^j)^q$

Also by taking the reverse images of u and v if necessary, we may assume $i>0$. In particular za is a prefix of u.

More generally, for all $0<r<n$ we have:

(3.2.2) $v^r a^i (za^{i+j})^\alpha z=a^i(u\, a^{i+j})^\beta u$ for some $0\leq\alpha<p-1$ and $0\leq \beta\leq q-1$

(3.2.3) $v^r=a^i(ua^{i+j})^\beta (za^{i+j})^{p-\alpha-1}za^j$ for some $0<\alpha\leq p-1$ and $0\leq \beta\leq q-1$

If the prefix za is $\{a,b^p\}$-covered then $z\in \{a,b^p\}^*$ and we have $h^2(a)\in \{a,h(a)\}^*$ (cf. (2.5)). Thus, we may assume that for some $t\in \{a,b^p\}^*$ the following holds:

$$u\in tx\Sigma^* \text{ where } x=b^r a \text{ for some } 0<r<p$$

and the prefix tx is a prefix of the prefix za of u.

In all cases, as in (2.12), if we set $x=x'a$ then $h(x')\in \{a,u\}^* u'$ for some prefix u' of u, and we have:

$$u'h(a)\in tx\Sigma^*$$

except if $j=0$, $\alpha=p-1$ (i.e., $rq\equiv 1 \bmod p$), $z=b^s$ and $s=r+\ell p$ for some $\ell\geq 0$ which is precisely the case that we consider next.

3.2.1 The special case $j=0$, $\alpha=p-1$, $z=b^s$ and $s=r+\ell p$ for some $\ell\geq 0$

We have $u=(b^s a^i)^{p-1}b^s$ and $v=(a^i b^s)^q$ with $s=r+\ell p$ for some $\ell\geq 0$. Also:

$$v^r=a^i(ua^i)^\beta z$$

Observe that the following holds:

(3.2.4) $v^r u=(a^i u)^\beta a^i b^{2s}(a^i b^s)^{p-1}$

and

(3.2.5) $v^{2r}=(a^i u)^{2\beta} a^i b^s a^i b^s$

We first consider the case $p=2$ and thus $r=1$ which as we said in the introduction, is one case where we must compute u_3 in order to check that the sequence is indeed locally catenative:

(3.2.6) *If for some* $k\geq 0$ *we have* $u=b^{2k+1}a^i b^{2k+1}$ *and* $v=a^i b^{2k+1}$ *then* $u_2\notin \{u_0,u_1\}^*$ *and* $u_3\in \{u_0,u_1,u_2\}^*$ *, i.e., the sequence is locally catenative.*

Proof:

We have $u_2=v^{2k+1}u^i v^{2k+1} \in (a^i u)^k a^i b^{4k+2}\Sigma^*$ which shows that $u_2 \notin \{a,u\}^*$. Compute
$v_2 u_2^i v_2 = (u^i v^{4k+2})^i u^i (v^{2k+1}u^i v^{2k+1}) = (u^i (a^i u)^{2k+1})^i u^i u_2$
We obtain:
$u_3=v_2^{2k+1}u_2^i v_2^{2k+1}=v_2^{2k}v_2 u_2^i v_2\, v_2^{2k}=(u^i u_2)^k v_2 u_2^i v_2\, (u^i u_2)^k \in \{u_0,u_1,u_2\}^*$ □

Returning to the remaining cases where $p>2$, we prove the following conjunction of predicates by induction on $m>0$:

- A_m: u_m is not U_{m-1}-covered
- B_m: all U_m-coverings of v_m^p meet v_m^p.
- C_m: $x_{m+1}=v_m^r u_m$ is not U_m-covered
- D_m: $v_m^s u_m$ is not U_m-covered
- E_m: u_m is not a prefix of some word in $U_{m-1}^+ u_{m+1}$ or $U_{m-1}^+ u_{m+2}$ and u_m is not a prefix of u_{m+1}

Proof:

A_1: trivial.
B_1: this follows from Lemma 2.4 .
C_1: this follows from (3.2.4).
D_1: this follows from B_1 and C_1.
E_1: since $u \in b^s a \Sigma^*$ the word u can not be a prefix of a^+u_2 or a^+u_3.

Assume $m>1$.
A_m: all U_{m-1}-coverings of $u_m \in v_{m-1}^s u_{m-1}\Sigma^*$ yield a U_{m-1}-covering of $v_{m-1}^s u_{m-1}$, contradicting D_{m-1}.
B_m: this follows from Lemma 3.1.1

C_m: we have $v_m^r u_m \in (u_{m-1}^i u_m)^\beta u_{m-1}^i v_{m-1}^{2s}(u_{m-1}^i v_{m-1}^s)^{p-1}$. By Lemma 3.1.1 all U_m-coverings meet $(u_{m-1}^i u_m)^\beta$ yielding a U_m-covering of $u_{m-1}^i v_{m-1}^{2s}(u_{m-1}^i v_{m-1}^s)^{p-1}$. Because of E_{m-1}, (3.2.3) yields a U_m-covering of:

(3.2.7) $$v_{m-1}^{2s}(u_{m-1}^i v_{m-1}^s)^{p-1}$$
$$=(u_{m-2}^i u_{m-1})^{2(\beta+\ell q)}\, u_{m-2}^i v_{m-2}^s u_{m-2}^i v_{m-2}^s (u_{m-1}^i v_{m-1}^s)^{p-1}$$

Observe that:
$$v_{m-1}^{2s}(u_{m-1}^i v_{m-1}^s)^{p-1} \in (u_{m-2}^i v_{m-2}^s)^+ v_{m-2}^s u_{m-2}^i v_{m-2}^s \Sigma^*$$
Now by the f.d.d. property, no occurrence $u_{m-2}^j u_{m-1} \in u_{m-2}^j v_{m-2}^s u_{m-2}^i \Sigma^*$ with $0 \le j<i$ is a prefix of $u_m \in u_{m-2}^i v_{m-2}^s u_{m-2}\Sigma^*$ and by E_{m-1}, u_{m-1} is not a prefix of u_m. Assume for some $0 \le \gamma \le 2(\beta+\ell q)$ we have a U_m-covering of the form:

(3.2.8) $$(u_{m-2}^i u_{m-1})^\gamma u_m=$$
$$(u_{m-2}^i u_{m-1})^\gamma v_{m-1}^s (u_{m-1}^i v_{m-1}^s)^{p-1} \in (u_{m-2}^i v_{m-2}^s)^+ v_{m-1}^s v_{m-2}^s u_{m-2}^i v_{m-2}^s \Sigma^*$$

Because of the f.d.d. property applied to u_{m-2}, v_{m-2} this implies $v_{m-1}^{2s}=(u_{m-2}^i u_{m-1})^\gamma v_{m-1}^s$, i.e., $v_{m-1}^s \in \{u_{m-2},u_{m-1}\}^*$, i.e., $v_{m-1}^r \in \{u_{m-2},u_{m-1}\}^*$ and thus $v^r \in \{a,u\}^*$, contradiction. This far we have proven that all U_m-coverings of (3.2.7) meet $(u_{m-2}^i u_{m-1})^{2(\beta+\ell q)}$ and is not followed by u_m, yielding a U_m-covering of $u_{m-2}^i v_{m-2}^s u_{m-2}^i v_{m-2}^{2s} u_{m-2}^i$. Now by the f.d.d. property, for all $0 \le j < i$ no occurrence $(u_{m-2}^{i-j}, u_{m-2}^j v_{m-2}^s u_{m-2}^i)$ is a prefix of:

$$u_m \in u_{m-2}^i v_{m-2}^s u_{m-2} \Sigma^*$$

which shows that the U_m-covering of (3.2.7) yields a U_m-covering of $v_{m-2}^s u_{m-2}^i v_{m-2}^{2s}$ that does not start with u_m. Because of C_{m-2}, $v_{m-2}^s u_{m-2}$ is a prefix of $yu_{m-1} \in yv_{m-2}^s u_{m-2} \Sigma^*$ or of $yu_m \in yv_{m-1}^s u_{m-1} \Sigma^* \subseteq yu_{m-2}^i v_{m-2}^s u_{m-2} \Sigma^*$ for some $y \in U_{m-2}^*$. In both cases this contradicts C_{m-1} except if $v_{m-2}^s u_{m-2}^i v_{m-2}^{2s}$ is a prefix of u_{m-1}. However since $p>2$ this word is comparable with $v_{m-2}^s u_{m-2}^i v_{m-2}^s u_{m-2}^i$, contradiction.

D_m: by B_m all U_m-coverings of $v_m^{r+\ell p} u_m$ meet $v_m^{\ell p}$, yielding a U_m-covering of $v_m^r u_m$ contradicting C_m.

E_m: because of $u_{m+2} \in u_m^i v_m \Sigma^*$ if u_m is a prefix of some word in $U_{m-1}^+ u_{m+2}$ we have that u_m is a prefix $U_{m-1}^+ u_m$ contradicting A_{m-1}. Furthermore, because of:

$$u_{m+1} \in u_{m-1}^i v_{m-1}^s u_{m-1} \Sigma^*$$

if $u_m \in v_{m-1}^s u_{m-1} \Sigma^*$ is a prefix of some word in $U_{m-1}^+ u_{m+1}$ then so is $v_{m-1}^s u_{m-1}$ contradicting C_{m-1}.□

3.2.2 The general case

Because of the previous case we have the following relation (as (2.12), cf. the outline of the proof):

$$h(x) = v^r u \in (a^i u a^j)^\beta a^i x \Sigma^*$$

We prove the following conjunction of predicates by induction on $m>0$:

- A_m: u_m is not U_{m-1}-covered
- B_m: all U_m-coverings of $v_m^p u_m$ and $v_m^p v_m$ meet v_m^p.
- C_m: $x_{m+1} = v_m^r u_m$ is not U_m-covered
- D_m: $z_{m+1} u_m$ is not U_m-covered
- E_m: u_m is not a prefix of some word in $U_{m-1}^+ u_{m+1}$

Proof:

A_1: trivial.

B_1: the prefix $v^p z \in \{a,z\}^*$ of $v^p u$ and the prefix $v^p a^i z \in \{a,z\}^*$ of $v^p v$ have $\{a,u\}$-coverings. Because of Lemma 2.4 they clearly meet v^p.

C_1: Because of (3.2.3) and Lemma 2.4, v^r has the unique $\{a,u\}$-covering of (3.2.2). If $v^r u$ is $\{a,u\}$-covered then $a^i z = z a^i$, contradiction.

D_1: If $z_2 u$ is $\{a,u\}$-covered then so is $t_2 v^r u$. Because of B_1 all $\{a,u\}$-coverings meet t_2 yielding thus a $\{a,u\}$-covering of $v^r u$, contradicting C_1.

E_1: if u is a prefix of $a^+ u_2$ then $u \in a\Sigma^*$, i.e., u_2 starts with u and thus u is a prefix of $a^+ u$, contradiction.

Assume $m>1$.
A_m: all U_{m-1}-coverings of u_m yield a U_{m-1}-covering of $z_m u_{m-1}$, contradicting D_{m-1}.

B_m: because of Lemma 3.1.1 $v_m^p u_m$ has a unique U_m-covering. Applying the Lemma 3.1.2 and D_{m-1} to the prefix $v_m^p u_{m-1}^i z_m u_{m-1}$ of $v_m^p v_m$ shows that all U_m-coverings meet v_m^p.

C_m: Because of (3.2.3), Lemma 3.1.2 and D_{m-1}, v_m^r can only be covered by:
$$u_{m-1}^i (u_m u_{m-1}^{i+j})^\beta u_m$$
By (3.2.2) if $v_m^r u_m$ is covered it implies $z_m u_{m-1}^i = u_{m-1}^i z_m$, i.e., $z \in a^+$, a contradiction .

D_m: $z_{m+1} u_m$ has a prefix $t_{m+1} v_m^r u_m$. Because of Lemma 3.1.2 and B_m, all U_m-coverings of $z_{m+1} u_m$ meet t_{m+1}, yielding a U_m-covering of $v_m^r u_m$ contradicting C_m.

E_m: assume u_m is a prefix of some word $y u_{m+1}$ with $y \in U_{m-1}^+$. If $u \in a\Sigma^*$ then u_m is a prefix of u_{m+1} and thus is a prefix of $y u_m$ violating A_m. If $u \in b\Sigma^*$ then $u_m \in z_m u_{m-1} \Sigma^*$ is a prefix of $u_{m+1} \in y v_m \Sigma^* \subseteq y u_{m-1}^i z_m \Sigma^*$. Since $i>0$ this implies that $z_m u_{m-1}$ is a prefix of $y u_{m-1}^i z_m$, i.e., of $y u_{m-1}^i z_m u_{m-1}$ showing that $z_m u_{m-1}$ is U_{m-1}-covered, contradicting D_{m-1}. □

3.3 Assumption: for some $n>0$ $v^n u\{u,v\}^* \cap \{a,u\}^* \neq \emptyset$

By Lemma A.3 of the Appendix, we have two possibilities:

$n>1$: $u=[(za^{2i})^{n-1} za^i]^p (za^{2i})^{n-1} z$ and $v=a^i z a^i$ with $i,p>0$ and $z \notin a^*$
i.e., u and v satisfy the equality:
$$(v^n u)^p v^n = (a^i u)^{p+1} a^i$$

or

$n=1$: $u=(za^i)^p z$ and $v=a^i (za^i)^q$ for some $q,p,i>0$ and $z \notin a^*$
i.e., u and v satisfy the equality:

(3.3.1) $$(vu)^p v = (a^i u)^{p+q} a^i$$

If $n>1$ the following holds for all $0 \leq j \leq p$:

(3.3.2) $$(a^i u)^{j+1} = (v^n u)^j v^n [(za^{2i})^{n-1} za^i]^{p-j-1} (za^{2i})^{n-1} z$$
and for all $0 \leq n' \leq n$ we have $(v^n u)^j v^{n'} = (a^i u)^j [a^i (za^{2i})^{n-1} z]^j (a^i z a^i)^{n'}$

If $n=1$ the following holds:

(3.3.3) for all $0<j \leq p$, $0<j' \leq p+q$, $(q+p+1)j=(p+1)j'+\lambda$ and $0<\lambda \leq p$
we have $(vu)^j = (a^i u)^{j'} (a^i z)^\lambda$ and $(a^i u)^{j'+1} = (vu)^j (a^i z)^{p+1-\lambda}$

(3.3.4) for all $0 \le j < p$, $0 \le j' < p+q$, $(q+p+1)j+q=(p+1)j'+\lambda$ and $0<\lambda \le p$
we have $(vu)^j v=(a^i u)^{j'}(a^i z)^{\lambda} a^i$ and $(a^i u)^{j'+1}=(vu)^j(a^i z)^{p+1-\lambda}$

Set $w=(b^n a)^p b^n$. If $u_2 \notin \{u_0,u_1\}^*$ then $u \notin \{a,w\}^*$. In fact a stronger result holds:

Claim : *If* $n>1$ *or* $i>1$ *then* za^2 *is a prefix of* u *and it is not covered by* $\{a,w\}$. *Otherwise, we may assume without loss of generality that* za *is not* $\{a,w\}$*-covered.*

Indeed, assume $n>1$ or $i>1$ (i.e., za^2 is a prefix of u) and za^2 is covered by $y_1 \ldots y_s$, $y_i \in \{a,w\}$ $i=1,\ldots,s$. Then $y_s=w$ because of $z \notin \{a,w\}^*$, implying w has a factor of the form a^2, impossible.

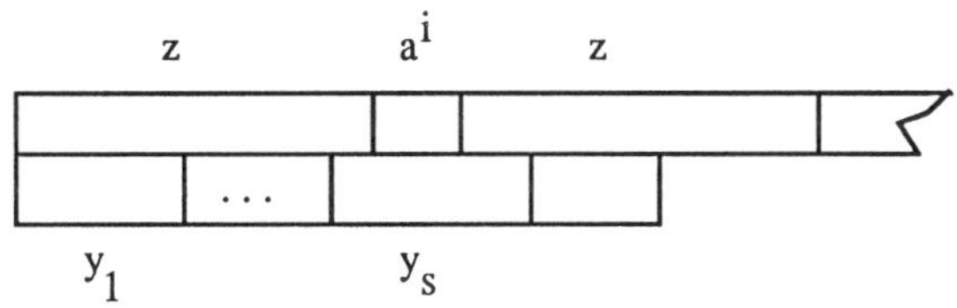

Now, if $n=1$ and $i=1$ then the reverse images $\tilde{u}$ and $\tilde{v}$ of u and v respectively satisfy the same equation (3.3.1). If $\tilde{z}a$ is not covered by $\{a,\tilde{w}\}$ then we are done. Otherwise, since $\tilde{w}=(ba)^p b=w$ the word u is a prefix and a suffix of the set $\{a,w\}^*$, i.e., it has two different $\{a,w\}$-interpretations yielding $z \in (ba)^k b$ for some $k \ge 0$. Now compute:

$$u=(za)^p z=(ba)^{p(k+1)+k} b=(wa)^p w$$

a contradiction. □

Let $x' \in b\Sigma^*$ be the proper prefix of w and let $c \in \Sigma$ be the letter such that $x=x'c$ is not a prefix of w and:

$$u \in tx\Sigma^* \text{ with } t \in \{a,w\}^*$$

Because of the previous claim, four different cases must be considered:

(3.3.5) $x=(b^n a)^j b^{n+1}$ $0 \le j < p$
and x occurs in the prefix z of u

(3.3.6) $x=(b^n a)^j b^{n'} a$ $0<n'<n$ and $0 \le j \le p$
and x occurs in the prefix za of u

(3.3.7) $x=(b^n a)^j a$ $0<j \le p$ and ($n>1$ or $i>1$)
and x occurs in the prefix za^2 of u

(3.3.8) $x=(ba)^j a$ $0<j \le p$ $n=1$ and $i=1$
and x occurs in the prefix za of u

For all $m>0$ we define:

$$c_m=h^m(c)$$

Because of the previous claim and equalities (3.3.2), (3.3.3) and (3.3.4), in all cases we have the following relation (as (2.12), cf. the outline of the proof):

$$h(x)=(a^iu)^ja^itxg \quad \text{where } txg \text{ is a prefix of } u$$

except if in the expression $h(x)=(vu)^ju=(a^iu)^{j'}(a^iz)^{\lambda}z\Sigma^*$ of (3.3.3), tx is not a prefix of the occurrence $(a^iz)^{\lambda}z$. This can only happen if $x=(ba)^ja$, $qj\equiv 1 \mod p+1$ (i.e., $\lambda=1$) and: $z\in w\{a,w\}^*(ba)^{j-1}b$ or $z\in w\{a,w\}^*(ba)^j$ or $z\in aw\{a,w\}^*(ba)^{j-1}b$. By taking the reverse images we only need consider the cases $z\in(ba)^{s(p+1)+j-1}b$ and $z\in(ab)^{s(p+1)+j}$ where $s\geq 0$, that we handle directly.

3.3.1 The special cases $u=(za^i)^pz$, $v=a^i(za^i)^q$, $i,p,q>0$, $qj\equiv 1 \mod p+1$ and $z\in(ba)^{s(p+1)+j-1}b$ or $z\in(ab)^{s(p+1)+j}$ for some $s\geq 0$

We first prove the following subcase (after renaming of the exponents) which as we said in the introduction, is one case where we must compute u_3 in order to check that the sequence is indeed locally catenative:

(3.3.9) *If* $u=za^iz$ *and* $v=a^i(za^i)^{2q+1}$ *with* $z=(ab)^{2k+1}$ *then* $u_2\notin\{u_0,u_1\}^*$ *and* $u_3\in\{u_0,u_1,u_2\}^*$*, i.e., the sequence is locally catenative.*

Proof:
Indeed, we have:

(3.3.10) $$vuv=(a^iu)^{2q+2}a^i$$

and

$$\begin{aligned}u_2&=(uv)^{2k+1}u^i(uv)^{2k+1}\\&=(uvuv)^kuvu^{i+1}v(uvuv)^k\\&\in(ua^i)^{k(2q+3)}u(a^iu)^qa^i(ab)^{4k+2}\Sigma^*\end{aligned}$$

which shows that $u_2\notin\{a,u\}^*$. Now compute:

(3.3.11) $$z_2z_2=(uv)^{4k+2}=(u(vuv))^{2k+1}\in(ua^i)^+$$

and

(3.3.12) $$\begin{aligned}v_2u_2^{i+1}v_2&=u^i(z_2u^i)^{2q+1}(z_2u^iz_2)^{i+1}u^i(z_2u^i)^{2q+1}\\&=(u^iu_2u^iz_2)^{q+1}u^i(z_2z_2u^i)^i(u^iu_2u^iz_2)^{q+1}\\&=(u^iu_2)^{q+1}u^i(z_2z_2u^i)^i(u^iu_2)^{q+1}\end{aligned}$$

In view of (3.3.11) we obtain:

$$v_2u_2^{i+1}v_2\in\{u_0,u_1,u_2\}^*$$

Thus (3.3.10) and (3.3.12) imply:

$$u_3=(u_2(v_2u_2v_2))^ku_2(v_2u_2^{i+1}v_2)(u_2(v_2u_2v_2))^k\in\{u_0,u_1,u_2\}^* \qquad \square$$

Now assume (3.3.9) does not hold. We prove the following conjunction of predicates by induction on m:

A_m: u_m is not U_{m-1}-covered
B_m: all U_m-coverings of $w_{m+1}u_m$ and $w_{m+1}v_m$ meet w_{m+1}
C_m: $x_{m+1}=(v_mu_m)^ju_m$ is not U_m-covered
D_m: $z_{m+1}u_m^2$ is not U_m-covered
E_m: v_mv_m is not U_m-covered

F_m: u_m is not a prefix of $U_{m-1}^+ u_{m+1}$ and $U_{m-1}^+ u_{m+2}$

Proof

A_1: obvious

B_1: the words w_2u and w_2v have a $\{a,u\}$-covering. The property follows from Lemma 2.4.

C_1: because of Lemma 2.4 and (3.3.3) all U_1-coverings of $x_2=(vu)^j u=(a^i u)^{j'} a^i z(za^i)^p z$ start with $(a^i u)^{j'+1}=(a^i u)^{j'}(a^i z)^{p+1}$ implying $za^i=a^i z$.

D_1: because of B_1, all U_1-coverings of $z_2u^2 \in t_2x_2\Sigma^*$ yield a U_1-covering of $x_2=h(x)$, contradicting C_1.

E_1: because of Lemma 2.3 (iv) all coverings of $vv=(a^i z)^q a^{2i}(za^i)^q$ imply that $(a^i z)^q \in \{a,u\}^*$, i.e., $v \in \{a,u\}^*$ contradiction.

F_1: if u is a prefix of a^+u_2 (resp. a^+u_3) then $u \in a\Sigma^*$, i.e., u_2 (resp. u_3) starts with u and thus u is a prefix of a^+u, contradiction.

Now assume $m>1$.

A_m: All U_{m-1}-coverings of u_m yield a U_{m-1}-covering of $z_m u_{m-1}^2$ contradicting D_{m-1}.

B_m: Because of Lemma 3.1.1, $w_{m+1}u_m \in \{u_{m-1},u_m\}^* u_m$ has a unique U_m-covering. Applying Lemma 3.1.2 and D_{m-1} to the prefix $w_{m+1}u_{m-1}^i z_m u_{m-1}^2$ of $w_{m+1}v_m$ shows that all U_m-coverings of $w_{m+1}v_m$ meet w_{m+1}.

C_m: by Lemma 3.1.1 a U_m-covering of $(v_m u_m)^j u_m \in (u_{m-1}^i u_m)^{j'} u_{m-1}^i z_m z_m u_{m-1}^i z_m \Sigma^*$ yields a U_m-covering of $u_{m-1}^i z_m z_m u_{m-1}^i z_m$. Furthermore, since u_{m-1} is not a prefix of u_m it yields a covering of $z_m z_m u_{m-1}^i z_m$. However, it does not start with $u_m \in z_m u_{m-1}^i z_m \Sigma^*$ else $z_m u_{m-1}^i = u_{m-1}^i z_m$, i.e., $z \in a^+$, contradiction.

1) $z \in (ba)^{s(p+1)+j-1}b$, i.e., $z_m=(u_{m-2}^i u_{m-1})^{s(p+q+1)+j''}(u_{m-2}^i z_{m-1})^{\lambda'} u_{m-2}^i$ with $0<j''\leq p+q$ Furthermore $0<\lambda'\leq p$ if $j>1$ and $\lambda'=q$ otherwise. Using F_{m-1}, assume that z_m has a covering of the form $gu_m \in gz_m u_{m-1}^i \Sigma^*$ where $g \in \{u_{m-2},u_{m-1}\}^*$ is a $\{u_{m-2},u_{m-1}\}$-prefix of $(u_{m-2}^i u_{m-1})^{s(p+q+1)+j''}$. Then gz_m is a prefix of $z_m z_m$. By Proposition 2.3. (iv) this implies that $z_m^+ \cap g^+ \neq \emptyset$, i.e., $z_2^+ \cap \{a,u\}^* \neq \emptyset$. We have $z_2^2 \in (a^i u)^{s(p+q+1)+j''}(a^i z)^{\lambda'} a^{2i} \Sigma^*$. If $j>1$ then $\lambda'<p+1$ which shows that this is impossible. Otherwise, $z_2 \in (a^i u)^+ v$, i.e., $z_2^2 \in (a^i u)^+ v a^i u \Sigma^*$ which can only be $\{a,u\}$-covered if $v \in \{a,u\}^*$, contradiction. Thus, all U_m-coverings of $z_m z_m u_{m-1}^i$ yield a U_m-covering of

(3.3.13) $\qquad (u_{m-2}^i z_{m-1})^{\lambda'} u_{m-2}^i v_{m-1} \in (u_{m-2}^i z_{m-1})^{\lambda'} u_{m-2}^{2i} (z_{m-1} u_{m-2}^i)^q.$

If $j=1$ then (3.3.13) becomes $v_{m-1}v_{m-1}$. Because of E_{m-1}, a U_m-covering starts with $yu_m \in yv_{m-1}\Sigma^*$ where $y \in U_{m-1}^+$. If y is a prefix of v_{m-1} then because of Proposition 2.3 (iv) the prefix $v_{m-1}v_{m-1}$ of $v_{m-1}z_m u_{m-1}^i$ is U_{m-1}-covered, else $v_{m-1}v_{m-1}$ is a prefix of yv_{m-1} thus of $yv_{m-1}v_{m-1}$ leading to the same contradiction.

If $j>1$ then because of the f.d.d. property, the covering does not start with:

$$u_{m-1}^k u_m \in u_{m-1}^k v_{m-1} v_{m-1} \Sigma^* \subseteq u_{m-1}^k (u_{m-2}^i z_{m-1})^{p+q+1} \Sigma^*$$

for some $0 \le k \le i$. Since u_{m-2} is not a prefix of u_{m-1} we obtain a U_m-covering of the word $z_{m-1}u_{m-2}^2$ of the form $yu_m \in yu_{m-2}^i z_{m-1}u_{m-2}^i \Sigma^*$ for some $y \in U_{m-1}^+$, contradicting D_{m-1}.

2) $z \in (ab)^{s(p+1)+j}$: we have:

$$z_m z_m u_{m-1}^i z_m = (u_{m-1}v_{m-1})^{2s(p+1)+2j} u_{m-1}^i z_m$$
$$= (u_{m-1}u_{m-2}^i)^{2s(p+q+1)+2j'} u_{m-1}^i z_{m-1}u_{m-1}^i z_{m-1}u_{m-1}^i z_m$$

Assume that $(u_{m-1}u_{m-2}^i)^{2s(p+q+1)+2j'}$ has a covering of the form $gu_m \in gz_m u_{m-1}^i \Sigma^*$. Then using F_{m-1}, we have $g \in (u_{m-1}u_{m-2}^i)^+$. If g is a prefix of z_m then by Proposition 2.3. (iv) we get: $z_m^+ \cap g^+ \neq \emptyset$, i.e., $z_2^+ \cap \{a,u\}^* \neq \emptyset$. This can only happen if:

$$g = (u_{m-1}u_{m-2}^i)^{t(p+q+1)} = (u_{m-1}v_{m-1})^{t(p+1)}$$

for some $0 < t \le s$. This implies that $u_{m-1}v_{m-1}$ is a prefix of $u_{m-1}^{i+1}v_{m-1}$ contradicting the f.d.d. property. Now if z_m is a prefix of g then u_{m-1} is a prefix of $u_{m-2}^i u_{m-1}$ contradicting A_{m-1}. As a consequence we obtain a U_m-covering of:

$$u_{m-1}^i z_{m-1}u_{m-1}^i z_{m-1}u_{m-1}^i z_m \in u_{m-1}^i z_{m-1}u_{m-1}^i z_{m-1}u_{m-1}^{i+1}v_{m-1}\Sigma^* .$$

Now $p>1$ because otherwise we are in the previous case. Thus there exists a U_m-covering of:

$$z_{m-1}u_{m-1}^i z_{m-1}u_{m-1}^{i+1}v_{m-1}$$

of the form $yu_m \in yu_{m-2}^i z_{m-1}u_{m-2}^i \Sigma^*$ for some $y \in U_{m-1}^+$, contradicting D_{m-1}.

D_m: because of B_m and Lemma 3.1.2, any U_m-covering of $z_{m+1}u_m^2 \in t_{m+1}x_{m+1}\Sigma^*$ meets t_{m+1}, i.e., yields a U_m-covering of x_{m+1} contradicting C_m.

E_m: because of (3.3.4), Lemma 3.1.2 and D_{m-1}, all U_m-coverings of $v_m v_m = (u_{m-1}^i z_m)^q u_{m-1}^{2i} z_m u_{m-1}^i \Sigma^*$ start with $(u_{m-2}^i u_{m-1})^{j'+1} \in v_m z_m u_{m-2}^i \Sigma^*$ implying $z_m u_{m-1}^i = u_{m-1}^i z_m$, i.e., $z \in a^*$, contradiction.

F_m: Assume u_m is a prefix of some word yu_{m+1} (resp. yu_{m+2}) with $y \in U_{m-1}^+$. If $u \in a\Sigma^*$ then u_m is a prefix of u_{m+1} (resp. u_{m+2}) and thus it is a prefix of yu_m violating A_m. Now assume, $u \in b\Sigma^*$. If $u_m \in z_m u_{m-1}^2 \Sigma^*$ is a prefix of yu_{m+1} it is a prefix of yv_m thus of $yu_{m-1}^i z_m u_{m-1}^i$ implying that $z_m u_{m-1}^2$ is U_{m-1}-covered, contradicting D_{m-1}. If u_m is a prefix of $yu_{m+2} \in yv_{m+1}\Sigma^* \subseteq yu_m\Sigma^*$ then it is a prefix of yu_m contradicting A_m. ◻

3.3.2 The general case

We prove the following conjunction of predicates by induction on m:

A_m: u_m is not U_{m-1}-covered
B_m: all U_m-coverings of $w_{m+1}u_m$ and $w_{m+1}v_m$ meet w_{m+1}
C_m: x_{m+1} is not U_m-covered
D_m: z_{m+1} if (3.3.5) holds (resp. $z_{m+1}u_m$ if (3.3.6) or (3.3.8) holds, $z_{m+1}u_m^2$ if (3.3.7) holds) is not U_m-covered
E_m: u_m is not a prefix of $U_{m-1}^+ u_{m+1}$

Proof

A_1: obvious

B_1: the words w_2u and w_2v have a $\{a,u\}$-covering. The property follows from Lemma 2.4.

C_1: because of (3.3.2) (resp. (3.3.3), (3.3.4) all U_1-coverings of x_2 start with $(a^iu)^{j+1}$ (resp. $(a^iu)^{j'+1}$). In all cases this implies $za^i=a^iz$.

D_1: because of B_1, all U_1-coverings of z_2 if (3.3.5) holds (resp. z_2u if (3.3.6) or (3.3.8) holds, z_2u^2 if (3.3.7) holds), yield a U_1-covering of $x_2=h(x)$ contradicting C_1.

E_1: if u is a prefix of a^+u_2 then $u\in a\Sigma^*$, i.e., u_2 starts with u and thus u is a prefix of a^+u, contradiction.

Now assume m>1.

A_m: All U_{m-1}-coverings of u_m yield a U_{m-1}-covering of z_m if (3.3.5) holds (resp. z_mu_{m-1} if (3.3.6) or (3.3.8) holds, $z_mu_{m-1}^2$ if (3.3.7) holds) contradicting D_{m-1}.

B_m: Because of Lemma 3.1.1, $w_{m+1}u_m\in\{u_{m-1},u_m\}^*u_m$ has a unique U_m-covering. Applying Lemma 3.1.2 and D_{m-1} to the prefix $w_{m+1}u_{m-1}^iz_m$ if (3.3.5) holds (resp. $w_{m+1}u_{m-1}^iz_mu_{m-1}$ if (3.3.6) or (3.3.8) holds, $w_{m+1}u_{m-1}^iz_mu_{m-1}^2$ if (3.3.7) holds) of $w_{m+1}v_m$ shows that all U_m-coverings of $w_{m+1}v_m$ meet w_{m+1}.

C_m:

z_m: by Lemma 3.1.2 and D_{m-1}, (3.3.2) if n>1 (resp. (3.3.4) if n=1) shows that the only possible U_m-covering of $(v_m^nu_m)^jv_m^n$ is $(u_{m-1}^iu_m)^{j+1}$ (resp. $(u_{m-1}^iu_m)^{j'+1}$). If x_{m+1} is U_m-covered, (3.3.2) (resp. (3.3.4)) implies $u_{m-1}^iz_m=z_mu_{m-1}^i$ contradicting $z\notin a^*$.

z_mu_{m-1}: by Lemma 3.1.2 and D_{m-1}, the condition (3.3.2) if n>1 (resp. (3.3.3) if n=1) shows that the only possible U_m-covering of $(v_m^nu_m)^jv_m^{n'}$ is $(u_{m-1}^iu_m)^{j+1}$ (resp. of $(v_mu_m)^j$ is $(u_{m-1}^iu_m)^{j'+1}$). If x_{m+1} is U_m-covered, then (3.3.2) (resp. (3.3.3)) implies $u_{m-1}^iz_m=z_mu_{m-1}^i$ contradicting $z\notin a^*$.

$z_mu_{m-1}^2$: by Lemma 3.1.2 and D_{m-1}, (3.3.2) if n>1 (resp. (3.3.3) if n=1) shows that the only possible U_m-covering of $(v_m^nu_m)^j$ is $(u_{m-1}^iu_m)^{j+1}$ (resp. $(u_{m-1}^iu_m)^{j'+1}$). If x_{m+1} is U_m-covered, (3.3.2) (resp. (3.3.3)) implies $u_{m-1}^iz_m=z_mu_{m-1}^i$ contradicting $z\notin a^*$.

D_m: Observe that we have:

(3.3.14) $\quad t_{m+1}\in\{u_{m-1},u_m\}^*u_m$

or

(3.3.15) $\quad t_{m+1}\in\{u_{m-1},u_m\}^*w_{m+1}$

We have $z_{m+1}\in t_{m+1}x_{m+1}\Sigma^*$ if (3.3.5) or (3.3.8) holds (resp. $z_{m+1}u_m\in t_{m+1}x_{m+1}\Sigma^*$ if (3.3.6) holds, $z_{m+1}u_m^2\in t_{m+1}x_{m+1}\Sigma^*$ if (3.3.7) holds) which by Lemma 3.1.1, if (3.3.14) holds or B_m if (3.3.15) holds, shows that any U_m-covering of z_{m+1} (resp. $z_{m+1}u_m$, $z_{m+1}u_m^2$) meets t_{m+1}, i.e., yields a U_m-covering of x_{m+1} contradicting C_m.

E_m: Assume u_m is a prefix of some word yu_{m+1} with $y\in U_{m-1}^+$. If $u\in a\Sigma^*$ then u_m is a prefix of u_{m+1} and thus is a prefix of yu_m violating A_m. If $u\in b\Sigma^*$ then $u_m\in z_m\Sigma^*$ if (3.3.5) or (3.3.8) holds (resp. $u_m\in z_m u_{m-1}\Sigma^*$ if (3.3.6) holds, $u_m\in z_m u_{m-1}^2\Sigma^*$ if (3.3.7) holds) is a prefix of yv_m thus of $yu_{m-1}^i z_m u_m$ implying that z_m (resp. $z_m u_{m-1}$, $z_m u_{m-1}^2$) is U_{m-1}-covered, contradicting D_{m-1}. ▫

3.4 Assumption: v is not {a,u}-covered

Set $u\in a^j b\Sigma^*$ and $v\in a^i b\Sigma^*$. There are two possibilities:

$v=tz$ where $t\in\{a,u\}^*u+1$, $z\in a^k b\Sigma^*$, $k<j$

or

$v\in ta^k bu'cw$ and $u\in a^j bu'dz$ where $t\in\{a,u\}^*u+1$, $u',w,z\in\Sigma^*$, $k\geq j$, $c,d\in\Sigma$, $c\neq d$

In the second situation the maximal prefix of v covered by {a,u} is $ta^k bu'$. This case is again split in two according to whether or not u and v are comparable.

3.4.1 The case v=tz where $t\in\{a,u\}^*u+1$, $z\in a^k b\Sigma^*$, $k<j$.

In this case the maximal prefix of v covered by {a,u} is ta^k. We prove the following conjunction of predicates by induction on m>0:

A_m: u_m is not U_{m-1}-covered
B_m: v_m is not U_m-covered
C_m: u_m is not a prefix of yu_{m+1}, $y\in U_{m-1}^+$

Proof
A_1 and B_1 are straightforward.
C_1: because of j>0, we have $u\in a^+\Sigma^*$. If u is a prefix of $a^+u_2\in a^+u\Sigma^*$, then $u\in a^+$.

Now assume, m>1.
A_m: by Lemma 3.1.1 all U_{m-1}-coverings of u_m yield a U_{m-1}-covering of v_{m-1} contradicting B_{m-1}.
B_m: this follows from Lemma 3.1.2 and B_{m-1}.
C_m: since j>0, u_m is a prefix of $yu_{m+1}\in yu_m\Sigma^*$ contradicting A_m. ▫

Case 3.4.2 : The case u,v are incomparable and $v \in ta^k bu'cw$ and $u \in a^j bu'dz$ where $t \in \{a,u\}^* u + 1$, $u',w,z \in \Sigma^*$, $k \geq j$, $c,d \in \Sigma$, $c \neq d$

For all $m>0$ we define:

$c_m = h^m(c)$
$d_m = h^m(d)$

We prove the following conjunction of predicates by induction on $m>0$:

A_m: u_m is not U_{m-1}-covered
B_m: v_m is not U_m-covered
C_m: u_m is not a prefix of yu_{m+1}, $y \in U_{m-1}^+$

Proof
A_1 and B_1 are obvious.
C_1: if u is a prefix of a^+u_2 then $u \in a\Sigma^*$, i.e., u_2 starts with u and thus u is a prefix of a^+u, contradiction.

Now assume $m>1$
A_m: by Lemma 3.1.1, all U_{m-1}-coverings of u_m yield a U_{m-1}-covering of v_{m-1} contradicting B_{m-1}.
B_m: by Lemma 3.1.2 and B_{m-1} all coverings of v_m start with $t_m u_{m-1}^{k-j} u_m$ which leads to c_m and d_m being comparable, contradiction.
C_m: if $j>0$, u_m is a prefix of $yu_{m+1} \in yu_m\Sigma^*$ contradicting A_m. Otherwise, $u_m \in v_{m-1}\Sigma^*$ and thus v_{m-1} is a prefix of $yu_{m+1} \in yv_m\Sigma^* \subseteq yu_{m-1}^j v_{m-1}\Sigma^*$ a contradiction to B_{m-1}. ▫

Case 3.4.3 : The case u is a prefix of v and $v \in ta^k bu'cw$ and $u \in a^j bu'dz$ where $t \in \{a,u\}^* u + 1$, $u',w,z \in \Sigma^*$, $k \geq j$, $c,d \in \Sigma$, $c \neq d$

Let us make the extra assumption that $t \in u^+$, $k=j$ and $dz=b$ implies $cw \notin a^+$. We set $bu'd \in x\Sigma^*$ for $x=ba$ or $x=bb$.
For all $m>0$ we define:

$c_m = h^m(c)$
$d_m = h^m(d)$

Observe that for all $m>0$, u_m is a prefix of v_m, i.e., $|u_m|<|v_m|$. Furthermore $v_m u_m$ is a prefix of x_m thus:

$u_m \in u_{m-1}^j v_{m-1} u_{m-1} \Sigma^*$
$v_m \in t_m u_{m-1}^k v_{m-1} u_{m-1} \Sigma^*$

We prove the following conjunction of predicates by induction on $m>0$:

A_m: u_m is not U_{m-1}-covered
B_m: if $dz=b$ then $v_m\Sigma^* \cap u_m^* = \emptyset$
C_m: $v_m u_m$ is not U_m-covered
D_m: u_m is not a prefix of yu_{m+1}, $y \in U_{m-1}^+$

Proof
A_1, B_1 and C_1 are clear.
D_1: if u is a prefix of a^+u_2 then $u \in a\Sigma^*$, i.e., u_2 starts with u and thus u is a prefix of a^+u, contradiction.

Now assume $m>1$.
A_m: by Lemma 3.1.1 all U_{m-1}-coverings of u_m yield a U_{m-1}-covering of $v_{m-1}u_{m-1}$ contradicting C_{m-1}.
B_m: by Lemma 3.1.1 the hypothesis $v_m\Sigma^* \cap u_m^* \neq \emptyset$ implies $t \in u^+$ and by the f.d.d. property $k=j$. Set $cw \in a^rb\Sigma^*$ for some $r>0$. We obtain $d_m=v_{m-1}$ is a prefix of $c_mw_m \in u_{m-1}^r v_{m-1}\Sigma^*$ contradicting B_{m-1}, via Proposition 2.1.
C_m: because of Lemma 3.1.2 and assertion C_{m-1} all U_m-coverings of:

$$v_m u_m = t_m\, u_{m-1}^k v_{m-1} u_m' c_m w_m\, u_m$$

start with $t_m\, u_{m-1}^{k-j} u_m = t_m\, u_{m-1}^{k-j} u_{m-1}^j v_{m-1} u_m' d_m z_m$
Since $c \neq d$, because of the f.d.d. property this can only occur if $dz=b$ or $dz \in a^+$.
If $dz=b$ then we have $cw \in a^rb\Sigma^*$ for some $r>0$. This implies that $v_{m-1}=d_m$ is a prefix of $u_{m-1}^r v_{m-1}$, i.e., that $v_{m-1}\Sigma^* \cap u_{m-1}^* \neq \emptyset$ contradicting B_{m-1}. If $dz \in a^p$ for some $p>0$ then $c_mw_m\, u_m$ starts with $v_{m-1}u_{m-1}$. By C_{m-1} this last occurrence of $v_{m-1}u_{m-1}$ is a prefix of $u_{m-1}^p y u_m \in u_{m-1}^p y\, u_{m-1}^j v_{m-1}\Sigma^*$, thus of $u_{m-1}^p y\, u_{m-1}^j v_{m-1} u_{m-1}$ for some $y \in U_{m-1}^*$. We obtain that $v_{m-1}u_{m-1}$ is U_{m-1}-covered contradicting C_{m-1}.
D_m: if $j>0$ then u_m is a prefix of $yu_{m+1} \in yu_m\Sigma^*$ which contradicts A_m. Otherwise, $u_m \in v_{m-1}u_{m-1}\Sigma^*$ is a prefix of $yu_{m+1} \in yv_m\Sigma^* \subseteq yu_{m-1}^i v_{m-1}u_{m-1}\Sigma^*$, i.e., $v_{m-1}u_{m-1}$ is a prefix of $yu_{m-1}^i v_{m-1}u_{m-1}$ contradicting C_{m-1}. ◻

We are thus left with the case: $u=xb$ and $v=(xb)^pxa^k$ $p>0$, $k>0$. By the previous study if the reverse image $\tilde{v}$ is not $\{a, \tilde{u}\}$-covered we are done since $\tilde{u}$ is not a prefix of $\tilde{v}$. Now if $\tilde{v}$ is $\{a, \tilde{u}\}$-covered then $a^k\tilde{x}b\tilde{x}$ and $a^ib\tilde{x}$ are comparable for some $i \geq 0$. This implies $\tilde{x}=(a^jb)^{r-1}a^j$ where $j=i-k$ and $r>1$, i.e., $\tilde{v}=a^k(a^jb)^{rp+r-1}a^j$ and $\tilde{u}=(ba^j)^r$. We are reduced to:

Case 3.4.4 : The special case $u=(ba^j)^r$ and $v=a^i(ba^j)^{rp+r-1}$ where $r>1$, $p>0$ and $i>j\geq 0$.

Observe that the following holds for all $n \geq 0$:

(3.4.1) $$vu^n=a^iu^{n+p}(ba^j)^{r-1}$$
$$vu^nba^j=a^iu^{n+p+1}$$

We prove the following conjunction of predicates by induction on $m>0$:

A_m: u_m is not U_{m-1}-covered
B_m: for all $n\geq 0$ all U_m-coverings of $v_m u_m^n$ start with $u_{m-1}^i u_m^{n+p}$
C_m: for all $n\geq 0$ $v_m u_m^n v_m$ is not U_m-covered
D_m: u_m is not a prefix of yu_{m+1}, $y \in U_{m-1}^+$
E_m: if $r=2$ then $v_m u_m^{i+j} v_m$ is not U_{m+1}-covered

Proof
A_1, B_1, C_1 and E_1 are easily verified.
D_1: since $u \in b\Sigma^*$, u can not be a prefix of a^+u_2.

Now assume $m>1$.
A_m: by Lemma 3.1.1 all U_{m-1}-coverings of u_m yield a U_{m-1}-covering of $v_{m-1}u_{m-1}^j v_{m-1}$ contradicting C_{m-1}.

B_m: by Lemma 3.1.1 this follows from (3.4.1).

C_m: by B_m all U_m-coverings of $v_m u_m^n v_m$ start with $u_{m-1}^i u_m^{n+p}$ yielding a U_m-covering of $(v_{m-1}u_{m-1}^j)^{r-1} v_m$. If $r>2$ then because of C_{m-1} and Lemma 3.1.2, this implies that the words v_{m-1} and u_{m-1} are comparable, contradicting. if $r=2$ then $v_{m-1}u_{m-1}^{j+i} v_{m-1}$ is U_m-covered contradicting E_{m-1}.

D_m: if $u_m \in v_{m-1}u_{m-1}^j v_{m-1}\Sigma^*$ is a prefix of $yu_{m+1} \in yv_m\Sigma^*$ then it is a prefix of:

$$yu_{m-1}^i v_{m-1} u_{m-1}^j v_{m-1}$$

contradicting C_{m-1}.

E_m: by the f.d.d. property no U_{m+1}-covering of

$$v_m u_m^{i+j} v_m \in u_{m-1}^i u_m^{i+j+p} v_{m-1} u_{m-1}^{i+j} v_{m-1} u_{m-1} \Sigma^*$$

starts with $u_{m+1} \in v_m u_m^j v_m u_m \Sigma^*$. Since $u_{m+1} \in v_m \Sigma^* \subseteq u_{m-1}\Sigma^*$ no occurrence of u_{m-1} can be covered by yu_{m+1} with $y \in U_{m-2}^+$. Furthermore by D_m the unique U_{m+1}-covering of an occurrence of u_m is u_m itself. Thus all coverings meet $u_{m-1}^i u_m^{i+j+p}$. Because of E_{m-1}, $v_{m-1}u_{m-1}^{i+j} v_{m-1}$ is a prefix of $yu_{m+1} \in yv_m\Sigma^* \subseteq yu_{m-1}^i v_{m-1} u_{m-1}^j v_{m-1}\Sigma^*$ with $y \in U_m^*$. Observe that because of B_{m-1} we can not have $|v_{m-1}| > |y|$. But $|v_{m-1}| \leq |y|$ implies that $v_{m-1}u_{m-1}^{i+j} v_{m-1}$ is a prefix of $yu_{m-1}^i v_{m-1} u_{m-1}^{i+j} v_{m-1}$ which contradicts E_{m-1}. ▫

3.5 Assumption: for some $n>0$ $v^n u$ is covered

By Lemma A.2 we have two possibilities:

1) $n=1$ and for some $z \in \Sigma^*$, $s \leq r$, we have: $v \in \{a, u\}^*(za^i)^s$, $u = (za^i)^r z$

or

2) $v=a^j z a^i$ $u=[(za^{i+j})^{n-1} za^i]^p (za^{i+j})^{n-1} z$ $i,j,p>0$, $n>1$ and $i \neq j$

What concerns condition 1), take the reverse images. If $\tilde{v}$ is not $\{a, \tilde{u}\}$-covered then we are reduced to the case 3.4. Otherwise, $\tilde{v}$ is a prefix of $(a^i z)^+$. If $\tilde{v}^2$ or $\tilde{v}\tilde{u}$ is -covered then $\tilde{v} \in (a^i z)^+$, i.e., $\tilde{v} \in \{a, \tilde{u}\}^*$ which reduces to case 3.8.

We can now assume that 2) holds. We have $j \neq 0$ since otherwise $v^+ \cap \{a,u\}^* \neq \emptyset$. We may further assume that $i \neq 0$. Indeed, if $i=0$, i.e., $v=a^j z$ and $u=[(za^j)^{n-1}z]^{p+1}$ $j,n,p>0$ then taking the reverse images, $\tilde{v}^{n+2}$ and $\tilde{v}^k u$ for all $k>0$ are not $\{a, \tilde{u}\}$-covered but $\tilde{v}^n$ and possibly $\tilde{v}^{n+1}$ are covered which is case 3.8.

We have $u \in a^* x \Sigma^*$ where x is of one of the following forms:

(3.5.1) $\quad x \in \{b^n ab, b^{n+1}\}$

(3.5.2) $\quad x \in \{b^k a \mid 0<k<n\}$

(3.5.3) $\quad x=b^n aa$

Furthermore, x occurs in the prefix z of u (resp. za, zaa).

The following holds for all $0<k<n$:

(3.5.4) $\quad a^j u \in v^k a^j z a^i \Sigma^*$

(3.5.5) $\quad a^j u \in v^n z a^{i+j} \Sigma^*$

(3.5.6) $\quad a^j u a^i u \in v^n u a^i z a^{i+j} \Sigma^*$

Set $x=x'c$ where c is the last letter of x. Then $h(x')=gu'$ for some $g \in \{a,u\}^*$ and some prefix u' of u. The following relations show, as in (2.12), that $u'h(c) \in a^i x \Sigma^*$.

(3.5.7) $\quad v^n u \in a^j u a^i z a^{i+j} \Sigma^*$

(3.5.8) $\quad v^2 \in a^j z a^{i+j} z \Sigma^*$

(3.5.9) $\quad vu \in a^j z a^i z \Sigma^*$

We prove the following conjunction of predicates by induction on m:

A_m: u_m is not U_{m-1}-covered
B_m: x_{m+1} is not U_m-covered
C_m: z_{m+1} in case (3.5.1) (resp. $z_{m+1}u_m$ in (3.5.2), $z_{m+1}u_m^2$ in (3.5.3)) is not U_m-covered
D_m: u_m is not a prefix of $U_{m-1}^+ u_{m+1}$

Proof

A_1 and B_1 are easily verified.

C_1: if z_2 (resp. $z_2 u$, $z_2 uu$) were U_1-covered so would x_2 be.

D_1: if u is a prefix of $a^+ u_2$ then $u \in a\Sigma^*$, i.e., u_2 starts with u and thus u is a prefix of $a^+ u$, contradiction.

Now assume $m>1$.

A_m: because of Lemma 3.1.1 any U_{m-1}-covering of u_m induces a U_{m-1}-covering of x_m.

B_m:

case 1: $x \in \{b^n ab, b^{n+1}\}$

If $x=b^n ab$ then because of C_{m-1}, (3.5.7) and Lemma 3.1.2 all U_m-coverings of $x_{m+1}=v_m u_m v_m = v_m u_m u_{m-1}^j z_m u_{m-1}^i \Sigma^*$ start with $u_{m-1}^j u_m u_{m-1}^i u_m \in v_m u_m u_{m-1}^i z_m u_{m-1}^{i+j} \Sigma^*$ violating the f.d.d. property, cf. Proposition 2.3 (iv). Now assume $x=b^{n+1}$. Because of (3.5.8) all U_m-coverings of $x_{m+1}=v_m^n u_{m-1}^j z_m u_{m-1}^i$ start with $u_{m-1}^j u_m = v_m^n z_m u_{m-1}^{i+j} \Sigma^*$. This implies $u_{m-1}^j z_m = z_m u_{m-1}^j$, i.e., $za^j=a^jz$ contradiction.

case 2: $x \in \{b^k a \mid 0<k<n\}$

Because of (3.5.8) if $k>1$, or (3.5.9) if $k=1$, all U_m-coverings of $x_{m+1}=v_m^k u_m \in v_m^k z_m u_{m-1}^{i+j} \Sigma^*$ start with:

$$u_{m-1}^j u_m = v_m^n z_m u_{m-1}^{i+j} \Sigma^* \subseteq v_m^k u_{m-1}^j z_m u_{m-1}^i \Sigma^*$$

implying $u_{m-1}^j z_m = z_m u_{m-1}^j$, i.e., $za^j=a^jz$ contradiction. .

case 3: $x=b^n aa$

Because of (3.5.7), all U_m-coverings of $x_{m+1}=v_m^n u_m u_m \in u_{m-1}^j u_m u_{m-1}^i z_m u_{m-1}^{i+j} \Sigma^*$ start with $u_{m-1}^j u_m u_{m-1}^i u_m \in v_m^n u_m u_{m-1}^i z_m u_{m-1}^{i+j} \Sigma^*$. This implies $u_{m-1}^i z_m = z_m u_{m-1}^i$, i.e., $za^i=a^iz$ contradiction.

C_m: if z_{m+1} (resp. $z_{m+1}u_m$, $z_{m+1}u_m^2$) were U_m-covered then by Lemma 3.1.2 so were x_{m+1} contradicting B_m.

D_m: assume u_m is a prefix of some word yu_{m+1} with $y \in U_{m-1}^+$. If $u \in a\Sigma^*$ then u_m is a prefix of u_{m+1} and thus is a prefix of yu_m violating A_m. If $u \in b\Sigma^*$ then $u_m \in z_m \Sigma^*$ (resp. $u_m \in z_m u_{m-1} \Sigma^*$, $u_m \in z_m u_{m-1}^2 \Sigma^*$) is a prefix of yv_m thus of $yu_{m-1}^i z_m u_m$ implying that z_{m+1}(resp. $z_{m+1}u_m$, $z_{m+1}u_m^2$) is U_{m-1}-covered, contradicting C_{m-1}.□

3.6 Assumption: v^n is $\{a,u\}$-covered for some $n>1$ and v has no prefix in a^*u

Because of Lemma A.4 we have to consider three cases. They all reduce to the already examined case 3.4.

3.6.1 The case $v=a^i(tpa^k)^r t$ and $u=(tpa^k)^r t v^{n-2} a^i tp$ for some $r>0$, $a^i tpa^k \neq pa^k ta^i$ and $t \notin a^*$ if $r=1$.

Take the reverse images $\tilde{u}=\tilde{p}\tilde{t}a^i\tilde{v}^{n-2}\tilde{t}(a^k\tilde{p}\tilde{t})^r$ and $\tilde{v}=\tilde{t}(a^k\tilde{p}\tilde{t})^r a^i$. We show that if $\tilde{v}$ is $\{a, \tilde{u}\}$-covered then $\tilde{v}^+ \cap \{a,\tilde{u}\}^* \neq \emptyset$.

Indeed, if $\tilde{v}$ is $\{a,\tilde{u}\}$-covered then there exists $j \geq 0$ such that $\tilde{v}=\tilde{t}(a^k\tilde{p}\tilde{t})^r a^i$ is a prefix of:

$$w= a^j\tilde{p}\tilde{t}a^i\tilde{t}(a^k\tilde{p}\tilde{t})^{r-1}a^k.$$

We have $|w| - |\tilde{v}| = j$. Looking at the suffixes $\tilde{t}a^i$ and $\tilde{t}a^k$ of $\tilde{v}$ and w respectively, shows that $\tilde{t} \in a^*$ or $k=i+j$. The first condition implies $r>1$ and thus by examing the prefixes $\tilde{t}a^k\tilde{p}$ and $a^j\tilde{p}$ of $\tilde{v}$ and w: $\tilde{t}=a^{k-j}$. Now taking the prefixes $\tilde{t}a^k\tilde{p}\,\tilde{t}a^k\tilde{p}$ and $a^j\tilde{p}\tilde{t}a^i\tilde{t}a^k\tilde{p}$ shows that $\tilde{v}$ is not $\{a,\tilde{u}\}$-covered. Now assume r=1 and k=i+j. We obtain:

$$\tilde{t}a^k\tilde{p}=\tilde{t}a^ia^j\tilde{p}=a^j\tilde{p}\tilde{t}a^i$$

Therefore, there exists $x\in\Sigma^*$ such that $\tilde{v}\in(a^jxa^i)^*$ and $a^j\tilde{u}a^i\in(a^jxa^i)^*$ as claimed. ▫

3.6.2 The case $v=a^i(ta^k)^rta^j$ and $u=(ta^k)^rta^jv^{n-2}a^it$ where $k\geq j$ and $k\neq i+j$

Take the reverse images $\tilde{u}$ and $\tilde{v}$. Then if $j>0$ $\tilde{v}$ is not $\{a,\tilde{u}\}$-covered and $\tilde{u}$ is not a prefix of $\tilde{v}$. If $j=0$ then because of $k\neq i+j$, $\tilde{v}$ is not $\{a,\tilde{u}\}$-covered and furthermore $\tilde{u}$ not a prefix of $\tilde{v}$. ▫

3.6.3 The case $v=(a^it)^p$ and $u=(ta^i)^qta^j$ $0<j\leq i$

Taking the reverse images $\tilde{u}$ and $\tilde{v}$, because of $i>0$ $\tilde{v}$ is not $\{a,\tilde{u}\}$-covered ▫

3.7 Assumption: v is $\{a,u\}$-covered and has a prefix in a^*u.

If the reverse image $\tilde{v}$ of v is not covered by $\{a,\tilde{u}\}$ where $\tilde{u}$ is the reverse image of u, then we are in case 3.4. Otherwise, this implies that $u=(za^k)^pz$ and $v=a^i(za^k)^qza^j$ for some $z\in\Sigma^*$ and some $0<k$, $0\leq i,j\leq k$, $0<p<q$. Since $v^+u\Sigma^*\cap\{a,u\}^*=\emptyset$ we have $j<k$ and since $v^+\cap\{a,u\}^*=\emptyset$ we have $k\neq i+j$, cf. Lemma A.1. Observe:

(3.7.1) vv and vu are not $\{a,u\}$-covered.

We set:

$za\in a^{\ell}x\Sigma^*$ where x=bb or x=ba and $\ell\geq 0$

(3.7.2) $v=t(a^kz)^{p'}a^j$ where $t\in a^iu(a^ku)^*$, $0<p'\leq p$

$vu\in t(a^kz)^{p'}a^jz\Sigma^*$

$vv\in t(a^kz)^{p'}a^{j+i}z\Sigma^*$

(3.7.3) $tu\in a^i(za^k)^qza^kz\Sigma^*$

Set $x'=b$ and $x=x'c$, i.e., c=a if x=ba and c=b if x=bb. Then by (3.7.2) we have $h(x')=tu'$ where u' is a prefix of u. As the condition (2.12) of the outline of the proof we obtain $u'h(c)\in a^{k+\ell}x\Sigma^*$ except if c=a, j=0, z=b and p'=1 which is treated directly here.

3.7.1 The special case $u=(ba^k)^pb$ and $v=a^i(ba^k)^qb$ with $q=(p+1)r$ for some $r>0$.

This implies:

$v=t(a^kb)$ where $t\in a^iu(a^ku)^{r-1}$

$vu\in ta^kbb\Sigma^*$

(3.7.4) $vv\in ta^kba^ib\Sigma^*$

We prove by induction on m that the following conjunction of predicates hold:

A_m: u_m is not U_{m-1}-covered
B_m: all U_m-coverings of v_m meet t_m
C_m: $v_m u_m$ and $v_m v_m$ are not U_m-covered
D_m: $z_{m+1}u_m$ is not U_m-covered
E_m: u_m is not a prefix of yu_{m+1} for some $y \in U_{m-1}^+$

Proof
A_1, B_1, C_1 and are easily verified.
D_1 : this follows from Proposition 2.4 and C_1.
E_1: if u is a prefix of a^+u_2 then $u \in a\Sigma^*$, i.e., u_2 starts with u and thus u is a prefix a^+u of , contradiction.

Now assume m>1.
A_m: u_m starts with $v_{m-1}u_{m-1}$. Because of Lemma 3.1.1 all U_{m-1}-coverings of u_m yield a U_{m-1}-covering of x_m violating C_{m-1}.

B_m: this is a consequence of (3.7.2) and Lemma 3.1.2.

C_m : if i>0 then because of (3.7.4) we have $v_m v_m \in t_m u_{m-1}^k v_{m-1} u_{m-1}^i v_{m-1}\Sigma^*$ which by Lemma 3.1.2 and C_{m-1} shows that any U_m-covering starts with $t_m u_{m-1}^k u_m$. Since $i \neq k$ this violates the f.d.d. property . Thus, $v_m v_m$ and $v_m u_m$ both start with $t_m u_{m-1}^k v_{m-1} v_{m-1} u_{m-1}$ and it suffices to show that this word can not be U_m-covered. Because of Lemma 3.1.1 all U_m-coverings meet t_m. By the f.d.d. property they further meet $t_m u_{m-1}^k$ yielding a U_m-covering of $w = v_{m-1}v_{m-1}u_{m-1}$. The same property shows that w is not covered by u_m thus because of C_{m-1} the covering starts with yu_m for some $y \in U_{m-1}^+$. If $|y| \geq |v_{m-1}|$ then $v_{m-1}v_{m-1}$ is a prefix of yv_{m-1} thus of $yv_{m-1}v_{m-1}$ violating C_{m-1}. If $|y| < |v_{m-1}|$ then yv_{m-1} is a prefix of $v_{m-1}v_{m-1}$ thus by the f.d.d. property we obtain $v_{m-1}^+ \cap y^+ \neq \varnothing$, i.e., $v^+ \cap \{a,u\}^+ \neq \varnothing$, contradiction.

D_m: if $z_{m+1}u_m$ is U_m-covered then so is $v_m u_m$ or $v_m v_m$ violating C_m.

E_m : if $u \in a\Sigma^*$ then u_m is a prefix of $yu_{m+1} \in yu_m\Sigma^*$ contradicting A_m. Otherwise, u_m is a prefix of yu_{m+1}, for some $y \in U_{m-1}^+$. Then we have $u_m \in z_m u_{m-1}\Sigma^*$ and further $u_{m+1} \in yv_m \Sigma^* \subseteq yu_{m-1}^i z_m u_{m-1}\Sigma^*$ violating D_{m-1}. □

3.7.2 General case

We prove by induction on m that the following conjunction of predicates hold:

A_m: u_m is not U_{m-1}-covered
B_m: $z_m\Sigma^* \cap u_{m-1}^+ = \varnothing$
C_m: $v_m\Sigma^* \cap u_m^+ = \varnothing$
D_m: x_{m+1} is not U_m-covered
E_m: $z_{m+1}u_m$ is not U_m-covered

F_m: u_m is not a prefix of yu_{m+1} for some $y \in U_{m-1}^+$

Proof
A_1: and B_1 are obvious.
C_1: this is a consequence of (3.7.2)
D_1: this is a consequence of (3.7.1)
E_1: if z_2u is covered so is x_2.
F_1: if u is a prefix of a^+u_2 then $u \in a\Sigma^*$, i.e., u_2 starts with u and thus u is a prefix of a^+u, contradiction.

Now assume $m>1$.
A_m: u_m has a prefix in $u_{m-1}^{\ell}x_m$. Because of Lemma 3.1.1 all U_{m-1}-coverings of u_m yield a U_{m-1}-covering of x_m violating D_{m-1}.
B_m: if $z \in a^{\ell}x\Sigma^*$ then $z_m\Sigma^* \cap u_{m-1}^+ \neq \emptyset$ implies $x_m\Sigma^* \cap u_{m-1}^+ \neq \emptyset$ contradicting D_{m-1}. Otherwise we have $z=a^{\ell}b$, implying $v_{m-1}\Sigma^* \cap u_{m-1}^+ \neq \emptyset$ contradiction to C_{m-1}.
C_m : because of $v_m \in u_{m-1}^i u_m u_{m-1}^k u_m \Sigma^*$, $v_m\Sigma^* \cap u_m^+ \neq \emptyset$ implies $u_m \in u_{m-1}^+$, contradicting A_m.

D_m : since the special case has been directly treated we may assume that vv (resp. vu) has a prefix in $ta^{k+\ell}bb$ (resp. $ta^{k+\ell}ba$) , i.e., $x_{m+1} \in t_m u_{m-1}^{k+\ell} x_m \Sigma^*$. Because of Lemma 3.1.2 and D_{m-1} all U_m-coverings of x_{m+1} start with $t_m u_{m-1}^k u_m$. If $x_{m+1}=v_m v_m$ (resp. $x_{m+1}=v_m u_m$) the condition (3.7.3) entails $u_{m-1}^{i+j} z_m$ (resp. $u_{m-1}^j z_m$) is comparable with $u_{m-1}^k z_m$ contradicting B_m.
E_m : if $z_{m+1}u_m$ is U_m-covered then so is x_{m+1}.
F_m : if $\ell>0$, u_m is a prefix of $yu_{m+1} \in yu_m\Sigma^*$ contradicting A_m. Otherwise, u_m is a prefix of yu_{m+1}, $y \in U_{m-1}^+$ then $u_m \in z_m u_{m-1}\Sigma^*$ and $u_{m+1} \in yv_m \Sigma^* \subseteq yu_{m-1}^i z_m u_{m-1}\Sigma^*$ violating E_{m-1}. ▫

3.8 Assumption: for some $n>0$, $i \geq 0$, v^n is a prefix of a^iu and v^{n+1} is not $\{a,u\}$-covered.

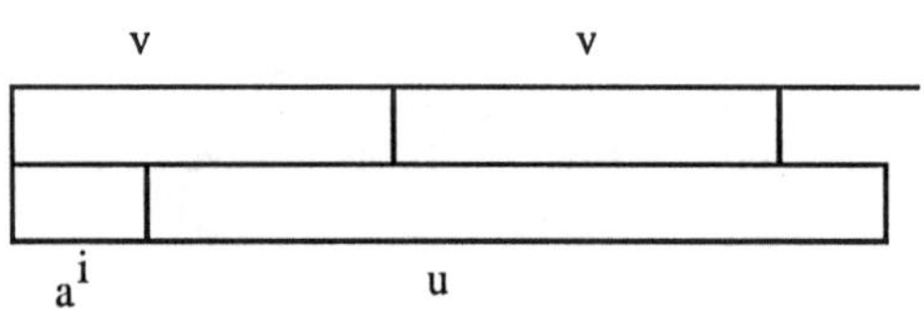

In particular:

$a^iu \in v^n\Sigma^*$

We set :

$u \in a^jb\Sigma^*$ with $j \geq 0$

We claim $u \in a^jx\Sigma^*$ holds where x is of one of the following forms:

(3.8.1) $x=b^k a$ for $0<k<n$

(3.8.2) $x=b^n a$

(3.8.3) $x=b^{n+1}$

Indeed, if this is not true then we have $u=a^j b^p$ for some $0<p<n+1$. This implies $n=1$ and $v=a^{i+j}b^q$ for some $0<q<p$. Taking the inverse images one sees that $\mathfrak{v}$ is not $\{a, \mathfrak{u}\}$-covered.

Now we consider 4 different cases according to whether or not $a^i u$ is a prefix of v^{n+1} or of $v^n u$.

3.8.1 The case $a^i u$ is a prefix of v^{n+1} and $v^n u$

Claim 1: $i=0$

Indeed, if this is not the case then the only prefixes common to u and v are in a^*:

$$a^i u \in v^n a^+$$

and $\mathfrak{v}$ is not $\{a, \mathfrak{u}\}$-covered. □

Claim 2: *without loss of generality we may assume that for some* $t,z \in \Sigma^*$ *and* $k \geq 0$ *we have* $u=v^n t$, $v=tz$ *and* $z \in a^k b^+ a\Sigma^*$

Indeed, if $z \in a^*$ then $v^{n+1} \in ua^+$. Now if $z \in a^* b^+$ then $\mathfrak{v}$ can be $\{a,\mathfrak{u}\}$-covered only if it is a prefix of $\mathfrak{u}$ which implies $tz=zt$, i.e., $vu=uv$, impossible. □

We prove the following conjunction of predicates by induction on $m>0$:

> A_m: u_m is not U_{m-1}-covered
> B_m: $v_m u_m$ is not U_m-covered
> C_m: u_m is not a prefix of some word in $U^+_{m-1} u_{m+1}$

Proof

A_1 and B_1 are trivial

C_1: if u is a prefix of $a^+ u_2$ then $u \in a\Sigma^*$, i.e., u_2 starts with u and thus u is a prefix $a^+ u$ of , contradiction.

Now assume $m>1$.

A_m: Set $u \in a^j b^r a \Sigma^*$. Because of Claim 2, $v^r_{m-1} u_{m-1}$ is a prefix of $v_{m-1} u_{m-1}$ and thus $u_m \in u^j_{m-1} v_{m-1} u_{m-1} \Sigma^*$ holds. Now by Lemma 3.1.1 all U_{m-1}-coverings of u_m yield a U_{m-1}-covering of $v_{m-1} u_{m-1}$, contradicting B_{m-1}.

B_m: because of Lemma 3.1.1 all U_m-coverings of $v_m u_m = u_m z_m t_m$ start with u_m which implies that $z_m t_m$ is U_m-covered. Because of Claim 2 $z_m t_m \in u^k_{m-1} v_{m-1} u_{m-1} \Sigma^*$ holds. Via B_{m-1}, Lemma 3.1.2 shows that $z_m t_m$ can only be U_m-covered by $u^{k-j}_{m-1} u_m$. Since vu is not $\{a,u\}$-covered this violates the f.d.d. property for u_{m-1} and v_{m-1}.

C_m: if $u \in a\Sigma^*$ then u_m is a prefix of u_{m+1} and thus is a prefix of yu_m violating A_m. If $u \in b\Sigma^*$ then because of Claim 2, $v_{m-1}u_{m-1}$ is a prefix of u_m thus of $yv_m \in yv_{m-1}u_{m-1}\Sigma^*$ implying that $v_{m-1}u_{m-1}$ is U_{m-1}-covered, contradicting B_{m-1}.□

3.8.2 The case a^iu is a prefix of v^nu but not of v^{n+1}

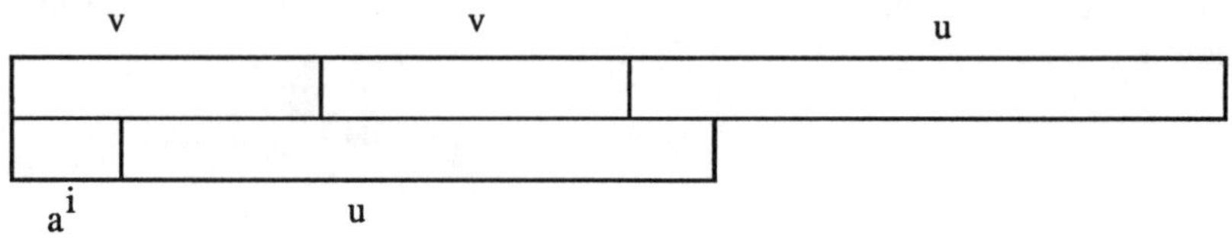

The hypothesis yields:

(3.8.4) $a^iu \in v^n tc\Sigma^*$ where $c,d \in \Sigma$, $c \neq d$ and td is a prefix of v.

This implies in particular $i>0$, i.e., u and v are incomparable, else v^{n+1} is a prefix of v^nu.

Claim 1: $v^nu=a^iuz$ *with* $z \in a^r w\Sigma^*$ $r \geq 0$ *and* $w=b^\ell a$ *for some* $0<\ell\leq n$ *or* $w=b^{n+1}$.

Indeed, arguing on the number of occurrences of the letter b in the equality $v^nu=a^iuz$ we have: $|v^n|_b=|z|_b$ which proves the claim if $|v|_b>1$. Now assume $|v|_b=1$ and $z \in a^*b^+$. If $z \notin a^*b$ then u ends in b^2 and thus $\tilde{v}$ can not be $\{a, \tilde{u}\}$- covered. We are left with $z \in a^*b$ and $n=1$. For some $k \geq 0$ we have $z=a^{k+j}b$, $v=a^{i+j}ba^k$ and $u \in (a^jba^k)^+a^jb$ which implies that vu is covered by a^iua^ku, contradiction □

3.8.2.1 The special case $v=a^ib$ and $u=ba\Sigma^*$

Here $x=ba$ and $n>1$ since $n=1$ implies $vu \in a^ib^2\Sigma^*$. If $n=2$, the reverse image $\tilde{v}$ is $\{a, \tilde{u}\}$-covered only if $u \in (ba^ib)^+$, i.e., if v^nu is $\{a,u\}$-covered. Thus we assume $n>2$. We prove by induction on m that the following conjunction of predicates holds:

A_m: u_m is not U_{m-1}-covered
B_m: all U_m-coverings of v_mv_m start with $u_{m-1}^iu_m$
C_m: $v_m v_m u_m$ is not U_m-covered
D_m: $v_m u_m$ is not U_m-covered

Proof

A_1,B_1,C_1 and D_1 are trivial.

Now assume $m>1$.

A_m: by Lemma 3.1.1, all U_{m-1}-coverings of u_m yield a U_{m-1}-covering of $v_{m-1}u_{m-1}$, contradicting D_{m-1}.

B_m: by Lemma 3.1.2 and D_{m-1}, all U_m-coverings of $v_m v_m \in u_{m-1}^i v_{m-1} u_{m-1} \Sigma^*$ start with $u_{m-1}^i u_m$.

C_m: by B_m, all U_m-coverings of:

$$v_m\, v_m u_m = u_{m-1}^i v_{m-1} u_{m-1}^i v_{m-1} u_m$$

start with $u_{m-1}^i u_m \in v_m^3 \Sigma^*$. This implies that u_m and v_m are comparable, contradiction.

D_m: if $v_m\, u_m \in u_{m-1}^i\, v_{m-1} v_{m-1} u_{m-1}^i v_{m-1}$ is U_m-covered then so is $v_{m-1} v_{m-1} u_{m-1}$ by Lemma 3.1.1, i.e., by C_{m-1} $v_{m-1} v_{m-1} u_{m-1}$ is a prefix of $y u_m \in y v_{m-1} u_{m-1} \Sigma^*$ for some $y \in U_{m-1}^+$. If $|y| \leq |v_{m-1}|$ then $y \in U_{m-2}^+$ and $v_{m-1}^+ \cap y^+ \neq \emptyset$ and therefore $v \in a^+$. Thus the word $v_{m-1} v_{m-1}$ is a prefix of $y v_{m-1}$, i.e., of $y v_{m-1} v_{m-1}$ which shows by B_{m-1} that y starts with $u_{m-2}^i u_{m-1} \in v_{m-1}^3 \Sigma^*$. This implies that v_{m-1} and u_{m-1} are comparable, contradiction.

□

3.8.2.2 The remaining cases

If (3.8.1) holds then we may assume:

(3.8.5) $v^2, v^k u \in a^{i+j} x \Sigma^*$

Indeed, this is clear if $k>1$. If $k=1$ this can only be violated if $v=a^i b$ and $u \in ba\Sigma^*$ which is 3.8.2.1. □

If (3.8.2) holds then we may assume:

$$v^n u \in a^{i+j} x \Sigma^*$$

This is clear if $n>1$. If $n=1$ and $vu \notin a^{i+j} x \Sigma^*$ then $j=0$, $v=a^i b^n$ and $u=b^p$ for some $p>n$ violating (3.8.2). □

If (3.8.3) holds then we have:

(3.8.6) $v \in a^{i+j} x \Sigma^*$

We prove by induction on m that the following conjunction of predicates holds:

A_m: u_m is not U_{m-1}-covered
B_m: x_{m+1} is not U_m-covered

Proof

A_1 and B_1 are obvious. Assume $m>1$.

A_m: because of Lemma 3.1.1 and $u_m \in u_{m-1}^j x_m \Sigma^*$ all U_{m-1}-coverings of u_m yield a U_{m-1}-covering of x_m, contradicting B_{m-1}.

B_m:

$x=b^k a$ $0<k<n$: by Lemma 3.1.2, B_{m-1} and (3.8.5) all U_m-coverings of the word $v_m^k u_m \in u_{m-1}^{i+j} x_m \Sigma^*$ start with $u_{m-1}^i u_m$. Since v_m^{k+1} is a prefix of $u_{m-1}^i u_m$ because of (3.8.4) this implies that v_{m-1} is a prefix of u_{m-1}, a contradiction.

$x=b^{n+1}$: by Lemma 3.1.2, B_{m-1} and (3.8.6) all U_m-coverings of the word $v_m^{n+1}\in v_m\Sigma^*$ $\subseteq u_{m-1}^{i+j}x_m\Sigma^*$ start with $u_{m-1}^{i}u_m$. This implies by (3.8.4) that v_{m-1} is a prefix of u_{m-1}, a contradiction.

$x=b^na$: assume $v_m^nu_m=u_{m-1}^iu_mz_m\in u_{m-1}^iu_mu_{m-1}^rw_m\Sigma^*$ is U_m-covered. If $w=x$ then by Lemma 3.1.2., $r\geq j$ and $v_m^nu_m$ is a prefix of $u_{m-1}^iu_mu_{m-1}^{r-j}u_m$, i.e., v^nu is a prefix of $a^iua^{r-j}u$, a contradiction.

If $w\neq x$ then by Lemma 3.1.3 all coverings meet $u_{m-1}^iu_mu_{m-1}^r$, i.e., w_m is U_m-covered. Then since w_m is not comparable with u_{m-1} or u_m, they all start with yu_{m-1} or with yu_m where $y\in U_{m-2}^+$. Observe that we have:

$$w_m\in u_{m-2}^{i+j}x_{m-1}\Sigma^*$$

except if $w=ba$, $j=0$ and $v=a^ib^n$. This last case can be ruled out since it yields for some $0\leq k<n-1$:

$$u\in[(b^na^i)^{n-1}b^n]^+(b^na^i)^kb^{n-1}$$

Taking the reverse image one sees that $\tilde{v}$ is not $\{a,\tilde{u}\}$-covered.

Assume some U_m-covering of w_m starts with yu_{m-1}. Now, $u_{m-2}^{i+j}x_{m-1}$ is a prefix of yu_{m-1}. Lemma 3.1.2 implies that $y=u_{m-2}^i$, i.e., w_m and $u_{m-2}^iu_{m-1}$ and thus w_2 and a^iu are comparable, contradicting the hypothesis that a^iu is not a prefix of v^{n+1}.

Assume the covering starts with yu_m and because of the previous verification $j=0$. This implies $w_m\in u_{m-2}^ix_{m-1}\Sigma^*$ is covered by $yu_m\in yv_{m-1}^nu_{m-1}\Sigma^*\subseteq yu_{m-2}^ix_{m-1}\Sigma^*$ showing that $u_{m-2}^ix_{m-1}$ is U_{m-2}-covered, contradicting B_{m-1}. □

3.8.3 The case a^iu is a prefix of v^{n+1} but not of v^nu

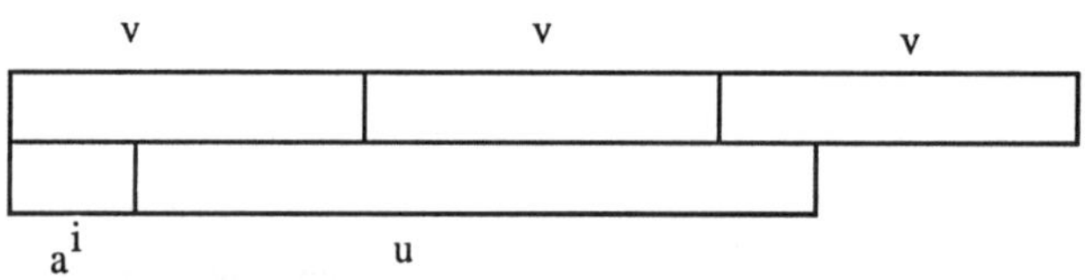

This implies in particular $i>0$, i.e., u and v are incomparable, else v^{n+1} is a prefix of v^nu.

Claim 1

(3.8.7) $v^{n+1}=a^iuz$ *with* $z\in a^rw\Sigma^*$ $r\geq 0$ *and* $w=b^\ell a$ *for some* $0<\ell\leq n$ *or* $w=b^{n+1}$.

Indeed, if $v\in\Sigma^*b$ then $\tilde{v}$ is $\{a,\tilde{u}\}$-covered only if it is a prefix of $\tilde{u}$ which implies $v^+\cap(a^iu)^+\neq\emptyset$. □

In particular we have:

(3.8.8) $v\in a^{i+j}x\Sigma^*$

We prove by induction on m that the following conjunction of predicates holds:

A_m: u_m is not U_{m-1}-covered
B_m: x_{m+1} is not U_m-covered

Proof
A_1 and B_1 are obvious. Assume m>1.

A_m: because of Lemma 3.1.1 and $u_m \in u_{m-1}^j x_m \Sigma^*$ all U_{m-1}-coverings of u_m yield a U_{m-1}-covering of x_m, contradicting B_{m-1}.
B_m:
$x=b^k a$, $0<k<n$:by Lemma 3.1.2, B_{m-1} and (3.8.8) all U_m-coverings of the word $v_m^k u_m \in u_{m-1}^{i+j} x_m \Sigma^*$ start with $u_{m-1}^i u_m$. Since v_m^{k+1} is a prefix of $u_{m-1}^i u_m$ because of (3.8.7) this implies that v_{m-1} is a prefix of u_{m-1}, a contradiction.

$x=b^n a$: by Lemma 3.1.2, B_{m-1} and (3.8.8) all U_m-coverings of the word $v_m^n u_m \in v_m \Sigma^* \subseteq u_{m-1}^{i+j} x_m \Sigma^*$ start with $u_{m-1}^i u_m$. This implies that v_{m-1} and u_{m-1} are comparable, a contradiction.

$x=b^{n+1}$: assume $v_m^{n+1} = u_{m-1}^i u_m z_m \in u_{m-1}^i u_m u_{m-1}^r w_m \Sigma^*$ is U_m-covered. If w=x then by Lemma 3.1.2., we get $r \geq j$ and v_m^{n+1} is a prefix of $u_{m-1}^i u_m u_{m-1}^{r-j} u_m$, i.e., v^{n+1} is a prefix of $a^i u a^{r-j} u$, a contradiction.

If $w \neq x$ then by Lemma 3.1.3 all coverings meet $u_{m-1}^i u_m u_{m-1}^r$, i.e., w_m is U_m-covered. Then since w_m is incomparable with u_{m-1} or u_m, they start with yu_{m-1} or with yu_m where $y \in U_{m-2}^+$. Observe that because of (3.8.8) we have:

$$w_m \in u_{m-2}^{i+j} x_{m-1} \Sigma^*$$

Assume some covering starts with yu_{m-1}. Now, $u_{m-2}^{i+j} x_{m-1}$ is a prefix of yu_{m-1}, i.e., it is U_{m-2}-covered. Lemma 3.1.2 implies $y = u_{m-2}^i$, i.e., w_m and $u_{m-2}^i u_{m-1}$ and thus w and $a^i u$ are comparable, contradicting the hypothesis that $a^i u$ is not a prefix of $v^n u$.

Assume the covering starts with yu_m and because of the previous verification j=0. This implies $w_m \in u_{m-2}^i x_{m-1} \Sigma^*$ is covered by $yu_m \in yv_{m-1}^{n+1} \Sigma^* \subseteq yu_{m-2}^i x_{m-1} \Sigma^*$ showing that $u_{m-2}^i x_{m-1}$ is U_{m-2}-covered, contradicting B_{m-1}. ▫

3.8.4 The case $a^i u$ is not a prefix of v^{n+1} or $v^n u$

Assume first i>0, i.e., u and v are incomparable. We directly treat three special cases.

3.8.4.1 The special case $v=a^i b$ and $u \in x\Sigma^*$ where $x=b^2$

We prove by induction on m that the following conjunction of predicates holds:

A_m: u_m is not U_{m-1}-covered
B_m: $x_{m+1} = v_m v_m$ is not U_m-covered
C_m: $v_m\ u_m$ is not U_m-covered

Proof

A_1 and B_1 are obvious. Assume m>1.

A_m: because of Lemma 3.1.1 all U_{m-1}-coverings of $u_m \in x_m\Sigma^*$ yield a U_{m-1}-covering of x_m, contradicting B_{m-1}.

B_m: by Lemma 3.1.3 if $v_m v_m = u^i_{m-1} v_{m-1} u^i_{m-1} v_{m-1}$ is U_m-covered then so is $v_{m-1}u_{m-1}$. Such a covering starts with yu_{m-1} or with $yu_m \in yv_{m-1}u_{m-1}\Sigma^*$ for some $y \in U^+_{m-2}$. In the first case, by C_{m-1}, $v_{m-1}u_{m-1}$ is covered by $yu_{m-1}y'u_m \in yu_{m-1}y'v_{m-1}\Sigma^*$ for some $y' \in U^*_{m-1}$, i.e., by consideration of lengths it is a prefix of $yu_{m-1}y'v_{m-1}$ thus of the word $yu_{m-1}y'v_{m-1}u_{m-1}$ contradicting C_{m-1}. In the second case $v_{m-1}u_{m-1} \in u^i_{m-2}v_{m-2}v_{m-2}\Sigma^*$ is a prefix of $yu_m \in yu^i_{m-2}v_{m-2}\Sigma^*$. Because of Lemma 3.1.1, u^i_{m-2} admits a unique U_{m-2}-covering thus $y \in u^+_{m-2}$, i.e., u_{m-2} and v_{m-2} are comparable, contradiction

C_m: by Lemma 3.1.2 and B_{m-1} all U_m-coverings of $v_m u_m \in u^i_{m-1} v_{m-1}v_{m-1}\Sigma^*$ start with $u^i_{m-1}u_m$, showing that vu and $a^i u$ are comparable, contradiction. ▫

3.8.4.2 The special case $v=a^ib$ and $u \in x\Sigma^*$ where x=ba

By taking the reverse images we may assume without loss of generality that v is a suffiix of u, i.e., that $|u_m|>|v_m|$ for all m>0. We prove by induction on m that the following conjunction of predicates holds:

> A_m: u_m is not U_{m-1}-covered
> B_m: $v_m u_m$ is not U_m-covered
> C_m: if n=1 then $v_m v_m$ is not U_m-covered, else it has the unique U_m-covering $u^i_{m-1}u_m$

Proof

For m=1 this is trivial. Assume m>1.

A_m: because of Lemma 3.1.1 all U_{m-1}-coverings of $u_m \in x_m\Sigma^*$ yield a U_{m-1}-covering of x_m, contradicting B_{m-1}.

B_m: because of Lemma 3.1.3 all U_m-coverings of the word $v_m u_m \in u^i_{m-1}v_{m-1}v_{m-1}u_{m-1}\Sigma^*$ meet u^i_{m-1} yielding a U_m-covering of $v_{m-1}v_{m-1}$. Now since u_{m-1} and v_{m-1} are not comparable, a covering of $v_{m-1}v_{m-1}$ starts with yu_{m-1} or with $yu_m \in yv_{m-1}u_{m-1}\Sigma^*$ for some $y \in U^+_{m-2}$.

In the first case, $|v_{m-1}v_{m-1}|<|yu_{m-1}|$ holds else $v_{m-1}v_{m-1}$ would be a prefix of $yu_{m-1}y'u_m \in yu_{m-1}y'v_{m-1}\Sigma^*$ for some $y' \in U^+_{m-1}$, thus of $yu_{m-1}y'v_{m-1}$ since the inequality $|u_{m-1}|>|v_{m-1}|$ holds, implying $v_{m-1}v_{m-1}$ has a different U_{m-1}-covering than $u^i_{m-2}u_{m-1}$. Therefore, because of C_{m-1}, we have n>1 and $y=u^i_{m-2}$, i.e., $v_{m-1}v_{m-1}u_{m-1}$ and $u^i_{m-2}u_{m-1}$, thus vvu and a^iu are comparable, contradiction. In the second case, if we have $|v_{m-1}v_{m-1}| \geq |yv_{m-1}|$ then because of Proposition 2.3 (iv) we have $v^+_{m-1} \cap y^+ \neq \emptyset$ contradicting C_{m-1}. Thus, $v_{m-1}v_{m-1}$ is a prefix of yv_{m-1}, i.e., of $yv_{m-1}v_{m-1}$, therefore $v_{m-1}v_{m-1}$ is U_{m-2}-covered, contradiction.

C_m: by Lemma 3.1.2 all U_m-coverings of $v_m v_m$ start with $u_{m-1}^i u_m$. If n=1 this would violate the f.d.d. property of v_{m-1} and u_{m-1}. ▫

3.8.4.3 The special case $v=a^{i+j}b\Sigma^+$ or $u\in a^+b\Sigma^*$

Then we claim that:

(3.8.9) $v^2 \in a^{i+j}x\Sigma^*$

(3.8.10) $v^k u\in a^{i+j}x\Sigma^*$ if (3.8.1) holds (resp. $v^n u\in a^{i+j}x\Sigma^*$ if (3.8.2) holds, $v\in a^{i+j}x\Sigma^*$ if (3.8.3) holds)

Indeed, assume n>1. Because of i>0, v^2 has a prefix in $a^+b\Sigma^*a\Sigma^*$ proving (3.8.9) and (3.8.10) if k>1. If k=1 or n=1 and if $v\notin a^{i+j}x\Sigma^*$ then $v=a^{i+j}b$. Therefore, $u\in a^+b\Sigma^*$, i.e., $v^k u\in a^{i+j}x\Sigma^*$ and $v^n u\in a^{i+j}x\Sigma^*$.

Under the conditions (3.8.9) and (3.8.10) we prove by induction on m that the following conjunction of predicates holds:

> A_m: u_m is not U_{m-1}-covered
> B_m: x_{m+1} is not U_m-covered
> C_m: u_m is not a prefix of yu_{m+1} for some $y\in U_{m-1}^+$

Proof

A_1 and B_1 are obvious.

C_1: if u is a prefix of a^+u_2 then $u\in a\Sigma^*$, i.e., u_2 starts with u and thus u is a prefix a^+u of, contradiction.

Now assume m>1.

A_m: because of Lemma 3.1.1 and $u_m\in u_{m-1}^j x_m\Sigma^*$ all U_{m-1}-coverings of u_m yield a U_{m-1}-covering of x_m, contradicting B_{m-1}.

B_m: by Lemma 3.1.2, B_{m-1} and (3.8.10) all U_m-coverings of $x_{m+1}\in u_{m-1}^{i+j}x_m\Sigma^*$ start with $u_{m-1}^i u_m$. This implies that v_{m-1} and u_{m-1} are comparable, a contradiction.

C_m: if j>1 then this violates A_m. If j=0 then because of (3.8.10), $x_{m+1}\in u_{m-1}^{i+j}x_m$ and thus x_m is a prefix of $u_{m-1}^{i+j}x_m$, contradicting B_{m-1}. ▫

From now on we assume i=0, i.e.:

(3.8.11) $u=v^n z$ where for some $c,d\in\Sigma$, $c\neq d$ and $w\in\Sigma^*$ we have $v=wc\Sigma^*$ and $z\in wd\Sigma^*$

We first rule out the particular case:

(3.8.12) $v=b^q$ and $u\in b^p a\Sigma^*$ with $p\geq q>0$.

Indeed, for all m>0 we have $u_m\in b^{r_m}a\Sigma^*$ with $r_m=p[q^{(m-1)}+..+q+1]$ which shows that all u_m are incomparable.

3.8.4.4 The case $v=a^jb^q$ and $u\in a^jb^pa\Sigma^*$ $0<q<p$, $j>0$

By taking the reverse images we may assume without loss of generality that v is a suffix of u, i.e., in particular that $|u_m|>|v_m|$ for all $m>0$. We prove by induction on m that the following conjunction of predicates holds:

A_m: u_m is not U_{m-1}-covered
B_m: v_mu_{m-1} is not U_m-covered
C_m: u_m is not a prefix of yu_{m+1} for some $y\in U_{m-1}^+$

Proof
A_1 and B_1 are obvious.
C_1: because of $u\in a\Sigma^*$, u_2 starts with u and thus u is a prefix of a^+u, contradiction.

Now assume $m>1$.
A_m: because of Lemma 3.1.1 and $u_m\in u_{m-1}^jv_{m-1}u_{m-2}\Sigma^*$ all U_{m-1}-coverings of u_m yield a U_{m-1}-covering of $v_{m-1}u_{m-2}$, contradicting B_{m-1}.
B_m: by the f.d.d. property, because of the relations $v_mu_{m-1}\in u_{m-1}^jv_{m-1}^qu_{m-2}^jv_{m-2}^pu_{m-2}\Sigma^*$ and $u_m\in u_{m-1}^jv_{m-1}^qu_{m-2}^jv_{m-2}^qu_{m-2}^j\Sigma^*$, v_mu_{m-1} is not a prefix of u_m. For the same reason, because of $v_mu_{m-1}\in u_{m-1}^jv_{m-1}^qu_{m-1}\Sigma^*$ a U_m-covering can not start with $u_{m-1}^k u_m\in u_{m-1}^{k+j}v_{m-1}^pu_{m-1}\Sigma^*$ where $0<k\leq j$, i.e., any U_m-covering of v_mu_{m-1} yields a U_m-covering of $v_{m-1}u_{m-2}$. Because of B_{m-1}, this U_m-covering is of the form yu_m for some $y\in U_{m-1}^+$. Thus, the word $v_{m-1}u_{m-2}$ is a prefix of $yu_m\in yu_{m-1}^jv_{m-1}\Sigma^*$ violating B_{m-1}.
C_m: since u_m is a prefix of u_{m+1} it is a prefix of yu_m violating A_m □

3.8.4.5 The remaining cases

Observe that $u\in a^jx\Sigma^*$ where $x=b^pa$ for some $p>0$. Indeed, if $u=a^jb^p$ then $n=1$ and $v=a^jb^q$ for some $0<q<p$ which shows that $\tilde{v}$ can not be $\{a, \tilde{u}\}$-covered. We claim:

(3.8.13) $v^pu\in a^jx\Sigma^*$

This is clear if $n>1$. Assume now $n=1$. If $j=0$ and $v\notin x\Sigma^*$ then this is (3.8.12). If $j>0$ this is the case 3.8.4.4.

We prove by induction on m that the following conjunction of predicates holds:

A_m: u_m is not U_{m-1}-covered
B_m: $v_m^pu_m$ is not U_m-covered
C_m: u_m is not a prefix of yu_{m+1} for some $y\in U_{m-1}^+$

Proof
A_1 and B_1 are obvious.
C_1: if u is a prefix of a^+u_2 then $u\in a\Sigma^*$, i.e., u_2 starts with u and thus u is a prefix a^+u of, contradiction.

Now assume $m>1$.

A_m: because of Lemma 3.1.1 and $u_m \in u_{m-1}^j x_m \Sigma^*$ all U_{m-1}-coverings of u_m yield a U_{m-1}-covering of x_m, contradicting B_{m-1}.

B_m: because of (3.8.13) all U_m-coverings of $v_m^p u_m$ start with u_m. Because of (3.8.11) we have $v_m^p u_m \in v_m^n w_m c_{m-1} \Sigma^*$ and $u_m \in v_m^n w_m d_{m-1} \Sigma^*$. If $c=a$ then because of $A_{m\ 1}$ and the f.d.d. property this implies $u \in v^n w b^k$ for some $k>0$, i.e., $v=wax$ and $u=(wax)^n wb^k$ for some $x \in \Sigma^*$ but then $\tilde{v}$ is not a prefix of $\tilde{u}$. If $c=b$ then because of the same property we must have $u=v^n wa$, i.e., $v=wbx$ for some $x \in \Sigma^*$ and again $\tilde{v}$ is not a prefix of $\tilde{u}$.

C_m: assume the word $u_m \in u_{m-1}^j x_m \Sigma^*$ is a prefix of some word $yu_{m+1} \in yu_m^j x_{m+1} \Sigma^*$ with $y \in U_{m-1}^+$. If $j>1$ then this violates A_m. If $j=0$ then because of (3.8.13) we have $x_{m+1} \in u_{m-1}^{i+j} x_m$ and thus x_m is a prefix of $u_{m-1}^{i+j} x_m$, contradicting B_{m-1}. □

4. APPENDIX

Biven two words $u,v \in \Sigma^*$, the purpose of this appendix is to determine under which conditions some elements of the semigroup $v\{u,v\}^*$ are $\{a,u\}$-covered and to define the relations between u and v that the existence of such elements induces. Because of the few words we work with (the two words u and v along with the very particular word reduced to the letter a) it can be expected that any relation between them imposes extremely strong restrictions on their form. As a consequences of the various Lemmas we then obtain the following exhaustive classification of the possible pairs (u,v) where, considering the cases in their order, the condition of a given case assumes the conditions of the previously cases do not hold:

1) $v^+ \cap \{a,u\}^* \neq \varnothing$
2) $v\{u,v\}^* u\{u,v\}^* v \cap \{a,u\}^* \neq \varnothing$
3) v is not $\{a,u\}$-covered
4) $v^n u$ is covered for some $n>0$
5) v^n is $\{a,u\}$-covered by $a^i u a^k u$ for some $i,k \geq 0$, $n>1$ and v has no prefix in $a^* u$
6) v has a prefix in $a^* u$.
7) v^n is a prefix of $a^i u$ for some $n>0$, $i \geq 0$ and v^{n+1} is not $\{a,u\}$-covered

In order to avoid triviality, throughout this appendix we assume that $u \notin a^*$, $v \notin \{a,u\}^*$ and $uv \neq vu$. We first study the conditions under which $v^+ \cap \{a,u\}^* \neq \varnothing$ holds.

Lemma A.1

Let $u, v \in \Sigma^*$ *satisfy* $v \notin \{a, u\}^*$. *Then* $v^+ \cap \{a,u\}^* \neq \varnothing$ *if and only if there exist* $z \in \Sigma^+$, $i+j>0$, $q>0$, $p>1$ *and* g.c.d.$(p,q)=1$ *such that:*

$u=(za^{i+j})^{p-1}z$ *and* $v=(a^i z a^j)^q$

Proof

Without loss of generality we may assume that $v^n \in \{a,u\}^*$ where $n>1$ is the smallest such non zero integer.

Case 1: v^n is $\{a,u\}$-primitive. Then by [LS] Theorem 5, $v^n=a^iua^j$ for some $i+j>0$, i.e., $v=a^iza^j$ and $u=(za^{i+j})^{n-1}z$.

Case 2: v^n is $\{a,u\}$-imprimitive. Then $v^n=x^r$ where x is $\{a,u\}$-primitive and $r>1$. By minimality of n we have $n\neq r$ and there exist $t\in\Sigma^*$, $q,p>1$, g.c.d.$(p,q)=1$, such that $v=t^q$ and $x=t^p$. Then by [LS] Theorem 5, we obtain $x=a^iua^j$ for some $i+j>0$, i.e., $t=a^iza^j$, $u=(za^{i+j})^{p-1}z$. □

Now we assume $v^+\cap\{a,u\}^*=\emptyset$ and investigate the elements in $v\{u,v\}^*\cap\{a,u\}^*$. Such elements have a prefix of the form v^nu for some $n>0$. More generally we enquire about the elements v^nu, for some $n>0$, that are $\{a,u\}$-covered. As anticipated above this condition already imposes dramatic restrictions on u and v:

Lemma A.2

Let u, v *be two words such that* $v^+\cap\{a,u\}^*=\emptyset$. *If for some* $n>0$ v^nu *is* $\{a,u\}$-*covered then one of the following two conditions applies*:

i) *for some* $i+j>0$, $n>1$, $p>0$, $z\in\Sigma^*$ *we have*: $v=a^iza^j$, $u=[(za^{i+j})^{n-1}za^i]^p(za^{i+j})^{n-1}z$

ii) $n=1$ *and for some* $z\in\Sigma^*$, $0<s\leq r$, $i\geq 0$, *we have*: $v\in\{a,u\}^*(za^i)^s$, $u=(za^i)^rz$

Proof

Let $x_1,\ldots,x_m$ be a $\{a,u\}$- covering of v^nu. Then for some $j\geq 0$ the following holds:

$x_{m-j-1}=x_m=u$ and $x_{m-j}=x_{m-j+1}=\ldots=x_{m-1}=a$ (cf. Figure)

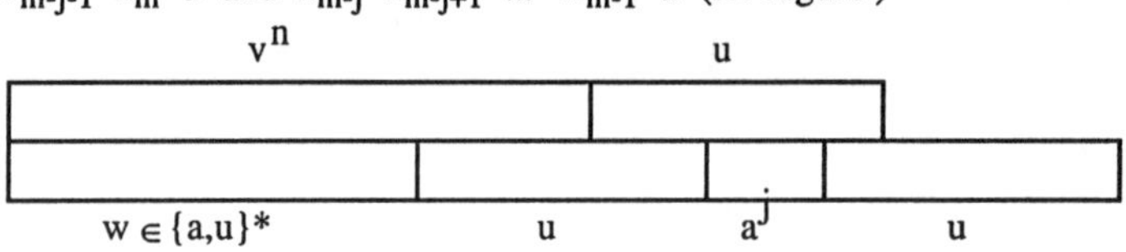

The covering induces an interpretation of the occurrence (v^n,u) of the form (s_1,a^j,p_1) where s_1 (resp. p_1) is a suffix (resp. prefix) of u. By Lemma 2.2 this implies $u=(za^j)^rz$ for some $z\in\Sigma^*$, $i\geq 0$, $r>0$. We obtain:

(A.1) $\quad v^n=w(za^j)^s$ for some $0<s\leq r$ and $w\in\{a,u\}^*$.

If $n=1$ then this is the case ii) so we assume from now on that $n>1$.

Case1: $v\in\{a,z\}^*$. Because of (A.1) and $v\notin\{a,u\}^*$, for some strict prefix u_1 of $u=(za^j)^rz$ we have $v\in\{a,u\}^*u_1$. Since v has a suffix in za^j, because of (2.9) applied to the suffixes we obtain:

(A.2) $\quad za^j$ is a $\{a,z\}$-suffix of v, i.e., $v\in\{a,z\}^*za^j$

Because of (2.8) and the hypothesis $v\notin\{a,u\}^*$ we have $u_1=(za^j)^k$ for some $0<k\leq r$ and $u_2=z(a^jz)^{r-k}$ where $u=u_1u_2$. Now u_2 is comparable with the second leftmost occurrence of v which shows that the $\{a,u\}$-covering of v starts with u.

We claim $v\in(za^j)^*$. Indeed, if v is a prefix of u then $v\in\{a,z\}^*za^j\cap(za^j)^*p$ for some prefix p of za^j. By (2.8) this yields $p=za^j$, $v\in(za^j)^*$. If u is a prefix of v we apply Lemma 2.5. In all cases we get and $v^+\cap\{ua^j\}^+\neq\emptyset$.

Now let $1<k\leq n$ be the least integer such that $v^k\in\{a,z\}^+$. Observe that v^k ends in za^j and that k divides n because of its minimality.

Case2: v^k is $\{a,z\}$-primitive. Then by [LS] Theorem 5, we have $v^k=a^iza^j$ for some $j\geq0$. If $k=n$ then for some z_1 we have $v=a^iz_1a^j$ and $z=(z_1a^{i+j})^{n-1}z_1$ which is i). Otherwise $v^{2k}=a^iza^{i+j}za^j$ is a prefix of v^n and therefore is covered by $\{a,u\}$ which implies $j=0$ and leads to $v^{k(r+1)}=ua^j$, contrary to the hypothesis.
Case3: v^k is $\{a,z\}$-imprimitive. Then $v^k=x^p$ for some $p>1$, where x is the $\{a,z\}$-root of v^k, i.e., $v=y^t$ and $x=y^q$ for some $t,q>1$ and $y\in\Sigma^*$. By [LS] Theorem 5, this means that $x=a^iza^j$, for some $i\geq0$ and furthermore $y=a^iz_1a^j$, where $z=(z_1a^{i+j})^{q-1}z_1$. If v^nu is covered then $v^n=a^iua^j$, contrary to the hypothesis. □

The previous Lemma allows us to characterize the words that belong to the intersection of the two semigroups $v\{u,v\}^*$ and $\{a,u\}^*$.

Lemma A.3 *Let* u, v *be two words such that* $v^+\cap\{a, u\}^*=\emptyset$. Then $v\{u,v\}^*\cap\{a,u\}^*\neq\emptyset$ *if and only if one of the following two conditions occurs:*
i) for some $n>1, p,i>0, z\in\Sigma^*$ *we have:* $v=a^iza^i$, $u=[(za^{2i})^{n-1}za^i]^p(za^{2i})^{n-1}z$
in which case we have:

$$(v^nu)^pv^n=(a^iu)^{p+1}a^i$$

ii) for some $p,q,i>0$, g.c.d $(q,p+1)=1$, $z\in\Sigma^*$ *we have:* $v=a^i(za^i)^q$ and $u=(za^i)^pz$
in which case we have:

$$(vu)^pv=(a^iu)^{p+q}a^i$$

Proof
Clearly the conditions are sufficient. Conversely condition i) of Lemma A.2 implies $i=j$. Consider condition ii) and let w be a word in $v\{u,v\}^*\cap\{a,u\}^*$. Because of $vu=ga^i(za^i)^s$ where $g\in\{a,u\}^*$, vu is covererd by ga^iu, i.e., w starts with vuv. Then the second occurrence v of w is a prefix of $(za^i)^{r-s}z\ \{a,u\}^*$, i.e., by Lemma 2.5 v is a prefix of $(za^i)^+$. Since it ends in za^i it impies $v\in(a^iz)^+a^i$ as claimed. Finally, if q and $p+1$ have a common divisor r then it suffices to pose $q=rq'$, $p+1=r(p'+1)$, $z'=(za^i)^rz$, $u=(z'a^i)^{p'}z$, $v=a^i(z'a^i)^{q'}$. □

The last case to be considered is the case when for some $n>0$ v^n is $\{a,u\}$-covered.

Lemma A.4 *Let* $u,v\in\Sigma^*$ *and assume* $v^+u\Sigma^*\cap\{a,u\}^*=v^+\cap\{a,u\}^*=\emptyset$. *If for some* $n>1$ v^n *is covered by* a^iua^ku *for some* $i,k\geq 0$, *and if* $v\notin a^*u\Sigma^*$, *then one of the following conditions holds*:

(A.3) $v=a^i(tpa^k)^rt$ *and* $u=(tpa^k)^rtv^{n-2}a^itp$ for some $r>0$, $a^itpa^k\neq pa^kta^i$ and $t\notin a^*$ if $r=1$.

(A.4) $v=a^i(ta^k)^rta^j$ *and* $u=(ta^k)^rta^j\, v^{n-2}a^it$ *where* $k\geq j$ *and* $k\neq i+j$

(A.5) $v=(a^it)^p$ *and* $u=(ta^i)^qta^j$ $0<j\leq i$ *for some* $p>0$, $q\geq 0$.

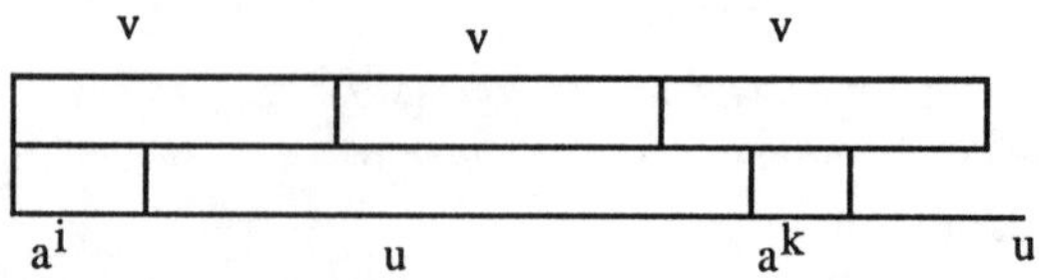

Proof

Considering the coverings of the leftmost and the rightmost occurrences of v in v^n we obtain:

$v=a^iu_1=u_2a^ku'$ where u_1 and u' are prefixes of u and u_2 is a suffix of u. Clearly, if u_1 is shorter than u' then we have $v\in a^*$ impossible. Thus setting, $u_1=xy$, $u_2=z$ and $u'=x$ we get:

$$v=a^ixy=za^kx \text{ and } u=xyv^{n-2}z$$

We are left with 3 different cases:

$|z|\leq|a^ix|$ and $|a^ix|\leq|za^k|$

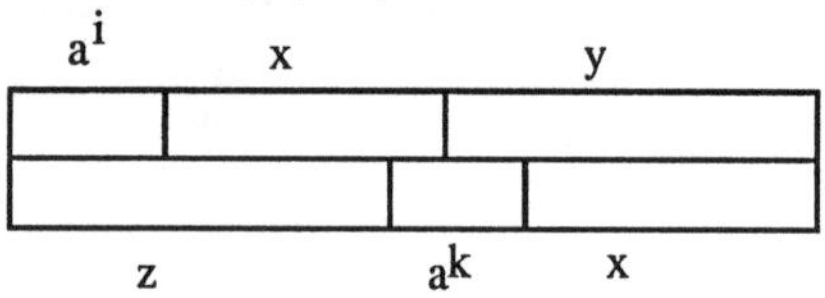

Then for some $t\in\Sigma^*$ and some $j,\ell\leq 0$ we have:

$x=ta^j$, $y=a^\ell ta^j$, $z=a^it$, $k=\ell+j$

Thus: $v=a^ixy=a^ita^kta^j$ and

$u=xyv^{n-2}z=ta^kta^jv^{n-2}a^it$

The hypothesis $k=i+j$ implies $v^{2n-1}=(a^iua^j)^2$. We may thus assume $k\neq i+j$.

$|a^ix|\leq|z|$

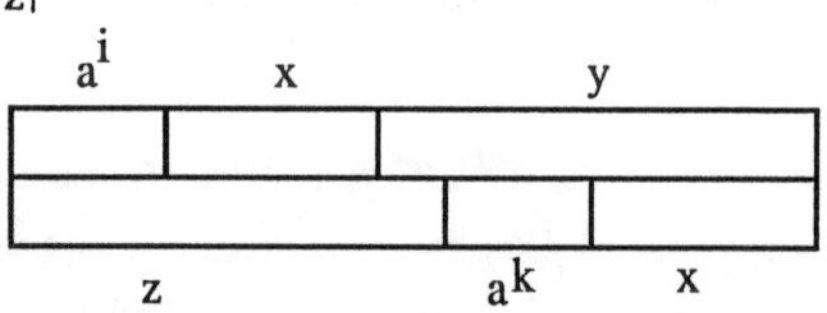

For some prefix y_1 of y we have $z=a^ixy_1$, $y=y_1a^kx$. Then:

$$v=a^ixy=a^ixy_1a^kx$$

and $u=xyv^{n-2}z=xy_1a^kxv^{n-2}a^ixy_1$.

This is (A.3) unless $a^ixy_1a^k=y_1a^kxa^i$ which implies:

$y_1\in a^i(ta^k)^*t$ and $x\in(ta^k)^*ta^{k-i}$ if $k\geq i$

or $y_1\in(a^it)^*a^{i-k}$ and $x\in(ta^i)^*t$ if $k<i$.

These two cases lead to (A.5).

$|a^i|\leq|z|$ and $|za^k|\leq|a^ix|$

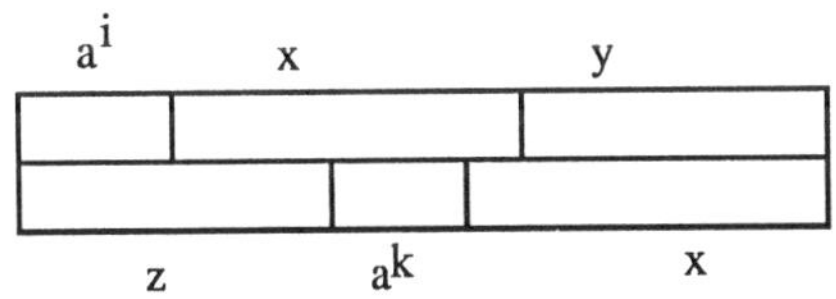

For some suffix z' of z we have $z=a^iz'$ and $xy=z'a^kx$. If $y=pa^kt$ with $z'=tp$ then $x\in(tpa^k)^*t$ which leads to solution (A.3) unless $a^itpa^k=pa^kta^i$ which leads to solution (A.5). If $y=a^jz'a^{k-j}$ for some $k\geq j$ then $x\in(z'a^k)^*z'a^j$ which yields solution (A.4). □

Acknowledgements. The author thanks Imre Simon for inviting him at the Universidade de São Paulo where some of this work was written.

5. REFERENCES

[Be] J. Berstel, Some recent results on Square-Free Words, Lecture Notes on Computer Science n°166, (1984), 14-25

[BP] J. Berstel & D. Perrin, Theory of Codes, Academic Press, 1985

[BL] E. Barbin & M. Lerest, Sur la combinatoire des codes à deux mots, Theoret. Comput. Sci., 41, (1985), 61-80

[CF] K. Culik II & I. Fris, The decidability of the equivalence problem for DOL-sequences, Inform. and Contr., 35, (1977), 20-39

[ER1] A. Ehrenfeucht & G. Rozenberg, Elementary homomorphisms and a solution of the DOL equivalence problem, Theoret. Comput. Sci., 7, (1978), 169-183

[ER2] A. Ehrenfeucht & G. Rozenberg, On the bound of two DOL-sequence equivalence problem, Theoret. Comput. Sci., 7, (1980), 169-183

[HL] T. Harju & M. Linna, On the periodicity of morphisms on free monoids, RAIRO Theoretical Informatics and Applications, 20, (1986), 47-54

[Ka1] J. Karhumäki, On the equivalence problem for binary DOL-systems, Inform. Contr., 50, (1981), 276-284

[Ka2] J. Karhumäki, A property of three element codes, LNCS 166, Springer Verlag, (1984), 305- 313

[Ka3] J. Karhumäki, About the intersection of two free monoids generated by two elements, Semigroup Forum, 29, (1984), 183-205

[La] J.-L. Lambert, The local catenativity of DOL-sequences in free commutative monoids is decidable in the binary case, submitted for publication

[Lo] Lothaire, Combinatorics in Words, Encyclopedia of Mathematics, Addison-Wesley, 1983

[LR] A. Lindenmayer & G. Rozenberg, Developmental Systems with Locally Catenative Formulas, Acta Informatica, 2, (1973), 214-248

[LS] A. Lentin & M.P. Schützenberger, A combinatorial problem in the theory of free monoids, Proc. University of North Carolina, (1967), 128-144

[Ni] M. Nielsen, On the decidability of some equivalence problems for DOL-systems, Inform. Contr., 25, (1974), 166-193

[Ru] K. Ruohonen, Remarks on locally catenative developmental sequences, Elektron. Informationsverarb. Kybernet., 14, (1978), 171-180

[Sa] A. Salomaa, DOL-equivalence: the problem of iterated morphisms, EATCS Bull., 4, (1978), 5-12

[Se] P. Séébold, About a family of binary morphisms whose stationary words are Sturmian, FCT'89, Springer LNCS 380, 384-399

[Sh] J. Shallit, A generalization of automatic sequences, Theoret. Comput. Sci, 61, (1988), 1-16

[Sp] J.C. Spehner, Quelques problèmes d'extension, de conjugaison et de présentation de sous-monoides d'un monoide libre, Thèse de Doctorat d'Etat, Paris, 1976

ALTERNATING ITERATION OF MORPHISMS AND THE KOLAKOVSKI SEQUENCE

K. Culik II[1], J. Karhumäki[2] and A. Lepistö[2]

[1]Department of Computer Science, University of South Carolina, Columbia, S.C. 29208

[2]Department of Mathematics, University of Turku, SF-20500 Turku, Finland

Abstract: Alternating iteration of morphisms is a natural extension of iteration of morphisms to define infinite words. We study properties of two infinite words thus obtained, one being the famous Kolakovski sequence.

Keywords: infinite words, Kolakovski sequence, alternating iteration of morphisms.

1 Introduction

One of the most important applications of Lindenmayer systems is to use the deterministic ones, in particular D0L systems, to generate infinite words. These words are generated by applying iteratively a (nonerasing) morphism h on a letter a satisfying "a is a prefix of $h(a)$", or equivalently, by applying iteratively h in a TAG-way on a, cf. [M]. These mechanisms are by no means new: The first one was used already by Thue at the beginning of this century, cf. e.g. [T], and the second one was introduced by Post in 1920's. However, they became well-known and popular only about fifteen years ago, mainly due to initiatives of Berstel, cf. [Be1] and [Be2]. Part of their popularity is because of their usefulness in computer graphics, cf. [Sm] and [PL].

An iteration of a morphism is a very simple mechanism to generate infinite words. Although this mechanism can generate complicated infinite words, such as the Thue–Morse word or the Fibonacci word, it is natural to look for more complicated mechanisms as well. An iteration of a deterministic *gsm* provides such a method, cf. [AG]. However, this is already so powerful that it is hard to give concrete examples of infinite words for which it is possible to prove that they are not obtainable by this method. An in-between case was introduced in [CK], where also a unified approach for all these different mechanisms was proposed.

The in-between case is based on alternating iteration of a finite number of morphisms. More presicely, let p be a natural number and h_i, for $i = 0, ..., p-1$, be morphisms of a free monoid Σ^*.

The sequence is generated by applying iteratively these morphisms as follows: In each iteration the letter which is ith (mod p) is rewritten by h_i. Clearly, this mechanism is a natural and proper extension of an iteration of a morphism. On the other hand it is a proper restriction of an iteration of a *dgsm*. Indeed, i can be identified with the state of a *dgsm*. In [CK] systems of this new mechanism were called D0L TAG systems with periodic control, and consequently words generated by them were called D0L TAG words with periodic control.

Concerning the families of infinite words generated by the above mechanisms it is easy to see that there are infinite D0L TAG words with periodic control which are not D0L words, that is generated by iterating a morphism. It is less obvious, but proved in [CK], that there are also infinite words obtained by iterating a *dgsm* such that they are not D0L TAG words with periodic control. This paper is devoted to further study of D0L TAG words with periodic control. We investigate a few such words, and shed light on the power of the mechanism by showing some properties of these particular words.

The contents of this note are as follows.

In Section 2 after fixing our notation and recalling the basic definitions we give three concrete examples of D0L TAG words with periodic control. One of these words, the Kolakovski sequence, is well-known from literature, cf. [K] and [D2], while another seems to be new.

In Section 3 we show that the Kolakovski word is not a D0L word, but fail to prove that it is not a coding of a D0L word, i.e. a CD0L word. Concerning our other infinite word we still fail to prove that it is not an infinite CD0L word, but are able to prove that its prefixes appearing in its generation by our D0L TAG system with periodic control do not form a coding of a D0L sequence of finite words.

Section 4 has two goals. First to point out several open problems connected to the sequences we have studied. Actually, the problems on Kolakovski sequence are not new, and are stated here only for the sake of completeness. Second we present some numerical calculations showing in one hand the very unpredictable behaviour of the Kolakovski word, and in the other hand supporting probable answers to most of the above open problems.

2 Preliminaries and examples

Let Σ be a finite alphabet, which in this note is most typically binary, say $\{1,2\}$. Let Σ^* (resp. Σ^+) denote the free monoid (resp. semigroup) generated by Σ. Elements of Σ^* are called (finite) words. The identity of Σ^* is called the empty word. For any word u let $|u|$, $first(u)$, and $last(u)$ denote its length, first symbol and last symbol, respectively. The left quotient of u by v

is denoted by $v^{-1}u$. A crucial notion for our considerations is that of a morphism $h : \Sigma^* \to \Sigma^*$. A *coding* c is a length preserving morphism, so that it maps each letter to another letter.

In addition to finite words we are dealing with infinite words over Σ as well. Formally, such words are mappings $\mathbf{N} \to \Sigma$, but we prefer to denote them as infinite sequences: $a_0a_1a_2...$ with $a_i \in \Sigma$. When there is no danger of confusion we call infinite words simply words.

Let $h : \Sigma^* \to \Sigma^*$ be a morphism. We call sequences of finite words of the form

$$w, h(w), h^2(w), ..., h^i(w), ... \tag{1}$$

D0L sequences, where $w \in \Sigma^+$. A *CD0L sequence* is a sequence of words which is obtained from a D0L sequence by mapping all the words by a fixed coding. If a nonerasing morphism h satisfies "a is a proper prefix of $h(a)$" then

$$\gamma = \lim_{n\to\infty} h^n(a)$$

is well defined element of Σ^ω, and we say that γ is obtained by *iterating h on a letter a.* Such infinite words are called *D0L words.* By a *CD0L word* we mean a coding of a D0L word.

Let p be a natural number and $h_0, ..., h_{p-1} : \Sigma^* \to \Sigma^*$ a finite collection of nonerasing morphisms. We use these morphisms to generate the sequence of finite words

$$w_0, w_1, w_2, \dots , \tag{2}$$

where $w_0 \in \Sigma^+$ and, for each $i \geq 0$, if $w_i = a_{i,0}...a_{i,t}$, with $a_{i,j} \in \Sigma$, then

$$w_{i+1} = \prod_{j=0}^{t} h_k(a_{i,j}) , \quad \text{with } k \equiv j (\text{mod } p). \tag{3}$$

Sequences thus obtained are called *p-alternating D0L sequences*, and by an *alternating D0L sequence* we mean a p-alternating D0L sequence for some $p \geq 1$. Clearly, 1-alternating D0L sequences are just D0L sequences.

Under the condition "a is a proper prefix of $h_0(a)$" an iterative application of morphisms $h_0, ..., h_{p-1}$ in the sense of (3) and starting at the word a yields a unique element of Σ^ω. Infinite words thus obtained are called *D0L words with periodic control* (in [CK] they were called D0L TAG words with periodic control). The requirement that we start from a single letter is for clarity only and not essential.

Now, we are ready to define two interesting examples of D0L words with periodic control. The first one is due to W. Kolakovski, cf. [K], and is quite famous of its intriguing properties, cf. [D2] or [Sh].

Example 1 (The Kolakovski word). Let $p = 2$ and define morphisms

$$h_a : \begin{array}{l} 1 \to 2 \\ 2 \to 22 \end{array} \qquad h_b : \begin{array}{l} 1 \to 1 \\ 2 \to 11 \end{array} .$$

Here, for clarity, we denote $h_0 = h_a$ and $h_1 = h_b$. We obtain by starting at 2 the infinite word

$$\mathrm{Kol} = 2211212212211\ldots . \tag{4}$$

The sequence (4) is referred to as the *Kolakovski sequence.* As is well-known, and easy to see, it is specified by the following rules:

(i) It starts with 2 ;

(ii) It consists of consecutive blocks of 1's and 2's ;

(iii) The length of the ith block is determined by the digit in the ith place of the sequence.

Consequently, Kol is considered as an example of a self-generating word.

As we saw the Kolakovski word is very easily definable as a D0L word with periodic control. It also has alternative, but related, definitions as follows. First it is obtained by iterating the following *dgsm* starting at 2:

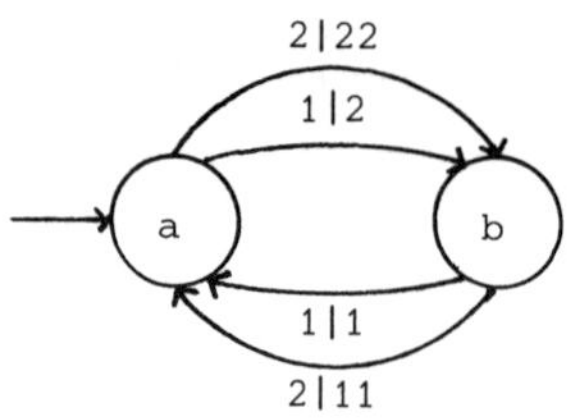

Second it is obtained, as was noted e.g. in [D1], by iterating the following mapping starting at 22:

$$\varphi : \begin{array}{l} 11 \to 21 \\ 12 \to 211 \\ 21 \to 221 \\ 22 \to 2211 \end{array} . \tag{5}$$

By (5) one might think that the Kolakovski sequence can be obtained by iterating a morphism in a four letter alphabet $\{11, 12, 21, 22\}$. This, however, is not true since the images of words 12 and 21 under the above mapping are not words in that alphabet. On the other hand, the representation (5) immediately gives a possibility to define Kol by iterating another *dgsm* over a four letter alphabet.

Example 2. Let again $p = 2$ and define morphisms

$$h_a : \begin{array}{l} 1 \to 11 \\ 2 \to 22 \end{array} \qquad h_b : \begin{array}{l} 1 \to 2 \\ 2 \to 1 . \end{array}$$

Now, the word generated starting at 2 is

$$\text{Alt} = 221112111112112111\ldots .$$

Clearly, the very same word is obtained by iterating the following *dgsm* starting at 2:

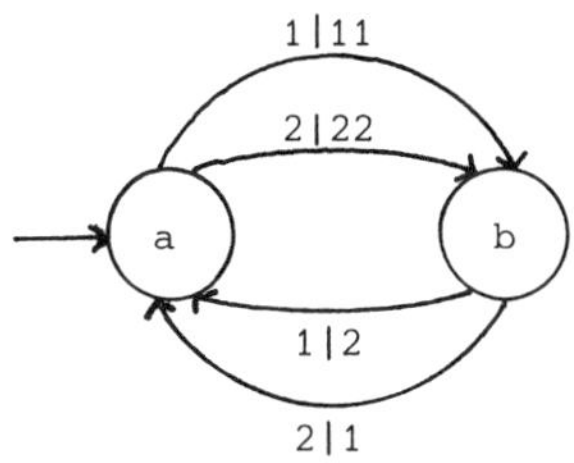

Let us denote by α_n the finite word produced after the nth iteration. Consequently,

$$\begin{aligned} \alpha_0 &= 2 \\ \alpha_1 &= 22 \\ \alpha_2 &= 221 \\ \alpha_3 &= 22111 \\ \alpha_4 &= 22111211 \\ \alpha_5 &= 221112111112 \\ \alpha_6 &= 221112111112112111 \\ &\vdots \end{aligned}$$

and the lengths of these words grow as: $1, 2, 3, 5, 8, 12, 18, \ldots$. Actually, for the lengths $|\alpha_n|$ we have a simple formula

$$\begin{aligned} |\alpha_0| &= 1 \\ |\alpha_{n+1}| &= |\alpha_n| + \left\lceil \frac{|\alpha_n|}{2} \right\rceil , \quad \text{for } n \geq 0 . \end{aligned}$$

Finally, as our last example we generalize the Kolakovski sequence to an arbitrary alphabet.

Example 3 (The Generalized Kolakovski sequence). Clearly, the rules (i)–(iii) in the definition of the Kolakovski sequence can be generalized to define the $\mathrm{Kol}_n \in \{1,...,n\}^\omega$ as follows:

(i) Kol_n starts with n ;

(ii) Kol_n consists of consecutive blocks of 1's, 2's, ..., n's ;

(iii) The length of the ith block is determined by the digit in the ith place of the sequence.

It is straightforward to see that Kol_n is a D0L word with periodic control. Indeed, for $n = 3$ it is obtained by choosing $p = 3$ and defining the morphisms by

$$h_a: \begin{array}{l} 1 \to 3 \\ 2 \to 33 \\ 3 \to 333 \end{array} \qquad h_b: \begin{array}{l} 1 \to 1 \\ 2 \to 11 \\ 3 \to 111 \end{array} \qquad h_c: \begin{array}{l} 1 \to 2 \\ 2 \to 22 \\ 3 \to 222 \end{array} \quad .$$

Consequently,

$$\mathrm{Kol}_3 = 33311122231233112233311 2... \quad .$$

3 Results

Here we prove two results on our infinite words Kol and Alt.

Theorem 1. The Kolakovski word Kol is not a D0L word.

Proof. Assuming that Kol is a D0L word it would contain (arbitrary long) repetitions at the beginning since it starts with two 2's. Let u be the shortest word satisfying $|u| > 1$ and

$$\mathrm{Kol} = uuw \text{ for some } w \text{ in } \{1,2\}^\omega . \tag{6}$$

We show that this leads to a contradiction.

Clearly, we can assume that

$$|u| \geq 13 . \tag{7}$$

Let φ be the mapping of Example 1 generating Kol and denote

$$C = \{21, 211, 221, 2211\} .$$

Now, each word in C start with 2 and ends with 1. Consequently, C is a code, and moreover, it has the property

$$v = v_1 12 v_2 \in C^+ \Rightarrow v_1 1, 2v_2 \in C^+ . \tag{8}$$

In particular, C has a finite deciphering delay equal to 1, cf. [BP], so that we have

$$(\varphi^{-1})^i(\text{Kol}) = \text{Kol}\ , \quad \text{for any} \ \ i \geq 0\ . \tag{9}$$

Since u starts with 22 it must end with 1. Otherwise Kol would contain a subword 222.

We have to consider separately the four different cases depending on the suffix of u.

I Assume that $u = u'2211$. Then, by (8), we conclude that

$$\varphi^{-1}(u) = 22...22\ ,$$

and hence the finite delay property of C together with (6) and (7) imply that Kol contains a subword 222, a contradiction.

II Assume that $u = u'211$. If the last symbol of u' is 2 we are in the case II. Otherwise we conclude, as above, that

$$\varphi^{-1}(u) = 22...12\ ,$$

which yields the same contradiction as case I.

III Assume that $u = u'21$ with $u' = u''1$. We conclude, assuming that $|u| \geq 7$, that

$$\text{Kol} = \varphi^{-1}(\text{Kol}) = \varphi^{-1}(u).\varphi^{-1}(uw) = 22...11.\varphi^{-1}(uw) = 22...11.2211...$$

Consequently, considerations of cases I and II apply to yield a contradiction.

IV Assume that $u = u'221$. We have two subcases to be considered. First the first symbol of w in (6) is 2. Then we have

$$\text{Kol} = \varphi^{-1}(\text{Kol}) = \varphi^{-1}(u)\varphi^{-1}(u)\varphi^{-1}(w)$$

which is a contradiction with the minimality of u. Second the first symbol of w is 1. Then we can write

$$\text{Kol} = \varphi^{-1}(\text{Kol}) = \varphi^{-1}(u')21\varphi^{-1}(u')22\varphi^{-1}(1^{-1}w)\ .$$

Hence the last symbol of $\varphi^{-1}(u')$ is 1 so that we can apply the considerations of III if we assume that $|\varphi^{-1}(u')| \geq 7$. This assumption is guaranteed by (7), completing the proof. □

The proof of Theorem 1 requires a few comments. We have defined, as in [D1] and [D2], the Kolakovski sequence starting with 22. However, sometimes it is defined starting with 1, that is as the sequence Kol′ = 1Kol. Our proof of Theorem 1 is specific for the sequence Kol, and does not apply, at least directly, for the sequence Kol′ (and hence neither for the sequence SKol, where S is a new symbol, cf. [CK]). In a list of problems on the Kolakovski sequence by Shallit, cf. [Sh], it was asked whether the Kolakovski sequence Kol′ is a D0L word. Consequently, we have given a solution to a very related question. This problem has been settled by Yu as well, cf. [Y].

As our second result we prove

Theorem 2. The sequence $(\alpha_n)_{n\geq 0}$ of Example 2 is not a CD0L sequence.

Proof. As we concluded in Example 2 the words α_n satisfy the recurrence

$$|\alpha_{n+1}| = |\alpha_n| + \left\lceil \frac{|\alpha_n|}{2} \right\rceil \tag{10}$$

for $n \geq 0$ and $|\alpha_0| = 1$. Now, the proof is based on the following straightforward observations:
(i) In each CD0L sequence $(w_n)_{n\geq 0}$ the sequence $(last(w_n))_{n\geq 0}$ is ultimately periodic ;
(ii) $last(\alpha_{n+1}) = last(\alpha_n)$ iff $|\alpha_n|$ is odd.

So it remains to be proved

Claim. The sequence (10) is not ultimately periodic in $\mathbf{Z}_2$.

Proof of Claim. Let us denote $a_n = |\alpha_n|$, for $n \geq 0$. Now, assume that the sequence $(a_n)_{n\geq 0}$ is ultimately periodic in $\mathbf{Z}_2$ with treshold t and period p, that is

$$a_{t+j+ip} \equiv a_{t+j+(i+1)p} \pmod 2 \tag{11}$$

for all $i \geq 0$ and $j = 0, ..., p-1$. In particular,

$$a_{t+j} \equiv a_{t+p+j} \pmod 2 \tag{12}$$

for all $j = 0, ..., p-1$. We show that (12) implies

$$a_t \equiv a_{t+p} \pmod{2^p} . \tag{13}$$

First, by (12), for each $j = 0, ..., p-1$ we can write

$$a_{t+j} = a_{t+p+j} + 2r_j \tag{14}$$

for suitably chosen numbers r_j. Consequently, for $j = 0, ..., p-1$,

$$\left\lceil \frac{3a_{t+j}}{2} \right\rceil = \left\lceil \frac{3a_{t+p+j}}{2} \right\rceil + 3r_j .$$

This, since $a_{n+1} = \lceil \frac{3}{2} a_n \rceil$ for $n \geq 0$, together with (12) yields

$$3r_j \equiv 0 \pmod 2$$

for $j = 0, ..., p-2$. Hence, the numbers $r_0, ..., r_{p-2}$ in (14) are even, and consequently we can replace (12) by

$$a_{t+j} \equiv a_{t+j+p} \pmod{2^2} \tag{12'}$$

for $j = 0, ..., p-2$. Repeating exactly the same argument replacing (12) by (12') we finally conclude (13).

Now, we are in the position to derive a contradiction. Clearly, not only p, but also any of its multiples is a period of the sequence $(a_n)_{n \geq 0}$ in $\mathbf{Z}_2$. Consequently, for any $n \geq 1$ we have

$$a_t \equiv a_{t+np} \pmod{2^{np}},$$

and

$$a_{t+p} \equiv a_{t+np} \pmod{2^{(n-1)p}}.$$

Hence, for all $n \geq 1$,

$$a_t \equiv a_{t+p} \pmod{2^{(n-1)p}}.$$

This is clearly impossible, since t and p are fixed and the sequence $(a_n)_{n \geq 0}$ is strictly monotonous. □

Theorem 2 still leaves it open whether Alt is a CD0L word, or in more general, whether there exist a D0L word with periodic control which is not a CD0L word, cf. [CK]. An example of a CD0L word which is not a D0L word with periodic control is given in [CK].

4 Concluding remarks and open problems

Let ω be an infinite word generated by a D0L TAG system G with periodic control and w_n, for $n \geq 0$, the prefix of ω defined after the nth iteration of G. We can assosiate with ω the following problems:

Problem 1. Is the word ω ultimately periodic?

Problem 2. Is the word ω a D0L word?

Problem 3. Is the sequence $(w_n)_{n \geq 0}$ a CD0L sequence?

Problem 4. Is the word ω a CD0L word?

Problem 5. Does the sequence $(|w_n|)_{n \geq 0}$ satisfy a linear recurrence?

Problem 6. Assume that $\omega \in \{1,2\}^\omega$, and let for $i = 1,2$, $|w_n|_i$ denote the number of i's in w_n. Is it true that $|w_n|_1 \sim |w_n|_2$, i.e. $\lim_{n\to\infty} \dfrac{|w_n|_1}{|w_n|_2} = 1$?

Problem 7. Does the limit $\lim_{n\to\infty} \dfrac{|w_{n+1}|}{|w_n|}$ exist?

These problems are very natural and by no means new. Indeed, Problems 2–7, together with some others, were formulated in [Sh] for the Kolakovski sequence. Moreover, Problem 4 is related to one of the main open problems in [CK]. In what follows we consider these problems for our particular sequences Kol and Alt.

Let us consider first the Kolakovski sequence. It was shown in [D1] that it is not ultimately periodic; the proof is not complicated. In Section 3 we proved that it is not a D0L word either. Our proof was straightforward, but specific to the sequence 221121... . Problems 3–7, in turn, seem to be open (although not independent of each other). It is likely that Problems 3, 4 and 5 have a negative answer, while Problems 6 and 7, which are equivalent, as is easy to see, have an affirmative answer. Computer calculations shown in Figures 1–2 strongly support these conjectures. In Figure 3 we present calculations similar to those in Figure 1 for the generalized Kolakovski word Kol_3.

For the sequence Alt we can say more about these problems. In Section 3 we gave a negative answer to Problem 3. In doing so we proved that the sequence $(|\alpha_n|)_{n\geq 0}$ is not ultimately periodic in $\mathbf{Z}_2$, and therefore cannot satisfy a linear recurrence in $\mathbf{Z}_2$. But then it cannot satisfy a linear recurrence in $\mathbf{Z}$ either, so that we have a negative answer to Problem 5 as well.

The fact that Alt is not a D0L word can be proved as Theorem 1. However, the proof is simpler since the representation corresponding to (5) is now

$$\begin{array}{lcl} 11 & \rightarrow & 112 \\ 12 & \rightarrow & 111 \\ 21 & \rightarrow & 222 \\ 22 & \rightarrow & 221\,. \end{array}$$

Consequently, the prefix 2211121 of Alt occurs as a subword in $\{112,111,222,221\}^*$ in the unique way: ...2211121... . This representation makes it also easy to conclude that Alt is not ultimately periodic. Problem 7 has an easy affirmative answer, too. The limit is 3/2 since $|\alpha_{n+1}| = 3/2|\alpha_n| + 0(1)$. The above presentation strongly indicates that also Problem 6 has an affirmative answer for the word Alt. However, a detailed argument is missing. Finally, Problem 4 for Alt remains open — our conjecture is "no", as for the Kolakovski sequence.

References

[AG] Autebert, J.M. and Gabarro, J.: Iterated GSM's and Co-CFL, Acta Informatica 26 (1989), 749–769.

[Be1] Berstel, J.: Sur les mots sans carré définis par un morphisme, Springer Lecture Notes in Computer Science 71 (1979), 16–25.

[Be2] Berstel, J.: Mots sans carré et morphismes iteres, Discret. Math. 29 (1980), 235–244.

[BP] Berstel, J. and Perrin, D.: Theory of Codes, Academic Press, New York (1986).

[CK] Culik II, K. and Karhumäki, J.: Iterative devices generating infinite words, manuscript (1991).

[D1] Dekking, F.M.: Regularity and irregularity of sequences generated by automata, Sem. Th. de Nombres de Bordeaux, exp. 9 (1979–1980).

[D2] Dekking, F.M.: On the structure of selfgenerating sequences, Sem. Th. de Nombres de Bordeaux (1980–1981).

[K] Kolakovski, W.: Self generating runs, Problem 5304, Amer. Math. Monthly 71 (1965).

[M] Minsky, M.L.: Computation: Finite and Infinite Machines, Prentice-Hall, Englewood Cliffs, N.J. (1967).

[PL] Prusinkiewicz, P. and Lindenmayer, A.: The Algorithmic Beauty of Plants, Academic Press, New York (1990).

[Sh] Shallit, J.: Open problems on the Kolakovski sequence, private communication.

[Sm] Smith III, A.R.: Plants, fractals and computer graphics, Computer Graphics 18 (1984), 1–10.

[T] Thue, A.: Über die gegenseitige Lage gleicher Teile gewisser Zeichenreihen, Videnskapsselskapets Skrifter. I. Mat.-naturv. Klasse, Kristiania (1912), 1–67.

[Y] Yu, S.: private communication.

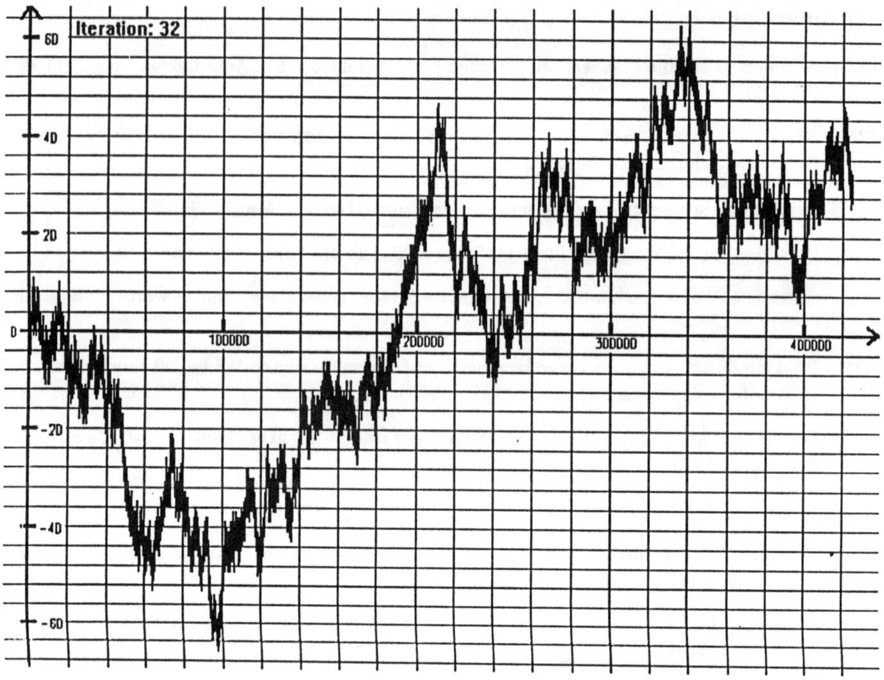

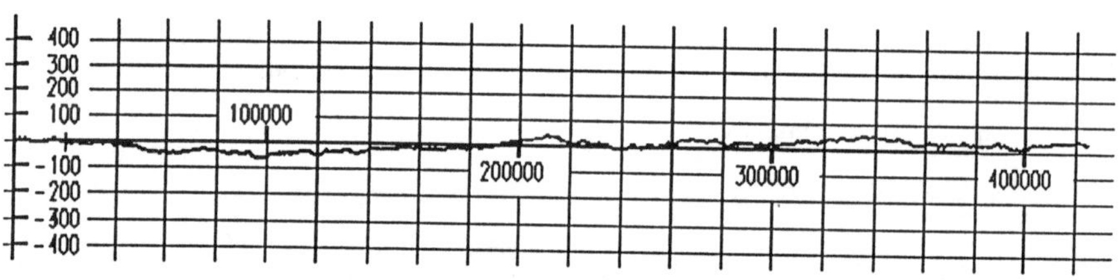

FIGURE 1

The diagrams show the value $|x|_1 - |x|_2$ for the prefixes x of the Kolakovski sequence up to the length 400 000 in two different scales.

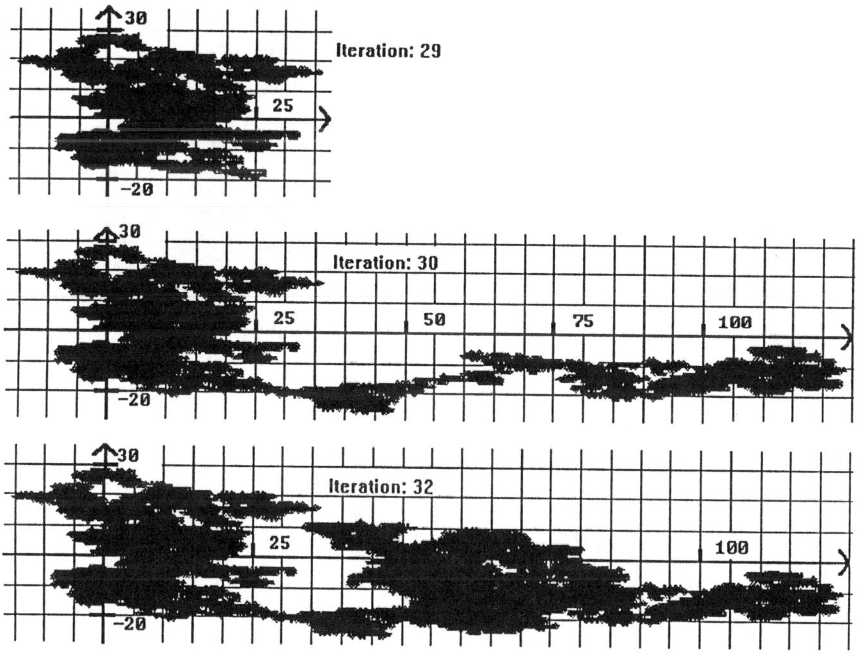

FIGURE 2

In this figure the Kolakovski sequence is divided into consecutive blocks of length two, i.e. to the blocks of 11, 12, 21 and 22 , and a relationship between the numbers of the occurrences of these blocks are shown. By (5) one can get an intuitive support for the fact that these numbers are in ratio 1:2:2:1 . In these figures we have combined the numbers of occurrences of 12 and 21 , so that any of their occurrences corresponds one fourth of the unit in the direction W . Similarly, 11 (resp. 22) corresponds one unit in the direction SSE (resp. NNE). The length of the prefix after 32 iterations is over 400 000 . The calculations indicate that the numbers of 11's and 22's are in ratio 1:1 , as well as the sum of the numbers of 11's and 22's and that of 12's and 12's are in ratio 1:2 .

It is also interesting to note that in the 30th iteration step the vector changes its value clearly, but returns considerably during the next two iteration steps.

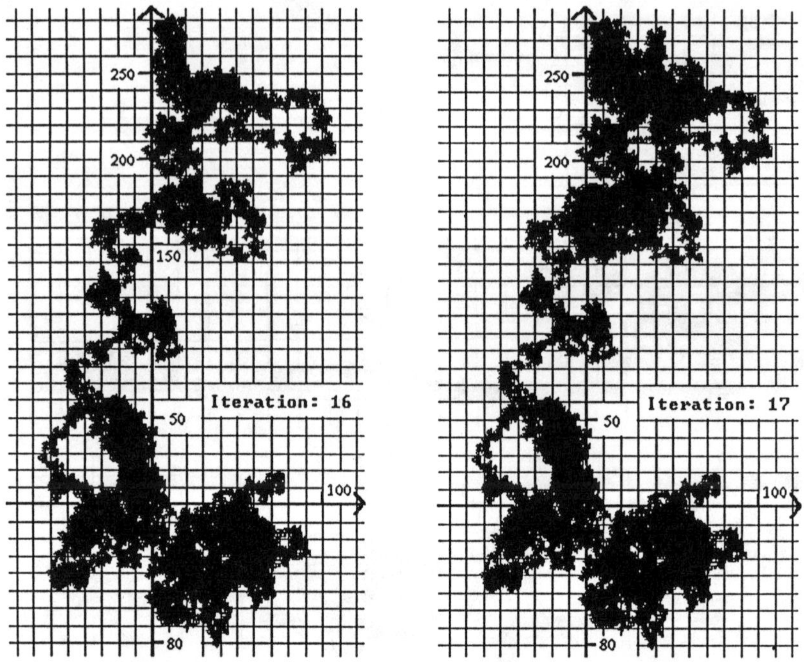

FIGURE 3

Figures are associated with the prefixes of the the generalized Kolakovski sequence Kol_3 after 16 and 17 iterations. Points present positions of the vectors $v = (|x|_1, |x|_2, |x|_3)$ for the prefixes x of Kol_3. Directions of the components of v are ENE, WNW and S, respectively, and the increment of the value $|x|_i$ by 1 corresponds one unit. The length of the prefix after 17 iterations is over 400 000, so that the vector v is all the time relatively very small.

Interaction Strength of DIL Systems

Youichi Kobuchi

Department of Electronics and Informatics, Faculty of Science and Technology, Ryukoku University, Seta, 520-21 Japan Bitnet : kobuchi@jpnrins

Abstract: Some DIL systems with certain property can be simulated by D0L systems. We have shown three such examples of D<1,0>L systems and obtained a few sufficient conditions for a D<1,1>L system to be simulated by a D0L system.

Keywords: Interaction strength, Simulation, DIL system, D0L system

1. Introduction

One way of measuring the strength of interaction in DIL systems is to use the code length needed to simulate a system under consideration through some D0L system if it is possible at all. The idea of simulating DIL systems with D0L systems has been elaborated on for some time. [1, 3, 4, 6]

In our previous works [4, 6], we defined an isomorphism between L schemes and classified the four cases of the isomorphism relations between D0L scheme S_0 and DIL scheme S as follows.

(1) Isomorphism between S_0 and S.

(2) Embedding S_0 into S : S_0 is isomorphic to a proper subscheme of S.

(3) Simulation of S by S_0 : A proper subscheme of S_0 is isomorphic to S.

(4) Isomorphism between proper subschemes of S_0 and S, respectively.

This note extends the results discussed in the analysis of simulation relation (case (3)) to the relation between D0L and DIL systems which falls in the last category in the above classification. (case (4)) We use D<1, 0>L systems as the simplest examples of DIL systems although definitions and theorems are written for D<1, 1>L systems. In this note, we have shown, in particular, that a D<1,0>L system used in Lindenmayer's original paper [5] can be simulated by a D0L system.

2. Definitions and relevant previous results

First, we give some necessary definitions for our later use although we assume the reader to be familiar with the basics of L system framework. (For the details, see [2].)

Let S = <Σ, P, g> be a D<k, l>L scheme where Σ is the alphabet of S, g is an environmental input symbol not in Σ ,and P is the set of deterministic generation rules with k left-side and l right-side context given as a function from $\Sigma_g{}^k \times \Sigma \times \Sigma_g{}^l$ into Σ^*, where $\Sigma_g = \Sigma \cup \{g\}$. Then, S defines the generation mapping f_S from Σ^* to Σ^* through the parallel application of the rules P as usual. An L subscheme of S, denoted as S(Γ) = <Γ, P, g>, is a closed part of the generations of S such that $\Gamma \subset \Sigma^*$ and $f_S(\Gamma) \subset \Gamma$. We denote the generation mapping of S(Γ), which is the restriction of f_S to Γ, as $f_{S(\Gamma)}$.

Definition. (Isomorphism between L subschemes)

Let X = <Γ_X, P_X, g_X> and Y = <Γ_Y, P_Y, g_Y> be two L subschemes. They are isomorphic to each other through a one-one mapping $h : \Gamma_X \rightarrow \Gamma_Y$, denoted as $X \Leftrightarrow^h Y$, if $f_Y(h(x)) = h(f_X(x))$ for every x in Γ_X.

In what follows, we consider D0L scheme S_0 = <Σ_0, P_0> and DIL scheme S = <Σ, P, g>, and treat the case where h is given by a λ-free homomorphism $\gamma : \Sigma_0 \rightarrow \Sigma^+$. As cited in Section 1, there are four types of ways in which isomorphisms exist between subschemes of S_0 and S. In the list, simulation of S by S_0 (case 3) means that some part of S_0 is isomorphic to a whole S. That is, $S_0(\Gamma) \Leftrightarrow^\gamma S$ for a λ-free homomorphism $\gamma : \Sigma_0 \rightarrow \Sigma^+$. It has been noted that [4, 6] the λ-free homomorphism γ can be restricted as $\gamma : \Sigma_0 \rightarrow \Sigma$ in this case. Thus the problem of case 3 is : Given a DIL scheme S, find conditions on S under which $S_0(\Gamma)$ and γ exist such that $S_0(\Gamma) \Leftrightarrow^\gamma S$. We review a few such necessary conditions for possible extension when S is a D<1,1>L scheme. Note that we are omitting some subtle points in our previous treatment in order to make this note readable. First, we define "the maximum derived string", which is just the longest string determined only by a given string, as follows.

Definition. (Maximum derived string)

For a D<1,1>L scheme S = <Σ, P, g>, the maximum derived string of $a_1a_2...a_m$ ($m \geq 2$) in Σ^+ is defined as follows.

$$\widetilde{f_{Sg}}(a_1a_2...a_m) = \beta_L\beta_2...\beta_{m-1}\beta_R$$

where β_L is the maximum length common suffix of $\{P(b,a_1,a_2) \mid b \text{ in } \Sigma \text{ or } b=g\}$, β_R is the maximum length common prefix of $\{P(a_{m-1},a_m,b') \mid b' \text{ in } \Sigma \text{ or } b'=g\}$, and $\beta_i = P(a_{i-1},a_i,a_{i+1})$ for i = 2,..., m-1.

Using the above concept, we have our central definition and a corresponding theorem.

Definition. (CD condition)

A D<1,1>L scheme $S = \langle \Sigma, P, g \rangle$ is said to fulfill the Critical Development condition (CD condition, for short) of order B if there exists an integer B (≥ 2) such that for every $a_1 a_2 \ldots a_B$ in Σ^B we have $|\widetilde{f_{Sg}}(a_1 a_2 \ldots a_B)| \geq B$.

Theorem 1.

If a D<1,1>L scheme $S = \langle \Sigma, P, g \rangle$ fulfills the CD condition of order B, then there exist a D0L scheme $S_0 = \langle \Sigma_0, P_0 \rangle$, a subscheme $S_0(\Gamma)$, and a homomorphism $\gamma : \Sigma_0 \rightarrow \Sigma^+$ such that $S_0(\Gamma) \Leftrightarrow^{\gamma} S$.

We have also had more general condition for a D<1,1>L scheme called CDE condition of order B which demands that any length B string does not diminish to a string whose length is less than two under any iterated application of the maximum derived string mapping. This CDE condition can be, with some additional condition related to the boundary effect, another sufficient condition for a D0L scheme to simulate it.

3. CD condition for systems

As an illustrative example, consider the following D<1,0>L scheme $S_1 = \langle \Sigma_1, P_1, g_1 \rangle : \Sigma_1 = \{0,1\}$, $P_1(0,0) = 0$, $P_1(0,1) = 10$, $P_1(1,0) = 1$, $P_1(1,1) = 0$, and $g_1 = 0$. This scheme does not satisfy the CD condition of any order. To see this, let $a_1 a_2 \ldots a_B = 0^B$ for an arbitrary integer B(≥ 2), then the maximum derived string $\widetilde{f_{Sg}}(a_1 a_2 \ldots a_B) = 0^{B-1}$. If we have a D<1,0>L system $G_1 = \langle S_1, w_0 \rangle$ where $w_0 = 1$, for example, then the above argument becomes futile. For a string 0^B where B is an integer greater than 2 cannot appear as a substring in any word generated by G_1. In fact, 0^3 cannot appear as a substring : Assume the contrary, then at least 0^3 or 1^3 must appear as a substring in a previous word. But 1^3 cannot appear in any derivation of this scheme, and 0^3 cannot appear in the axiom $w_0 = 1$. Any strings of length 3 except 0^3 and 1^3, on the other hand, satisfy the inequality for B = 3 required in the CD condition. Then we can apply the method described in our previous work to find the following D0L system which simulates this D<1,0>L system G_1.

Example 1. (D0L system $G_0 = \langle \Sigma_0, P_0, W_0 \rangle$ that simulates G_1)

There are 18 states in Σ_0 expressed in the form $\langle a_1\, a_2\, \underline{a_3}\, a_4\rangle$ where $a_1\, a_2\, a_3\, a_4$ is in $g^*\{0,1\}^+g^*$. Rule P_0 is given as follows where $x \rightarrow y$ means $P_0(x) = y$.

$\langle g\,g\,\underline{1}\,g\rangle \rightarrow \langle g\,g\,\underline{1}\,0\rangle\langle g\,1\,\underline{0}\,g\rangle$

$\langle g\,g\,\underline{1}\,0\rangle \rightarrow \langle g\,g\,\underline{1}\,0\rangle\langle g\,1\,\underline{0}\,1\rangle$

$\langle g\,1\,\underline{0}\,g\rangle \rightarrow \langle 1\,0\,\underline{1}\,g\rangle$

$\langle g\,1\,\underline{0}\,1\rangle \rightarrow \langle 1\,0\,\underline{1}\,1\rangle\langle 0\,1\,\underline{1}\,0\rangle$

$\langle 1\,0\,\underline{1}\,g\rangle \rightarrow \langle 1\,1\,\underline{0}\,g\rangle$

$\langle 1\,0\,\underline{1}\,1\rangle \rightarrow \langle 1\,1\,\underline{0}\,0\rangle$

$\langle 0\,1\,\underline{1}\,0\rangle \rightarrow \langle 1\,0\,\underline{0}\,1\rangle$

$\langle 1\,1\,\underline{0}\,g\rangle \rightarrow \langle 0\,0\,\underline{1}\,g\rangle$

$\langle 1\,1\,\underline{0}\,0\rangle \rightarrow \langle 0\,0\,\underline{1}\,0\rangle$

$\langle 1\,0\,\underline{0}\,1\rangle \rightarrow \langle 0\,1\,\underline{0}\,1\rangle\langle 1\,0\,\underline{1}\,0\rangle$

$\langle 0\,0\,\underline{1}\,g\rangle \rightarrow \langle 0\,1\,\underline{0}\,g\rangle$

$\langle 0\,0\,\underline{1}\,0\rangle \rightarrow \langle 0\,1\,\underline{0}\,1\rangle$

$\langle 0\,1\,\underline{0}\,1\rangle \rightarrow \langle 1\,0\,\underline{1}\,1\rangle\langle 0\,1\,\underline{1}\,0\rangle$

$\langle 1\,0\,\underline{1}\,0\rangle \rightarrow \langle 1\,1\,\underline{0}\,1\rangle$

$\langle 0\,1\,\underline{0}\,g\rangle \rightarrow \langle 1\,0\,\underline{1}\,g\rangle$

$\langle 1\,1\,\underline{0}\,1\rangle \rightarrow \langle 0\,0\,\underline{1}\,1\rangle\langle 0\,1\,\underline{1}\,0\rangle$

$\langle 0\,0\,\underline{1}\,1\rangle \rightarrow \langle 0\,1\,\underline{0}\,0\rangle$

$\langle 0\,1\,\underline{0}\,0\rangle \rightarrow \langle 1\,0\,\underline{1}\,0\rangle$

Note that a state $\langle a_1\, a_2\, \underline{a_3}\, a_4\rangle$ corresponds to the state a_3 in D<1,0>L system G_1, and the axiom for this D0L system W_0 is $\langle g\,g\,\underline{1}\,g\rangle$.

Some of the earlier generation steps are compared in Fig. 1.

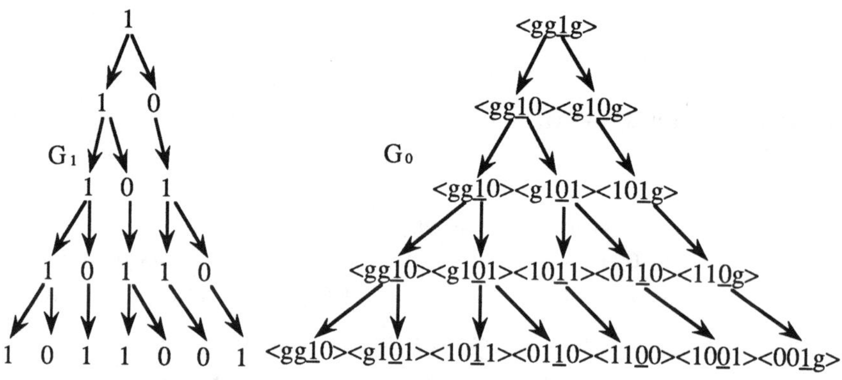

Fig. 1 Derivation steps of G_1 and those of simulating G_0.

The idea illustrated above yields the following extension of Theorem 1.

<u>Theorem 2.</u>

If a D<1,1>L system $G = \langle\Sigma, P, g, w_0\rangle$ fulfills the following CD condition of order B (≥ 2), then there exist a D0L system $G_0 = \langle\Sigma_0, P_0, W_0\rangle$, and a homomorphism $\gamma : \Sigma_0 \rightarrow \Sigma^+$ such that $G_0 \Leftrightarrow^{\gamma} G$.

CD condition for a D<1,1>L system G whose underlying scheme is S :

A D<1,1>L system G satisfies the Critical Development condition of order B if there exists an integer B (≥ 2) such that $|\widetilde{f_{S_g}}(a_1a_2...a_B)| \geq B$ for every $a_1a_2...a_B$ in $Sub_B(G)$ where $Sub_B(G)$ is the set of length B subwords of any word in L(G).

4. Lindenmayer's example and CDE condition for systems

Now we examine another D<1,0>L system $G_2 = <\Sigma_2, P_2, g_2, w_0>$ which was introduced in Lindenmayer's original paper [5]: $\Sigma_2 = \{0,1\}$, $P_2(0,0) = 0$, $P_2(0,1) = 11$, $P_2(1,0) = 1$, $P_2(1,1) = 0$, $g_2 = 0$, and $w_0 = 1$. Although this system G_2 is only slightly different from the system G_1 in its outlook, the interaction strength from our viewpoint is far stronger as is shown below.

First, note that the underlying scheme $S_2 = <\Sigma_2, P_2, g_2>$ also does not fulfill the CD condition of any order because the maximum derived string of 0^B is 0^{B-1} for any integer $B \geq 2$. Then we go through the following series of observation for possible substrings in the words of $L(G_2)$.

(1) $(10)^3$ cannot appear as a substring:

Assume that 1010 appears somewhere. Then possible predecessor to produce the leftmost 1 is 10 or 01. But 100 and 101 both fail to produce 1010. The only possible predecessor is then 01100 but we cannot have 101010 by adding any letters to the right of 01100.

(2) $(01)^3$ cannot appear as a substring:

Assume that 0101 appears. Then the only possible predecessor is 11001, but this produces 01011.

(3) 1^7 cannot appear as a substring:

Assume that 1^7 appears. Then it should come from $(10)^3$ or $(01)^3$ which is impossible by (1) and (2).

(4) 0^6 cannot appear as a substring:

0^6 can be obtained only by 0^7 or 1^7 , which is not possible because of (3) and by the fact that the axiom is 1.

Now we are going to check whether the CD condition for system G_2 hold when B = 7. Any possible string in $\{0,1\}^7$ (that is, except 0^7, 10^6, and 1^7 by the above consideration) can be classified into two classes : Class A where each component string contains 01, or Class B where the strings have the form 1^u0^v such that $u + v = 7$, $u \geq 2$, and $v \geq 1$. Every string in Class A trivially meets the CD condition for B = 7. On the other hand, every string in Class B does not satisfy the CD condition. Take 1^60, for example. We have a sequence of strings through the maximum derived string operation as follows.

1111110 $\rightarrow$ 000001 $\rightarrow$ 000011 $\rightarrow$ 000110 $\rightarrow$ 001101 $\rightarrow$ 0110111.

The last string belongs to Class A. Similarly, we have the following sequences for the other strings in Class B.

1111100 $\rightarrow$ 000010 $\rightarrow$ 000111 $\rightarrow$ 001100 $\rightarrow$ 011010 $\rightarrow$ 1101111.

1111000 $\rightarrow$ 000100 $\rightarrow$ 001110 $\rightarrow$ 011001 $\rightarrow$ 1101011.

1110000 $\rightarrow$ 001000 $\rightarrow$ 011100 $\rightarrow$ 110010 $\rightarrow$ 010111 $\rightarrow$ 1111100.

1100000 $\rightarrow$ 010000 $\rightarrow$ 111000 $\rightarrow$ 00100 $\rightarrow$ 01110 $\rightarrow$ 11001 $\rightarrow$ 01011 $\rightarrow$ 111110 $\rightarrow$
00001 $\rightarrow$ 00011 $\rightarrow$ 00110 $\rightarrow$ 01101 $\rightarrow$ 110111 $\rightarrow$ 011100 $\rightarrow$ 110010 $\rightarrow$
010111 $\rightarrow$ 1111100.

Thus, every string in Class B ends up a string in Class A or Class B. These facts ensure that the following modified CDE condition holds for substrings which appear in $L(G_2)$ when B = 7.

Definition. (Modified CDE condition for D<1,1>L system)

A D<1,1>L system G whose underlying scheme being S satisfies the modified Extended Critical Development condition (CDE condition, for short) of order B if there exists an integer B (≥ 2) such that for every $a_1a_2...a_B$ in $Sub_B(G)$ there exists an integer n for which $|\widetilde{f_{S_g}}^n(a_1a_2...a_B)| \geq B$.

By a similar coding as in Theorem 2, we have

Theorem 3.

If a D<1,1>L system $G = <\Sigma, P, g, w_0>$ fulfills the modified CDE condition of order B (≥ 2), then there exist a D0L system $G_0 = <\Sigma_0, P_0, W_0>$, and a homomorphism $\gamma : \Sigma_0 \rightarrow \Sigma^+$ such that $G_0 \Leftrightarrow^{\gamma} G$.

Thus we have shown that the Lindenmayer's D<1,0>L system $G_2 = <\Sigma_2, P_2, g_2, w_0>$ can be simulated by a D0L system. Since this D0L system has a big alphabet, we use still another example to show actual coding process as follows.

Consider a D<1,0>L system $G_3 = <\Sigma_3, P_3, g_3, w_0>$ where $\Sigma_3 = \{0,1\}$, $P_3(0,0) = 01$, $P_3(0,1) = 0$, $P_3(1,0) = 0$, $g_2 = 0$, and $w_0 = 0$. Note that $P_2(1,1)$ is not defined because substring 11 does not appear in any string derived by G_3. Take any length 4 string in $\Sigma_3{}^*$ which does not contain 11 as a substring. If it contains 00, then the maximum derived string has length more than or equal to 4. The other possibilities are 1010 and 0101. Both of them yield 000 which in turn generates 0101. Thus we have seen that G_3 fulfills the modified CDE condition and can be simulated by the following DOL system.

Example 2. (DOL system $G_0 = \langle \Sigma_0, P_0, W_0 \rangle$ that simulates G_3)

$\langle g\,g\,g\,\underline{0}\,g\rangle \to \langle g\,g\,g\,\underline{0}\,1\rangle\langle g\,g\,0\,\underline{1}\,g\rangle$	$\langle 0\,0\,0\,\underline{1}\,0\rangle \to \langle 1\,0\,1\,\underline{0}\,0\rangle$
$\langle g\,g\,g\,\underline{0}\,1\rangle \to \langle g\,g\,g\,\underline{0}\,1\rangle\langle g\,g\,0\,\underline{1}\,0\rangle$	$\langle 0\,0\,1\,\underline{0}\,1\rangle \to \langle 0\,1\,0\,\underline{0}\,0\rangle$
$\langle g\,g\,0\,\underline{1}\,g\rangle \to \langle g\,0\,1\,\underline{0}\,g\rangle$	$\langle 0\,1\,0\,\underline{1}\,0\rangle \to \langle ?\,0\,0\,\underline{0}\,0\rangle$
$\langle g\,g\,0\,\underline{1}\,0\rangle \to \langle g\,0\,1\,\underline{0}\,0\rangle$	$\langle 1\,0\,1\,\underline{0}\,g\rangle \to \langle ?\,0\,0\,\underline{0}\,g\rangle$
$\langle g\,0\,1\,\underline{0}\,g\rangle \to \langle 0\,1\,0\,\underline{0}\,g\rangle$	$\langle 1\,0\,1\,\underline{0}\,0\rangle \to \langle ?\,0\,0\,\underline{0}\,0\rangle\langle 0\,0\,0\,\underline{0}\,1\rangle$
$\langle g\,0\,1\,\underline{0}\,0\rangle \to \langle 0\,1\,0\,\underline{0}\,0\rangle\langle 1\,0\,0\,\underline{0}\,1\rangle$	$\langle ?\,0\,0\,\underline{0}\,0\rangle \to \langle 0\,1\,0\,\underline{1}\,0\rangle\langle 1\,0\,1\,\underline{0}\,1\rangle$
$\langle 0\,1\,0\,\underline{0}\,g\rangle \to \langle 0\,0\,0\,\underline{1}\,g\rangle$	$\langle ?\,0\,0\,\underline{0}\,g\rangle \to \langle 0\,1\,0\,\underline{1}\,g\rangle$
$\langle 0\,1\,0\,\underline{0}\,0\rangle \to \langle 0\,0\,0\,\underline{1}\,0\rangle\langle 0\,0\,1\,\underline{0}\,1\rangle$	$\langle 0\,0\,0\,\underline{0}\,1\rangle \to \langle 0\,1\,0\,\underline{1}\,0\rangle$
$\langle 1\,0\,0\,\underline{0}\,1\rangle \to \langle 0\,1\,0\,\underline{1}\,0\rangle$	$\langle 1\,0\,1\,\underline{0}\,1\rangle \to \langle ?\,0\,0\,\underline{0}\,0\rangle$
$\langle 0\,0\,0\,\underline{1}\,g\rangle \to \langle 1\,0\,1\,\underline{0}\,g\rangle$	$\langle 0\,1\,0\,\underline{1}\,g\rangle \to \langle ?\,0\,0\,\underline{0}\,g\rangle$

There are 20 states in S_0 and $W_0 = \langle g\,g\,g\,\underline{0}\,g\rangle$. The symbol ? means "don't know."

5. Concluding remarks

We have shown by examining some examples of D<1, 0>L systems that they can be simulated by DOL systems. The coding method is basically the same as before [4, 6], and our effort has gone into checking whether the possible substrings of certain length can maintain the length under the maximum derived string mapping.

All the three examples happen to have regenerative property as defined in [3]. That is, we can write as a formula $w_t = w_{t-1}\, z_t$ $(t = 1,2,...)$ for these D<1, 0>L systems (z_t varies from example to example) where w_t is a string generated at step t. It seems that these regenerative structures are a sign of weak interaction as can be suggested by locally catenative DIL systems [3].

Acknowledgements

Most of the work reported here was carried out while the author was visiting the Theoretical Biology Group at Utrecht University during the summer of 1987. Hospitality shown to him by A. Lindenmayer and the other group members is greatly appreciated.

References

1. Culik, K. and Karhumäki, J.: Interactive L Systems with Almost Interactionless Behaviour. *Inform. Contr.* **43**, 83-100. (1979)
2. Herman, G.T. and Rozenberg, G.: *Developmental Systems and Languages.* North-Holland, Amsterdam. 1975
3. Kobuchi, Y. : Two Characterization Theorems of Locally Catenative Developmental Systems. *Inform. Processing Lett.* **6**, 4, 120-124. (1977)
4. Kobuchi, Y. and Seki, S. : Cell-Cell Interactions in Developmental Systems. in "Proceedings, International Symposium on Mathematical Topics in Biology" Kyoto 1978. 207-217.
5. Lindenmayer, A. : Mathematical Models for Cellular Interaction in Development. Parts I and II. *J. Theoret. Biol.* **18** 280-299. and 300-315.(1968)
6. Seki, S. and Kobuchi, Y. : On DIL Schemes Simulated by D0L Schemes. *JCSS* **21**, 1, 30-45. (1980)

Inductive Inference of 0L Languages*

Takashi YOKOMORI

Department of Computer Science and Information Mathematics
University of Electro-Communications
1-5-1 Chofugaoka, Chofu, Tokyo 182, JAPAN

Abstract. This paper studies the problem of identifying the family of 0L languages, in which the identifiability in the limit from positive data is the main concern.

We show that the family of P0L languages is not identifiable in the limit from positive data. Thus, in general, the family of 0L languages is not identifiable in the limit from positive data. It is also shown that the family of PD0L languages is identifiable in the limit from positive data. Further, we show that there is a consistent, responsive and conservative algorithm for identifying the family of PD0L languages, if we make an appropriate choice of the representation space for PD0L languages.

Keywords: inductive inference, 0L languages, identification in the limit

1 Introduction

In the study of inductive inference of formal languages, Gold [7] showed that the family of languages containing all finite sets and one infinite set is not identifiable in the limit from positive data only. This fact was shocking in a sense because a simple implication is that even the family of regular languages, one of the most prominent families in the formal language theory, is not identifiable in the limit from positive data.

Angluin [1] has given several conditions for the family of languages to be identifiable in the limit from positive data, and presented some examples of identifiable families. At present, there are not a few families of languages shown to be identifiable in the limit from positive data.

---

* Supported in part by Grants-in-Aid for Scientific Research No.02650261 from the Ministry of Education, Science and Culture, Japan.

Speaking of the inductive inference of Lindenmayer language families, the following type of inference model is a main framework: Given a finite set of finite sequences of strings, the goal is to decide if there exists an appropriate set (family) of developmental systems in which a correct system(grammar) for the given data exists, and (2) select a particular system from the set if exists.([5], [8]) Some results on the identification of L-systems in "in the limit" problem setting can be also found in [8].

There obviously exists a biological motivation of this inference problem which usually reads as follows: A developmental biologist interested in a particular plant has a large number of *observed experimental data.* His task is to explain the observed data by some formal model, and to find a concrete mechanism with which the explanation is achieved.

The observed experimental data are usually assumed to be given as a finite set of finite sequences of strings from a fixed alphabet. In the standard framework of grammatical inference, this corresponds to a finite set of finite *positive samples* that are provided together with some sort of *derivational information.*

In this paper, we are concerned with the inductive inference problem for the families of 0L languages, in which the inference model of "identification in the limit from positive data" is adopted. When compared with the results in a traditional model mentioned above, the identifiablity results in this model provide us with more implicative results in twofold senses ; the correctness of the conjecture identified in the limit is assured for all data from the target language including observed ones, and no derivational (historical) information on data is required.

We shall show that the family of P0L languages is not identifiable in the limit from positive data. Thus, in general, the family of 0L languages is not identifiable in the limit from positive data. It is also shown that the family of PD0L languages is identifiable in the limit from positive data. Further, we show that there is a consistent, responsive and conservative algorithm for identifying the family of PD0L languages, if we make an appropriate choice of the representation space for PD0L languages.

2 Definitions

2.1 Basic Definitions and Notations

We assume the reader to be familiar with the rudiments of automata and formal language theory as well as L-system theory. (For notions and notations not stated here, see, e.g., [10], [15].)

Let Σ be a fixed finite alphabet and Σ^* be the set of all strings over Σ. Further, let $\Sigma^+ = \Sigma^* - \{\lambda\}$, where λ is the null string. By $lg(u)$ we denote the length of string u. A *language over* Σ is a subset of Σ^*. The cardinality of a set S is denoted by $|S|$. By $\mathbf{N}$ we denote the set of non-negative integers. A *finite substitution* h on Σ is a mapping defined by : for $\forall a \in \Sigma$, $h(a)$ is a finite subset of Σ^*.

Let $\Sigma = \{a_1, ..., a_m\}$. A *Parikh mapping* φ on Σ^* is a mapping from Σ^* to $\mathbf{N}^m$ defined by $\varphi(w) = (\#_{a_1}(w), ..., \#_{a_m}(w))$, where $\#_{a_i}(w)$ denotes the number of symbols a_i appearing in w. The relation $\preceq$ on $\mathbf{N}^m$ is defined as follows:

for $v_i = (x_{i,1}, ..., x_{i,m})(i = 1, 2)$ in $\mathbf{N}^m$, $v_1 \preceq v_2$ iff $1 \leq \forall j \leq m : v_{1,j} \leq v_{2,j}$.

Further, $v_1 \prec v_2$ iff $v_1 \preceq v_2$ and $v_1 \neq v_2$.

2.2 Language Family and Its Representations

An *indexed family of nonempty languages* is an infinite sequence $L_1, L_2, L_3, ...$, where each L_i is a nonempty language. An *indexed family of nonempty recursive languages* is an indexed family of nonempty languages $L_1, L_2, L_3, ...$, such that there exists an effective procedure to compute the membership function

$$f(i,w) = \begin{cases} 1 & (\text{ if } w \in L_i) \\ 0 & (\text{ otherwise }). \end{cases}$$

Let $\mathcal{F} = L_1, L_2, L_3, ...$ be an indexed family of nonempty recursive languages to be identified over a fixed alphabet Σ. Then, we consider a *class of representations* $\mathcal{R}$ with the following properties :

1. $\mathcal{R}$ is a recursive language (over some fixed alphabet),

2. for $\forall L \in \mathcal{F}$, there exists $r \in \mathcal{R}$ such that r represents L(denoted by $L(r) = L$).

(In this case, we say that $\mathcal{R}$ is a class of representations for $\mathcal{F}$.)

2.3 Identification in the Limit

For a given $r \in \mathcal{R}$, a *presentation* of the language $L(r)$ is any infinite sequence of examples (or data) such that every $w \in \Sigma^*$ occurs at least once in the sequence with its sign (+: when $w \in L(r)$, or −: otherwise), and no other examples (incorrectly labeled) appear in the sequence. A *positive presentation* of $L(r)$ is any infinite sequence of examples such that every $w \in L(r)$ occurs at least once in the sequence and no other examples not in $L(r)$ appear in the sequence.

Let r be a representation in $\mathcal{R}$ representing a given L(i.e., $L = L(r)$). An algorithm A is said to *identify a language L in the limit using $\mathcal{R}$* iff for any presentation of L, the infinite sequence of representations r_i in $\mathcal{R}$ produced by A satisfies the property that there exists a representation r' in $\mathcal{R}$ such that for all sufficiently large i, the i-th conjecture (representation) r_i is identical to r' and $L(r') = L(r)$. An indexed family of languages $\mathcal{F}$ is *identifiable in the limit using $\mathcal{R}$* iff there exists an algorithm A that, given an L in $\mathcal{F}$, identifies L in the limit using $\mathcal{R}$.

When $\mathcal{R}$ is explicit or not important, we simple say that $\mathcal{F}$ is idenfifiable in the limit. $\mathcal{F}$ is *identifiable in the limit from positive data* iff it is identifiable in the limit and only positive presentations are used.

2.4 Consistent, Responsive and Conservative Algorithms

Let A be an inference algorithm for identifying $\mathcal{F}$ in the limit from positive data. We say that A is *consistent on* $\mathcal{F}$ provided that for every $L \in \mathcal{F}$ and every positive presentation of L, the conjecture produced by A contains all the input data read in up to that point in the computation. We say that A is *responsive* on $\mathcal{F}$ provided that for every $L \in \mathcal{F}$ and every positive presentation of L, between any two distinct steps of computation of A requiring input data on the presentation, there exists a step in which A produces a conjecture. We say that A is *conservative* on $\mathcal{F}$ provided that for every $L \in \mathcal{F}$ and every positive presentation of L, each conjecture r_i from A is equal to the last conjecture r_j

made by A unless $L(r_j)$ fails to contain some input data read in prior to the point of the conjecture r_i.

Intuitively, an inference algorithm A is "consistent" if every conjecture is compatible with all the data read in so far. A is "responsive" if it makes one conjecture in response to each input before requesting another input, and A is "conservative" if it never changes its conjecture unless the conjecture fails to explain all the data read in so far.

2.5 Known Results

We summarize the known results on the identifiability in the limit from positive data, which play a central role to obtain the main resutls.

Condition A: An indexed family of nonempty languages *satisfies Condition A* iff there exists an effective procedure which on any input $i \geq 1$ enumerates a set of strings T_i such that

(*i*) T_i is finite,
(*ii*) $T_i \subseteq L_i$, and
(*iii*) for all $j \geq 1$, if $T_i \subseteq L_j$, then L_j is not a proper subset of L_i.

Theorem 1 ([1]) *An indexed family of nonempty recursive languages is identifiable in the limit from positive data iff it satisfies Condition A.*

Condition B: An indexed family of nonempty recursive languages *satisfies Condition B if* for every $i \geq 1$, there exists a finite subset T_i of L_i such that for every $j \geq 1$, if $T_i \subseteq L_j$, then L_j is not a proper subset of L_i.

Note that Condition B is simply Condition A without the requirement of effective enumerability of T_i.

Theorem 2 ([1]) *If an indexed family of nonempty recursive languages is identifiable in the limit from positive data, then it satisfies Condition B.*

3 Main Results

3.1 OL Systems and Languages

A *0L system* is a triple $G = (\Sigma, h, w)$, where Σ is an alphabet, h is a finite substitution on Σ(into the set of subsets of Σ^*), and w, refered to as the *axiom*, is an element of Σ^+. The *language* of G is defined by $L(G) = \bigcup_{i \geq 0} h^i(w)$. In this case, we say that G generates $L(G)$([15]).

Let $G = (\Sigma, h, w)$ be a 0L system. A symbol $a \in \Sigma$ is *useful* if there exists a string in $L(G)$ containing a. G is termed *reduced* if any $a \in \Sigma$ is useful.

A 0L system $G = (\Sigma, h, w)$ is termed *propagating*(or, shortly a *P0L system*) if for no $a \in \Sigma$, $h(a)$ contains λ. G is called *deterministic*(or, shortly a *D0L system*) if for every $a \in \Sigma$, $|h(a)| = 1$. Further, by *PD0L system* we denote a propagating D0L system.

Let G be a triple $(\Sigma, h, \{w_1, ..., w_p\})$, where Σ, h are the same as PD0L system, $\{w_1, ..., w_p\}$ is a finite set of strings in Σ^+. G is termed *FPD0L system*. The language of G is defined by $L(G) = \bigcup_{i=1}^{p} L(G_i)$, where $G_i = (\Sigma, h, w_i)$ is a PD0L system.

A language is termed a 0L(resp. P0L, D0L, PD0L, FPD0L) *language* if it equals $L(G)$ for some 0L(resp. P0L, D0L, PD0L, FPD0L) system G.

By $\mathcal{L}(X)$ we denote the family of X languages, where X runs through $\{0L, P0L, D0L, PD0L, FPD0L\}$.

Let $G = (\Sigma, h, w)$ be a D0L system. G is *conservative* if for every $i \geq 0$, alph$(h^i(w))= \Sigma$. Further, G is *growing* if for every $i \geq 0$, $\varphi(h^i(w)) \prec \varphi(h^{i+1}(w))$.

In the process of identifying some family of languages in the limit, it would be much better if the family is identifiable using only positive data. However, we shall first show that even the family of P0L languages is not identifiable in the limit from positive data.

Theorem 3 *$\mathcal{L}$(P0L) is not identifiable in the limit from positive data.*

Proof. From Theorem 2, it suffices to show that Condition B fails.

Consider a P0L language $L = \{b\} \cup \{a^n | n \geq 2\}$ which is generated by a P0L system $G = (\{a, b\}, h, b)$, where $h(a) = \{a, a^2\}, h(b) = \{a^2\}$. Let T be any nonempty finite subset of L.

Suppose that $T = \{a^{n_1}, ..., a^{n_p}\}$(where, $2 \leq n_1 < \cdots < n_p$.) Then, consider a P0L system $G_T = (\{a, b\}, h_T, b)$, where $h_T(a) = \{a\}, h_T(b) = \{a^{n_1}, ..., a^{n_p}\}$. Since $L(G_T) = \{b\} \cup T$, it is clear that $T \subseteq L(G_T) \subset L$ holds.

Suppose that $T = \{b\} \cup T'$, where either $T' = \{a^{n_1}, ..., a^{n_p}\}$ $(2 \leq n_1 < \cdots < n_p)$, or $T' = \emptyset$. Then, consider, again, the P0L system $G_T = (\{a, b\}, h_T, b)$ given above, where if T' is empty, then h_T is defined as an identity mapping. In either case, $T \subseteq L(G_T) \subset L$ holds. Thus, Condition B fails. □

Thus, it is proved that the family $\mathcal{L}(0L)$ is not identifiable in the limit from positive data. On the other hand, since it is known that any indexed family of recursive languages is identifiable in the limit from not only positive but also *negative* data([7]), we obviously obtain:

Theorem 4 *$\mathcal{L}$(0L) is identifiable in the limit.*

Here we are interested in the identifiability of the 0L language family from only positive data. In the sequel, we shall show that the family of PD0L languages is identifable in the limit from positive data. For that purpose, we need some preparations.

The morphic equivalence problem for a family $\mathcal{F}$ is the problem of deciding, given a language $L \in \mathcal{F}$ and two morphisms h and g, whether or not h and g are equivalent on L, i.e., whether or not $h(x) = g(x)$ holds for all $x \in L$.

In reference to the problem above, the following notion is quite important as well as interesting. A finite subset F of a language $L \subseteq \Sigma^*$ is a *test set* for L with respect to morphisms if for any two morphisms h and g, they agree on L iff they agree on F.

It was proved that each language possesses a test set([3]). Especially, concerning D0L languages, the possession of a test set is effective. That is,

Lemma 5 ([3]) *Each D0L language L possesses effectively a test set T_L.*

This is a crucial result for our next theorem. Before presenting the next theorem, we need another lemma.

Lemma 6 *Let $G = (\Sigma, h, w)$ be a PD0L system such that $L(G)$ is infinite. Then, there effectively exist integers m_G, $n_G (1 \leq m_G \leq n_G)$ such that*

1. *$L(G) = F_G \cup \bigcup_{j=0}^{m_G-1} L(G_j)$, where F_G: finite set depending on m_G, n_G*
2. *for all $0 \leq j \leq m_G - 1$ each $G_j = (\Sigma_j, h_j, w_j)$ is conservative and growing*
3. *F_G, $L(G_j)$ $(0 \leq j \leq m_G - 1)$ are all disjoint.*

Proof. The proof is essentially due to Theorem 4.4 in [14].

Let n_G be the smallest natural number such that there exists an m_G, $1 \leq m_G \leq n_G$ with the property that

$$\text{alph}(h^{n_G}(w)) = \text{alph}(h^{n_G - m_G}(w)).$$

Clearly, it holds that $\forall i, j \geq 0, 0 \leq j < m_G$; $\text{alph}(h^{n_G + m_G i + j}(w)) = \text{alph}(h^{n_G - m_G + j}(w))$. Hence, we have a set of PD0L systems: $(0 \leq j \leq m_G - 1)$

$$G_j = (\Sigma_j, h_j, w_j)$$

where $\Sigma_j = \text{alph}(h^{n_G + j}(w))$, $h_j = h^{m_G}_{\Sigma_j}$(restriction of h^{m_G} to Σ_j), $w_j = h^{n_G - m_G + j}(w)$. Finally, let $F_G = \{w, h(w), ..., h^{n_G - m_G - 1}(w)\}$. Then, it is easy to see that the conditions are all satisfied. □

[Note.] For $0 \leq \forall i \leq n_G - m_G - 1$, let $\Sigma_F(i) = \text{alph}(h^i(w))$. Then, $\Sigma_F(i)(0 \leq \forall i \leq n_G - m_G - 1)$ and $\Sigma_j(0 \leq \forall j \leq m_G - 1)$ are all distinct.

Example 1 Consider a PD0L system $G = (\{a, b, c, d\}, h, a)$, where

$$h(a) = b, \quad h(b) = cd$$
$$h(c) = db, \quad h(d) = d.$$

Then, $L(G) = \{a, b\} \cup \{d^n c d^{n+1} | n \geq 0\} \cup \{d^n b d^n | n \geq 1\}$. Since

$$\text{alph}(h^0(a)) = \{a\}, \quad \text{alph}(h^1(a)) = \{b\},$$
$$\text{alph}(h^2(a)) = \{c, d\}, \quad \text{alph}(h^3(a)) = \{b, d\},$$

and for $\forall i > 0$,

$$\text{alph}(h^2(a)) = \text{alph}(h^{2+2i}(a)) = \{c, d\}, \text{ and}$$
$$\text{alph}(h^3(a)) = \text{alph}(h^{3+2i}(a)) = \{b, d\},$$

we have that $n_G = 4$ and $m_G = 2$.

Let $F_G = \{a, b\}$, and let G_0 and G_1 be defined as follows:

$$G_0 = (\{c, d\}, h_0, cd), \text{ where } \quad h_0(c) = dcd, \quad h_0(d) = d$$
$$\text{and } G_1 = (\{b, d\}, h_1, dbd), \text{ where } \quad h_1(b) = dbd, \quad h_1(d) = d.$$

(Note that each G_i is conservative and growing.) Then, we have a decomposition of $L(G)$:

$$L(G) = F_G \cup L(G_0) \cup L(G_1).$$

Thus, $L(G)$ is represented as the disjoint union of a finite set and several PD0L languages.
□

We are now in a position to show the next theorem.

Theorem 7 *$\mathcal{L}$(PD0L) is identifiable in the limit from positive data.*

Proof. We shall prove that Condition A is satisfied which, from Theorem 1, completes the proof.

Let L be any PD0L language generated by a PD0L system $G = (\Sigma, h, w)$. Suppose that L is finite. The finiteness problem for D0L systems is decidable(e.g.,[9]). Then, letting $T = L$, we have that for any PD0L language L', $T \subseteq L'$ implies that L' is not a proper subset of L. Hence, we may assume that $L(= L(G))$ is infinite.

Now, let

$$L(G) = F_G \cup \bigcup_{j=0}^{m_G-1} L(G_j)$$

be a decomposition of $L(G)$ by Lemma 6, where $F_G = \{w, h(w), ..., h^{n_G-m_G-1}(w)\}$, and for $0 \leq \forall j \leq m_G - 1$, $G_j = (\Sigma_j, h_j, w_j)$, $\Sigma_j = \text{alph}(h^{n_G+j}(w))$, $h_j = h^{m_G}_{\Sigma_j}$(restriction of h^{m_G} to Σ_j), $w_j = h^{n_G-m_G+j}(w)$. (Note that each G_j is conservative and growing.)

Now, consider a finite set T effectively defined by

$$T = F_G \cup \bigcup_{j=0}^{m_G-1} T_{G_j},$$

where T_{G_j}, a test set of $L(G_j)$, contains at least two strings. The existence of such a test set is effectively assured by Lemma 5.

Suppose that $T \subseteq L'(= L(H))$ for some PD0L system $H = (\Sigma, h', x)$. (We shall prove that L' is not a proper subset of L.)

Let

$$L(H) = F_H \cup \bigcup_{j=0}^{m_H-1} L(H_j),$$

be a decomposition of $L(H)$ by Lemma 6, where $F_H = \{x, h(x), ..., h^{n_H-m_H-1}(x)\}$, and for $0 \leq \forall j \leq m_H - 1$, $H_j = (\Sigma'_j, h'_j, x_j)$, $\Sigma'_j = \text{alph}(h'^{n_H+j}(x))$, $h'_j = h'^{m_H}_{\Sigma'_j}$(restriction of h'^{m_H} to Σ'_j), $x_j = h'^{n_H-m_H+j}(x)$.

We consider and examine the following five conditions:

[C1]: $F_G = F_H$
[C2]: $m_G = m_H$
[C3]: $n_G = n_H$
[C4]: $\{\Sigma_j | 0 \leq j \leq m_G - 1\} = \{\Sigma'_{j'} | 0 \leq j' \leq m_H - 1\}$
[C5]: (for $0 \leq \forall j \leq m_G - 1$) $\exists p$(permutation) on $\{0, 1, ..., m_G - 1\}$ such that $L(G_j) = L(H_{p(j)})$.

Further, for each $i = 1, 2, ..., 5$, by Not[Ci] we denote that [Ci] does not hold.

Claim 1: Not[C1] implies that $L(H)$ is not a proper subset of $L(G)$.

(Proof) Suppose that $F_G \neq F_H$: (1) Now assume that $F_G \subset F_H$ and let $x' \in F_H - F_G$. If $x' \notin \bigcup_j L(G_j)$, then it is obvious that there exists $x' \in L(H) - L(G)$. Otherwise, let j be an integer such that $x' \in L(G_j)$. There exists $w_j \in T - F_G$ such that w_j is the axiom of G_j and alph(w_j)=alph(x'). Since $T \subseteq L(H)$, w_j is in $L(H)$.

Here, suppose that $w_j \in F_H$. Then, since T contains at least two strings from each $L(G_j)$, there exists $x_j(\neq w_j)$ in $L(H)$ such that alph(x_j)=alph(w_j). If $x_j \in F_H$, then we have a contradiction, because for all $x \in F_H$, alph(x) is distinct each other. If $x_j \notin F_H$, then this also leads to a contradiction, because for all $x \in L(H) - F_H$, alph(x) is distinct from alph(w_j)(where $w_j \in F_H$).

Suppose tht $w_j \notin F_H$. Then, since $T \subseteq L(H)$, there exists j' such that $x_{j'}$ is the axiom of $H_{j'}$ and $w_j \in L(H_{j'})$. This implies that alph(x')=alph(w_j)=alph($x_{j'}$), contradicting to that alph(x') is distinct from the alphabet of any other string in in $L(H)$.

(2) Assume that $F_H \subset F_G$ and let $w' \in F_G - F_H$. Then, since $w' \in L(H)$, there must exist $\exists j$ such that $w' \in L(H_j)$. The alphabet of H_j is alph(w') and H_j is conservative, while alph(w') is only for w' of F_G and never appears as an alphabet of any other string of $L(G)$. Hence, we conclude that there exists $z \in L(H_j) - L(G)$ with the property that alph(z)=alph(w'). (Note that each $L(H_j)$ contains at least two strings.)

Claim 2: [C1] and Not[C2] imply that $L(H)$ is not a proper subset of $L(G)$.

(Proof) Suppose that $m_G \neq m_H$ and $F_G = F_H$. Since $T \subseteq L(H)$, $m_G < m_H$. This implies that there exists $x' \in L(H) - L(G)$ such that alph(x')$\neq \Sigma_j$, for $0 \leq \forall j \leq m_G - 1$.

Claim 3: [C1] and [C2] imply [C3]. This is because $n_G \neq n_H$ and $m_G = m_H$ imply that $|F_G| \neq |F_H|$, and hence, $F_G \neq F_H$.

Claim 4: [C1] and [C2] and [C3] imply [C4].

(Proof) Let $m = m_G = m_H$, then $T = F_G \cup \bigcup_{j=0}^{m-1} T_{G_j}$. Since $F_G = F_H$ and $T \subseteq L(H)$, $\bigcup_{j=0}^{m-1} T_{G_j} \subseteq \bigcup_{j=0}^{m-1} L(H_j)$. For each $j : 0 \leq j \leq m-1$, let y_j be any string in T_{G_j}. Then, there exists $j' : 0 \leq j' \leq m-1$ such that $y_j \in L(H_{j'})$. Since G_j and $H_{j'}$ are conservative, alph(y_j)=$\Sigma_j = \Sigma'_{j'}$. Thus, the conclusion follows from that $m_G = m_H$ and all Σ_j (or $\Sigma'_{j'}$) are distinct.

Claim 5: [C1]$\sim$ [C4] imply [C5].

(Proof) Let $m = m_G = m_H$, then from the assumption we may assume that for each $0 \leq j \leq m-1$, there exists $j' = p(j)$ such that $T_{G_j} \subseteq L(H_{j'})$, where $G_j = (\Sigma_j, h_j, w_j)$ and $H_{j'} = (\Sigma_j, h'_{j'}, w_j)$. (Note that since $H_{j'}$ is growing as well as conservative, the axiom of $H_{j'}$ is the minimum elememt w_j in T_{G_j}.) Recall $T_{G_j} \subseteq L(H_{j'})$ and T_{G_j} is a test set for $L(G_j)$. Then, from the property of the test set, we have that for $\forall i \geq 0$ $h_j^i(w_j) = h'^i_{j'}(w_j)$, which imply that $L(G_j) = L(H_{j'})$.

Finally, we note that [C1] $\sim$ [C5] imply that $L(G) = L(H)$. Thus, eventually [C1] and [C2] imply that $L(G) = L(H)$. This completes the proof. □

Thus, it is shown that $\mathcal{L}(PD0L)$ is identifiable in the limit from positive data. However, the algorithm implied from Theorem 7, that is, Theorem 1 is based on the enumerative method, and therefore, is inefficient in nature.

In the next section, we shall present a non-enumerative algorithm for idenfifying the family of PD0L languages.

3.2 Identification Algorithm: A_1

Let $G = (\Sigma, h, w)$ be a PD0L system such that $L(G)$ is infinite. Recall a finite set T of $L(G)$ constructed in the proof of Theorem 7:

$$T = F_G \cup \bigcup_{j=0}^{m_G-1} T_{G_j},$$

where $F_G = \{w, h(w), ..., h^{n_G-m_G-1}(w)\}$ and for $0 \le \forall j \le m_G - 1$, T_{G_j} is a test set for $L(G_j)$. If $L(G)$ is finite, then let $T = L(G)$. (T is usually called "*finite telltale set*" of $L(G)$.)

Suppose that a finite telltale set T is given. We now describe a procedure for obtaining a PD0L system G_T, equivalent to G, from a given T.

Procedure(T;G_T):

Construct a PD0L systems $G_T = (\Sigma, h_T, w_T)$ in the following manner:

(1) Compute

$$\begin{aligned} \text{ALPH}(T) &= \{\Sigma' \subseteq \Sigma |\ \text{alph}(x) = \Sigma',\ \text{for some } x \in T\}, \\ \text{ALPH}(T)_p &= \{\Sigma' \subseteq \Sigma | \exists x, x' \in T \text{ s.t. alph}(x) = \text{alph}(x') = \Sigma' \text{ and } x \neq x'\},\ \text{and} \\ \text{ALPH}(T)_s &= \text{ALPH}(T) - \text{ALPH}(T)_p. \end{aligned}$$

(2) We note that m_G is computed as follows: $m_G = |\text{ALPH}(T)_p|$. Further, let

$$\begin{aligned} F &= \{x \in T | \text{alph}(x) \in \text{ALPH}(T)_s\} \\ &= \{x_1, ..., x_t\}. \end{aligned}$$

(3) For $\forall \Sigma_j \in \text{ALPH}(T)_p = \{\Sigma_0, ..., \Sigma_{m_G-1}\}$, let

$$\begin{aligned} T_{\Sigma_j} &= \{w \in T | \text{alph}(w) = \Sigma_j\} \\ &= \{w_{j,1} (= w_j), ..., w_{j,t_j}\}, \end{aligned}$$

where all elements are linearly ordered with respect to $\preceq$. (Clearly, F and each T_{Σ_j} correspond to F_G and a test set for $L(G_j)$, respectively, of the original system G.)

For $0 \le \forall j \le m_G - 1$, consider

$$\text{Hom}(T_{\Sigma_j}) = \bigcap_{1 \le k \le t_j - 1} \text{Hom}(w_{j,k}, w_{j,k+1}),$$

where, $\text{Hom}(x, y) = \{f | f :\ \text{homomorphism on alph}(x) \text{ s.t. } f(x) = y\}$.

For each j, take any f_{Σ_j} from $\text{Hom}(T_{\Sigma_j})$ with the property that if $a \in \Sigma_j \cap \Sigma_{j'}$ for $\exists j'(\neq j)$, then

$$f_{\Sigma_j}(a) = f_{\Sigma_{j'}}(a).$$

Then, let

$$f_{\Sigma_p} = \bigcup_j f_{\Sigma_j}, \quad \text{where } \Sigma_p = \bigcup_j \Sigma_j.$$

(4) Let's define

$$\begin{aligned} \text{short}(F) &= \{x \in F | lg(x) : \text{ shortest } \}, \\ \text{first}(T) &= \{w_j | \text{ the first element of } T_{\Sigma_j}\}, \\ \text{short}(T) &= \{w_j \in \text{first}(T) | lg(w_j) : \text{ shortest}\}, \\ \Sigma_s &= \bigcup_{x \in F} \text{alph}(x), \text{ and} \\ u &= |\text{short}(F)|. \end{aligned}$$

(4.1) In order to define h_T, we can utilize the fact that the intended h_T must satisfy the property:

$$\forall a \in \Sigma_p, \quad h_T^{m_G}(a) = f_{\Sigma_p}(a).$$

Now, find a homomorphism f_p on Σ_p such that there exists a permutation p on $\{0, ..., m_G - 1\}$ with the property that

1. for $0 \leq \forall i < m_G - 1$, $f_p(w_{p(i)}) = w_{p(i+1)}$,

2. $f_p(w_{p(m_G-1)}) = f_{\Sigma_p}(w_{p(0)})$,

where first(T)=$\{w_0, ..., w_{m_G-1}\}$

Then, let's define, for $\forall a \in \Sigma_p$,

$$h_T(a) = f_p(a).$$

Clearly h_T is compatible with f_{Σ_p} over Σ_p. Hence, all we have to do is to determine the values $h_T(a)$ for all $a \in \Sigma - \Sigma_p$.

(4.2) Now, find a homomorphism f_s on Σ_s such that there exists a permutation p' on $\{1, ..., t\}$ with the property that

1. the first u elements of $\{x_{p'(i)} | 1 \leq i \leq t\}$ comprises short(F),

2. for $1 \leq \forall i \leq t - 1$, $f_s(x_{p'(i)}) = x_{p'(i+1)}$,

3. $f_s(x_{p'(t)}) = w_j$(for $\exists w_j \in$ short(T)),

4. for $\forall a \in \Sigma_p \cap \Sigma_s$, $f_s(a) = f_p(a)$.

Let $w_T = x_{p'(1)}$ be the axiom of G_T. Further, let's define, for $\forall a \in \Sigma - \Sigma_p$,

$$h_T(a) = f_s(a).$$

From the way of constructing h_T, we have a PD0L system $G_T = (\Sigma, h_T, w_T)$ equivalent to G. (The end of Procedure(T;G_T)) □

Remark. In the third stage (3) above, one may adopt an *algebraic method* for obtaining all homomorphisms satisfying a given D0L sequence of strings proposed in [4] which provides a considerablly efficient procedure, compared with a standard combinatorial method presented here.

We note that a procedure Procedure(T;G_T) works for all T containing a finite telltale set T_L of $L(= L(G))$, i.e., for all T containing T_L, a PD0L system G_T obtained through Procedure(T;G_T) is equivalent to the original G for L.

Thus, we have :

Lemma 8 *Given a PD0L system G, let T_L be a finite telltale set of $L = L(G)$ and let T be any finite set such that $T_L \subseteq T \subset L(G)$. Then, Procedure($T$;$G_T$) always succeeds and G_T is equivalent to G.*

Example 2 Let's take a PD0L system $G = (\{a, b, c, d\}, h, a)$ considered in the previous example, where

$$h(a) = b, \quad h(b) = cd$$
$$h(c) = db, \quad h(d) = d.$$

Then, suppose that a finite telltale set

$$T = \{a, b, cd, dbd, dcd^2, d^2bd^2, d^2cd^3, d^3bd^3\}$$

is given. The construction of G_T is performed as follows.

(1) Let

$$\begin{aligned} \text{ALPH}(T) &= \{\{a\}, \{b\}, \{c, d\}, \{b, d\}\}, \\ \text{ALPH}(T)_p &= \{\{c, d\}, \{b, d\}\}, \text{ and} \\ \text{ALPH}(T)_s &= \{\{a\}, \{b\}\}. \end{aligned}$$

(2) Compute m_G and F as follows:

$$\begin{aligned} m_G &= |\text{ALPH}(T)_p| = 2, \\ F &= \{a, b\}. \end{aligned}$$

(3) First, construct

$$\begin{aligned} T_{\{c,d\}} &= \{cd, dcd^2, d^2cd^3\} \\ T_{\{b,d\}} &= \{bdb, d^2bd^2, d^3bd^3\}. \end{aligned}$$

Then, compute

$$\text{Hom}(T_{\{c,d\}}) = \{f_{\{c,d\}}\},$$
$$\text{where } f_{\{c,d\}}(c) = dcd, \ f_{\{c,d\}}(d) = d.$$

Similarly,

$$\text{Hom}(T_{\{b,d\}}) = \{f_{\{b,d\}}\},$$
$$\text{where } f_{\{b,d\}}(b) = dbd, \ f_{\{b,d\}}(d) = d.$$

Hence, we have: $f_{\{b,c,d\}} = f_{\{b,d\}} \cup f_{\{c,d\}}$.

(4) We now have that

$$\begin{aligned} \text{short}(F) &= \{a,b\}(= F), \\ \text{first}(T) &= \{cd, dbd\}, \\ \text{short}(T) &= \{cd\}, \\ \Sigma_s &= \{a,b\}, \quad \text{and} \\ u &= 2. \end{aligned}$$

(4.1) For $c, d \in \Sigma_p$, the following relations hold:

$$\begin{aligned} f_p(cd) &= dbd, \text{ and} \\ f_p(dbd) &= f(cd) \\ &= dcd^2. \end{aligned}$$

1) If $f_p(c) = d$, then $f_p(d) = bd$ and since $f_p^2 = f$, $f_p^2(d) = f_p(bd) = cdbd \neq d(= f(d))$(contradiction). 2) If $f_p(c) = db$ and hence $f_p(d) = d$, then $f_p(d)f_p(b)f_p(d) = dcdd = f(cd)$. This is compatible with f. That is, $f_p(c) = db, \quad f_p(d) = d$ and hence,

$$h_T(c) = db, \quad h_T(d) = d.$$

(4.2) There are two possibilities : i) $a \Rightarrow f_s(a) = b \Rightarrow f_s^2(a) = cd \in \text{short}(T)$, ii) $b \Rightarrow f_s(b) = a \Rightarrow f_s^2(b) = cd \in \text{short}(T)$. (Note that $f_s^2 = f$.) However, from $f(b) = dbd$, the case ii) fails. Hence, w_T(the axiom of G_T) must be "a" and it must hold that $f_s(a) = b, \quad f_s(b) = cd$, that is, we define:

$$h_T(a) = b, \quad h_T(b) = cd.$$

Thus, we obtain a PD0L system $G_T = (\Sigma, h_T, a)$ equivalent (in fact, as far as this example is concerned, isomorphic) to G. □

We now present an identification algorithm A_1. The identification protocol for A_1 consists of only the positive presentation of an unknown PD0L language U over the fixed alphabet Σ.

We shall show that A_1 given in Figure 1 below eventually identifies in the limit a PD0L system G_T such that $U = L(G_T)$.

Lemma 9 *Let $G_{T_0}, G_{T_1}, ..., G_{T_i}, ...$ be the sequence of PD0L systems produced by A_1. Then, there exists $r \geq 0$ such that for $\forall i \geq 0$, $L(G_{T_r}) = L(G_{T_{r+i}}) = U$.*

Proof. Consider sufficiently large integer $N_0 > 0$ such that $T_U \subseteq T_{N_0}$, where T_U is a finite telltale set of U. (Since T_U is finite, this is always possible.) From Lemma 8 it follows that for $\forall i \geq 0$, $U = L(G_{T_U}) = L(G_{T_{N_0}}) = L(G_{T_{N_0+i}})$. □

Note. Procedure(T;G_T) in A_1 does not always end up with success. That is, the task for constructing G_{T_i} might fail for some of T_is not containing a finite telltale set of U.

Theorem 10 *Given any unknown PD0L language U, the algorithm A_1 identifies in the limit a PD0L system G_T such that $U = L(G_T)$.*

Input: a positive presentation of U
Output: a sequence of PD0L systems G_T

Procedure
Initialize $T = \emptyset$;
repeat (forever)
read the next positive example w ;
let $T := T \cup \{w\}$;
call **Procedure**(T;G_T);
if Procedure(T;G_T) succeeds,
then output G_T ;
else output $G_T = (\Sigma, \mathrm{Id}, w)$;
(where Id: identity over Σ)

Figure 1. The Identification Algorithm A_1

We remark that although the alrogithm A_1 is constructive, it is too naive to run fast. (Note that A_1 is responsive, but neither consistent nor conservative.)

In what follows, we shall show that there exists a *consistent, responsive* and *conservative* algorithm for identifying the family of PD0L languages if we do *not care about* the choice of the representation space for the family.

3.3 Another Algorithm: A_2

A 4-tuple $S = (\Sigma, h, \{w_1, ..., w_p\}, F)$ is termed *quasi-FPD0L* system if F is a finite set and $G = (\Sigma, h, \{w_1, ..., w_p\})$ is an FPD0L system. By $L(S)$ we denote $F \cup L(G)$. A language L is termed *quasi-FPD0L* if there exists a quasi-FPD0L system S such that $L = L(S)$.

The next lemma almost immediately follows from the definition and Lemma 6:

Lemma 11 *Let $G = (\Sigma, h, w)$ be a PD0L system. Then, there effectively exists quasi-FPD0L system S such that $L(G) = L(S)$.*

Proof. Similarly to the proof for Lemma 6, it follows that there effectively exist integers m_G, n_G such that

1. $L(G) = F_G \cup \bigcup_{j=0}^{m_G - 1} L(G_j)$,
2. $F_G = \{w, h(w), ..., h^{n_G - m_G - 1}(w)\}$,
3. for all $0 \leq j \leq m_G - 1$, each $G_j = (\Sigma, h^{m_G}, w_j)$ is a PD0L system.

(Note that each G_j is not necessarily reduced.) Let's define

$$S = (\Sigma, h^{m_G}, \{w_0, ..., w_{m_G-1}\}, F_G),$$

then we have that $L(G) = L(S)$ holds. □

Thus, instead of PD0L systems, we are able to adopt the family of quasi-FPD0L systems as the representation space for the family of PD0L languages

Let U be an unknown PD0L language such that $U = L(G)$ for some PD0L system. Suppose that a finite set T of the unknown U is given. (Note that T is not necessarily a finite telltale set of U.) We shall describe a procedure for obtaining a quasi-FPD0L system S_T from a given T.

By **Procedure(*Seq*;*Hom*)** we denote the procedure which, given $Seq =< w, w' >$, a sequence of two strings, produces Hom, the set of all homomorphisms f such that $f(w) = w'$.

We now describe the procedure.

Procedure(T;S_T):
Construct a quasi-FPD0L systems $S_T = (\Sigma_T, h_T, F_T^a, F_T)$ in the following manner:

(1) Compute

$$\begin{aligned} \Sigma_T &= \text{alph}(T), \\ \text{ALPH}(T) &= \{\Sigma' \subseteq \Sigma |\ \text{alph}(x) = \Sigma',\ \text{for some } x \in T\}, \\ \text{ALPH}(T)_p &= \{\Sigma' \subseteq \Sigma | \exists x, x' \in T \text{ s.t. alph}(x) = \text{alph}(x') = \Sigma' \text{ and } x \neq x'\}, \\ \text{ALPH}(T)_s &= \text{ALPH}(T) - \text{ALPH}(T)_p. \end{aligned}$$

(2) Compute $m = |\text{ALPH}(T)_p|$. Then, for $\forall \Sigma_j \in \text{ALPH}(T)_p = \{\Sigma_0, ..., \Sigma_{m-1}\}$, let

$$\begin{aligned} T_{\Sigma_j} &= \{w \in T | \text{alph}(w) = \Sigma_j\} \\ &= \{w_{j,1}(= w_j), ..., w_{j,t_j}\}, \end{aligned}$$

where all elements are linearly ordered with respect to $\preceq$.
Further, let

$$F_T^a = \{w_0, ..., w_{m-1}\}.$$

(3) Case (3.1)$(0 \leq \forall j \leq m-1)[t_j \geq 2]$: Let

$$Seq_j =< w_{j,1}, w_{j,2} >.$$

Then, call **Procedure(Seq_j;Hom_j)**.
Case (3.1²)$(\forall j)[Hom_j \neq \emptyset]$: For each j and each $h_j \in Hom_j$, check whether or not $T_{\Sigma_j} \subseteq h_j^*(w_j)$.
Case (3.1³) If so, then check whether or not for all j there exists $h_j \in Hom_j$ with the property that if $a \in \Sigma_j \cap \Sigma_{j'}$ for $\exists j'(\neq j)$, then $h_j(a) = h_{j'}(a)$.
Case (3.1⁴) If so, then take those h_j and define

$$h_T(a) = \begin{cases} \bigcup_j h_j(a) & (\forall a \in \Sigma_p(=\bigcup_j \Sigma_j)) \\ a & (a \in \Sigma_T - \Sigma_p). \end{cases}$$

Otherwise, i.e., neither Cases (3.1^2), (3.1^3), nor (3.1^4): Follow Case (3.2).
Case (3.2)(=not Case (3.1)): Define, for $\forall a \in \Sigma_T$,

$$h_T(a) = a \text{ (i.e., identity mapping).}$$

(4) Finally, let

$$F_T = \{w \in T | w \notin \bigcup_{n \geq 0} h_T^n(F_T^a)\}.$$

(The end of Procedure(T;S_T)) □

From the way of constructing S_T and Lemma 11, it easily follows:

Lemma 12 *Let T be a finite set of $L(G)$ for a given PD0L system G and let S_T be a result from Procedure(T;S_T). Then, (1) it holds that $T \subseteq L(S_T)$, and (2) if T contains a finite telltale set of $L(G)$, then $L(G) = L(S_T)$ holds.*

We now present an identification algorithm A_2 in Figure 2 which is obtained from A_1 by replacing Procedure(T;G_T) with Procedure(T;S_T) together with a small modification.

Input: a positive presentation of U
Output: a sequence of quasi-FPD0L systems S_T

Procedure
  Initialize $T = \emptyset$;
  Let $S_\emptyset = (\Sigma_\emptyset, \mathrm{Id}, \emptyset, F_\emptyset)$;
  repeat (forever)
    read the next positive example w ;
    let $T' := T \cup \{w\}$;
    if $w \notin L(S_T)$ **then** call **Procedure**(T';$S_{T'}$);
    else let $S_{T'} =: S_T$;
    output $S_{T'}$;
    let $T := T'$;

Figure 2. The Identification Algorithm A_2

In Figure 2, the time complexity of A_2 is dominated by Procedure(T';$S_{T'}$). In most steps, Procedure(T';$S_{T'}$) is actually implemented in an efficient way, because, as a matter of course, we can utilize the computation results of the former quasi-FPD0L system S_T produced in the previous step, and only the computation on the current example w is partially required for obtaining the new system $S_{T'}$.

The correctness of the algorithm A_2 immediately follows from Lemma 12 and the fact that for any input sample data T, A_2 never produces a system S_T which overgeneralizes the target language U. Further, as easily seen, A_2 is consistent, responsive and conservative.
Thus, we have:

Lemma 13 *Given any unknown PD0L language U, the algorithm A_2 identifies in the limit a quasi-FPD0L system S_T such that $U = L(S_T)$.*

Theorem 14 *$\mathcal{L}$(PD0L) is identifiable in the limit from positive data using the family of quasi-FPD0L systems. Further, the algorithm is consistent, responsive and conservative on $\mathcal{L}$(PD0L).*

4 Discussions

4.1 Related Works

As mentioned in Section 1, there have been reported a number of work by Herman and Walker([8])and by Feliciangeli and Herman([5]), investigating the syntactic inference problem for variety of L-language families. They deal with the problem of (1) finding an algorithm which decides for any finite set of sequences of strings whether or not there exists a system of certain type such that each of the sequences is a part, or a regularly sampled part, or a randomly sampled part of a derivation by the system, and (2) giving an algorithm which, whenever the answer for (1) is positive, will produce an appropriate system.

Ducet proposes an improved method for the syntactic inference problem of D0L-sequences by presenting algorithms based on an algebraic approach rather than combinatorial one([4]). It is obvious that some of these results are applicable to the current problem in this paper in order to find a collection of D0L schemata consistent with given data.

Rather lately, Jürgensen and Lindenmayer investigate the inference problem for D0L systems from sequences of branching structures in which structural information with cell lineages are available([12]).

In reference to inductive inference algorithms for formal languages, there are numerous work reported in the recent active research on the computational or algorithmic learning theory. (e.g., [2], [6], [11]) To author's knowledge, however, there is no work concerning the identification of 0L languages families in "in the limit" framework, and a little is known about the *efficient* identifiability in the limit, in particular, from positive data. As for inconsistent algorithms, it is quite recently demonstrated in [13] that they surely have the advantage of the consistent algorithms in the inference problem of pattern languages.

4.2 Conclusions

We have shown that the family of P0L languages is not identifiable in the limit from positive data. Thus, in general, the family of 0L languages is not identifiable in the limit from positive data. Also, it has been shown that the family of PD0L languages is identifiable in the limit from positive data. Unfortunately, the algorithm does not bear the property of "consistency" which is one of the desirable features of an inference algorithm. Then, we have shown that there is a consistent, responsive and conservative algorithm for identifying the family of PD0L languages in the limit, if we make an appropriate choice of the representation space for hypotheses.

An open problem is whether or not the family of D0L languages is identifiable in the limit from positive data. Also, the question remains open whether or not there exists a consistent, responsive and conservative algorithm for identifying the family of D0L(PD0L) languages in the limit from positive data using *D0L(PD0L) systems*, respectively. We conjecture that the former will be solved in the affirmative, while the latter in the negative.

For further future research, it is also interesting and important to explore the possibility of *efficient* algorithms for solving the inference problems of L-language families. More generally, the ternary relationship between the identifiability, the consistency and the efficiency of the inference algorithm remains open to be investigated.

References

[1] D. Angluin. Inductive inference of formal languages from positive data. *Information and Control*, 45:117–135, 1980.

[2] S. Arikawa, S.Goto, S.Ohsuga and T.Yokomori(Eds.). *Algorithmic Learning Theory* (Proceedings of ALT'90). Ohmsha and Springer-Verlag, 1990.

[3] K. Culik II and J.Karhumäki. A new proof for the D0L sequence equivalence problem and its implications. In G.Rozenberg and A.Salomaa, editors, *The Book of L*, pages 63–74. Springer-Verlag, 1986.

[4] P. Doucet. The syntactic inference problem for D0L-sequences. In *L-Systems,* Lecture Notes in Computer Science 15, Springer-Verlag, pages 146–161, 1974.

[5] H. Feliciangeli and G.T.Herman. Algorithms for producing grammars from sample derivations: A common problem of formal language theory and developmental biology. *Journal of Computer and System Sciences*, 7:97–118, 1973.

[6] M. Fulk and J.Case(Eds.). *Computational Learning Theory: Proceedings of COLT'90.* Morgan Kaufmann, 1990.

[7] E. M. Gold. Language identification in the limit. *Information and Control*, 10:447–474, 1967.

[8] G.T.Herman and A.Walker. The syntactic inference problem as applied to biological systems. In B.Meltzer and D.Michie, editors, *Machine Intelligence Vol.7*, pages 341–356. Edingurgh University Press, 1972.

[9] G.T.Herman and G.Rozenberg. *Developmental Systems and Languages.* North-Holland, Amsterdam, 1975.

[10] M. Harrison. *Introduction to Formal Language Theory.* Addison-Wesley, Reading, MA, 1978.

[11] P. Jantke(Ed.). *Analogical and Inductive Inference.* Lecture Notes in Artificial Intelligence 397, Springer-Verlag. 1989.

[12] H. Jürgensen and A. Lindenmayer. Inference algorithms for developmental systems with cell lineages. *Bulletin of Mathematical Biology*, 49:93–123, 1987.

[13] S. Lange and R.Wiehagen. Polynomial-time inference of arbitrary pattern languages. *New Generation Computing*, 8:361–370, 1991.

[14] M. Nielsen. On the decidability of some equivalence problems for D0L-systems. *Information and Control*, 25:166–193, 1974.

[15] G. Rozenberg and A. Salomaa. *The Mathematical Theory of L Systems.* Academic Press, 1980.

A normal form for structurally equivalent E0L grammars

Valtteri Niemi

Mathematics Department, University of Turku, SF-20500 Turku, Finland

Abstract: A normal form for structurally equivalent E0L grammars is constructed. Two E0L grammars are structurally equivalent iff the respective normal form grammars are isomorphic. This result gives also a new proof for the decidability of structural equivalence in the E0L case.

Keywords: E0L systems, equivalence problems, normal forms

1 Introduction

The notion of equivalence between two objects is of great importance in all mathematical sciences. In formal language theory various equivalence problems have motivated huge amount of investigation. In the case of rewriting systems there are two basic notions of equivalence: in the weaker one only the resulting languages are considered while in the stronger one also the generation processes are taken into account.

Let us take an example. Classical results (see, e.g., [6] or [9]) show that it is *undecidable* whether or not two context-free grammars G_1 and G_2 are *equivalent* (or *language* equivalent or *weak* equivalent), that means $L(G_1) = L(G_2)$. On the other hand (see [1], [4], [8]), it is *decidable* whether G_1 and G_2 are *structurally equivalent*, that means also the derivation trees generating equal words are of equal shape. As expected, E0L grammars resemble context-free ones also in this context. The (language) equivalence between two E0L grammars G_1 and G_2 is undecidable (see, e.g., [5]) while the structural equivalence is decidable (see [7]).

In this paper we are interested in structural equivalence of E0L grammars. In the context-free case decidability of structural equivalence has been proved by two, essentially different, strategies. Grammatical methods are applied in [1] and [4] while automata-theoretic methods, namely tree automata, were in use in [8]. The latter idea is gives a simpler proof but the former one gives more than just the decidability of the structural equivalence problem. It establishes a *normal form* for structural equivalence of context-free grammars. Each context-free grammar can be effectively reduced into a structurally equivalent grammar in a standard form in such way that for two structurally equivalent context-free grammars G_1 and G_2 the respective reduced grammars are *isomorphic.*

Ottmann and Wood (see [2] and [3]) attacked structural equivalence of E0L grammars along the lines of the grammatical tradition. Salomaa and Yu [7] continued the automata-theoretic tradition and, by introducing so-called *height-counting* tree automata, they were able to show that structural equivalence is, indeed, decidable also in the case of E0L grammars. We join the former branch of study. It is shown in the paper that it is possible to effectively construct a *normal form* also for E0L grammars. The main result of this paper shows that for any two structurally equivalent grammars G_1 and G_2 the respective normal form grammars are isomorphic. As a byproduct we obtain a new proof for the decidability of structural equivalence of E0L grammars.

The paper is organized as follows. Some definitions and preliminary observations are presented in section 2. The crucial construction is presented in section 3. The isomorphism result is proved and a normal form grammar with infinite nonterminal alphabet is found in section 4. Section 5 contains our normal form for E0L grammars with respect to structural equivalence, while in section 6 we illustrate all constructions via an example. Some concluding remarks are finally made in section 7.

2 Preliminaries

As regards basics of theory of grammars and L systems we refer to , e.g., [5], [6], [9]. We discuss here basic definitions only briefly.

An E0L grammar is a quadruple $(\mathcal{N}, \Sigma, \mathcal{P}, \mathcal{S})$ where the elements of $\mathcal{N}$ are called *nonterminals*, elements of Σ *terminals*, $\mathcal{P}$ is a finite set of productions $X \to w$ with $X \in \mathcal{N}$, $w \in \mathcal{N}^+ \cup \Sigma^+$ and $\mathcal{S} \subseteq \mathcal{N}$ is the set of *start symbols*.

Similarly as in [3], our definition is not the usual one. There are three changes:

- the grammar is *propagating*: there are no productions with empty word as right-hand side.
- the grammar is *synchronized*: there are no productions with a terminal letter as left-hand side.
- we allow several start symbols.

In the end of this section we show how the usual case can be transformed to our case in the context of structural equivalence.

Let $\alpha = A_1 \dots A_n$ be a word over the nonterminal alphabet while each of the words $\beta_1, \dots, \beta_n$ contains either solely nonterminals or solely terminals. If further $A_i \rightarrow \beta_i$ ($i = 1, \dots, n$) is a valid production we say that α is rewritten as $\beta = \beta_1 \dots \beta_n$, denoted by $\alpha \Rightarrow \beta$. A sequence of the type $\alpha = \beta_0 \Rightarrow \beta_1 \Rightarrow \beta_2 \Rightarrow \dots \Rightarrow \beta_d = \beta$ is called a *derivation*. The transitive closure of the relation $\Rightarrow$ is denoted by $\Rightarrow^+$ and the reflexive transitive closure is denoted by $\Rightarrow^*$, as usual.

The language generated by G is denoted by $L(G)$ and it is defined by

$$L(G) = \{w \in \Sigma^* | (\exists S \in \mathcal{S}) S \Rightarrow^+ w\}$$

Unless stated otherwise when speaking about derivations we mean derivations starting from a start symbol.

Note that the empty word can never be derived. It is possible that the same word can result from several different derivations. Each derivation can be further structured with a *derivation tree*: nodes of the derivation tree are letters and in each rewriting step some daughters are added to each leaf of the tree in such a way that if the production $A \rightarrow \beta$ is used then the node A has letters of β as daughters. Every terminating derivation tree has *uniform depth*, in other words it is *balanced*. Balanced derivation trees with a start symbol as root are called *sentential* derivation trees. The words derived by such trees are *sentential forms*.

Two E0L grammars G_1 and G_2 are *equivalent* if $L(G_1) = L(G_2)$. Furthermore, G_1 and G_2 are *structurally equivalent* if

1. they are equivalent,

2. for each terminating derivation tree in G_1 there is a structurally similar derivation tree in G_2 that generates the same terminal word (the two trees differ only as regards the nonterminals in the inner nodes).

Because of mainly technical reasons we want to put some restrictions on our grammars. Some of the restrictions were already mentioned in the beginning of this section. Next we show that none of these restrictions will cause any loss of generality in our argumentation concerning structural equivalence.

1. No start symbol occurs on the right-hand side of any production. For this purpose we introduce, for each start symbol S, a new start symbol S', duplicate all productions of the type $S \rightarrow w$ to obtain productions $S' \rightarrow w$ and finally take the status as a start symbol away from the old start symbol S. It is straight-forward to show that the modified grammar is structurally equivalent to the original one.

2. Terminals cannot be rewritten. This can be handled by changing each terminal a into a nonterminal A and adding a production $A \to a$. All derivations change respectively. There is a slight difference between the derivation trees of the new grammar and those of the original one (in the former one there is always one more derivation step which turns the new nonterminals into terminals) but this correspondence is one-to-one and the modification does not affect structural equivalence of grammars.

3. The grammars are reduced: all nonterminals can be derived from some start symbol and, on the other hand, all nonterminals generate words which contain at least one terminal letter. This can be assumed, since only nonterminals of this type can appear in terminating derivation trees.

4. The grammars are propagating, in other words they are EP0L grammars. It is well-known (see, e.g., [5]) that this modification does not affect the family of *languages* generated by grammars (apart from the loss of the empty word λ). However, it *does* affect the family of derivation trees. Fortunately, any use of an erasing production $N \to \lambda$ is tracable when derivation trees are considered.

   We introduce a new *terminal* letter, say λ_0, and a new nonterminal Λ. A production $N \to \lambda$ is replaced by three productions $N \to \Lambda\Lambda$, $\Lambda \to \Lambda|\lambda_0$. After this modification exactly the same grammars are structurally equivalent as before. It is important to notice that the new nonterminal Λ is really doubled in the production $N \to \Lambda\Lambda$. This is done in order to mark the exact place where the erasing production was applied. A replacement by the production $N \to \Lambda$ would discard all differences between *erasing chains*, i.e. derivations of the type $N_1 \Rightarrow N_2 \Rightarrow \ldots \to N_k \Rightarrow \lambda$.

5. The grammars are *invertible*: there are never two productions with equal right-hand sides. It is shown in [3] that for any E0L grammar G an invertible structurally equivalent grammar G' can be constructed (the subset construction method is used).

6. We allow more than one start symbol. This is needed in the aforementioned construction of an invertible grammar. However, this modification is quite unessential, since a new unique start symbol can always be introduced to replace all the old ones and a structurally equivalent grammar results. Moreover, because of the restriction 1, an invertible grammar remains invertible even if all start symbols are identified.

7. A specific modification is needed in section 4. We do not allow chains in the

beginning of derivations, that means derivations of the type $S \Rightarrow A_1 \Rightarrow A_2 \Rightarrow \ldots \Rightarrow A_k \Rightarrow w \Rightarrow \ldots$. Hence, every production with a start symbol on left-hand side has at least two letters on right-hand side. This modification, of course, affects the structure of derivation trees.

However, clearly the length set of such beginning chains is *ultimately periodic*. Hence, cutting away chains in the beginning of derivations leads us to a finite number of new grammars, each of which is associated to some subset of the length set of all chains. By using totally isolated new nonterminals, a new grammar, structurally equivalent to the old one, can be constructed effectively and in a uniquely determined manner from the bunch of grammars with no beginning chains.

3 Constructions

Let us now begin with an invertible EP0L grammar G. Construct the following set of letters in G.

$B_0 = \Sigma$ (= terminal alphabet of G).

$B_1 = \{N \mid \text{there exists a production } N \to \alpha \text{ in } G \text{ such that } \alpha \in B_0^+\}$.

$B_{i+1} = \{N \mid \exists\, N \to \alpha \text{ in } G : \alpha \in B_i^+\}$ for $i = 1, 2, \ldots$

(Since the alphabet of G is finite, there are integers t and p such that $B_t = B_{t+p}$. However, we do not use this fact for some time.)

It is clear now that a nonterminal N belongs to B_i iff there exists a (balanced) derivation tree whose height is i, root is N and leaves are terminals.

Let us next modify the grammar G to obtain an infinite grammar G' in the following way. We separate the sets B_i by introducing new copies of nonterminals to satisfy the condition $B_i \cap B_j = \emptyset$ for $i \neq j$. Hence, the nonterminal alphabet of G' is

$$\mathcal{N}_{G'} = \{N^{(i)} | N \in \mathcal{N}_G, i = 1, 2, \ldots\}.$$

Indeed, we separate also such sets B_i, B_j $(i \neq j)$ that are originally equal. This means we get an infinite nonterminal alphabet.

The terminal alphabet of G' is the same as in G. For each production $N \to \alpha$ in G there exist in G' all productions $N^{(i)} \to \alpha^{(i-1)}$ such that $N \in B_i$ and $\alpha \in B_{i-1}^+$. (By the terminology of *grammar form theory*, G' is an *interpretation* of G but we do not need this information in the sequel.)

Now for each terminating derivation in G' there exists a corresponding terminating derivation in G which is obtained by removing the indices from nonterminals. On the other hand, it is quite clear, by the definition of the sets B_i, that we can also add indices to each terminating derivation in G to obtain a corresponding derivation in G'. Thus, G and G' are structurally equivalent. Furthermore, G' is invertible, since otherwise we could see by removing indices that G is not invertible. Also, terminals are not rewritten, start symbols do not occur on right-hand side of any productions and there are no beginning chains.

It is possible that G' is not reduced any more. In that case we just drop the extra nonterminals that do not participate in any terminating derivations.

For each nonterminal $N^{(i)}$ all terminating derivation trees with the root $N^{(i)}$ are of height i.

The second step in our construction is to reduce the grammar G' according to so-called *context equivalence.* This notion was introduced in the context-free case in [1] and Ottmann and Wood [3] applied it in the E0L case.

We denote by α a specific sentential derivation tree where all leaves are nonterminals except one which is replaced by an empty space. Moreover, all letters in the inner nodes are also left open. We call such α a *context.*

We say that a nonterminal A *appears in the context* α if A can be placed into the empty leaf space and also the inner nodes can be filled in by nonterminals in such a way that a valid derivation tree results. Two nonterminals A and B are *context equivalent,* denoted $A \equiv B$, if they can appear in exactly the same contexts. We say an E0L grammar is *context reduced* if two context equivalent nonterminals are always equal.

The following theorem is a modification of Theorem 4.4 in [3] for the case of infinite grammars. The proof is also a straight-forward modification of the original proof and we omit it here.

Theorem 1. Let $G = (\mathcal{N}, \Sigma, \mathcal{P}, \mathcal{S})$ be an invertible E0L grammar (with $\mathcal{N}$ possibly infinite). Then there exists a context reduced E0L grammar $\bar{G} = (\bar{\mathcal{N}}, \Sigma, \bar{\mathcal{P}}, \bar{\mathcal{S}})$ which is structurally equivalent to G.

Let us now continue with our grammar G'. In every sentential derivation tree of G' all nonterminals on the same level have the same index. Also, in a sentential derivation tree there are at least two letters on each level except the first one (, since there are no beginning chains). On the first level only a start symbol can appear, and by the fact that start symbols never appear later in any derivations, it can be seen that all start symbols constitute a single context equivalence class. These facts imply that two nonterminals symbols can be context equivalent only if they have the same index i or they both are start symbols.

In particular, for G' the context reduced version $\bar{G}'$ is invertible. This can be seen in the same way as in the proof of Theorem 4.4 of [3]. The only difference is that in [3] a specific notion of so-called *n-completeness* was needed to assure that each sentential derivation tree can be extended by one parallel derivation step. However, in the case of G' this is automatically guaranteed.

Furthermore, it is clear that in $\bar{G}'$ no terminals are rewritten, there are no productions with the start symbol on right-hand side, there are no beginning chains and all nonterminals occur in some terminating derivation.

4 Characterization of structural equivalence

We are now prepared for the following characterization of structural equivalence of E0L grammars G_1 and G_2:

Theorem 2. Two invertible E0L grammars G_1 and G_2 are structurally equivalent iff $\bar{G}'_1$ and $\bar{G}'_2$ are *isomorphic.*

Proof. The implication from right to left is trivial, since we have seen that G_i is structurally equivalent to G'_i and, furthermore, by Theorem 1, G'_i is structurally equivalent to $\bar{G}'_i$ $(i = 1, 2)$.

To prove the converse implication let us choose an arbitrary nonterminal $N^{(s)}$ in $\bar{G}'_1$. Thus, by the fact that in $\bar{G}'_1$ all nonterminals appear in some terminal derivation, there is a terminating derivation tree in $\bar{G}'_1$ of the following form:

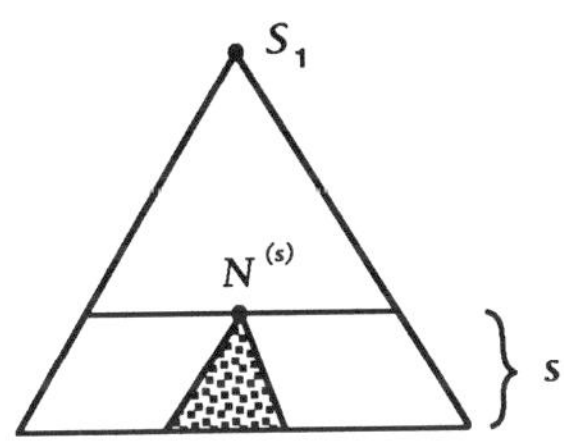

It follows from our assumption that $\bar{G}'_1$ and $\bar{G}'_2$ are structurally equivalent. Hence, there is a derivation tree in $\bar{G}'_2$ of the form

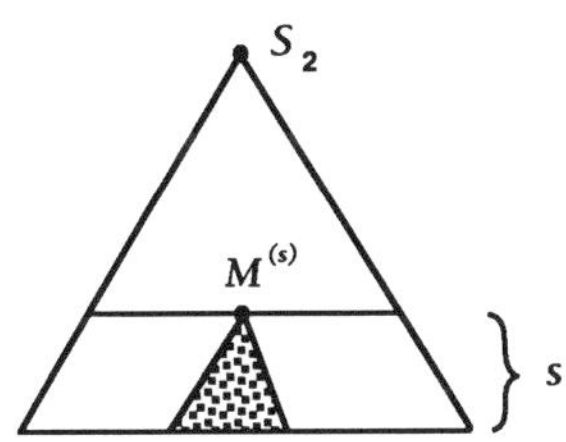

Now we define a mapping f from $\mathcal{N}_{\bar{G}_1'}$ to $\mathcal{N}_{\bar{G}_2'}$ by the rule $N^{(s)} \mapsto M^{(s)}$. The mapping f is then extended to cover also terminals by applying identity mapping in their case.

First, we show f is well-defined. Indeed, suppose there are other trees (first one in $\bar{G}_1'$ and the other in $\bar{G}_2'$) :

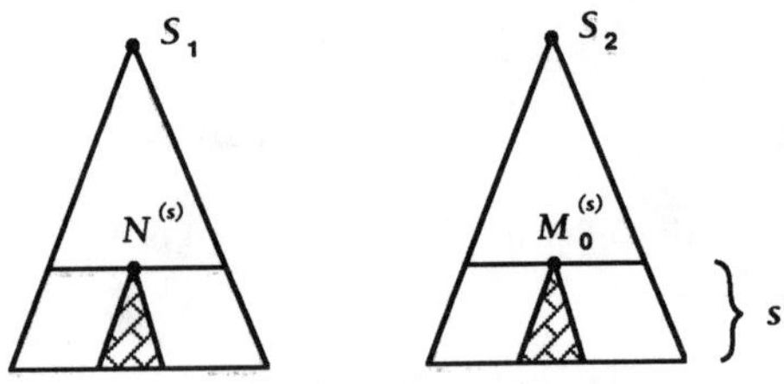

Let us now choose an arbitrary context α in which $M^{(s)}$ appears in $\bar{G}_2'$:

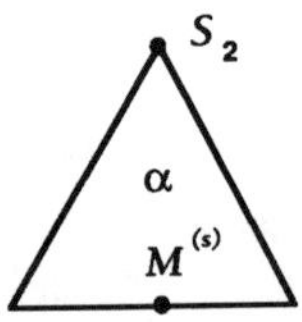

By the definition of G' this sentential form can be terminated in exactly s steps :

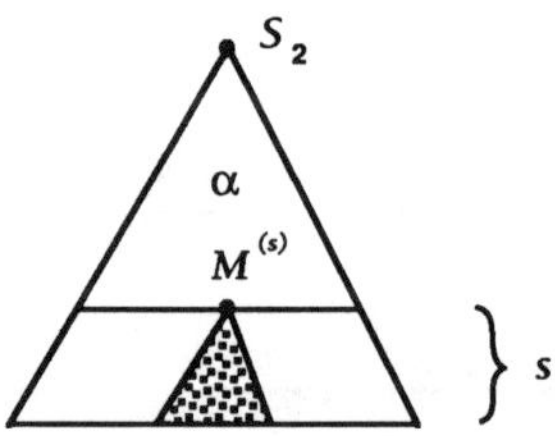

Since $\bar{G}_1'$ is structurally equivalent to $\bar{G}_2'$ and $\bar{G}_1'$ is invertible there exists a tree of the following form in $\bar{G}_1'$:

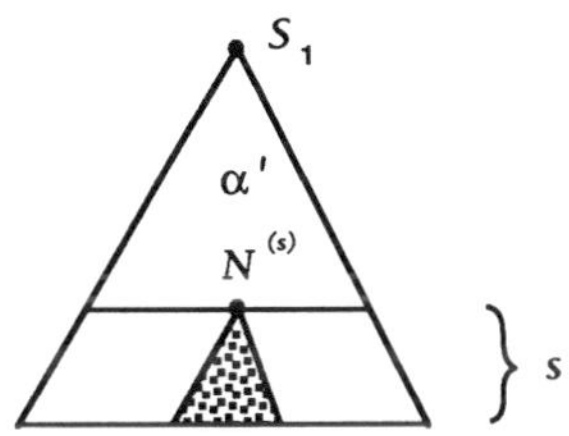

Then there exists also the following tree in $\bar{G}'_1$:

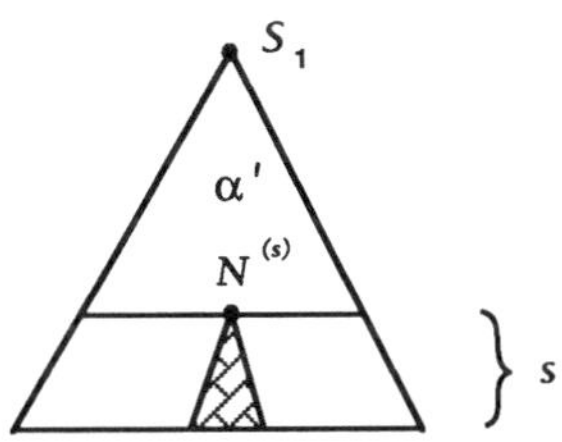

and, again, by structural equivalence of $\bar{G}'_1$ and $\bar{G}'_2$, together with invertibility of $\bar{G}'_2$, there exists the following tree in $\bar{G}'_2$:

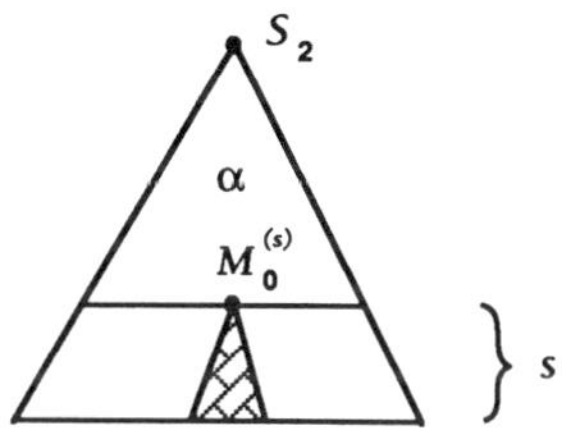

Thus, $M_0^{(s)}$ also appears in the context α. By symmetry it follows that $M^{(s)}$ and $M_0^{(s)}$ are context equivalent and, since $\bar{G}'_2$ is context-reduced, $M^{(s)} = M_0^{(s)}$.

Secondly, f is bijective. This is guaranteed already by the symmetry in its definition.

The third task is to show f is a morphism, i.e. it preserves the structure of productions. Let us consider a production $N^{(i)} \to \alpha^{(i-1)}$ in $\bar{G}'_1$ and some derivation where it occurs (there must be such a derivation, since $\bar{G}'_1$ is reduced):

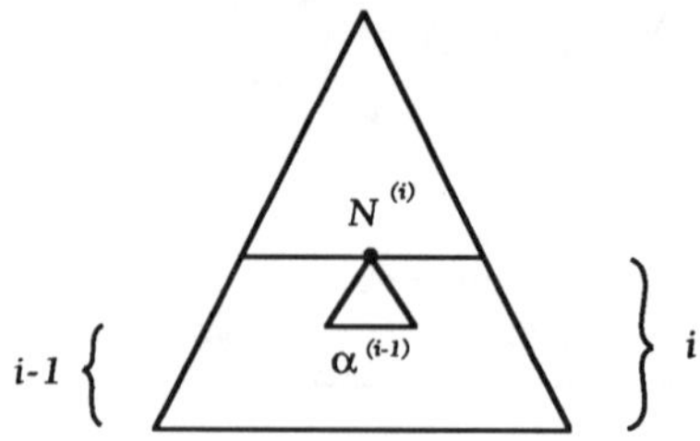

Then, by the structural equivalence of $\bar{G}_1'$ and $\bar{G}_2'$ there is a tree of the following type in $\bar{G}_2'$:

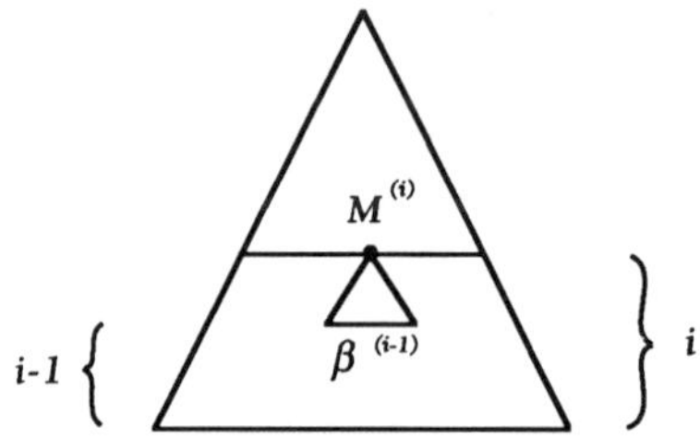

By definition of f it is clear that $f(N^{(i)}) = M^{(i)}$ and, also, $f(\alpha^{(i-1)}) = \beta^{(i-1)}$. Hence, $f(N^{(i)}) \to f(\alpha^{(i-1)})$ is a production in $\bar{G}_2'$.

Thus, we have shown that f defines an isomorphism from $\bar{G}_1'$ to $\bar{G}_2'$. □

As regards decidability of structural equivalence there are still some problems. Is it possible to construct $\bar{G}'$ effectively, and is it possible to check the isomorphism between $\bar{G}_1'$ and $\bar{G}_2'$ effectively (remember that the grammars are infinite)?

Let us investigate the first problem. We already pointed out in the beginning of the previous section that there are only finitely many different sets B_i in G, implying there exist integers t and p such that $B_{t+p} = B_t$. We call the integer t a *threshold* and the integer p a *period* of the grammar G. It is now clear that after i has reached the threshold level, the construction process of new nonterminals $N^{(i)}$ is similar every pth time. In this sense the construction process is effective.

In order to make G' reduced we dropped out unnecessary nonterminals. How can we do this effectively in an infinite grammar? A little bit similarly as in the case of sets B_i we can find in the original grammar G a family $\{C_i\}$ of such sets of nonterminals that can constitute the entire leaf set of some sentential derivation tree. Consider now an arbitrary nonterminal N in some set B_i. The nonterminal $N^{(i)}$ is qualified to the reduced version of G' only if there is a set C_j such that $N \in C_j \subseteq B_i$. Indeed, if $N^{(i)}$ appears in some terminating derivation then this must hold. On the other hand, if the condition is satisfied then there exists a terminating derivation in which $N^{(i)}$ appears on the ith level from the bottom, by definitions of B_i and C_j. By the ultimate periodicity of the sequence $\{B_i | i = 1, 2, \ldots\}$ it is clear that a nonterminal $N^{(i)}$ is dropped out iff all nonterminals $N^{(i+kp)}$ are dropped out for $i \geq t, k \geq 1$.

It also follows from the periodicity of G' that the context equivalence relation is ultimately periodic. For each context in which some nonterminal $N^{(i)}$ with $i \geq t$ appears there exist respective contexts for every $N^{(i+kp)}$ $(k = 1, 2, \ldots)$. We just add kp to all indices in the derivation tree. Therefore, for all $N^{(i)}, M^{(i)} \in B_i; N^{(i+kp)}, M^{(i+kp)} \in B_{i+kp}; i \geq t, k \geq 1$:

$$N^{(i)} \equiv M^{(i)} \text{ iff } N^{(i+kp)} \equiv M^{(i+kp)}.$$

This fact implies there are only finitely many equivalence relations in $\bar{G}'$ that are possible candidates for the context equivalence relation. Applying Lemma 4.2 of [3] we can infer that only derivations shorter than a certain bound (which depends only on $|\mathcal{N}_G|$) have to be considered in order to determine which candidate is the right one. This means, in particular, that $\bar{G}'$ is effectively constructable. (Of course, we cannot never list down all productions of an infinite grammar but if a batching of productions with the help of the denotation "$i + kp, k = 1, 2, \ldots$" is allowed, then the listing is finite.)

We proceed into the other problem: how to check effectively whether $\bar{G}'_1$ and $\bar{G}'_2$ are isomorphic? The periodicity gives an answer also to this problem. If the two grammars really are isomorphic then there must be two integers i and j $(i \neq j)$ such that

1. $|i - j|$ is a multiple of *both* the period of $\bar{G}'_1$ and that of $\bar{G}'_2$,
2. the bijections $\mathcal{N}^{(i)}_{\bar{G}'_1} \longrightarrow \mathcal{N}^{(i)}_{\bar{G}'_2}$ and $\mathcal{N}^{(j)}_{\bar{G}'_1} \longrightarrow \mathcal{N}^{(j)}_{\bar{G}'_2}$ are equal (or, more exactly, correspond to the same mapping between the original grammars G_1 and G_2).

Hence, isomorphism must be checked only up to some j that depends only on cardinalities of the original grammars. Of course, the upper bound is quite high, since in principle the least common multiple of the periods of G_1 and G_2 could be of the order $2^n \cdot 2^n$, while the number of possible bijections is of the order $n!$ (n is the maximal cardinality of nonterminal sets in G_1 and G_2).

Thus, we have proved the following result.

Theorem 3. There is an effective method to decide whether or not two E0L grammars G_1 and G_2 are structurally equivalent.

5 A normal form

The grammar $\bar{G}'$ serves as a kind of a normal form for an E0L grammar G with respect to structural equivalence. However, this is not quite satisfactory, since $\bar{G}'$ is an infinite grammar. Fortunately, it is possible to find a *finite* normal form grammar structurally equivalent to G.

Let us, once again, recall the property of ultimate periodicity in $\bar{G}'$. Instead of introducing new and new nonterminals for each $i = 1, 2, \ldots$ we could identify these sets periodically. That means we do not introduce new nonterminals $\{N^{(t+p)}\}$ at the level $t+p$ but use the old nonterminals $\{N^{(t)}\}$ of the threshold level t instead. In this manner we obtain a finite grammar G''. It is clear that all derivations in $\bar{G}'$ have structurally similar counterparts in G''. On the other hand, in *terminating* derivations of G'' the indices at each level must be equal. It follows that for all terminating derivations of G'' there exist structurally similar counterparts in $\bar{G}'$, since indices at higher levels than $t+p-1$ could be changed. Altogether, G'' and $\bar{G}'$ are structurally equivalent.

The form of the grammar G'' is heavily dependent on the form of the original grammar G. We would like to "squeeze" out this nondeterminism and obtain a unique grammar (modulo isomorphism) for all structurally equivalent grammars. First step toward this goal is to minimize the sum $t+p$. We begin with the infinite normal form grammar $\bar{G}'$. For each level i in $\bar{G}'$ we try to map bijectively all nonterminals $\{N^{(i)}\}$ to each lower level. Each such mapping gives a candidate for a grammar G_2 that is structurally equivalent to G but has a lower $t+p$-value.

This process leads to a grammar G''' that is structurally equivalent to G and has minimal values for $t+p$ and t (in this order). The grammar G''' can be found effectively, since the value of $t+p$ is bounded from above by the specific value of $t+p$ in the original grammar G. The grammar G''' is still not determined uniquely (modulo isomorphism). Indeed, it is possible that there are two bijections from a certain upper level to a certain lower level in G'' which both give a grammar structurally equivalent to G. An example of this is given in the next section. This means there exist two structurally equivalent grammars G_1 and G_2 with the same values of t and p such that the bijection of nonterminals giving the isomorphism (of Theorem 2) from $\bar{G}_1'$ to $\bar{G}_2'$ is *different* on the levels t and $t+p$, although the nonterminal sets in question are the same in these two levels. Nonterminals are *permuted* on the level $t+p$.

To obtain a unique normal form (modulo isomorphism, of course) we do not reduce anything but, on the contrary, we increase the number of nonterminals. First we check which permutations are possible in the aforementioned sense. More specifically, we check for each permutation of the set B_{t+p} in G''' whether or not a structurally equivalent grammar results if we replace all nonterminals of B_{t+p} by their permuted counterparts in the left-hand sides of such productions whose right-hand sides belong to $(B_{t+p-1})^+$. This checking is effective, since there are only finitely many permutations.

For each "successful" permutation we pick up its ***order***, i.e. the smallest number of times that a permutation must be performed before the original order is reached again. Calculate l = least common multiple of all such orders of permutations. Replace then the original period p by $l \cdot p$. This means more and more nonterminals are introduced until, finally, on the level $t + lp$ we choose $B_{t+lp} = B_t$. In this way we obtain a grammar $G^\square$ that is structurally equivalent to G. Moreover, the following theorem holds.

Theorem 4. For two structurally equivalent grammars G_1 and G_2 the grammars $G_1^\square$ and $G_2^\square$ are isomorphic. Thus, $G^\square$ constitutes a normal form for G with respect to the structural equivalence.

Proof. Only the implication from left to right needs a justification. By Theorem 2, there exists a sequence of bijections f_i between the sets B_i in G_1'''' and G_2'''' that preserves the production structure, i.e. $N^{(i)} \rightarrow \alpha^{(i-1)}$ is a production of G_1'''' iff $f_i(N^{(i)}) \rightarrow f_{i-1}(\alpha^{(i-1)})$ is a production of G_2''''.

What is needed is to show that the two bijections f_t and f_{t+lp} are equal. We can construct the bijections f_i recursively starting form f_0 = identity mapping $i : \Sigma \longrightarrow \Sigma$, because of the facts that the production structure is preserved and the grammars are invertible. Each bijection f_i determines completely the next one f_{i+1}.

We already saw above that it is possible to identify the whole alphabeta of G_1'''' and G_2''''. The only difference between these two is that there possibly is a nontrivial permutation p on the nonterminal set B_t such that $f_{t+p} = f_t \circ p$. (Note that this does not cause any contradiction with our claim that $G_1^\square$ and $G_2^\square$ are isomorphic, since if a nontrivial p exists then $B_t \cap B_{t+p} = \emptyset$ in both grammars.)

Going up from f_{t+p} to f_{t+2p} is exactly the same process as going up from f_t to f_{t+p}. Hence, $f_{t+2p} = f_{t+p} \circ p = f_t \circ p^2$, and inductively it can be seen that $f_{t+kp} = f_t \circ p^k$ for $k = 1, 2, \ldots$. In particular, $f_{t+lp} = f_t \circ p^l$. Since l is, by definition, a multiple of the order of p, it follows that $p^l = i$ and $f_{t+lp} = f_t$. □

6 An example

Let us go through all the construction steps of the normal form grammar in the case of the following E0L grammar:

$$G: \quad S \rightarrow AB|BA, A \rightarrow A|a, B \rightarrow B|b.$$

The language $L(G)$ is simply $\{ab, ba\}$ and both these words have derivations of the length i for $i = 2, 3, \ldots$

A typical derivation in G:

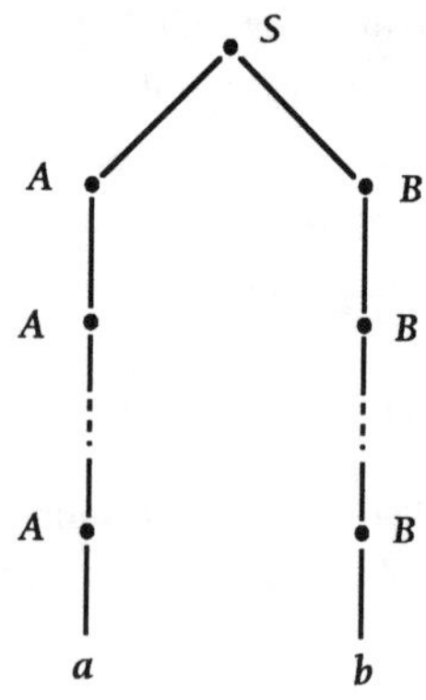

Let us construct the grammar G' :

$$B_0 = \{a, b\}$$
$$B_1 = \{A, B\}$$
$$B_2 = \{A, B, S\} = \mathcal{N}_G = B_3 = B_4 = \dots$$

Therefore, we have productions

$$\begin{array}{lll} A^{(1)} \to a, & B^{(1)} \to b; & \\ A^{(2)} \to A^{(1)}, & B^{(2)} \to B^{(1)}, & S^{(2)} \to A^{(1)}B^{(1)}|B^{(1)}A^{(1)}; \\ & \dots & \\ A^{(i+1)} \to A^{(i)}, & B^{(i+1)} \to B^{(i)}, & S^{(i+1)} \to A^{(i)}B^{(i)}|B^{(i)}A^{(i)}; \\ & \dots & \end{array}$$

It is easy to see that the only nontrivial context equivalence class constitutes of all start symbols. Let us identify them by $S^{(i)} = S$ to obtain $\bar{G}'$.

Coming back to finite grammars we construct G'' by identifying $A^{(k)} = A^{(2)}$ and $B^{(k)} = B^{(2)}$ for $k \geq 3$. Let us further erase the superscript (1) and replace the superscript (2) by '.

$$\begin{aligned} G'' : \quad & S \to A'B'|B'A'|AB|BA \\ & A' \to A'|A \ , \ B' \to B'|B \\ & A \to a \ , \ B \to b \end{aligned}$$

We have $t = 2, p = 1$. It is not possible to reduce t to one, since there are no derivations with length one. Therefore, G''' can be chosen to be equal to G''.

Next phase is to study permutations in $B_t = B_2 = \{S, A', B'\}$. Clearly, S must map into itself, hence the only possible permutation is the transposition $A' \leftrightarrow B'$. It turns out that this gives a structurally equivalent grammar:

$$\begin{aligned} G_0 : \quad & S' \to A'B'|B'A'|AB|BA \\ & A' \to B'|A \ , \ B' \to A'|B \\ & A \to a \ , \ B \to b \end{aligned}$$

The order of a transposition is two, and thus, the l.c.m. of orders of all successful permutations is also two.

Therefore, the normal form $G^{\square}$ is the following:

$$\begin{aligned} G^{\square} \ : \quad & S' \to A''B''|B''A''|A'B'|B'A'|AB|BA \\ & A'' \to A' \ , \ B'' \to B' \\ & A' \to A''|A \ , \ B' \to B''|B \\ & A \to a \ , \ B \to b \end{aligned}$$

For instance, in $G^{\square}$ the derivation of the word ab with length 5 is

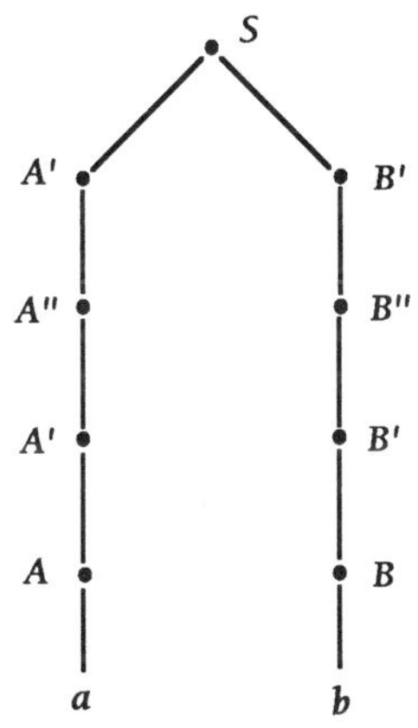

It is worth noting that also the grammar

$$G_2 : \quad S \to AB|BA \ , \ A \to B|a \ , \ B \to A|b$$

is structurally equivalent to G (and also to all the other grammars). This fact explains why the normal form of such a simple grammar as G is a bit complicated. There is no nontrivial equivalence relation between the nonterminals of G such that reducing the grammar according to that relation would preserve structural equivalence. On the other hand, G_2 is an example of a nonisomorphic grammar that is structurally equivalent to G but has as few nonterminals and productions as G.

7 Conclusions

We have seen how an arbitrary E0L grammar G can be transformed into a structurally equivalent normal form grammar $G^{\square}$. For two structurally equivalent grammars the respective normal forms are isomorphic. This gives a new method for deciding whether or not two E0L grammars are structurally equivalent.

There might exist other normal forms. The main disadvantage of $G^{\square}$ is the great number of nonterminals due to the multiplication of the period p. Also, $G^{\square}$ is not minimal with respect to t or p. As a basic open problem it may be asked whether there are normal forms that are simpler than $G^{\square}$ or minimal with respect to some measure. On the other hand, it might be possible to find such a measure that $G^{\square}$ is minimal with respect to it.

References

[1] McNaughton, R., Paranthesis grammars, *J. Assoc. Comput. Mach.* **14**, 1967, 490-500.

[2] Ottmann, T. and Wood, D., Defining Families of Trees with E0L Grammars, *Discrete Appl. Math.*, to appear.

[3] Ottmann, T. and Wood, D., Simplifications and E0L grammars, *in this volume.*

[4] Paull, M. and Unger, S., Structural equivalence of context-free grammars, *J. Comput. Systems Sci.* **2**, 1968, 427-463.

[5] Rozenberg, G. and Salomaa, A., *The Mathematical Theory of L Systems*, Academic Press, New York, 1980.

[6] Salomaa, A., *Formal Languages*, Academic Press, New York, 1973.

[7] Salomaa, K. and Yu, S., Decidability of Structural Equivalence of E0L Grammars, *Theoret. Comput. Sci.*, to appear.

[8] Thatcher, J. W., Tree automata: an informal survey, in Aho, A. V. (ed.): *Currents in the Theory of Computing*, Prentice-Hall, Englewood Cliffs, NJ, 1973, 143-172.

[9] Wood, D., *Theory of Computation*, John Wiley & Sons, New York, 1987.

Simplifications of E0L Grammars

Thomas Ottmann[1] and Derick Wood[2]

[1]Institut für Informatik, Universität Freiburg, Rheinstraße 10-12, D-7800 Freiburg, Germany
[2]Data Structuring Group, Department of Computer Science, University of Waterloo, Waterloo, Ontario N2L 3G1, Canada.

Abstract: We establish some simplification results for E0L grammars that preserve their structure. In the context-free case the simplifications we consider are sufficient to solve the structural equivalence problem, but unfortunately this does not hold in the E0L case. Therefore, we provide a number of examples that illustrate why the structural equivalence problem is considerably different for E0L grammars.

Keywords: E0L grammar; simplifications; normal forms; structural equivalence.

1 Introduction

Context-free grammars and E0L grammars are two popular means of generating languages by iterative rewriting. In both cases the set of symbols is divided into terminal and nonterminal symbols and words consisting of only terminal symbols belong to the language generated by a grammar. The major difference between the two generating devices is that rewriting is carried out in *parallel* (and synchronously) for E0L grammars and *sequentially* (and asynchronously) for context-free grammars. Two grammars are *equivalent* if they generate the same language. The (language) equivalence problem is undecidable for context-free and for E0L grammars. Even very restricted cases of this problem turn out to be undecidable.

To each derivation in a context-free and E0L grammar we can associate a syntax tree. It contains the sentence symbol at the root and each branching node is labeled with a terminal or nonterminal symbol. The labeling is determined by the productions used in the derivation. Syntax trees of E0L grammars are labeled trees of uniform depth; syntax trees of context-free grammars can have arbitrary shapes. Considering syntax trees leads to the notion of *structural equivalence* between grammars. Two structurally equivalent grammars not only generate the same terminal words (or language) but also structure these words in the same manner. It has been shown by McNaughton [1] in 1967 that structural equivalence is decidable for context-free grammars. Surprisingly, the structural equivalence problem for E0L grammars has not attracted much interest. However a preliminary version of this paper motivated Salomaa and Yu [4] to show, by completely different methods, that structural equivalence is decidable for E0L grammars. Their results leave open, however, the problem of characterizing when two E0L grammars are structurally equivalent. For example, two context-free grammars are structurally equivalent if and only if their simplified versions (á la McNaughton) are isomorphic. It is this kind of result that we seek and it this kind of result that we have not yet found, although we have been able to show that the simplifications discussed in this paper are sufficient to solve the structural equivalence problem for a restricted class of E0L grammars. We illustrate a number of differences between the case of context-free grammars and the case of E0L grammars, show that some special cases of the structural equivalence problem for E0L grammars can be reduced to isomorphism, and prove some simplification results for E0L grammars that preserve their structure. As we will see, parallel rewriting makes structural equivalence a challenging basic problem for E0L grammars which is considerably different from the corresponding problem for context-free grammars.

We assume that the reader is familiar with the basic notions for context-free grammars [5, 6]. In Section 2 we provide the necessary definitions for E0L grammars and some special cases of the structural equivalence problem are solved, via isomorphism, in Section 3. Section 4 introduces the notion of context equivalence and we show how to effectively reduce an E0L grammar with respect to this equivalence relation and we close with some concluding remarks in Section 5.

2 Definitions and examples

An *E0L grammar* G is a quadruple (N, Σ, P, S), where N is an alphabet of *nonterminals*, Σ is an alphabet of *terminals*, $P \subseteq N \times (N^+ \cup \Sigma^+)$ is a finite set of *productions*, and $S \in N$ is the *sentence symbol.*

This definition differs from the classical one [3] in two respects. First, only nonterminals have productions; the grammar is *synchronized.* Second, the right hand sides of productions cannot be the empty word; the grammar is *propagating.* These modifications do not affect the languages generated by E0L grammars (apart from the loss of the empty word). Throughout this paper we use upper case letters to denote nonterminals and lower case letters to denote terminals.

Rewriting is defined in the usual way. Let α be a nonempty word over N; that is, $\alpha = A_1 \dots A_n$, where A_i is in N, $1 \leq i \leq n$ and $n = |\alpha|$. α can be rewritten as $\beta = \beta_1 \dots \beta_n$, for some $\beta_i \in N^+ \cup \Sigma^+$, $1 \leq i \leq n$, if $A_i \to \beta_i$ is in P, $1 \leq i \leq n$. We usually denote this by $\alpha \Rightarrow \beta$. We write $\alpha \Rightarrow^d \beta$ to denote that α yields β in d steps, for $d \geq 0$, if: $d = 0$ and $\alpha = \beta$; $d = 1$ and $\alpha \Rightarrow \beta$; or $d > 1$ and there exists γ in N^+ such that $\alpha \Rightarrow \gamma$ and $\gamma \Rightarrow^{d-1} \beta$. We write $\alpha \Rightarrow^+ \beta$ if $\alpha \Rightarrow^d \beta$, for some $d \geq 1$, and we write $\alpha \Rightarrow^* \beta$, if $\alpha \Rightarrow^d \beta$ for some $d \geq 0$. We say that $\alpha \Rightarrow^+ \beta$ and $\alpha \Rightarrow^* \beta$ are *derivations.* A derivation starting with the sentence symbol is called a *sentential derivation.*

Note that only purely nonterminal words can be rewritten. This is the reason for only allowing right hand sides of productions to be either completely nonterminal or completely terminal.

The *language generated by* G is denoted by $L(G)$ and is defined by

$$L(G) = \{x : x \in \Sigma^+ \text{ and } S \Rightarrow^+ x\}.$$

We say that an E0L grammar $G = (N, \Sigma, P, S)$ is *reduced* if the sentence symbol generates a terminal word and each nonterminal appears in at least one sentential derivation of a terminal word. From now on we tacitly assume that E0L grammars are reduced if not explicitly stated otherwise.

With each derivation $S \Rightarrow^+ \alpha$, $\alpha \in N^+ \cup \Sigma^+$, we can associate a *derivation tree* $tr(S \Rightarrow^+ \alpha)$. This is a uniform depth tree whose internal nodes are labeled with nonterminal symbols and whose external nodes are labeled with the symbols in α in left-to-right order. It also satisfies the following condition: For all internal nodes u, if u has r children $u_1, \ldots, u_r$, for some $r \geq 1$, then $L(u) \to L(u_1) \ldots L(u_r)$ is in P, where $L(v)$ denotes the label of node v.

A syntax tree whose root is labeled with the sentence symbol and whose leaves are labeled with terminal symbols is called a *sentential syntax tree.* Concatenating the labels at the frontier of a sentential syntax tree in left-to-right order gives a terminal word x generated by the grammar. Note that a derivation $S \Rightarrow^d x$ for $x \in \Sigma^+$, has a sentential syntax tree $tr(S \Rightarrow^d x)$ of height d.

Two E0L grammars G and G' are (language) *equivalent* if $L(G) = L(G')$. Two E0L grammars G and G' are *structurally equivalent* if they are equivalent and, moreover, their sentential syntax trees are structurally identical; that is, they differ only in the labels of their internal nodes. Because a terminal word generated by an E0L grammar may have more than one derivation the latter condition in the above definition of structural equivalence is to be understood as follows: For each sentential syntax tree T generating a terminal word $x \in L(G)$ there is a structurally identical syntax tree T' in G' generating the same word $x \in L(G')$, and vice versa. In Section 4 we will reduce structural equivalence to equivalence by the use of parenthesized versions of grammars.

An E0L grammar is *unambiguous* if each terminal word generated by the grammar has exactly one derivation tree. An E0L grammar $G = (N, \Sigma, P, S)$ is *invertible* if no two productions in P have the same right hand side. An invertible grammar may be ambiguous; however, there cannot be two different syntax trees with the same structure for the same terminal word if the grammar is invertible.

An E0L grammar $G = (N, \Sigma, P, S)$ is called *nonterminal complete*, or *n-complete*, for short, if for each nonterminal $X \in N$ there is at least one production $X \to \alpha$ in P, where $\alpha \in N^+$.

We will illustrate these notions by a number of examples in which the sentence symbol is always denoted by S. In each case the given E0L grammar has no useless nonterminals; that is, it is reduced.

Example 2.1 Let G_1 be given by

$$S \to AA,\ A \to AB,\ A \to a,\ B \to b.$$

Then, $L(G_1) = \{aa, abab\}$. The E0L grammar G_1 is invertible and unambiguous.

Example 2.2 Let G_2 be given by

$$S \to AS|a,\ A \to AA|a.$$

Then, $L(G_2) = \{a^{2^i} | i \geq 0\}$. G_2 is not invertible, because there are two productions with right hand side a. However, it can easily be seen that each $x \in L(G_2)$ has exactly one syntax tree. Hence, G_2 is unambiguous.

Example 2.3 The grammar G_3 defined by

$$S \to SS|a$$

is invertible, unambiguous, and structurally equivalent to G_2.

Example 2.4 The E0L grammar G_4 defined by

$$S \to AA|a,\ A \to S$$

is invertible and equivalent to both G_2 and G_3. However, G_4 is not structurally equivalent to either G_2 or G_3.

Example 2.5 Let G_5 be given by the productions

$$S \to AA, \ A \to A|a.$$

Then, $L(G_5) = \{aa\}$. G_5 is invertible and ambiguous, indeed there are infinitely many syntax trees for aa.

Note that the grammar G_1 is not n-complete, but $G_2, \ldots, G_5$ are n-complete.

Theorem 2.1 *Let $G = (N, \Sigma, P, S)$ be an E0L grammar. Then, an n-complete structurally equivalent E0L grammar $G' = (N', \Sigma, P', S)$ can be effectively constructed from G such that invertibility is maintained; that is, G' is invertible if G is invertible.*

Proof: N' and P' are obtained from N and P, respectively, by adding "looping" nonterminals and productions as follows. For each nonterminal $X \in N$ such that there is no production $X \to \alpha$ in P with $\alpha \in N^+$, choose a new nonterminal X', add X' to N' and the productions $X \to X'$, $X' \to X'X'$ to P. Obviously, G' is n-complete. G' is structurally equivalent to G because the derivations of terminal words are not affected by the new productions. Finally, all new productions have pairwise distinct right hand sides which are different from the right hand sides of the productions in P. Therefore, G' is invertible if G is invertible. Observe that G' is not reduced because the looping nonterminals do not occur in any sentential derivation of a terminal word. □

Invertibility allows us to reconstruct the labeling of a syntax tree of a given structure bottom up: If the labels of the leaves are given we can uniquely reconstruct the labels at all internal nodes. Requiring invertibility does not affect the languages generated by E0L grammars as the following theorem shows.

Theorem 2.2 *Let $G = (N, \Sigma, P, S)$ be an E0L grammar. Then, an invertible structurally equivalent E0L grammar $G' = (N', \Sigma, P', S')$ can be effectively constructed from G.*

Proof: This is a straightforward modification of the proof given in Ottmann and Wood [2] for the case where $\Sigma = \{a\}$ and a set of sentence symbols instead of a single sentence symbol is allowed. Define N' to be the set $\{X \subseteq N : X \neq \emptyset\}$ and S' the set $\{X : X \subseteq N \text{ and } X \cap S \neq \emptyset\}$. Given a word α' over N' we say a word α over N *corresponds* to α' if $|\alpha| = |\alpha'|$ and each nonterminal symbol in α belongs to the set of nonterminal symbols appearing at the same position in α'.

The set P' of productions is defined as follows.

(i) P' contains a production $X \to \alpha'$, for $\alpha' \in (N')^+$, if and only if

$$X = \{A : A \in N, A \to \alpha \in P \text{ and } \alpha \text{ corresponds to } \alpha'\}$$

(ii) P' contains a production $X \to w$, for $w \in \Sigma^+$, if and only if

$$X = \{A : A \to w \in P\}.$$

It is not difficult to prove that G' is invertible and G and G' are structurally equivalent. □

Note that an E0L grammar with only one nonterminal is always invertible.

3 The one and two nonterminal cases

McNaughton's proof of the decidability of the structural equivalence problem for context-free grammars [1] is based on the following idea. For each context-free grammar a structurally equivalent and simplified grammar is constructed; a simplified grammar is obtained by appropriately identifying nonterminals which play the same role in all derivations. Simplification drastically reduces the class of structurally equivalent context-free grammars; any two simplified structurally equivalent context-free grammars are isomorphic. This provides a basis for the decidability result. To test whether two context-free grammars are structurally equivalent we simplify them and test whether the resulting grammars are isomorphic.

An obvious approach to solving the structural equivalence problem for E0L grammars is to proceed in a similar manner. This suggests imposing additional requirements on E0L grammars which preserve structural equivalence but restrict the variety of grammars having these properties. Invertibility is an example of such a property. Restricting the number of nonterminals is another obvious choice. In this section we show that both restrictions can force an E0L grammar to be uniquely determined up to isomorphism.

Theorem 3.1 *Let $G = (N, \Sigma, P, S)$ be an E0L grammar with $N = \{S\}$. If $G' = (N', \Sigma, P', S')$ is structurally equivalent to G and invertible, then G and G' are isomorphic.*

Proof: Because G has only one nonterminal its productions must be of the following forms

$$S \rightarrow S^{i_1}|\cdots|S^{i_k},$$

where $i_j \geq 1$, for $1 \leq j \leq k$, and

$$S \rightarrow w_1|\cdots|w_l,$$

where $w_j \in \Sigma^+$, for $1 \leq j \leq l$. Clearly, G' must also have productions $S' \rightarrow w_j$, $1 \leq j \leq l$, and cannot have productions $A \rightarrow w_j$, where $A \in N'$ and $A \neq S'$, otherwise G' would not be invertible. Now consider the syntax tree of Figure 1. Because G and G' are structurally equivalent, there is a syntax tree of the same structure in G' which generates the same terminal word. This is only possible if $S' \rightarrow S'^{i_j}$ is in P'. Because of invertibility no other nonterminal can generate S'^{i_j}. Thus, the productions in P and P' are isomorphic. □

Theorem 3.1 can be extended to the case where both E0L grammars have exactly two nonterminals.

Theorem 3.2 *Let $G = (\{S, A\}, \Sigma, P, S)$ and $G' = (\{S', A'\}, \Sigma, P', S')$ be two structurally equivalent invertible E0L grammars. Then, G and G' are isomorphic.*

Proof: Let us first consider productions with a terminal right hand side. Clearly $S \rightarrow w$, $w \in \Sigma^+$, is a production in P if and only if $S' \rightarrow w$ is a production in P'. For, otherwise G and G' would not be structurally equivalent. If $S \rightarrow w$ is a production in P, we cannot have a production $A' \rightarrow w$ in P', because invertibility implies that w can be generated in one step from S'. Similarly, if $S' \rightarrow w$ is in P' we cannot have a production $A \rightarrow w$ in P. Thus, the productions in P and P' with terminal right hand sides and the start symbol as left hand side are isomorphic.

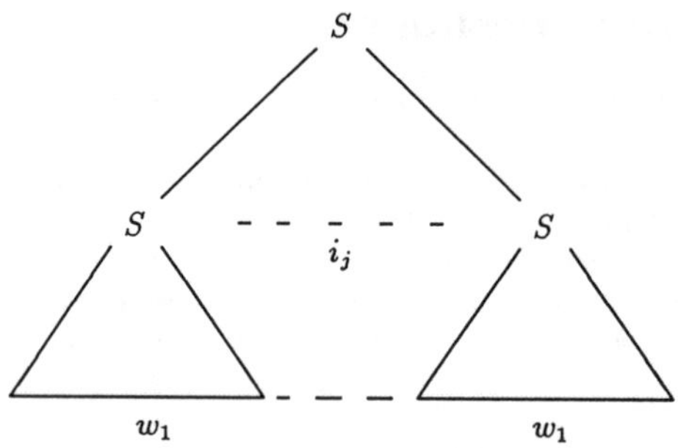

Figure 1: Syntax tree with one nonterminal

Assume that $A \to w$ is in P, $w \in \Sigma^+$. Then, $S \to w$ is not in P. Consider a syntax tree in which this production is applied. Because a syntax tree of the same structure can be generated in G', $A' \to w$ must be a production in P'; $S' \to w$ cannot be a production in P', because in this case, w would be generated by a height one syntax tree in G', but not in G. This argument shows that the productions with terminal right hand sides in G and G' are isomorphic.

Let us now consider productions with nonterminal right hand sides. Assume $S \to X_1 \ldots X_k$, $X_j \in \{S, A\}$, $1 \le j \le k$, is a production in P. Then, there exist $w_1, \ldots, w_k \in \Sigma^+$ such that $X_j \Rightarrow^d w_j$, for $1 \le j \le k$ and some $d \ge 1$. Now reconstruct the labels of the syntax tree with frontier $w_1 \ldots w_k$ in G'. We obtain $X'_j \Rightarrow^d w_j$, $1 \le j \le k$, and $X'_j = S'$ if and only if $X_j = S$, and $X'_j = A'$ if and only if $X_j = A$, for $1 \le j \le k$. For, otherwise a sentential syntax tree in G of height d would not have a structurally equivalent counterpart in G' and vice versa. Applying this argument once again we conclude that $S' \to X'_1 \ldots X'_k$ is in P'. By symmetry, this shows that P and P' must have isomorphic productions with the start symbol as left hand side and nonterminals as right hand sides.

By similar arguments we see that P and P' must also have isomorphic productions with the other nonterminal as left hand side and nonterminals as right hand side. A production $A \to X_1 \ldots X_k$ of this form must occur in a terminal derivation in G. Retracing this derivation in G' shows that the corresponding, isomorphic production must occur in G' and vice versa. □

Theorems 3.1 and 3.2 cannot be extended. If we allow more nonterminals we can have invertible, structurally equivalent E0L grammars which are not isomorphic. First, we give an example where both E0L grammars have a different number of nonterminals.

Example 3.1 Recall the E0L grammar G_5 of Example 2.5 which is given by the productions

$$S \to AA, \; A \to A|a.$$

A structurally equivalent but nonisomorphic grammar G_6 with three nonterminals is given by the productions

$$S \to AA|BB, A \to B|a, B \to A.$$

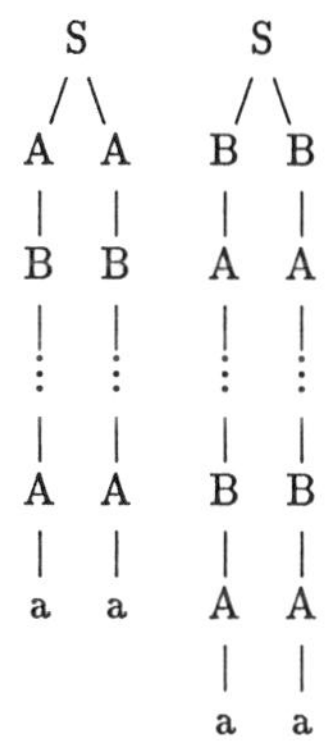

Figure 2: Syntax trees of even and odd heights in G_6

G_6 is also an invertible E0L grammar which generates just one word aa by syntax trees of even and odd heights as shown by Figure 2.

Note that G_5 and G_6 are not structurally equivalent if considered as context-free grammars. For, when parallel rewriting is replaced by (asynchronous) sequential rewriting, we can generate syntax trees of different structures with both grammars. The terminal leaves of syntax trees generated by G_6 always have an even depth difference whereas in G_5 we can generate a syntax tree where the terminal leaves have a depth difference of one. For example, the following syntax tree in G_5

has no structural equivalent counterpart in G_6, if both grammars are considered to be context-free grammars.

Example 3.2 The E0L grammar G_7 with three nonterminals defined by

$$S \to AA|BB, \ A \to a, \ B \to A|B$$

is structurally equivalent to G_6, but not isomorphic to G_6. It is easy to obtain an overview of the sentential syntax trees generated by G_7. The sentential syntax tree

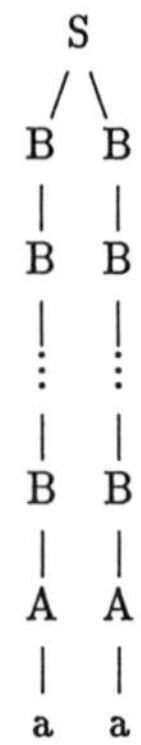

Figure 3: Syntax tree of height greater than two in G_7

is the only sentential syntax tree of height two in G_7. A sentential syntax tree of height $h > 2$ is of the form shown by Figure 3.

It is easy to see that there is no isomorphism between the productions of G_6 and G_7. Thus, there exist structurally equivalent, nonisomorphic E0L grammars with the same number of nonterminals. In G_6 and G_7 the roles played by the nonterminals A and B are different. We cannot have an A whenever we have a B and vice versa.

Example 3.3 Let the invertible E0L grammar G_8 be given by the productions

$$S \to AA|BB|CC, \; B \to A, \; C \to B|C, \; A \to a.$$

If we map both labels B and C to B, each syntax tree in G_8 becomes a valid syntax tree in G_7. In this case, G_7 can be considered as an homomorphic, but not as an isomorphic, image of G_8. It is obvious how to extend the sequence G_5, G_7, G_8 to obtain an infinite sequence of invertible, structurally equivalent, pairwise nonisomorphic E0L grammars, namely, H_k for $k \geq 1$ is given by

$$\begin{aligned} S &\to X_1X_1|X_2X_2|\ldots|X_kX_k, \\ X_k &\to X_{k-1}|X_k, \\ X_{k-1} &\to X_{k-2}, \; \ldots, \; X_2 \to X_1, \\ X_1 &\to a. \end{aligned}$$

Example 3.4 Consider the E0L grammars G_9 and G_{10} defined as follows:

$$G_9 \ : \ S \rightarrow AA, \ A \rightarrow A|a|b.$$

$$\begin{aligned} G_{10} \ : \ & S \rightarrow AA|AB|BA|BB \\ & A \rightarrow B|a \\ & B \rightarrow A|b. \end{aligned}$$

$L(G_9) = L(G_{10}) = \{aa, ab, ba, bb\}$; moreover, G_9 and G_{10} are structurally equivalent. Intuitively, in G_{10} A and B play essentially the same roles. It is possible to identify these nonterminals, yielding the simpler grammar G_9 in this case.

In the next section we define context equivalence and show how to effectively reduce an E0L grammar according to this equivalence relation; G_{10} is not context reduced.

4 Context equivalence

The structure of syntax trees of grammars can be encoded via parentheses. Introducing parenthesized versions of E0L grammars allows us to reduce structural equivalence to equivalence.

Given an E0L grammar $G = (N, \Sigma, P, S)$, the *parenthesized version* $G_{()}$ of G is the E0L grammar $G_{()} = (N \cup \{L, R\}, \Sigma \cup \{(,)\}, P_{()}, S)$, where L and R are two new nonterminals and "(" and ")" are new terminal symbols. Matching nonterminal L-R pairs represent matching pairs of parentheses. For $\alpha \in N^+$, $X \rightarrow L\alpha R$ is a production in $P_{()}$ if and only if $X \rightarrow \alpha \in P$ and, for $w \in \Sigma^+$, $X \rightarrow (w)$ is in $P_{()}$ if and only if $X \rightarrow w \in P$. Furthermore, $P_{()}$ contains the additional productions $L \rightarrow L|($ and $R \rightarrow R|)$. Rewriting and other related notions from Section 2 are extended to parenthesized versions of grammars in the obvious way. Of course, if $L(G) \subset \Sigma^+$, then $L(G_{()}) \subseteq (\Sigma \cup \{(,)\})^+$. Observe that for a word $x \in L(G_{()})$ all terminal symbols from Σ in x are surrounded by the same number of matching pairs of parentheses. This number is equal to the length of the derivation of x in $G_{()}$ from the start symbol. Obviously, two E0L grammars G and G' are structurally equivalent if and only if their corresponding parenthesized versions generate the same language $L \subseteq (\Sigma \cup \{(,)\})^+$.

A parenthesized nonterminal *context* of an E0L grammar $G = (N, \Sigma, P, S)$ is a sentential form of the parenthesized version $G_{()}$ in which one occurrence of a nonterminal symbol from N is replaced by an underscore. We call such a word $\alpha \in (N \cup \{L, R\} \cup \{\_\})^+$ simply a *context*. Given a context α and a nonterminal A, $\alpha[A]$ denotes the word obtained by replacing the underscore in α with A.

Let $S \Rightarrow^d \alpha[A]$ in $G_{()}$. $\alpha[A]$ can be identified with a sentential syntax tree of height d in G from which all labels except for the labels at its frontier have been removed. The sequence of labels at the frontier of this tree is denoted by $frontier(\alpha[A])$. Clearly, $S \Rightarrow^d frontier(\alpha[A])$ in G if and only if $S \Rightarrow^d \alpha[A]$ in $G_{()}$. Given an E0L grammar $G = (N, \Sigma, P, S)$ and two nonterminals A and B, we say that A and B are *d-context equivalent*, denoted by $A \equiv^d B$, if for all contexts α, we have in the parenthesized version $G_{()}$ of G

$$S \Rightarrow^d \alpha[A] \text{ if and only if } S \Rightarrow^d \alpha[B].$$

This definition in particular implies that $A \equiv^0 B$ is equivalent to ($A = S$ if and only if $B = S$). A and B are said to be *context equivalent*, denoted by $A \equiv B$, if $A \equiv^d B$, for all integers $d \geq 0$. Observe that $A \equiv S$ for some nonterminal A implies $A = S$. This reflects the fact that at the root of each syntax tree in G we can have only one nonterminal, the sentence symbol S, but no other nonterminal. We say an E0L grammar is *context reduced* if whenever two nonterminals are context equivalent they are equal.

Example 4.1 Grammar G_{10} of Example 3.4 is not context reduced. It is easy to prove by induction on d that A and B are d-context equivalent, for all integers $d \geq 0$.

Given an E0L grammar $G = (N, \Sigma, P, S)$, context equivalence of nonterminals can also be characterized in a different manner (cf. the case of context-free grammars [1]). For each context α and integer $d \geq 0$, we define

$$N_d^\alpha = \{X : S \Rightarrow^d \alpha[X] \text{ in } G_{()}\}$$

to be the set of nonterminals *d-distinguishable in context* α. The definition implies that $N_0^\alpha = \{S\}$, for all α, that is, $\{S\}$ is 0-distinguishable, for all α. A set M of nonterminals is said to be *d-distinguishable* if there is a context α such that $M = N_d^\alpha$. M is *strictly d-distinguishable* if M is d-distinguishable but not d'-distinguishable for any $d' < d$. M is *distinguishable* if there is a $d \geq 0$ such that M is d-distinguishable.

Theorem 4.1 *Let $G = (N, \Sigma, P, S)$ be an E0L grammar and $A, B \in N$. Then, $A \equiv B$ if and only if for all distinguishable sets $M \subseteq N$, A is in M if and only if B is in M.*

Proof: *if:* If $V = \{S\}$, then ($A \in V$ if and only if $B \in V$) immediately implies ($A = S$ if and only if $B = S$). Therefore, we have $A \equiv^0 B$. Now consider an arbitrary context α and an integer $d \geq 1$. We have to show

$$S \Rightarrow^d \alpha[A] \text{ if and only if } S \Rightarrow^d \alpha[B].$$

Consider the set N_α^d of nonterminals d-distinguishable in context α, namely,

$$\{X : S \Rightarrow^d \alpha[X]\}.$$

By definition N_α^d is d-distinguishable, therefore distinguishable. By assumption this set contains A if and only if it contains B. This immediately gives the hypothesis.
only if: Let $A \equiv B$. This implies by definition ($A = S$ if and only if $B = S$) and, for all contexts α, and for all $d \geq 1$, ($S \Rightarrow^d \alpha[A]$ if and only if $S \Rightarrow^d \alpha[B]$). If $M = \{S\}$, then we immediately have $A \in M$ if and only if $B \in M$. Hence, let M be a distinguishable subset of N and $M \neq \{S\}$. Then, there exists a context α and an integer $d \geq 1$ such that $M = N_d^\alpha$. In particular $A \equiv B$ implies ($S \Rightarrow^d \alpha[A]$ if and only if $S \Rightarrow^d \alpha[B]$); hence, ($A \in M$ if and only if $B \in M$). □

It is clear that there exist only finitely many different distinguishable subsets of nonterminals for a given grammar. If the distinguishable sets can be effectively determined the equivalence relation $\equiv$ can also be computed effectively. The question is whether or not we have to consider arbitrary long derivations in order to find all distinguishable sets of nonterminals. The following lemmas show that this is not necessary.

Lemma 4.2 *Let $G = (N, \Sigma, P, S)$ be an invertible n-complete E0L grammar and $M \subseteq N$ strictly d-distinguishable. Then, there exists an $M' \neq M$, $M' \subseteq N$, which is strictly $(d-1)$-distinguishable.*

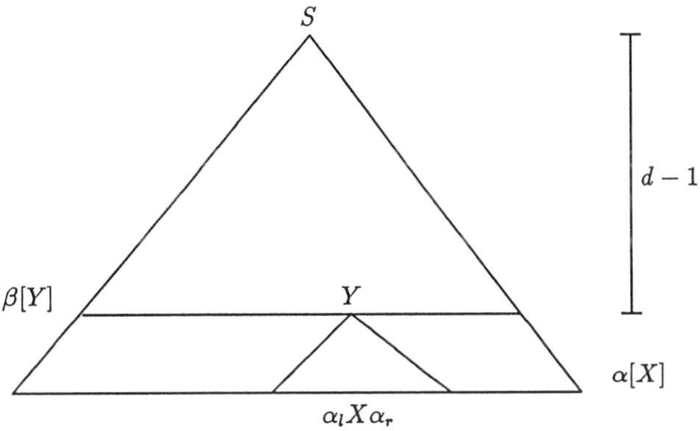

Figure 4: Syntax tree of height $d > 0$ for $frontier(\alpha[X])$

Proof: Let $M = N_d^\alpha = \{X : S \Rightarrow^d \alpha[X]\}$, for some $d > 1$ and for some context α such that, for each $d' < d$ and each context β, $M \neq N_{d'}^\beta$. Consider a syntax tree for $frontier(\alpha[X])$ in G of height $d > 1$, for some $X \in M$; see Figure 4.

All nonterminals in M which may occur instead of X at the same position on level d in $frontier(\alpha[X])$ are generated by nonterminals which occur at the same position on level $d-1$. Because G is invertible, all elements of $frontier(\alpha[X])$ on level d have a uniquely determined predecessor on level $d-1$. Thus, the context α uniquely determines a context β. Replace each innermost $L - R$ pair surrounding a right hand side of a production in P by its left hand side and, moreover, replace the innermost $L - R$ pair containing the underscore in α by an underscore in β. Let us assume that the innermost $L - R$ pair in α containing the underscore is of the form $L\alpha_l\_\alpha_r R$. Then, it is clear that

$$M = N_d^\alpha = \{X : Y \to L\alpha_l X \alpha_r R \text{ is a production in } P_{()} \text{ and } Y \in N_{d-1}^\beta\}.$$

We claim that N_{d-1}^β is strictly $(d-1)$-distinguishable. Assume that this is not true. This implies that there is a context β' and a $d' < d-1$ such that

$$N_{d-1}^\beta = N_{d'}^{\beta'} = \{Y : S \Rightarrow^{d'} \beta'[Y] \text{ in } G_{()}\}.$$

Because G is n-complete, we can add one parallel derivation step to the derivation of length d' in $G_{()}$ of the context β'. Each nonterminal in β' is replaced by the right hand side of a corresponding production in $P_{()}$ and the underscore is replaced by $L\alpha_l\_\alpha_r R$. This yields a context α'; see Figure 5. We obtain $M = N_d^\alpha = N_{d'+1}^{\alpha'}$, where $d'+1 < d$; this contradicts the assumption that M is strictly d-distinguishable. Hence we have proved our claim that N_{d-1}^β is strictly $(d-1)$-distinguishable. Therefore, the lemma holds with $M' = N_{d-1}^\beta$. □

As an immediate consequence of Lemma 4.2 we infer that in $G_{()}$ only derivations of length $d = 2^n$, $n = |N|$, have to be considered in order to determine all distinguishable

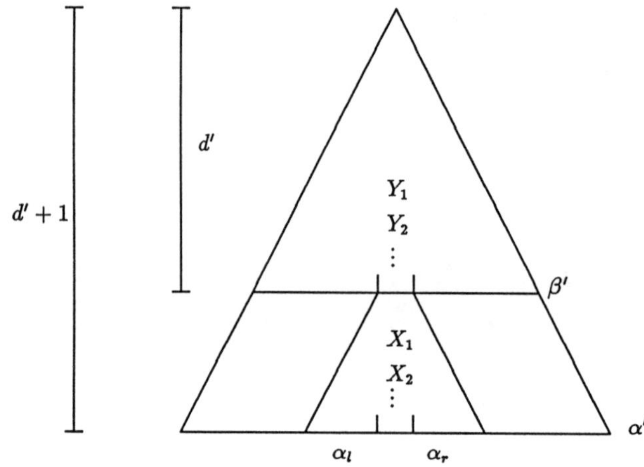

Figure 5: Definition of context α'

sets of nonterminals. The argument used in the proof of Lemma 4.2, however, requires that each derivation of a nonterminal word in the n-complete E0L grammar G can be extended by one parallel derivation step. The following lemma shows that it is not necessary to assume that an E0L grammar is n-complete in order to infer that only finitely many derivations have to be considered in order to determine the distinguishable sets of nonterminals.

Lemma 4.3 *Let $G = (N, \Sigma, P, S)$ be an invertible E0L grammar with $n = |N|$ and let $M \subseteq N$, $|M| = m$, be a strictly d-distinguishable set of nonterminals. Then, $d < (n^m + 1)(2^{n^m} + 1)$.*

Proof: Assume the contrary, that is, $M = N_d^\alpha$, for some context α and some minimal $d \geq (n^m + 1)(2^{n^m} + 1)$, and

$$M = N_d^\alpha = \{X : S \Rightarrow^d \alpha[X] \text{ in } G_{()}\} = \{A_1, \ldots, A_m\}.$$

Because G is invertible we can uniquely associate a labeled tree of height d and structure α to each $A_j \in M$, $1 \leq j \leq m$. Reconstruct the labels at all internal nodes in $\alpha[A_j]$ bottom up to obtain a sentential syntax tree for $frontier(\alpha[A_j])$ in G, $1 \leq j \leq m$. All associated syntax trees in G for $frontier(\alpha[X])$, $X \in M$, of height d have the same structure. They all have the label S at the root and the same labels at the frontier with only one exception, the position of the underscore in α. There $A_1, \ldots, A_m$ appear. However, the labels of internal nodes may be different. We call the root-to-leaf path connecting the root and the position of the underscore in α the *designated* path. With each internal node in α we can associate an m-tuple $(X_1, \ldots, X_m)$ of nonterminals as follows. $(X_1, \ldots, X_m)$ is associated to a node p if and only if X_j occurs in the syntax tree for $frontier(\alpha[A_j])$ of height d

and structure α at node p, $1 \leq j \leq m$. Because there are only n^m different m-tuples of nonterminals and $d \geq (n^m + 1)(2^{n^m} + 1)$ there must be at least $2^{n^m} + 1$ different levels in α such that the same m-tuple is associated to the nodes on the designated path on these levels. Because there are at most 2^{n^m} different subsets of m-tuples of nonterminals there must be at least two levels in α such that not only the same m-tuple of nonterminals is associated to the nodes on the designated path on these levels but also the sets of m-tuples of nonterminals associated to the other nodes on these two levels are identical. This implies that there are integers d_1, d_2, and d_3 such that $d_1 + d_2 + d_3 = d$ and (i), (ii), and (iii) hold.

(i) For each j, $1 \leq j \leq m$, the labels of the nodes on the designated path on levels d_1 and $d_1 + d_2$ below the root in the syntax tree of $frontier(\alpha[A_j])$ of height d and structure α are identical.

(ii) For each j, $1 \leq j \leq m$, there are nonterminal words w_j and w'_j such that

$$S \Rightarrow^{d_1} w_j \Rightarrow^{d_2} w'_j \Rightarrow^{d_3} frontier(\alpha[A_j]),$$

where w_j and w'_j are words over the same set N_j of nonterminals. N_j is obtained by taking the set of j-th components in the set of all m-tuples of nonterminals on levels d_1 or $d_1 + d_2$ in α.

(iii) Let p and q be nodes on levels d_1 and $d_1 + d_2$ in α, respectively, to which the same m-tuple of nonterminals is associated. Then for all j, $1 \leq j \leq m$, the syntax tree of $frontier(\alpha[A_j])$ of height d and structure α has identical nonterminals at the nodes p and q.

Therefore, we can cut out the part in between levels d_1 and d_1+d_2 in α as follows. Replace each subtree of height $d_2 + d_3$ with root on level d_1 by the subtree of height d_3 with root on level $d_1 + d_2$ with the same m-tuple of nonterminals associated to its root. Because the sets of m-tuples of nonterminals associated to the nodes on these two levels are identical we will find for each node on level d_1 an appropriate node on level $d_1 + d_2$. In particular, the subtree with root node on the designated path on level d_1 is replaced by the subtree with root node on the designated path on level $d_1 + d_2$. This situation is illustrated by Figure 6. In this way we obtain a context α' such that $N_d^{\alpha} = N_{d_1+d_2}^{\alpha'} = M$. Because $d_1 + d_2 < d$ we have a contradiction. □

Let $G = (N, \Sigma, P, S)$ be an invertible E0L grammar. We will now show how to reduce G according to the equivalence relation "$\equiv$".

Theorem 4.4 *Let $G = (N, \Sigma, P, S)$ be an invertible E0L grammar. We can effectively construct a context reduced E0L grammar $\bar{G} = (\bar{N}, \Sigma, \bar{P}, \bar{S})$ which is structurally equivalent to G. The construction maintains invertibility, if the given E0L grammar G is n-complete.*

Proof: For each $X \in N$ the equivalence class of X with respect to $\equiv$ will be denoted by $\bar{X}$. Define $\bar{G} = (\bar{N}, \Sigma, \bar{P}, \bar{S})$, where

$$\bar{N} = \{\bar{X} : X \in N\},\ \bar{P} = \{\bar{A} \to \bar{\alpha} : A \to \alpha \in P\},$$

and $\bar{\alpha}$ denotes the word over the alphabet $\bar{N} \cup \Sigma$ which is obtained by replacing each nonterminal by its equivalence class with respect to $\equiv$; we say $\bar{\alpha}$ *corresponds* to α. Because

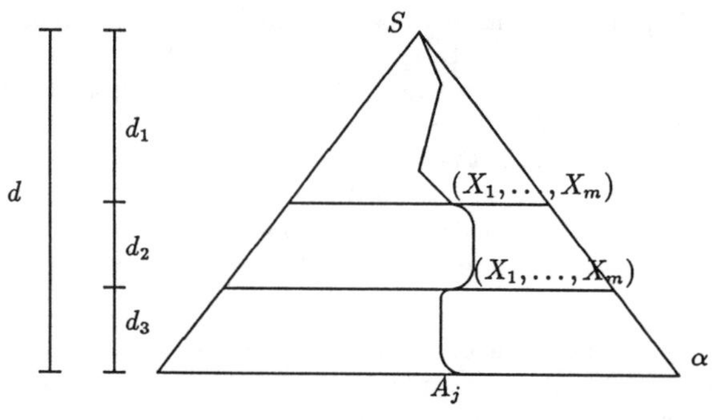

Figure 6: Levels with identical sets of m-tuples

of Lemmas 4.2 and 4.3, $\bar{G}$ can be effectively constructed from G. Note that $\bar{\alpha} = \bar{\beta}$ and $\alpha \neq \beta$ is possible, if α and β contain different but context equivalent nonterminals. In other words, the same word in $\bar{G}$ may correspond to different words in G. However, $\alpha = \beta$ if $\bar{\alpha} = \bar{\beta}$ and both α and β are terminal words.

It is clear that each derivation $S \Rightarrow^* \alpha$ in G can be transformed into a derivation of the same length and with a syntax tree of the same structure in $\bar{G}$. Each derivation step in G becomes a derivation step in $\bar{G}$ by taking corresponding words. This yields a derivation $\bar{S} \Rightarrow^* \bar{\alpha}$ in $\bar{G}$. We will now show that conversely for each derivation in $\bar{G}$ there is a derivation of the same length and structure in G. More precisely, we will show by induction on d, $d \geq 0$, that the following claim holds.

Claim *Let $\bar{S} \Rightarrow^d \bar{\alpha}$ in $\bar{G}$, $\bar{\alpha} \in (\bar{N}^+ \cup \Sigma^+)$, and let γ be an arbitrary word such that $\bar{\gamma} = \bar{\alpha}$. Then, $S \Rightarrow^d \gamma$ in G and both syntax trees are structurally identical.*

Proof of Claim: The proof is by induction on the length of derivations. We consider first a derivation of length 0. Let $\bar{S} \Rightarrow^0 \bar{\alpha}$ in $\bar{G}$ and γ such that $\bar{\gamma} = \bar{\alpha}$. Then, $\bar{\alpha} = \bar{S}$ and $\gamma = A$, where $A \equiv S$. By the definition of $\equiv$ this implies $A = S$. Hence, we have $S \Rightarrow^0 \gamma$ in G.

Consider a derivation $\bar{S} \Rightarrow^d \bar{\beta} \Rightarrow \bar{\alpha}$ of length $d+1$ in $\bar{G}$. We first show that there is a γ that satisfies the claim and, then, in the next paragraph we show that the claim holds for all γ such that $\bar{\gamma} = \bar{\alpha}$. Let $\bar{\beta} = \bar{B}_1 \ldots \bar{B}_k$, for some k and $B_1, \ldots, B_k \in N$. Then, $\bar{\alpha}$ is obtained by parallel rewriting all $\bar{B}_j$ according to productions $\bar{B}_j \rightarrow \bar{a}_j$ in $\bar{P}$; that is $\bar{\alpha} = \bar{\alpha}_1 \ldots \bar{\alpha}_k$, for some $\alpha_1, \ldots, \alpha_k$. By the definition of P, $\bar{B}_j \rightarrow \bar{\alpha}_j$ are productions in $\bar{P}$ only if $B_j \rightarrow \alpha_j$ are productions in P. $\bar{\beta}$ corresponds to $B_1 \ldots B_k$; hence, by the inductive assumption, $B_1 \ldots B_k$ can be derived in G by a derivation of length d with a syntax tree of the same structure as the syntax tree of $\bar{S} \Rightarrow^d \bar{\beta}$ in $\bar{G}$. Rewriting $B_1, \ldots, B_k$ in parallel by $\alpha_1, \ldots, \alpha_k$ yields a word γ such that $\bar{\gamma} = \bar{\alpha}$ and γ can be derived in $(d+1)$ steps in G with a syntax tree which is of the same structure as the syntax tree for $\bar{\alpha}$ in $\bar{G}$. Thus, we know that $\bar{S} \Rightarrow^{d+1} \bar{\alpha}$ in $\bar{G}$ implies that there *exists* a word γ such that $\bar{\gamma} = \bar{\alpha}$ and $S \Rightarrow^{d+1} \gamma$ in G and the two syntax trees have the same structure.

Now assume that γ' is a word which also corresponds to $\bar{\alpha}$; that is, $\bar{\gamma}' = \bar{\alpha}$. We have to show that $S \Rightarrow^{d+1} \gamma'$ in G and the syntax tree is of the same structure as the one in the derivation of $S \Rightarrow^{d+1} \gamma$. We may assume that both γ and γ' are nonterminal words. (For, otherwise, they must be identical). Let $\gamma = X_1 \ldots X_l$; then $\bar{\gamma} = \bar{\gamma}'$ implies $\gamma' = X_1' \ldots X_l'$, where $X_j \equiv X_j'$, for $1 \leq j \leq l$. By the definition of the relation $\equiv$ we infer that with γ also the following sequence of words is derivable in $(d+1)$ steps in G by syntax trees of identical structure:

$$\begin{aligned} \gamma &= X_1 X_2 \ldots X_l, \\ & X_1' X_2 \ldots X_l, \\ & \vdots \\ & X_1' \ldots X_j' X_{j+1} \ldots X_l \\ & \vdots \\ & X_1' \ldots X_l' = \gamma' \end{aligned}$$

This completes the induction and the proof of the claim. □

The claim in particular implies that the same terminal words are derivable in G and $\bar{G}$ with sentential syntax trees of identical structures. Therefore, G and $\bar{G}$ are structurally equivalent.

Next, we show that $\bar{G}$ is context reduced. Assume that $M, M' \subseteq N$ are two nonterminals in $\bar{G}$ and $M \equiv M'$. Choose $X \in M$ and $X' \in M'$. Let α be an arbitrary context for G. $\bar{\alpha}$ corresponds to α and is a context for $\bar{G}$. Moreover $\bar{\alpha}[M]$ corresponds to $\alpha[X]$ and $\bar{\alpha}[M']$ corresponds to $\alpha[X']$. As we have shown above, corresponding words are derivable in G and $\bar{G}$ by derivations of the same length and with syntax trees of the same structure. Therefore, we have, for all d,

$$S \Rightarrow^d \alpha[X] \text{ in } G_{()} \text{ if and only if } \bar{S} \Rightarrow^d \bar{\alpha}[M] \text{ in } \bar{G}_{()},$$

and

$$S \Rightarrow^d \alpha[X'] \text{ in } G_{()} \text{ if and only if } \bar{S} \Rightarrow^d \bar{\alpha}[M'] \text{ in } \bar{G}_{()}.$$

Hence, we can infer from $M \equiv M'$ that $X \equiv X'$. This, in turn, implies that $\bar{X} = M = M' = \bar{X}'$. Thus, we see that context equivalent nonterminals in $\bar{G}$ must be identical; that is, $\bar{G}$ is context reduced.

Finally, we show that the construction maintains invertibility, if the given E0L grammar is n-complete; that is we show that $\bar{G}$ is invertible if G is invertible and n-complete. Let $\bar{X} \to \bar{w}_1$ and $\bar{Y} \to \bar{w}_2$ be two productions in $\bar{G}$ with equal right hand sides; that is $\bar{w}_1 = \bar{w}_2$, and let these productions correspond to two productions $X \to w_1$ and $Y \to w_2$ in G. If w_1 and w_2 are both terminal words, they must be identical and, moreover, $X = Y$ because G is invertible. Hence, we may assume that w_1 and w_2 are nonterminal words. Consider an arbitrary context α such that $S \Rightarrow^d \alpha[X]$ in G. Because G is n-complete we can extend this derivation by one parallel derivation step. We replace X by w_1 and rewrite all other nonterminals by nonterminal words arbitrarily. We obtain some $\beta[w_1]$ such that $S \Rightarrow^d \alpha[X] \Rightarrow \beta[w_1]$ in $G_{()}$. Consider $\gamma_1 = \mathit{frontier}(\beta[w_1])$. Then, we know that $S \Rightarrow^{d+1} \gamma_1$ and $\beta[w_1]$ describes the structure of this syntax tree in G. As we have just shown, γ_1 can be derived in G if and only if $\bar{\gamma}_1$ can be derived in $\bar{G}$ with syntax trees of identical structure if they exist. Replacing w_1 by w_2 in $\beta[w_1]$ gives a word $\beta[w_2]$ in

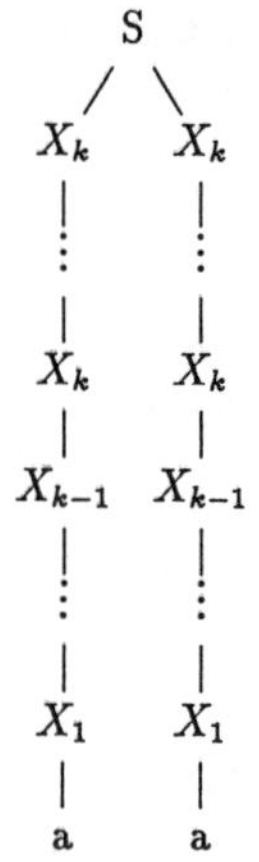

Figure 7: Syntax tree of height $dk + r + 1$ in $G(k)$

$G_{()}$ such that, for $\gamma_2 = frontier(\beta[w_2])$, we have $\bar{\gamma}_2 = \bar{\gamma}_1$, because $\bar{w}_1 = \bar{w}_2$ and all other letters in γ_1 and γ_2 are identical. Hence, γ_1 can be derived in G if and only if γ_2 can be derived in G and both syntax trees have the same structure if they exist. Because G is invertible, we can infer that in $G_{()}$ $S \Rightarrow^d \alpha[Y]$ holds. Thus, we see that for each context α and for each d, $S \Rightarrow^d \alpha[X]$ if and only if $S \Rightarrow^d \alpha[Y]$ in $G_{()}$. This implies $X \equiv Y$ and, therefore, $\bar{X} = \bar{Y}$. This shows that there cannot be two different productions in $\bar{G}$ with equal right hand sides. □

Remark: The proof of Theorem 4.4 can easily be carried over to the case of E0L grammars with more than one sentence symbol. It is in this form that the theorem is used in Ottmann and Wood [2].

Eliminating context equivalent nonterminals by reducing a grammar according to the relation $\equiv$ does *not* uniquely determine an E0L grammar up to isomorphism. There exist structurally equivalent invertible context reduced E0L grammars which are not isomorphic. Consider the grammar G_6 of Examples 3.1 given by the productions

$$S \rightarrow AA|BB, A \rightarrow B|a, B \rightarrow A.$$

Clearly, $A \not\equiv S$ and $B \not\equiv S$. Consider the context $\alpha = L\_AR$; then, we have $S \Rightarrow^1 \alpha[A]$ in $G_{6,()}$ but $S \Rightarrow^1 \alpha[B]$ does *not* hold; hence, $A \not\equiv B$. This shows that G_6 is context reduced. The structurally equivalent grammar G_5 of Example 2.5 given by the productions

$$S \rightarrow AA, \ A \rightarrow A|a$$

is also context reduced but not isomorphic to G_6. We can even have structurally equivalent invertible, context reduced grammars with the same number of nonterminals. Consider the E0L grammar G_7 of Example 3.2. It is easy to see that G_7 is context reduced. As

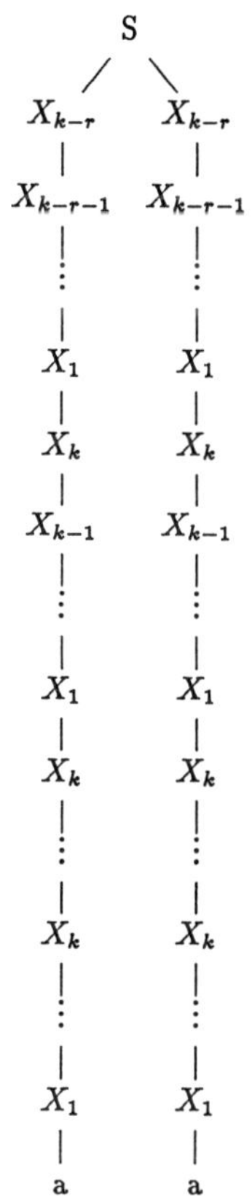

Figure 8: Syntax tree of height $dk + r + 1$ in $G'(k)$

we already know, G_7 is invertible and structurally equivalent to G_6. This example can be generalized to provide an infinite sequence of grammars. Consider the E0L grammar $G(k)$ with $k+1$ nonterminals given by the productions

$$S \to X_1X_1|X_2X_2|\dots|X_kX_k,$$

$$X_k \to X_{k-1},\dots,X_2 \to X_1,$$

$$(*)X_k \to X_k,$$

$$X_1 \to a.$$

Now, $G(k)$ is structurally equivalent to the grammar $G'(k)$, which has the same productions except for the production marked by (*). Instead of this production $G'(k)$ contains $X_1 \to X_k$. It is easy to check that both grammars are invertible, context reduced, structurally equivalent but not isomorphic.

The syntax trees of height $dk+r+1$ in $G(k)$, $d \geq 1$, $0 \leq r < k$, are displayed in Figure 7. Figure 8 shows the syntax trees of the same height $dk+r+1$ in $G'(k)$.

5 Conclusions

We have demonstrated that structural equivalence of E0L grammars is reducible to isomorphism in some special cases. However, we have been unable to obtain a general isomorphism result in contradistinction to the case of context-free grammars. We have imposed additional properties on E0L grammars that preserve structural equivalence in order to achieve a new normal form. In the case of context-free grammars this has the effect that the resulting grammar is uniquely determined up to isomorphism, but this does not hold for E0L grammars. Thus, the basic open question resulting from our investigation is: Is there an effectively decidable equivalence relation between the nonterminals of an E0L grammar such that reducing the grammar according to this relation preserves structural equivalence and yields a grammar which is uniquely determined up to isomorphism?

References

[1] R. McNaughton. Parenthesis grammars. *Journal of the ACM*, 14:490–500, 1967.

[2] Th. Ottmann and D. Wood. Defining families of trees with E0L grammars. *Discrete Applied Mathematics*, to appear.

[3] G. Rozenberg and A. Salomaa. *The Mathematical Theory of L Systems.* Academic Press, New York, 1980.

[4] K. Salomaa and S. Yu. Structural equivalence of E0L grammars is decidable. *Theoretical Computer Science*, to appear.

[5] A. Salomaa. *Formal Languages.* Academic Press, New York, 1973.

[6] D. Wood. *Theory of Computation.* John Wiley & Sons, New York, 1987.

The Complexity of the Emptiness Problem for EOL Systems

Klaus-Jörn Lange[1] and Michael Schudy[2]

[1]Institut für Informatik, Technische Universität München, Arcisstr. 21, D-8000 München 2
[2]Deutsche Airbus GmbH, Kreetslag 10, D-2103 Hamburg 95

Abstract: The emptiness problem for EOL systems is shown to be computationally equivalent to the word problem of RPAC automata which are augmented with a LOGSPACE working tape and a two-way input head.

Keywords: L systems, complexity theory, formal languages

1 Introduction

Given a device generating or accepting a formal language, it is usual in formal language theory to consider the problems of membership, equivalence, and emptiness. While the former two appear to be naturally motivated, the latter one seems to be quite artificial at first glance. This impression changes when looking at the emptiness problem from the view of complexity theory, where it provides an abundance of completeness results. Thus, the emptiness problem is **NL**-complete for right-linear grammars, **P**-complete for context-free grammars, **NP**-complete for EDOL systems, and **PSPACE**-complete for ETOL systems.

Most results of this type are special cases of a general relationship between the emptiness problem of one-way automata and the membership problem of the corresponding two-way automaton (see [2].) In this paper a corresponding result is shown for EOL systems and a corresponding automata model of Rozenberg, the *RPAC* automaton [7]. In this case, the techniques of [2] do not apply. For one thing, the *push-down array of counters* (PAC) is not an abstract storage type in the sense of Engelfriet [3] (, unless we allow 'nondeterministic' functions to model the push-down instruction). In addition, there seems to be no possibility to describe a storage type *restricted PAC*. Finally, the family of RPAC automata apparently does not allow for the normal forms used in [2]. Hence, techniques more specific for L systems have to be used.

The paper is organized as follows. The next section contains a review of RPAC automata and their equivalence to EOL systems. An extended version of a RPAC automaton is introduced and a certain normal form is derived. It will be indicated why these relations also hold with respect to LOGSPACE constructability. The subsequent section will introduce the notion of an *NARPAC*($\log n$) automaton and the equivalence of its word problem to the emptiness problem of EOL systems. Finally, a discussion concludes this work.

The constructions contained in this paper are carried out in full detail in [10]. A preliminary abstract appeared as [6].

2 Automata recognizing EOL languages

We assume the reader to be familiar with some standard notation of formal language theory and of complexity theory, see e.g. [4]. In the same way we assume familarity with OL languages as contained in [8]. In particular, we will consider EOL systems. An EOL system $G = (V, T, \sigma, S)$ consists of a finite set V of *nonterminal symbols*, a set $T \subseteq V$ of *terminal symbols*, an *axiom* $S \in V$ and a finite non-empty substitution $\sigma : V^* \to 2^{V^*}$. A pair $B \to v$, $B \in V$, $v \in \sigma(B)$, is called a *production*. Considering EOL systems as generalisations of context-free grammars, extensions of push-down automata have been developed which recognize exactly the EOL language. Depending in whether we simulate leftmost or rightmost derivations, two different types of automata have been introduced: Following the top-down approach, van Leeuwen ([12]) developed the *(LFR-)UCS-PD* automaton. In this paper we will deal with the corresponding bottom-up model, the *(R)PAC automaton* invented by Rozenberg in [7].

A *PAC automaton*[1] is a hexatuple $A = (T, \Gamma, Q, q_0, F, \delta)$ consisting of a finite set T of *input symbols*, a finite set $\Gamma \supseteq T$ of *push-down symbols*, a finite set Q of *states*, a distinguished *central state* $q_0 \in Q$, a subset $F \in Q$ of *final* (or accepting) states, and a *transition relation* $\delta \subseteq (Q \times \Gamma \times \Gamma \times \{q_0\}) \cup (Q \times \Gamma \times \{\lambda\} \times (Q - \{q_0\})) \cup (\{q_0\} \times \lambda \times \Gamma \times \{q_0\})$. Here, λ denotes the empty word.

A PAC automaton may be thought of as a push-down automaton equipped with counters at each cell of the push-down tape, which guarantee that all leaves of the simulated derivation tree have the same height, i.e., that we simulate an EOL derivation. But this kind of automaton has to be severely restricted in order to recognize nothing but context-free L languages. The main requirement is to distinguish a central state q_0, which is entered after each step filling the push-down store. A pop step is only allowed if the counters attached with the two topmost push-down cells contain the same value. In addition, a pop step always enters a noncentral state. Finally, it is possible to modify the topmost push-down element, but these *overwrite* steps increment the counter of the topmost cell by one.

A *configuration* of A is a triple $(w, \gamma, q) \in \mathbf{K} := T^* \times (\Gamma \times \mathbb{N}_0)^* \times Q$. Here w is (a rest of) the input word still to be read, γ the current content of PAC-store, and q the state entered by A. If $s_1 = (a_1 a_2 \dots a_n, \langle z_1, t_1 \rangle \langle z_2, t_2 \rangle \dots \langle z_m, t_m \rangle, q) \in \mathbf{K}$ we say that $s_1 \vdash_A s_2$ (s_1 reaches s_2) if one of the following conditions is matched:

- s_1 reaches $s_2 = (a_1 a_2 \dots a_m, \langle z_1, t_1 \rangle \dots \langle z_{m-1}, t_{m-1} \rangle \langle z, t_m + 1 \rangle, q_0)$ by an *overwrite step* if $m \geq 1$ and $(q, z_m, z, q_0) \in \delta$,

- s_1 reaches $s_2 = (a_1 a_2 \dots a_m, \lambda, q')$ by a *clear step* if $m = 1$ and $(q, z_1, \lambda, q') \in \delta$,

---

- s_1 reaches $s_2 = (a_1a_2\dots a_m, \langle z_1,t_1\rangle\dots\langle z_{m-1},t_{m-1}\rangle, q')$ by a *pop-up step* if $m \geq 2$, $t_m = t_{m-1}$, and $(q, z_m, \lambda, q') \in \delta$,

- s_1 reaches $s_2 = (a_1a_2\dots a_m, \langle z_1,t_1\rangle\dots\langle z_m,t_m\rangle\langle z,t\rangle, q_0)$ by a *push-down step* if $q = q_0$, $t \in \mathbb{N}$, and $(q_0, \lambda, z, q_0) \in \delta$ or

- s_1 reaches $s_2 = (a_2\dots a_m, \langle z_1,t_1\rangle\dots\langle z_m,t_m\rangle\langle a_1,0\rangle, q_0)$ by a *read step* if $m \geq 1$ and $q = q_0$.

Let $\overset{*}{\vdash}_A$ be the transitive and reflexive closure of $\vdash_A$. The *language accepted by A* is the set $\{w \in T^* \mid (w, \lambda, q_0) \overset{*}{\vdash}_A (\lambda, \lambda, q) \text{ for some } q \in F\}$.

The push-down instruction is used to simulate λ-productions within a derivation: Since λ is rewritten by λ, the bottom-up simulation of a λ-production $N \to \lambda$ has to locate nondeterministically the left-hand side N at a position of arbitrary height within the derivation tree. That is why the topmost element of configuration s_2 reached by a push-down step gets an arbitrary counter value $t \in \mathbb{N}$. This correct notion of a push-down step was first introduced by Rozenberg and Vermeir in [9]. It is argued in [9] that this operation could be simulated by a modified push-down step which uses $t = 1$ followed by a sequence of overwrites in order to increase the value t. But one has to be careful in using this construction, since it is possible to show that the question '$\lambda \in L(A)$' is NP-hard for PAC automata using a push-down instruction which pushes symbols with the counter value $t = 1$, only. On the other hand, this question '$\lambda \in L(A)$' is solvable in polynomial time for both EOL systems and PAC automata with that version of the push-down instruction used here.

The family of EOL languages is closed under nonerasing inverse homomorphisms, but not under arbitrary inverse homomorphisms. Thus they do not form an AFL. The least AFL containing the EOL languages is the family **RMOL** (see [1]).

Theorem 2.1 [[9]] The family of languages accepted by PAC automata coincides with **RMOL**.

A PAC automaton A is said to be *k-restricted* if in any computation of A, i.e. sequence of configurations related by $\vdash_A$, of length k, the central state is entered at least once. (Thus a k-restricted PAC automaton may pop at most $k-1$ symbols before returning to the central state). A PAC automaton is *restricted* if it is k-restricted by some $k \in \mathbb{N}$. We will use the terms of R_kPAC and RPAC languages. When simulating derivations of EOL systems the number k corresponds to the length of the longest right-hand side of any production.

Theorem 2.2 [[7]] A language is acceptable by an RPAC automaton if and only if it is an EOL language.

Since any EOL language is generatable by an EOL_2 system, i.e. an EOL system where the length of each right-hand side is bounded by two, we get the coincidence of the families of RPAC and R_2PAC languages, using the constructions of [7].

properties: i.e. if $n = |\langle G\rangle|$ is the length of the coding $\langle G\rangle$ of an EOL system $G = (V, T, \sigma, S)$, then the cardinality of V is bounded by n. In addition, the length $|\langle\langle N\rangle\rangle|$ of a coding $\langle\langle N\rangle\rangle$ is bounded by $\log(n)$ for each $N \in V$. Similar remarks hold for the coding $\langle A\rangle$ of a PAC automaton $A = (T, \Gamma, Q, q_0, F, \delta)$. In particular, if A is restricted, then it is n-restricted where $n = |\langle A\rangle|$.

The proof of Theorem 2.2 in [7] is constructive, but the construction of an EOL system G equivalent to a given RPAC automaton A is not computable in polynomial time since there exist automata, such that the constructed grammars have exponential size compared to the size of the automata. Hence we will give a modified construction which will be LOGSPACE computable. For lack of space we can only indicate the essential ideas. The construction is carried out in full detail in [10].

The main idea of the construction in [7] of an EOL system to a given RPAC automaton $A = (T, \Gamma, Q, q_0, F, \delta)$ is to regard the elements of Γ as nonterminals and to introduce productions $N \rightarrow N_1 N_2 \ldots N_k$ for each sequence of push-down symbols which can be popped up in this order according to δ. Here N is the symbol resulting from the subsequent overwrite step. Although k is bounded by $n := |\langle A\rangle|$ since A is restricted, we could get in this way up to $c^{O(n)}$ productions. Here c is the cardinality of Γ. Therefore the first step will be to construct within LOGSPACE to each restricted PAC automaton an equivalent 2-restricted one. But still the subsequent construction has to be modified in order to reduce the complexity. This will be done here for "REPAC" automata which might be regarded as RPAC automata enhanced with the possibility of having several central states and of performing 'change state' steps which simply change the state without affecting the push-down store or the input.

A *EPAC* automaton is a heptuple $A = (T, \Gamma, Q, Q_0, F, q_0, \delta)$, where T, Γ, Q, and F have the meaning as in the definition of a PAC automaton above. In addition, $Q_0 \subseteq Q$ is the set of *central states*, $q_0 \in Q_0$ is the *starting state*, and the *transition relation* δ is now a hexatuple $\delta = (\delta_{CS}, \delta_{OW}, \delta_{CL}, \delta_{PU}, \delta_{PD}, \delta_{RD})$, where

$$\begin{aligned}
\delta_{CS} &\subseteq (Q_0 \times (\Gamma \cup \{\lambda\}) \times Q_0) \cup ((Q \setminus Q_0) \times (\Gamma \cup \{\lambda\}) \times (Q \setminus Q_0)),\\
\delta_{OW} &\subseteq Q \times \Gamma \times \Gamma \times Q_0,\\
\delta_{CL} \cup \delta_{PU} &\subseteq Q \times \Gamma \times (Q \setminus Q_0),\\
\delta_{PD} &\subseteq Q_0 \times \Gamma \times Q_0, \text{ and}\\
\delta_{RD} &\subseteq Q_0 \times T \times Q_0.
\end{aligned}$$

While the notion of a 'configuration' remains unchanged, we have to extend the step relation $\vdash_A$ in the following way. A configuration $s_1 = (w := a_1 a_2 \ldots a_n,\ \gamma := \langle z_1, t_1\rangle \ldots \langle z_m, t_m\rangle, q)$ is said to reach a configuration s_2 in one step, abbreviated by $s_1 \vdash_A s_2$, if one of the following conditions is satisfied:

- s_1 reaches $s_2 = (w, \gamma, q')$ by a *change state step* if $(q, z_m, q') \in \delta_{CS}$ and $m \geq 1$ or if $(q, \lambda, q') \in \delta_{CS}$ and $m = 0$,

- s_1 reaches $s_2 = (w, \langle z_1, t_1\rangle \ldots \langle z_{m-1}, t_{m-1}\rangle\langle z, t_m{+}1\rangle, q')$ by an *overwrite step* if $m \geq 1$ and $(q, z_m, z, q') \in \delta_{OW}$,

- s_1 reaches $s_2 = (w, \langle z_1, t_1\rangle \dots \langle z_{m-1}, t_{m-1}\rangle, q')$ by a *pop-up step* if $m \geq 2$, $t_m = t_{m-1}$, and $(q, z_m, q') \in \delta_{PU}$,
- s_1 reaches $s_2 = (w, \gamma\langle z, t\rangle, q')$ by a *push-down step* if $t \in \mathbb{N}$ and $(q, z, q') \in \delta_{PD}$, or
- s_1 reaches $s_2 = (a_1 a_2 \dots a_{n-1}, \gamma\langle a_n, 0\rangle, q')$ by a *read step* if $n \geq 1$ and $(q, a_m, q') \in \delta_{RD}$.

Again we let $\overset{*}{\vdash}_A$ denote the transitive and reflexive closure of $\vdash_A$ and set

$$L(A) = \{w \in T^* \mid (w, \lambda, q_0) \overset{*}{\vdash}_A (\lambda, \lambda, q) \text{ for some } q \in F\}.$$

It should be remarked here that change state steps do not allow to switch between and noncentral states. As for PAC automata we introduce restricted and k-restricted EPAC automata, denoted as REPAC and R_kEPAC automata.

Theorem 2.3 There is a LOGSPACE computable function f which maps encodings of REPAC automata to encodings of R_2EPAC automata such that $L(A) = L(A')$ if $\langle A'\rangle = f(\langle A\rangle)$.

Idea of proof: Main idea is to replace a sequence of k pop-up steps and one overwrite by a sequence of k pairs of one pop-up followed by one overwrite which works if we reorganize the structure of counter values. E.g. consider a sequence of configurations $(w, \langle B,t\rangle\langle C,t\rangle\langle D,t\rangle, q) \vdash (\text{POP}-\text{UP})\,(w, \langle B,t\rangle\langle C,t\rangle, q_1) \vdash (\text{POP}-\text{UP})\,(w, \langle B,t\rangle, q_2) \vdash (\text{OVERWRITE})\,(w, \langle N, t+1\rangle, q_0)$, which means something like $N \to BCD$ seen from the grammar side. This computation would be simulated by $(w, \langle B, t\cdot k + (k-1)\rangle\langle C, t\cdot k + (k-2)\rangle\langle D, t\cdot k + (k-2)\rangle, q) \vdash (\text{POP}-\text{UP})\,(w, \langle B, t\cdot k+(k-1)\rangle\langle C, t\cdot k + (k-2)\rangle, q_1) \vdash (\text{OVERWRITE})\,(w, \langle B, t\cdot k + (k-1)\rangle\langle C', t\cdot k + (k-1), q_0^1) \vdash (\text{POP}-\text{UP})\,(w, \langle B, t\cdot k + (k-1)\rangle, q_2) \vdash (\text{OVERWRITE})\,(w, \langle N, (t+1)\cdot k, q_0)$.

To do so, the push-down symbols of the R_2EPAC automaton to be constructed are 'levelled', i.e. supplied with an index i, $0 \leq i < k$, where k is the bound of the restriction of the EPAC. Since k is input dependent we use $k = |\langle A\rangle|$, where A is the given REPAC automaton. Now all pop-up sequences of A have to be 'virtually' extended to have maximal length $k-1$. In addition, the level of the symbol N resulting from sequence is nondeterministically put into some level between 1 and $k-1$, if the preceeding sequence of pop-up and overwrite pairs ends correctly on level $k-1$. This works as long as not all symbols occurring in a pop-up sequence are generated by push-down steps. But if this should be the case, we may treat N as like a result of a push-down step and no wrong word is accepted by the constructed R_2EPAC automaton. □

We will now construct within LOGSPACE to each R_2EPAC automaton an equivalent EOL system. If we look at the proofs in [7] we see that the construction of a RPAC automaton to a given EOL system works in LOGSPACE. But, as mentioned above, the EOL grammar constructed to a given RPAC automaton even does not have polynomial size. If, however, the considered automata are restricted by some fixed constant, say, if they are all 2-restricted, then this construction can be done in LOGSPACE. In the

Theorem 2.4 There is a LOGSPACE computable function g which maps encodings of R_2EPAC automata to encodings of EOL systems such that $L(A) = L(G)$ if $\langle G\rangle = g(\langle A\rangle)$.

Idea of proof: There are two points in which EPAC and a PAC automata differ. First, the extended PAC automaton has several central states, and second it can perform 'changestate'-moves which do not affect the push-down store, in particular, do not increment the value of its topmost element. To cope with the first difference it is enough to use a bottom up analogue of the well-known triple construction as it is known from context-free grammars. The second one needs a little bit more work. A first idea would be to relate in the right-hand sides of the productions neighbouring nonterminals $[q', C, r][p, B, q]$ by the condition that state q' can be reached from state q with B as topmost element by pure changestate moves. Unfortunately, this kind of construction would need to solve some reachability problem. Since this is NL-complete, we can only expect to do this in polynomial time, but not in logarithmic space. The solution is to add nonterminals $[B, q, q', i]$ which derive λ in a subderivation of height i (or otherwise blocks) if state q' is reachable from state q with B as the topmost element in exactly i steps. That is, we integrate this reachability problem within the productions of the constructed grammar instead of solving it. Since i is bounded by n, the length of the given encoding of the R_2EPAC automaton, this leads to a construction similar to the level technique used in Theorem 2.3. That is, every overwrite/pop-up-cycle of the given automaton is simulated by a subderivation of height n. □

It could be remarked that the constructed grammar always contains λ-production, regardless whether or not the simulated automaton contains push-down instructions. If we want to keep this equivalence between EPOL systems and push-down free REPAC automata, we would have to use polynomial time instead of logarithmic space reducibilities.

3 EOL emptiness and RPAC automata with two-way input

It is well-known that both the emptiness problem for context-free grammars and the membership problem for push-down automata with two-way input are **P**-complete and thus computationally equivalent with respect to LOGSPACE reducibilities. We are now going to show that this equivalence holds true, if we replace context-free grammars by EOL systems and push-down automata by the corresponding automata for EOL languages, the RPAC automata.

In the following let $\emptyset \neq$ **EOL** $:= \{\langle G\rangle \mid G$ is an EOL system and $L(G) \neq \emptyset\}$ be the encoding of the emptiness problem for EOL systems. Analoguously, we build sets like $\emptyset \neq$ **PAC** or $\emptyset \neq$ **R_2EPAC** . By the results of the previous section we know that $\emptyset \neq$ **EOL** , $\emptyset \neq$ **RPAC** , $\emptyset \neq$ **R_2PAC** , $\emptyset \neq$ **REPAC** , and $\emptyset \neq$ **R_2EPAC** all are computationally equivalent. We will now characterize their common complexity by NARPAC($\log n$) automata, that is RPAC automata with two-way input which are augmented by logarithmically space bounded working tapes. We will not give an explicit

the input tape and from the working tape. In particular, it is possible to get information from the input tape without pushing a symbol onto the PAC store. In this way we derive NAPAC($\log n$), NARPAC($\log n$), and $\mathrm{NAR_2PAC}(\log n)$ automata. The class of languages accepted by these automata are denoted by **NAPAC**($\log \boldsymbol{n}$), **NARPAC**($\log \boldsymbol{n}$), and $\mathbf{NAR_2PAC}(\log \boldsymbol{n})$. We then have $\mathbf{P} = \mathbf{NAPDA}(\log \boldsymbol{n}) \subseteq \mathbf{NARPAC}(\log \boldsymbol{n}))$. Obviously, these classes are closed under LOG reducibilities.

The following subsections will characterize the complexity of $\emptyset \neq$ **EOL** in terms of **NARPAC**($\log \boldsymbol{n}$)-completeness.

3.1 Hardness

Theorem 3.1 $\mathbf{NARPAC}(\log \boldsymbol{n}) \leq_{log} \emptyset \neq \mathbf{REPAC}$

Idea of proof: Let $L \subseteq T^*$ be accepted by an NARPAC($\log n$) automaton A. For each $w \in T^*$ we will now construct a REPAC automaton A_w fulfilling $w \in L(A) \Leftrightarrow L(A_w) \neq \emptyset$.

This is done in a way very similar to the context-free/push-down-case. The input tape containing w and the working tape are simulated by the finite control of A_w. But there is one essential difference: all steps of A which fill the stack, i.e., all push-down instructions and all read instructions, are simulated by read instructions of A_w. For push-down steps of A this is done by reading the symbol which is to be pushed, from the input tape of A_w. That means that A_w reads something like a 'stack protocol' of a successful computation of A on w. Obviously, a k-restriction of A implies a k-restriction of A_w. Thus, we have $w \in L \Leftrightarrow \langle A_w \rangle \in \emptyset \neq$ **REPAC** . □

Since $\emptyset \neq \mathbf{REPAC} \leq_{log} \emptyset \neq \mathbf{EOL}$ and the latter one is a subproblem of $\emptyset \neq$ **ETOL** which is **PSPACE**-complete we have

Corollary 3.2 $\mathbf{NARPAC}(\log \boldsymbol{n}) \subseteq \mathbf{PSPACE}$.

3.2 Completeness

In order to show the **NARPAC**($\log \boldsymbol{n}$)-completeness of $\emptyset \neq$ **EOL** we have to solve $\emptyset \neq$ **EOL** or some computationally equivalent program by an NARPAC($\log n$)-automaton.

Theorem 3.3 $\emptyset \neq \mathbf{EOL} \in \mathbf{NARPAC}(\log \boldsymbol{n})$.

Idea of Proof: A first step is to transform EOL systems into a normalform, where the lengths of all right-hand sides are bounded by two. This is possible to do within LOGSPACE. If we would now handle systems over some fixed alphabet V, only, we could solve this "$\boldsymbol{V}^* - \emptyset \neq \mathrm{EOL}_2$" problem on an $\mathrm{NAR_2PAC}$ automaton by simulating a successful derivation of the input grammar in bottom-up mode: Guess productions, find their right-hand sides on top of the PAC store, and *overwrite* them by the corresponding left-hand sides. Here it is important to have a two-way input head! Problems arise if we handle arbitrary alphabets which are coded over some fixed alphabet. In this case an original right-hand side of length k is encoded into $O(\log |\langle G \rangle| \cdot k)$ symbols. Thus, if

constant. The solution to this problem is again to use the construction that worked in theorems 2.3 and 2.4. A sequence of m pop-up moves is replaced by a sequence of m pairs of pop-up and overwrite instructions. Again this requires a careful handling of the counter-management within the PAC store. In particular, after finishing a pop-up/overwrite-sequence of a right-hand side of a production we have nondeterministically to guess whether the corresponding left-hand side will be the first or second symbol of the right-hand side of some production. Details may be found again in [10]. □

Corollary 3.4 $\emptyset \neq$ **EOL** is **NARPAC**($\log \boldsymbol{n}$)-complete.

Since $\emptyset \neq$ **EOL** is located between $\emptyset \neq$ **EDOL** and $\emptyset \neq$ **ETOL** which are **NP**-resp. **PSPACE**-complete. We have

Corollary 3.5 **NP** $\subseteq$ **NARPAC**($\log \boldsymbol{n}$) $\subseteq$ **PSPACE**.

The above constructions also show that the *general membership problem* **GP(EOL)** := $\{\langle G, n\rangle \mid w \in L(G), G$ is an EOL system $\}$ is **NARPAC**($\log \boldsymbol{n}$)-complete corresponding to the **P**-completeness of **GM(CF)**.

In the context-free case the construction has the property $L(A_w) \neq \emptyset \Leftrightarrow L(A_w) = \{\lambda\}$, i.e. it even shows the **P**-completeness of the question whether or not the empty word is generated by a given grammar. This equivalence $L(A_w) \neq \emptyset \Leftrightarrow L(A_w) = \{\lambda\}$ no longer holds for the REPAC automaton constructed in Theorem 3.3. And in fact the problem '$\lambda \in L(G)$' can be solved in polynomial time for a given EOL system or a given REPAC automaton. This problem is closely related to the choice of an appropriate shape of the push-down instruction as discussed in the previous section. In this connection it could be remarked, that the REPAC automaton constructed in theorem 3.3 is push-down free.

4 Discussion and other questions

If we look at the corresponding **NP**-completeness result of $\emptyset \neq$ **EDOL** in [5] we see that the EDOL system which is constructed to a given formula and which generates an nonempty language exactly if F is satisfiable, has the property that every right-hand side of a production has length one. If we denote systems of this kind as EOL_1 and EDOL_1 systems, we see $\emptyset \neq$ **EDOL** $\in$ **NP** $\leq_{log}$ $\emptyset \neq$ $\textbf{EDOL}_1$.

Thus we know that there is a (LOGSPACE computable) transformation from EDOL to EDOL_1 systems which preserves (non)emptiness. If we would have a corresponding result for EOL systems we would have **NARPAC**($\log \boldsymbol{n}$) = **NP** since $\emptyset \neq$ $\textbf{EOL}_1$ is in **NP**.

We close this section with a collection of open questions.

- Is it possible to use within our results reducibility notions weaker than LOG reducibilities? In particular, is it possible to use NC^1 or AC^0 reductions?
- A further relationship often encountered between one-way and two-way automata

time restriction. E.g. Sudborough showed in [11] **LOGCFL** = $\mathbf{NAPDA}_{pt}(logn)$ and **LOGDCFL** = $\mathbf{DAPDA}_{pt}(logn)$. Question: Does correspondingly $\mathbf{NARPAC}_{pt}(logn)$ = LOG(**EOL**) = **LOGCFL** hold?

- Is it possible to derive similar results for the corresponding top-down automaton of van Leeuwen (see [12])?

References

[1] K. II Culik and J Opatrny. Macro 0l systems. *Internat. J. Comput. Math.*, 4a:327–342, 1975.

[2] J. Dassow and K.-J. Lange. Complexity of automata with abstract storages. Preprint 2/91, Technische Universität Magdeburg, Jan. 1991. accepted at FCT91.

[3] J. Engelfriet. Context-free grammars with storages. Techn. Report 86-11, University of Leiden, Department of Computer Science, 1986.

[4] J. Hopcroft and J. Ullman. *Introduction to Automata Theory, Language, and Computation.* Addison-Wesley, Reading Mass., 1979.

[5] N. Jones and S. Skyum. Complexity of some problems concerning L systems. *Math. Systems Theory*, 13:29–43, 1979.

[6] K.-J. Lange and M. Schudy. A further link between formal languages and complexity theory. *EATCS Bull.*, 33:67–71, 1987.

[7] G. Rozenberg. On a family of acceptors for some classes of developmental languages. *Internat. J. Comput. Math.*, 4a:199–228, 1974.

[8] G. Rozenberg and A. Salomaa. *The Mathematical Theory of L Systems.* Academic Press, New York, 1980.

[9] G. Rozenberg and D. Vermeir. On acceptors of iteration languages. *Internat. J. Comput. Math.*, 7a:3–19, 1979.

[10] M. Schudy. Augmentierte Zählerkellerautomaten, April 1990. In German.

[11] I. Sudborough. On the tape complexity of deterministic context-free languages. *J. Assoc. Comp. Mach.*, 25:405–414, 1978.

[12] J. van Leeuwen. Notes on pre-set pushdown automata. In G. Rozenberg and A. Salomaa, editors, *L Systems*, pages 177–188, 1974.

A0L- and CFG- Size of Languages

Alica Kelemenová[1], Miriam Removčíková[2]

[1]Department of Informatics, Computer Center Slovak Academy of Sciences,
Dúbravská cesta 9, CS 842 35 Bratislava, CSFR
[2]Martinčekova 30, 821 01 Bratislava, CSFR

Abstract: For a given language the size of minimal underlying A0L systems and the size of minimal underlying CF grammars are compared.

Keywords: A0L system, CF grammar, number of auxiliary symbols, rules and length of grammar / system.

1.Introduction

Descriptional (syntactical) complexity of languages belongs to the classic of formal language theory. (See [**1**], [**2**] for the first overview papers on that topic). It deals with the effective description of languages, where effectivity is characterized by some basic parameters of devices describing languages, e.g. by parameters of grammars, machines or expressions.

In the present paper we shall study adult languages of 0L systems (A0L systems for short). Their introduction were motivated biologically. There is an interesting result in [**3**] that adult languages of 0L systems and CF languages are identical.

One basic part of descriptional complexity theory deals with comparison of effective language descriptions by different descriptive objects. Naturally, equivalent adult 0L systems and CF grammars can differ in size. The aim of this paper is to discuss that distinction in detail. For languages we express precisely relations between minimal size of CF grammars and minimal size of A0L systems, where the size of grammars and systems are characterized by the number of auxiliary symbols, by the number of rules and by total length of rules and starting nonterminal / axiom.

In Section 2 we summarize basic definitions, notations and preliminary results. In Section 3 we state that number of auxiliary symbols equals in minimal A0L systems and in minimal CF grammars for given language, while difference between the number of productions in minimal A0L system and the number of productions in minimal CF grammar equals to the number of letters in the alphabet of language. In Section 4 quadratic dependence between the length of minimal CF grammars and length of minimal A0L system is presented.

2.Preliminaries

We assume that the reader is familiar with introduction to formal language and L system theory [**5**]. We fix notations of basic notions as follows.

We shall use $G = (N, T, P, S)$ for a CF grammar, where N is a finite set of nonterminals, T is a finite set of terminals, P is a set of productions of form $A \to \alpha$, where A is in N and α is in $(N \cup T)^*$, S in N is a starting symbol. The set V being $N \cup T$ is the alphabet of G.

Derivation in CF grammar proceeds sequentially and we use $\Rightarrow_G$ for derivation step in G, $\Rightarrow_G^*$ for derivation in G and $L(G)$ to denote a language generated by G (i.e.$L(G) = \{w \in T^* : S \Rightarrow_G^* w\}$). $Alph(L)$ is the set of all letters occurring in words of L.

In contrast with CF grammars, L systems are generating devices with parallel mode of derivation and with single (nondivided) alphabet.

DEFINITION 2.1. *An 0L system is a 3 tuple $H = (\Sigma, R, w_0)$, where Σ is a finite alphabet, R is a finite set of productions of the form $a \to \alpha$, where a is in Σ and α is in Σ^*. A derivation step in 0L system $H = (\Sigma, R, w_0)$ is a binary relation $\Rightarrow_H$ on Σ^*. We write $x_1 \ldots x_n \Rightarrow_H \alpha_1 \ldots \alpha_n$ iff $x_i \to \alpha_i$ is in R for $i = 1, \ldots, n$. By $\Rightarrow_H^*$ we denote the reflexive and transitive closure of $\Rightarrow_H$.*

DEFINITION 2.2. *An adult language of 0L system $H = (\Sigma, R, w_0)$ is the language $L_A(H) = \{w : w_0 \Rightarrow_H^* w$ and if $w \Rightarrow_H w'$ then $w' = w\}$. The adult alphabet Σ_A of H is defined by $\Sigma_A = alph(L_A(H))$, i.e. it consists exactly of the letters of $L_A(H)$.*

We shall use A0L system for 0L system with adult mode of language determination. The class of languages determined by CF grammars and by A0L systems will be denoted by $\mathcal{L}(CF)$ and $\mathcal{L}(A0L)$, respectively.

THEOREM 2.1. [5] $\mathcal{L}(A0L) = \mathcal{L}(CF)$.

We fix $\mathcal{L}$ to denote $\mathcal{L}(CF) = \mathcal{L}(A0L)$.

Size of grammar/system is indoubtedly one of basic complexity characteristics of grammar/system. To discuss size of grammars/systems we shall use number of its auxiliary symbols Var, number of its rules $Prod$ and total length of grammars/systems $Symb$.

DEFINITION 2.3. *Let $G = (N, T, P, S)$ be a CF grammar and $H = (\Sigma, R, w_0)$ be an A0L system.*

Then

$$Var\ G = |N| \qquad Var\ H = |\Sigma - \Sigma_A|$$

$$Prod\ G = |P| \qquad Prod\ H = |R|$$

$$Symb\ G = 1 + \sum_{\substack{A \to \alpha \\ in\ P}} (2 + |\alpha|) \qquad Symb\ H = |w_0| + \sum_{\substack{a \to \alpha \\ in\ R}} (2 + |\alpha|)$$

For K being $Var, Prod$ or $Symb$ we put

$K_{CF}(L) = min\{K(G) : L(G) = L$ and G is a CF grammar$\}$

$K_A(L) = min\{K(H) : L_A(H) = L$ and L is an A0L system$\}$

Note 2.1. In our definition of *Symb* we take into the consideration also the lenght of initial symbol and axiom. This was not necessary in previous papers dealing with CF grammars only, where $|S| = 1$.

For complexity measure K a CF grammar G is a K-minimal grammar of $L(G)$ if $K(G) = K_{CF}(L(G))$ and analogically an A0L system H is K-minimal if $K(H) = K_A(L(H))$.

3. Note on variables and productions

For technical reasons it will be useful to discuss A0L systems with initial nonadult symbol S, i.e. systems with $w_0 = S$ and $S \notin \Sigma_A$, abbreviated by $\mathrm{A0L}_S$-systems.

A complexity measure determined by $\mathrm{A0L}_S$ systems will be denoted by K_{A_S}. A slight influence of that restriction to the complexity of languages is presented in the next result.

Lemma 3.1. *Let L be in $\mathcal{L}$. Then*

$$Var_A\ L \leq Var_{A_S}\ L \leq Var_A\ L + 1$$
$$Prod_A\ L \leq Prod_{A_S}\ L \leq Prod_A\ L + 1$$
$$Symb_A\ L \leq Symb_{A_S}\ L \leq 3 + Symb_A\ L.$$

Relations between $K_{A_S}(L)$ and $K_{CF}(L)$ for the number of variables and number of symbols come out immediately using algorithms from [**5**].

THEOREM 3.1. Let L be in $\mathcal{L}$. Then $Var_{A_S} L = Var_{CF} L$ and $Prod_{A_S} L = |alph(L)| + Prod_{CF} L$.

SKETCH OF PROOF: Let $G = (N, T, P, S)$ be a Var-minimal ($Prod$-minimal) CF grammar of L. Then $H = (N \cup T, R, S)$, where $R = P \cup \{a \to a : a \in T\}$ is an $\mathrm{A0L}_S$ system and $L(H) = L(G)$. So

$$Var_{A_S} L \leq Var_{CF} L, \qquad Prod_{A_S} L \leq Prod_{CF} L + |alph(L)|. \tag{1}$$

Let $H = (\Sigma, R, S)$ be a Var-minimal ($Prod$-minimal) $\mathrm{A0L}_S$ system of L. Let h_A be a homomorphism on Σ with values in Σ^* such that

$$h_A(a) = \begin{cases} w_a & \text{for } a \in \Sigma_A, \text{ where } w_a \text{ satisfies } a \Rightarrow^* w_a \Rightarrow w_a \\ a & \text{otherwise} \end{cases}$$

Then we put $G = (\Sigma - \Sigma_A, \Sigma_A, R_0, S)$, where $R_0 = \{a \to h_A(\beta) : a \notin \Sigma_A, a \to \beta \text{ is in } R\}$. So $L(G) = L(H)$ and

$$Var_{CF} L \leq Var\ G = |\Sigma - \Sigma_A| = Var_{A_S} L;$$

$$Prod_{CF} L \leq Prod\ \ G = Prod_{A_S} L - |alph(L)|. \tag{2}$$

(1) and (2) together prove the Theorem.

Corollary 3.1. $Var_A L \leq Var_{CF} L \leq Var_A L + 1$
$Prod_A L \leq Prod_{CF} L + |alph(L)| \leq Prod_A L + 1$

Corollary 3.2. *For every n there is a language L_n in $\mathcal{L}$ such that $Prod_{CF} L_n = 1$ and $Prod_A L_n = n$.*

PROOF. *It is enough to consider* $L_n = \{a_1 a_2 \dots a_n\}$.

4. Total length of grammars and systems

In this section we shall deal with the relation between $Symb_{CF}L$ and $Symb_A L$.

THEOREM 4.1. *Let* L *be in* $\mathcal{L}$. *Then* $Symb_{A_S} L \leq 4\, Symb_{CF} L - 9$.
Moreover there are languages L_n *in* $\mathcal{L}$ *for* $n \geq 1$ *such that* $Symb_{A_S} L_n = 4\, Symb_{CF} L_n - 9$.

PROOF: Using similar transformation of CF grammar to $A0L_S$ system as in the proof of Theorem 3.1 we get $Symb_{A_S} L \leq Symb\ G + 3\,|T| = Symb_{CF} L + 3\,|T|$. Since $3 + |T| \leq Symb_{CF} L$ we have $Symb_{A_S} L \leq 4\, Symb_{CF} L - 9$.
Let $L_n = \{a_1 \dots a_n\}$ for $n \geq 1$. Then $Symb_{CF} L_n = 3 + n$ and $Symb_{A_S} L_n = 4n + 3$ gives $Symb_{A_S} L_n = 4\, Symb_{CF} L_n - 9$.

The transformation of A0L systems to CF grammars presented in the previous section is not optimal with respect to complexity measure *Symb*. We illustrate this by example.

Example 4.1. Let $H = (\{S, a, b, c, d\}, R, S)$ be an A0L system with rules $R = \{S \to a, a \to ab^3, b \to c^3, c \to d^3, d \to \varepsilon\}$. Then $L_A(H) = \{ab^3c^9d^{27}\}$ and $Symb\ H = 22$.
Using the method presented in previous section we obtain the grammar $G = (\{S\}, \{a, b, c, d\}, \{S \to ab^3c^9d^{27}\}, S)$ and $Symb\ G = 43$. But the grammar G_1, with rules $S \to ab^3A^3B^3$, $A \to c^3$, $B \to C^3$, $C \to d^3$ is more affective for $L = \{ab^3c^9d^{27}\}$, since $Symb\ G_1 = 28$.

Note that the transformation from previous section enables and exponential increase of *Symb*.

We present here a new method for transformation of A0L systems to equivalent CF grammars. It produces more effective CF grammars with respect to complexity measure *Symb* as the method used for *Var* and *Prod*.

Following [5] for an adult letter a of A0L system $H = (\Sigma, R, w_0)$ we have

1^0 There is exactly one rule $a \to \alpha$ in R. Moreover $\alpha \in \Sigma_A^*$ and $|\alpha|_a \leq 1$. ($|\alpha|_a$ denotes the number of occurrences of a in α)

2^0 There is exactly one word w_a such that $a \Rightarrow^* w_a \Rightarrow w_a$ and $w_a \in \Sigma_A^*$.

3^0 $w_a \neq \varepsilon$ iff $|\alpha|_a = 1$ and $w_a = \varepsilon$ iff $|\alpha|_a = 0$.

We denote $\Sigma_1 = \{a \in \Sigma_A : a \to \alpha \in R, |\alpha|_a = 1\}$

$\Sigma_2 = \{a \in \Sigma_A : a \to \alpha \in R, |\alpha|_a = 0\}$

$\Sigma_3 = \Sigma - \Sigma_A$

Evidently $\Sigma_1 \cup \Sigma_2 = \Sigma_A$.

In the next result a property of *Symb*-minimal $A0L_S$ system is specified.

Lemma 4.1. Let $H = (\Sigma, R, S)$ be a *Symb*-minimal $A0L_S$ system. If $c \to u$ is in R for c in Σ_3 then $u \in (\Sigma - \Sigma_2)^*$.

PROOF: Suppose that $H_0 = (\Sigma, R_0, S)$ is a *Symb*-minimal $A0L_S$ system and for some c in Σ_3 and b in Σ_2 a production $c \to ubv$ is in R. Then for system $H = (\Sigma, R, S)$, where $R = R_0 - \{c \to ubv\} \cup \{c \to uv\}$ we have $Symb\ H < Symb\ H_0$. For $L(H) = L(H_0)$ this leads to contradiction with *Symb*-minimality of H_0.

The equality $L(H_0) = L(H)$ comes out from the fact that $w_{xby} = w_{xy}$ for arbitrary $x, y \in \Sigma^*$ and $b \in \Sigma_2$, where w_u denotes an adult word derivable from u.

We continue by some denotation.

Let $u = x_1 x_2 \dots x_n$ for $x_1, \dots, x_n \in \Sigma$ and let $y \geq 1$ be a natural number. By U and U^j we denote a capital copy of u and j-th capital copy of u, respectively, i.e. $U = X_1 X_2 \dots X_n$, $U^j = X_1^j X_2^j \dots X_n^j$. For technical reason we put $U^0 = u$.

Let $H = (\Sigma, R, S)$ be an $A0L_S$ system and let $m = card\ \Sigma_2' - 1$, where $\Sigma_2' = \{b \in \Sigma_2 : b \to u \text{ is in } R \text{ and } u \neq \varepsilon\}$.

By $\mathcal{A}(H)$ we denote a CF grammar $G = (N, \Sigma_A, P, S)$, where

$N = \{X : x \in \Sigma_1 \cup \Sigma_3\} \cup \{X^j : 1 \leq j \leq m, x \in \Sigma_2\}$

$P = \{A \to U : a \to u \in R, a \in \Sigma_3\} \cup \{A \to U^m U^{m-1} \dots U^1 uavV^1 \dots V^{m-1} V^m : a \to uav \in R, a \in \Sigma_1\} \cup \{A^j \to U^{j-1} : 1 \leq j \leq m, a \to u \in R, a \in \Sigma_2\}$.

Lemma 4.2. Let H be an *Symb*-minimal $A0L_S$ system and $\mathcal{A}(H)$ be a CFgrammar $G = (N, \Sigma_A, P, S)$ defined above. Then $L(G) = L_A(H)$.

PROOF: Assume that following propositions are proved.

PROPOSITION 1: Let X be the capital version of x in Σ_1 and $L(X) = \{w \in \Sigma_A^* : X \Rightarrow_G^* w\}$. Then $L(X) = \{w_x\}$.

PROPOSITION 2: If $w_{x_1} \dots w_{x_n}$ is in $L_A(H)$ and $x_1, \dots, x_n$ are in Σ_1 then $X_1 \dots X_n$ is a sentential form in G, where X_i is the capital version of x_i.

To prove the lemma

(a) assume $w \in L_A(H)$. Then $w = w_{x_1} \dots w_{x_n}$ for some $x_1, \dots, x_n$ in Σ_1. By Proposition 2, $S \Rightarrow_G^* X_1 \dots X_n$ and by Proposition 1, $X_1 \dots X_n \Rightarrow_G^* w_{x_1} \dots w_{x_n}$, i.e. $L_A(H) \subseteq L(G)$.

(b) assume $w \in L(G)$. According to Lemma 4.1 for $A \to U$ in P, where A is the capital copy of $a \in \Sigma - \Sigma_A$, U is over capital copies of letters in $\Sigma_1 \cup \Sigma_3$. So for D being the derivation of w in G such that nonterminals corresponding to $\Sigma - \Sigma_A$ are rewritten before the others we obtain sentential form $X_1 \dots X_n$ over capital copies of Σ_1 producing w. By Proposition 1 $w = w_{x_1} \dots w_{x_n}$ and so $w \in L_A(H)$. This gives $L(G) \subseteq L_A(H)$.

PROOF OF PROPOSITION 1: There is exactly one production in G for X being the capital version of x in Σ_1 and for $X^j, j = 1, \dots, m$ being the j-th copy of x in Σ_2. Therefore $L(X)$ is a singleton. In G we derive w, $\{w\} = L(X)$.

$X \Rightarrow_G U^m \dots U^1 uxvV^1 \dots V^m \Rightarrow_G^* u_m \dots u_1 uxvv_1 \dots v_m = w$, where $U^i \Rightarrow_G^* u_i$ for $i = 1, \dots, m$.

By the definition of P for a in Σ_2 we have

$A^i \Rightarrow_G^* x \quad iff \quad a \Rightarrow_H^i x$.

Therefore $u \Rightarrow_H^i u_i$ and $w = w_x$.

PROOF OF PROPOSITION 2: Suppose that $w = w_{x_1} \dots w_{x_n}$ is in $L_A(H)$ and T be a derivation tree of w in H. Let y_i be the predecessor of x_i in T such that $y_i \neq x_i$. Then

y_i and all predecessors of y_i in T are in $\Sigma - \Sigma_A$.
Let $R_0 \subseteq R$ be the set of productions $y_i \to u_i x_i v_i$, $i = 1, \ldots, n$ and productions used in derivation tree T to reach vertices $y_1, \ldots, y_n$ from S. For $z \to u$ in R_0 we have $u \in (\Sigma_1 \cup \Sigma_3)^*$. Then capital versions of R_0 are in P and $S \Rightarrow_G^* X_1 \ldots X_n$.

THEOREM 4.2. $Symb_{CF} L < \frac{1}{3} Symb_A^2 L$.

PROOF: Let $H = (\Sigma, R, S)$ be a *Symb*-minimal A0L$_S$ system for L. Let $G = \mathcal{A}(H)$. Since $\Sigma_1, \Sigma_2, \Sigma_3$ are pairwise disjoint and $\Sigma = \Sigma_1 \cup \Sigma_2 \cup \Sigma_3$, we have $Symb\ H = 1 + Symb\ R_1 + Symb\ R_2 + Symb\ R_3$, where $Symb\ R_n$ denotes the sum of lengths of productions $a \to \alpha$ for a in Σ_n. Moreover $3m < Symb\ R_2$.
For productions in P we have
$Symb\{A \to U : a \to u \in R, a \in \Sigma_3\} = Symb\ R_3$
$Symb\{A \to U^m \ldots U^1 uavU^1 \ldots V^n : a \to uav \in R, a \in \Sigma_1\} = m\, Symb\, R_1 - 3m|\Sigma_1|$
$Symb\{A^j \to U^{j-1} : 1 \le j \le m, a \to u \in R, a \in \Sigma_2\} = m\, Symb\, R_2$
$Symb_{CF} L \le Symb\ G = 1 + m\ Symb\, R_1 + m\ Symb\, R_2 + Symb\, R_3 - 3m\, |\Sigma_1| < m\, Symb\, H < \frac{1}{3} Symb_A^2 L$.

Note 4.1. $\mathcal{A}(H)$ is not a *Symb*-minimal CF grammar for L since it contains nonterminals producing empty word only. They can be eliminated. Nevertheless they have no influence to quadratic increase of *Symb* as we illustrate in the next example.

Example 4.2. Suppose that H is an A0L$_S$ system with productions $S \to a$, $a \to ab_1^3, b_n \to \varepsilon$ and $b_i \to b_{i+1}^3$ for $i = 1, \ldots, n-1$. Let $G = \mathcal{A}(H)$ and let G_e be a CF grammar obtained from G by the elimination of all nonterminals producing ε word only. Then $Symb\, H = 5n + 7$, $Symb\, G = 5n^2 - 10n + 12$ and $Symb\, G_e = \frac{5}{2}n^2 + \frac{13}{2}n + 7$.

References

1. Gruska, J.: Descriptional complexity of context free languages. In: Proceedings of MFCS' 73, High Tatras, 71-83, 1973
2. Gruska, J.: Descriptional complexity (of languages) - A short survey. In: Proceedings of MFCS' 76 (A. Mazurkiewicz, ed.) Lecture Notes in Computer Science, Vol. 45, pp. 65-80. Berlin-Heidelberg-New York: Springer-Verlag 1976
3. Herman, G.T. and Walker, A.: Context free languages in biological systems. Intern. J. of Computer Math. Section A, 369-391 (1975)
4. Lindenmayer, A.: Models for multicellular development: Characterization, inference and complexity of L-systems. In: Trends, Techniques and Problems in Theoretical Computer Science (A. Kelemenová, J. Kelemen, eds.) Lecture Notes in Computer Science, Vol. 281, pp. 138-168. Berlin-Heidelberg-New York-London: Springer-Verlag 1987
5. Rozenberg, G., Salomaa, A.: The mathematical theory of L systems. Berlin-Heidelberg-New York: Springer-Verlag 1986

A Comparison of Two Controlled Rewriting Mechanisms for Table Lindenmayer Systems

Mária Pastorová

Department of Informatics, Faculty of Mathematics and Physics, Comenius University, Mlynská dolina, 842 15 Bratislava, Czechoslovakia

Abstract: Two mechanisms of controlled rewriting in table Lindenmayer systems, namely graphs and finite automata, are considered. It is shown that the finite automata concept of controlled rewriting is strongly more powerful than the graph concept of controlled rewriting according to the language generation. Moreover, a precise, natural description of the subclass of finite automata corresponding exactly to the graph controlled rewriting is found.

Keywords: TOL systems, DTOL systems, control graph, finite automaton, prefix automaton, language family.

1. Introduction

Already in 1975, Ginsburg and Rozenberg [4] introduced the concept of table Lindenmayer (TOL) systems with a rewriting mechanism controlled by the regular sets. This approach has been also used by other authors, e.g. Dassow. As it is known, the regular set corresponds to the finite automaton which can be represented by a so-called state diagram, i.e. a finite directed graph. On the other hand, graphs can be considered as a relatively general approach to controlled rewriting. This approach was introduced by Hromkovič and Kupková in [1, 6] and investigated by Dassow and Pastorová in their recent papers [2, 7].

The existence of two different rewriting mechanisms gives rise to a natural question what relation is between the families of languages generated by the two rewriting mechanisms mentioned above. In this paper it is shown that the finite automata concept of controlled rewriting is strongly more powerful than the graph concept of controlled rewriting according to the language generation. By the definitions and notations introduced in Section 2, this fact can be written as $\mathcal{L}(\mathrm{GTOL}) \subsetneq \mathcal{L}(\mathrm{FATOL})$, $\mathcal{L}(\mathrm{GDTOL}) \subsetneq \mathcal{L}(\mathrm{FADTOL})$ and is proved in Section 3. In Section 4 the exact correspodence between the graph controlled rewriting and the class of finite automata closed to prefixes is proved.

2. Basic definitions

Table OL (TOL) system introduced in [5] is a L-system whose development is described by some tables of rules. The TOL system is given by a triple $(V, \mathcal{P}, x)$, at every moment of time only one table P from the set $\mathcal{P}$ of developmental rule tables can be used. If these tables are deterministic, one deals with deterministic TOL (DTOL) systems. In this case each element of the alphabeth V is exactly once situated on the left hand side of rewriting rules in each table P.

Definition 1. A GTOL (GDTOL, respectively) system is quadruple $S = (V, \mathcal{P}, G, x)$, where $(V, \mathcal{P}, x)$ is a TOL (DTOL) system and $G = (U, H, h, U_0)$ is a labelled directed finite graph with the vertex set U and the edge set H satisfying the condition $\forall u \in U \; \exists (u, v) \in H$, further $h\colon U \to \mathcal{P}$ is a (not necessarily bijective) mapping, and $U_0 \subseteq U$ is the set of initial vertices (denoted by circles in figures).

The set $\mathcal{R} = \{\, h(u);\; u \in U \,\}$ is the set of initial tables.

Note that, according to Definition 1, each graph G must contain at least one cycle.

According to this definition, the GDTOL system is a particular case of GTOL system.

Definition 2. Let $S = (V, \mathcal{P}, G, x)$ be a GTOL system. Let $v = a_1 a_2 \dots a_n \in V^+$ be a word over V and $w \in V^*$. We shall say that v can be rewritten to w (w is derived from v) in S, if there exists a walk – a finite sequence $c = \{u_1\}_{i=0}^{k}$ $(k > 0)$ of vertices from U such that $(u_i, u_{i+1}) \in H$ $(i = 1, 2, \dots k-1)$ and

$$v \underset{h(u_1)}{\Longrightarrow} w_1 \underset{h(u_2)}{\Longrightarrow} w_2 \Longrightarrow \dots \underset{h(u_k)}{\Longrightarrow} w_k = w.$$

This will be denoted $v \overset{c}{\Longrightarrow} w$.

The language generated by a GTOL system $S = (V, \mathcal{P}, G, x)$ is defined as the set

$$L(S) = \{x\} \cup \{w \in V^*; x \overset{c}{\Longrightarrow} w,\; c \in C(S)\} \text{ where}$$

$C(S)$ is the union of all walks in S. We say that two {GTOL} systems S_1, S_2 are equivalent iff

$$L(S_1) = L(S_2).$$

Denote by C_n the set of all walks of length $n \geq 0$ in S starting from an initial vertex, and by $T(S)$ the set of table sequences, $T(S) = \cup_{n=1}^{\infty} T_n$ where

$$T_1 = \{\{P\}; h(u) = P, \{u\} \in C_0\},$$

$$T_n = \left\{\{P_i\}_{i=0}^{n-1}; \{u_i\}_{i=0}^{n-1} \in C_{n-1}, h(u_i) = P_i \quad (i = 0, 1, \dots, n-1)\right\}.$$

Definition 3. A FATOL (FADTOL, respectively) system S is a quadruple $S = (V, \mathcal{P}, A, x)$ where $(V, \mathcal{P}, x)$ is a TOL (DTOL) system and A is a finite deterministic automaton over the alphabet $\mathcal{P}$.

Let $v \in V^*, w \in V^*$, we shall say that v can be rewritten to w in S, if there exists $p = P_1 P_2 \dots P_k \in L(A)$ for some $k \geq 1$ ($L(A)$ is the language generated by A) and strings $w_i \in V^*$ $(i = 1, 2, \dots k)$ that

$$v \underset{P_1}{\Longrightarrow} w_1 \underset{P_2}{\Longrightarrow} w_2 \Longrightarrow \dots w_{k-1} \underset{P_k}{\Longrightarrow} w_k = w.$$

This fact will be shortly denoted as $v \underset{P_1 P_2 \dots P_k}{\Longrightarrow} w$ or $v \overset{p}{\Longrightarrow} w$.

The language generated by a FATOL system S is defined as the set $L(S) = \{x\} \cup \{w \in V^*; x \overset{p}{\Longrightarrow} w,\; p \in L(A)\}$.

Definition 4. Let $A = (Q, \Sigma, \delta, q_0 F)$ be a finite deterministic automaton where Q is a nonempty state set, Σ is a nonempty imput alphabet, δ is a transition function, q_0 is a initial state, $F \subseteq Q$ is a set of finite states. Then A will be called a prefix automaton, if $L(A)$ is closed according to prefixes, i.e. if for each $x, y \in \Sigma^*$, $xy \in L(A)$ implies $x \in L(A)$.

The FATOL (FADTOL, respectively) system with a prefix finite automaton we shall denotes as FA$_P$TOL (FA$_P$DTOL).

Recall, that one of representation of a finite automaton is the state diagram, i.e. the directed graph whose vertices correspond to states of automaton, the edges denote possible transitions between states and the edges are evaluated by symbols of the alphabet. Initial and finite states are signed particularly (e.g. by unevaluable arrows enterring or leaving the state).

3. Theorems about the FATOL and FADTOL systems

Now, we shall show that for every language generated by some GTOL system it is possible to find a FATOL system generating the same language, however not reversely. Denote by $\mathcal{L}(A)$ the class of languages generated by a family of systems A.

Lemma 1. There holds $\mathcal{L}(\text{GTOL}) \subseteq \mathcal{L}(\text{FATOL})$.

Proof. Let $S = (V, \mathcal{P}, G, x)$ be an available GTOL system, where $G = (U, H, h, U_0)$ and $\mathcal{P} = \{P_1, P_2, \ldots P_k\}$, $k \geq 1$. We construct a FATOL system $S' = (V, \mathcal{P}, A, x)$ with the property $L(S) = L(S')$. The nondeterministic finite automaton $A' = (Q, \mathcal{P}, \delta, Q_0, F)$ from graph G can be construct by the natural way: $Q = U$, $Q_0 = U_0$, $F = Q$, $\delta \colon Q \times \mathcal{P} \to 2^Q$, where $\delta(u_i, P_i) = \{u_j; (u_i, u_j) \in H, h(u_i) = P_i\}$. It will be shown that the set $T(S)$ is equivalent to $L(A')$. Indeed, there is $p = P_{j_0} P_{j_1} \ldots P_{j_{n-1}} \in L(A')$ iff $p \in \mathcal{P}^+$ and there exists some sequence $\{u_{j_i}\}_{i=0}^{n}$ $(n > 0)$ with $u_{j_i} \in Q$ $(i = 1, \ldots, n)$, $u_{j_0} \in Q_0$ that $u_{j_{i+1}} \in \delta(u_{j_i}, P_{j_i})$ $(i = 0, 1, \ldots, n-1)$. This is equivalent to $u_{j_0} \in U_0, (u_{j_i}, u_{j_{i+1}}) \in H, h(u_{j_i}) = P_{j_i}$ $(i = 0, 1, \ldots, n-1)$, $n > 0$ and further to $\{P_{j_i}\}_{i=0}^{n-1} \in T_n$, $n > 0$. Hence $\{P_{j_i}\}_{i=0}^{n-1} \in T(S)$, $n > 0$.

There exists such a deterministic finite automaton A to the automaton A' that $L(A) = L(A')$. Since the set $T(S)$ is equivalent to $L(A)$, according to the definitions of languages defined by GTOL and FATOL systems, respectively $L(S) = L(S')$ holds. Therefore $\mathcal{L}(\text{GTOL}) \subseteq \mathcal{L}(\text{FATOL})$. $\square$

Theorem 1. $\mathcal{L}(\text{GDTOL}) \subsetneq \mathcal{L}(\text{FADTOL})$.

Proof. The proof of the previous Lemma did not utilize the property that the tables are nondeterministic. Therefore, by the same way it can be proved $\mathcal{L}(\text{GDTOL}) \subseteq \mathcal{L}(\text{FADTOL})$. Now, we construct a FADTOL system $S = (V, \mathcal{P}, A, x)$ that $L(S) \notin \mathcal{L}(\text{GDTOL})$. Let $V = \{a, b\}$, $\mathcal{P} = \{P_1, P_2, P_3, P_4, P_5\}$, $x = a$ and $P_1 = \{(a, a^3), (b, b)\}$, $P_2 = \{(a, a^5), (b, b)\}$, $P_3 = \{(a, a^7 b), (b, b)\}$, $P_4 = \{(a, a), (b, a^2 b)\}$, $P_5 = \{(a, a), (b, a^2)\}$ and the state diagram corresponding to the nondeterministic finite automaton A', which is equivalent to the deterministic finite automaton A, is of the form

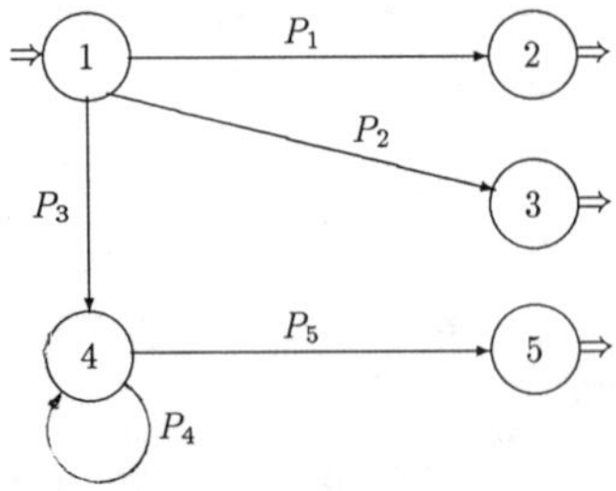

Obviously $L(A) = L(A') = \{P_1, P_2\} \cup P_3 P_4^* P_5$ is not prefix and $L(S) = \{a^{2k-1}; k \in N\} - \{a^7\}$. However, such language contains infinitely many words whose length is prime number and therefore, according to [6], cannot be generated by no GDTOL system. □

Before proving the analogical theorem for GTOL systems, we formulate two lemmas.

Lemma 2. Let a, b be some coprime integers,at least one even. Then for each integer $n > 0$ the integers $a+b$ and $a^{2n} + b^{2n}$ are coprime.

Proof. As one can easily verify, we have

$$(a^{2n} + b^{2n}) = (a+b)(a^{2n-1} - a^{2n-2}b + \cdots + ab^{2n-2} - b^{2n-1}) + 2b^{2n}.$$

The numbers $a+b$ and $2b^{2n}$ are coprime according to our assumptions. Thus, the sum in the equation above and $a+b$ are coprime, as well. □

Lemma 3. Let a, b be some coprime integers, at least one even. Then, for all integers i and j such that $j < 2^i$, the integer $a^j + b^j$ is not a divisor of $a^{2^i} + b^{2^i}$.

Proof. If j is odd then $a+b$ divides $a^j + b^j$, but by Lemma 2, $a+b$ does not divide $a^{2^i} + b^{2^i}$. This implies the assertion of Lemma. If j is even, in these considerations the integers a, b can be replaced by a^2, b^2 (or a^4, b^4 etc.). □

Theorem 2. $\mathcal{L}(\text{GTOL}) \subsetneq \mathcal{L}(\text{FATOL})$.

Proof. According to Lemma 1 $\mathcal{L}(\text{GTOL}) \subseteq \mathcal{L}(\text{FATOL})$ holds. Let $L = \{a\} \cup \{a^{2^i+3^i}; i = 0, 1, \dots\}$. We shall show that $L \in \mathcal{L}(\text{FATOL})$ and $L \notin \mathcal{L}(\text{GTOL})$.

$L \in \mathcal{L}(\text{FATOL})$: Let $S = (\{a, b\}, \mathcal{P}, A, a)$, $\mathcal{P} = \{P_1, P_2, P_3\}$, $P_1 = \{(a, ab), (b, b)\}$, $P_2 = \{(a, a^3), (b, b^2)\}$, $P_3 = \{(a, a), (b, a)\}$.

Let A denote a deterministic finite automaton which is equivalent to a nondeterministic finite automaton with the following state diagram

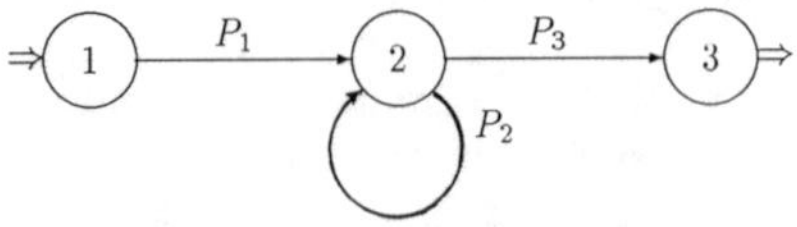

Obviously $L(A) = \{P_1P_3\} \cup P_1P_2^*P_3$. The system S is FADTOL system and $L(S) = L$. Hence, $L \in \mathcal{L}$(FATOL).

$L \notin \mathcal{L}$(GTOL): Asumme there exists a GTOL system $S' = (V', P', G, x)$ with such a graph $G = (U, H, h, U_0)$ that $L(S') = L$.

Since only the symbol a can occur on the right hand side of any rule, $V' - \{a\}$. It is obvious that $x = a$ because $\varepsilon \notin L$. Hereby, for arbitrary $P \in \mathcal{P}'$, $(a, \varepsilon) \notin P$.

Since $a^2 \in L$, there exists such a sequence $c = \{u_i\}_{i=0}^n \in C(S')$, $n \geq 0$ that $a \overset{c}{\Longrightarrow} a^2$. Denote $D = \{u; (u,u) \in H\}$, $D' = \{h(u); u \in D\}$. There must hold $D' = \{(a,a)\}$, because the opposite case $(a, a^k) \in D'$, $k > 1$ implies $a^{2k} \in L$, a contradiction.

Thus, consider words of the form a^{s_i} where $s_i = 2^{2^i} + 3^{2^i}$ $(i = 1, 2, \dots)$. Since the number of such words is infinite they cannot be derived immediately from the axiom by using some of initial tables. (Note that the number of initial tables is finite and each initial table contains a finite number of rules.)

We shall prove that any such word cannot be derived from another $v \in L$, $v \neq x$ using the only rule of type (a, a^k), $k > 1$. Indeed, the opposite case means that for each given i there exists k, j that

$$(a^k)^{2^j + 3^j} = a^{s_i}, \qquad \text{e. i.} \quad k(2^j + 3^j) = 2^{2^i} + 3^{2^i}.$$

However, in contradiction to Lemma 3, for $k > 1$ we get $j < 2^i$.

Therefore, there exists a table

$$P = \{(a, a^{k_1}), (a, a^{k_2}), \dots, (a, a^{k_m})\} \qquad (m \geq 2),$$

as well as the vertex v in graph G evaluated by $h(v) = P$.

Let $k_1 < k_2 < \dots < k_m$. Denote $E = \{2^i + 3^i; i = 0, 1, \dots\}$. Consider $a^q \in L$, $q \in E$ such that $a \overset{c}{\Longrightarrow} a^q$, $c = \{u_i\}_{i=0}^r \in C_r$, $(u_r, v) \in H$. Then the words $a^{k_1 q}, a^{k_2 q}, \dots, a^{k_m q}$ belong to $L(S')$, e.i., $k_1 q, \dots, k_m q \in E$. The same can be said about the words $a^{k_1(q-1)} a^{k_2}, a^{k_1(q-2)} a^{2k_2}, \dots, a^{k_1} a^{k_2(q-1)}$.

If we denote by α_1 the exponent of a in the first word, then for $q \in E$ we have $q \geq 2$ means $\alpha_1 = k_1(q-1) + k_2 \neq k_2$, and therefore $k_1 q < \alpha_1 = k_1 q + (k_2 - k_1) < k_2 q$.

Analogically if the exponent in the second word is denoted by $\alpha_2 = k_1 q + 2(k_2 - k_1)$, evidently $\alpha_1 < \alpha_2 \leq k_2 q$.

Now, we shall prove that if $k_1 q \in E$, $\alpha_1 \in E$, then $\alpha_2 \notin E$, hence $L(S') \neq L$. Suppose the opposite. Let $k_1 q, \alpha_1, \alpha_2 \in E$, i.e., there exists such $i, j, i < j$ that $k_1 q = 2^i + 3^i, \alpha_1 = 2^j + 3^j$. With regard to definition of α_1, α_2 we have

$$\alpha_2 - \alpha_1 = k_2 - k_1, \quad \alpha_1 - k_1 q = k_2 - k_1, \text{ hence}$$

$\alpha_2 = 2\alpha_1 - k_1 q = 2(2^j + 3^j) - (2^i + 3^i)$. However, the least integer in E greater than α_1 is some integer $t = 2^{j+1} + 3^{j+1}$. Since $2^{j+1} > 2 \cdot 2^j - 2^i$, $3^{j+1} > 2 \cdot 3^j - 3^i$, there is $t > \alpha_2 > \alpha_1$ what contradicts to the assumption $\alpha_2 \in E$.

For this reason $m = 1$. Therefore, each vertex of graph G can be evaluated only by a table containing the only rule. Together with the above consideration, this proves that $L \notin \mathcal{L}$(GTOL). □

4. Characterization of GTOLs in terms of FATOLs

We have shown that FATOLs have more generation power than GTOLs in the previous assertion. Now, the question arises whether we are able to characterize precisely $\mathcal{L}(\text{GTOL})$ as a proper subclass of $\mathcal{L}(\text{FATOL})$ in terms of FATOLs.

Theorem 3. $\mathcal{L}(\text{GTOL}) = \mathcal{L}(\text{FA}_\text{P}\text{TOL})$.

Proof. At first we shall prove the inclusion $\mathcal{L}(\text{GTOL}) \subseteq \mathcal{L}(\text{FA}_\text{P}\text{TOL})$. Let $S = (V, \mathcal{P}, G, x)$ be an arbitrary GTOL system and $S' = (V, \mathcal{P}, A, x)$ the corresponding FATOL system from the proof of Lemma 1. We shall show that S' is a FA_PTOL system, i.e. that the deterministic finite automaton A is a prefix automaton. That is true if and only if the corresponding nondeterministic finite automaton A' overthere defined is also a prefix one. It is known that $L(A')$ corresponds to $T(S)$. Let $p = P_{r_1}P_{r_2}\dots P_{r_n} \in L(A')$, $n \geq 2$. Then $\{P\}_{i=1}^k \in T_k$ $(k = 1, \dots, n)$. That means $P_{r_1}P_{r_2}\dots P_{r_n} \in L(A')$ $(k = 1, \dots n)$. Hence, A' is a prefix automaton.

Now, we shall prove the reverse inclusion. Let $S = (V, \mathcal{P}, A, x)$ be an arbitrary FA_PTOL system and let $\mathcal{P} = \{P_1, \dots, P_n\}$, $A = \{\mathcal{P}, Q, \delta, q_0, F\}$. Further, let $S' = (V, \mathcal{P}, G, x)$ be a GTOL system with the graph $G = (U, H, h, U_0)$ constructed in the following way :

$$\begin{aligned} U &= \{1, 2, \dots, n\}, \\ H &= \{(i,j); \exists\, q, q', q'' \in F\ \delta(q, P_i) = q' \text{ and } \delta(q', P_j) = q''\}, \\ h(i) &= P_i \qquad (i = 1, \dots, n), \\ U_0 &= \{i;\ \exists\, q \in F\ \delta(q_0, P_i) = q\}. \end{aligned}$$

One can easily prove that the class $L(A)$ is equivalent to the class $T(S')$. Since A is a prefix automaton, there is

$$\begin{aligned} &p = P_{r_1}P_{r_2}\dots P_{r_k} \in L(A) \quad (k \geq 1) \\ \text{iff}\quad & P_{r_1}P_{r_2}\dots P_{r_j} \in L(A) \quad (j = 1, \dots, k) \quad (k \geq 1) \\ \text{iff}\quad & \delta(q_{j-1}, P_{r_j}) = q_i \in F \quad (j = 1, \dots, k) \quad (k \geq 1) \\ \text{iff}\quad & (r_{j-1}, r_j) \in H \quad (j = 1, \dots, k) \quad (k \geq 1) \\ \text{iff}\quad & \{r_j\}_{j=0}^k \in C_k \quad (k \geq 1) \\ \text{iff}\quad & \{h(r_j)\}_{j=0}^k \in T_{k+1} \quad (k \geq 1). \end{aligned}$$

Since the class $L(A)$ is equivalent to $T(S')$, $L(S') = L(S)$. □

Note. Regarding that in the proof above we have handled only the class $T(S)$, i.e. sequences of tables without utilizing they are nondeterministic, the following consequence is obvious.

Theorem 4. $\mathcal{L}(\text{GDTOL}) = \mathcal{L}(\text{FA}_\text{P}\text{DTOL})$.

Example 1. The language L introduced in the proof of Theorem 1 has the property $L \in \mathcal{L}(\text{FADTOL})$, $L \notin \mathcal{L}(\text{GDTOL})$. Then, Theorem 4 implies $L \notin \mathcal{L}(\text{FA}_\text{P}\text{DTOL})$. However, at the same time $L \in \mathcal{L}(\text{GTOL})$ because there exists a GTOL system

$$S = (\{a\}, \{P\}, G, a), \text{ where } P = \{(a, a^3), (a, a^5)\}$$

and $G = (\{u_0\}, \{(u_0, u_0)\}, h(u_0) = P, \{u_0\})$, that $L(S) = L$.

In the proof of Theorem 2 we have dealt with language L such that $L \in \mathcal{L}(\text{FADTOL})$, $L \notin \mathcal{L}(\text{GTOL})$. Now we show a language with the opposite property.

Example 2. Consider a language $L_1 = \{a^{10(10-i)+i};\ i = 0,\dots,10\}$. Then $L_1 \notin \mathcal{L}(\text{FADTOL})$ but $L_1 \in \mathcal{L}(\text{GTOL})$ because there exists a GTOL system $S_1 = (\{a\},\{P_1,P_2\},G,a^{10})$ where $P_1 = \{(a,a),(a,a^{10})\}$, $P_2 = \{(a,a)\}$, $G = (\{u_1,u_2\}, \{(u_1,u_2),(u_2,u_2)\},h,\{u_1\}), h(u_1) = P_1, h(u_2) = P_2$ that $L(S) = L_1$.

Let A, B be families of systems. Denote by $A \rightarrow B$ if $\mathcal{L}(A) \subseteq \mathcal{L}(B)$, and $A \Longleftrightarrow B$ if $\mathcal{L}(A) = \mathcal{L}(B)$. The fact $\mathcal{L}(A) - \mathcal{L}(B) \neq 0$ and simultancously $\mathcal{L}(B) - \mathcal{L}(A) \neq 0$, denote $A - - - B$. Using this notation, the relation between some families of OL systems can be visualized by the diagram on Fig. 1,as it is proved in [4–6] and Theorems above.

In [6] it is shown that the language families $\mathcal{L}(\text{GTOL})$, and $\mathcal{L}(\text{GDTOL})$ are not closed under union, intersection with regular languages, ε-free homomorphism, inverse homomorphism, and complement. The analogical theorem is true for language families $\mathcal{L}(\text{FATOL})$, $\mathcal{L}(\text{FADTOL})$, $\mathcal{L}(\text{FA}_\text{P}\text{TOL})$, $\mathcal{L}(\text{FA}_\text{P}\text{DTOL})$.

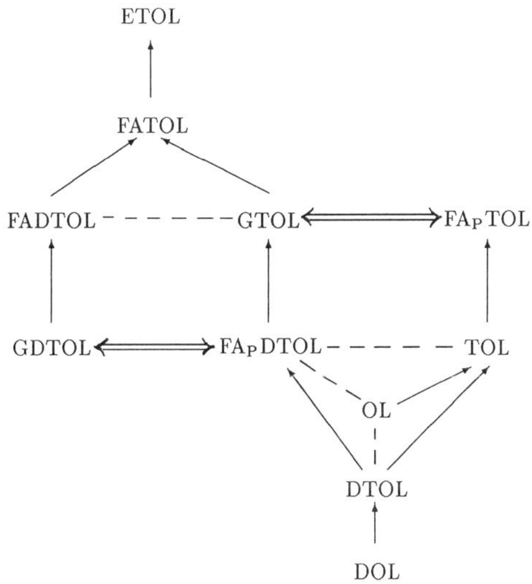

Fig. 1

1. Braunsteinerová, E., Hromkovič, J.: Graphic controlled tabled Lindenmayer systems. In: Proc. 3rd Internat. Meeting of Young Comp. Scientists, pp. 53–59. Hungarian Academy of Sciences 1984
2. Dassow, J.: On the sequence equivalence problem for OL systems with controlled derivations. Computers and Artificial Intelligence 7, 97–106 (1988)
3. Dassow, J., Fest, U.: On regulated L systems. Rostock Math. Kolloq. 25, 99–118 (1984)
4. Ginsburg, S., Rozenberg, G.: TOL schemes and control set. Inform. Control 27, 109–125 (1975)
5. Herman, G.T., Rozenberg, G.: Developmental systems and languages. Amsterdam: North-Holland 1975
6. Hromkovič , J., Kupková , E.: Graph-controlled table Lindenmayer systems. J. Inf. Process. and Cybern. EIK 24, 481–493 (1988)
7. Pastorová , M.: Growth functions of DTOL systems controlled by Bouquet graph. Computers and Artificial Intelligence 3, 299–311 (1990)

3. Images from the Path:

Computer Graphics, Pictures and Animation

L-systems: from formalism to programming languages

Przemyslaw Prusinkiewicz and James Hanan

Department of Computer Science, University of Regina, Saskatchewan, Canada S4S 0A2

Abstract: Originated as a mathematical formalism, L-systems have become a compact notation for specifying fractals, graftals, and plant models, and have acquired features, such as parameters, found in high-level programming languages. This paper discusses parametric L-systems as a model of parallel computation, and as a basis for a class of programming languages. In the latter context, extensions to parametric L-systems are proposed and a strikingly simple implementation is described.

Keywords: parametric L-systems, parallel computational models, programming languages, concurrency, algorithms.

1 Introduction

Soon after L-systems were introduced to describe developmental processes in plants [15], they were embraced by computer scientists as an exciting research topic in formal language theory [25]. This paper suggests that L-systems can be viewed from other perspectives as well, notably as a model of parallel computation, and as a basis for a class of programming languages.

The presentation begins with a justification for adding new constructs to the original definition of L-systems (Section 2). The resulting concept of parametric L-systems is described in Section 3. It leads in a natural way to an unorthodox computational model with self-replicating processors, presented in Section 4. Section 5 considers L-systems as the basis for a class of concurrent programming languages, and introduces extensions to parametric L-systems, such as local and global variables. Section 6 tailors L-systems to a single-thread programming language. Parallel application of productions, the distinctive feature of L-systems, is compromised, but the view of computation as a sequence of derivation steps is retained and leads to the concept of function evaluation using a queue rather than a stack. The proposed extensions to L-systems are summarized in Section 7. A simple, practical method for implementing these L-systems in the C programming language is outlined in Section 8. Finally, the main theses of the paper are recaptured in Section 9.

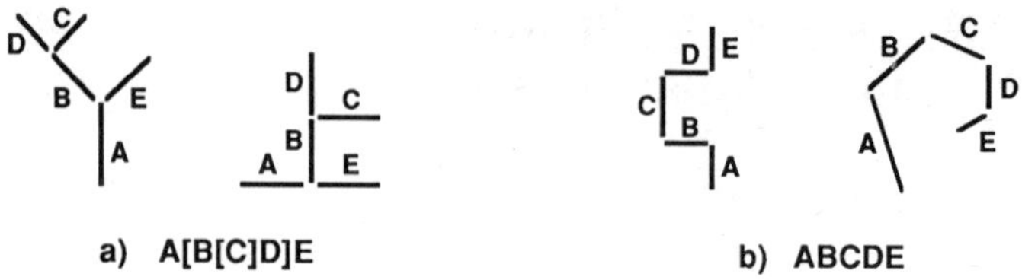

Figure 1: Intuitive interpretations of bracketed strings

2 The need for extensions

According to the original definition, L-systems operate on strings with brackets. For example, the string **A[B[C]D]E** represents a topological tree with letters denoting edges and matching pairs of brackets enclosing branches. A parallel string-rewriting mechanism is used to simulate changes in this tree over time.

By examining a branching structure, such as that of a plant, it is easy to derive the corresponding bracketed string. On the other hand, graphical interpretation of a given string is ambiguous because bracketed strings contain no geometric information. Different interpretations of the same string are shown in Figure 1.

Originally, the graphical interpretation of bracketed strings was performed by a draftsperson who incorporated knowledge of the modeled structures into the drawings. Frijters and Lindenmayer [8], and Hogeweg and Hesper [10] proposed the first algorithms intended to make drawing more automatic. No modifications of the L-systems under consideration are required and in this sense the geometric interpretation is *external* to the L-systems. Drawing is controlled by a predefined set of rules applied uniformly to all components of the structure. The same rules are used to interpret any string, making some structures impossible to specify. For example, the rule stating that branches should be issued in alternating directions, to the left and right, does not allow for structures with two consecutive branches oriented the same way.

To overcome these limitations, Prusinkiewicz [20, 21] proposed an alternative method for interpreting L-systems, related to earlier results of Szilard and Quinton [27]. According to this method, the alphabet of an L-system is extended to incorporate symbols with geometric interpretation. A string generated by the L-system is viewed as a sequence of commands for a LOGO-style turtle [1, 18], where **F** instructs the turtle to trace a line, and symbols + and − turn the turtle to the left and right (Figure 2, a and b). Although geometric information becomes an *internal* part of L-systems, their original definition still applies because productions operate in the same way on all symbols, irrespective of their interpretation.[1]

The turtle interpretation outlined above requires all lines and angles to be expressed as integer multiples of globally defined units. Consequently, many simple figures cannot be represented exactly (Figure 2b). Furthermore, variables ranging over an interval, such

[1]The only exception occurs in the context-sensitive case, where the insertion of symbols with geometric interpretation may change context for other symbols. To overcome this effect, selected symbols are ignored while searching for context [24, page 32].

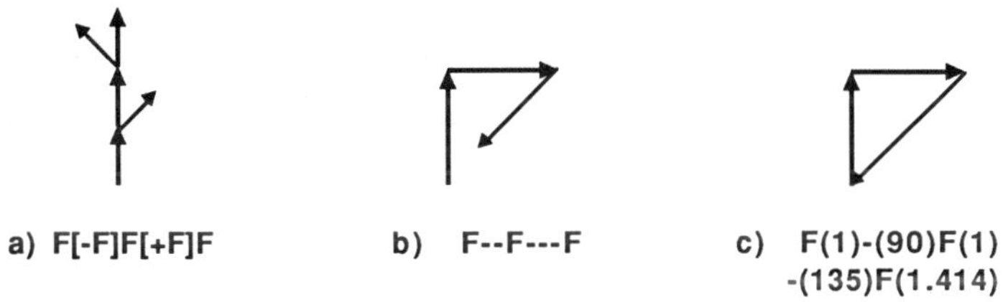

Figure 2: Turtle interpretation of strings without parameters (a, b), and with parameters (c). The right isosceles triangle cannot be represented exactly (b) unless numerical parameters are introduced (c).

as the concentration of a morphogen in a growing structure, must be quantized. This may increase the number of L-system symbols and productions beyond manageability. As a solution to such problems, Lindenmayer proposed the association of numerical *parameters* with L-system symbols [16]. He illustrated this idea by referring to the diffusion of nitrogenous compounds in a growing filament, and to the continuous development of a branching structure. Both problems were revisited in later papers [5, 11]. The actual simulations were carried out by a program called CELIA (an acronym for CEllular Linear Iterative Array simulator) [3, 5, 6, 9]. CELIA assumes that productions without parameters are listed in an input data file, using the notation *predecessor* $\rightarrow$ *successor*. In contrast, productions with parameters are specified as FORTRAN subroutines linked with the main body of the simulation program [6]. This inconsistency does not foster a unified, formal view of L-systems with parameters.

More recently, definitions of L-systems operating on symbols with parameters were proposed independently by Shebell [26], Chien [4], and Prusinkiewicz and Hanan [23] (see also [24, pages 40–50]). The next section summarizes the latter approach.

3 Parametric L-systems

Parametric L-systems operate on *parametric words*, which are strings of *modules* consisting of *letters* with associated *parameters*. The letters belong to an *alphabet* V, and the parameters belong to the set of *real numbers* $\Re$. A module with letter $A \in V$ and parameters $a_1, a_2, ..., a_n \in \Re$ is denoted by $A(a_1, a_2, ..., a_n)$. Every module belongs to the set $V \times \Re^*$, where $\Re^*$ is the set of all finite sequences of parameters. The sets of all strings of modules and all nonempty strings of modules are denoted by $(V \times \Re^*)^*$ and $(V \times \Re^*)^+$, respectively.

The real-valued *actual* parameters appearing in the strings are contrasted with the *formal* parameters used in the specification of L-system productions. Formal parameters may occur in logical and arithmetic *expressions*. If Σ is a set of formal parameters, $C(\Sigma)$ denotes a logical expression with parameters from Σ, and $E(\Sigma)$ denotes an arithmetic expression with parameters from the same set. Expressions of both types consist of formal parameters and numeric constants, combined using the arithmetic operators $+$, $-$, $*$, $/$, the remainder operator %, the exponentiation operator $\wedge$, the relational operators $=$, $<$, $<=$, $<$, $>$, $>=$, the logical operators !, &, | (not, and, or), and parentheses (). Standard rules for constructing syntactically correct expressions with proper operator precedence are observed. The sets of logical and arithmetic expressions with parameters from Σ are denoted by $\mathcal{C}(\Sigma)$ and $\mathcal{E}(\Sigma)$. The empty string is considered a logical expression that evaluates to *true*.

A *parametric OL-system* is defined as an ordered quadruplet $G = \langle V, \Sigma, \omega, P \rangle$, where:

- V is the *alphabet* of the system,
- Σ is the *set of formal parameters*,
- $\omega \in (V \times \Re^*)^+$ is a nonempty parametric word called the *axiom*, and
- $P \subset (V \times \Sigma^*) \times \mathcal{C}(\Sigma) \times (V \times \mathcal{E}(\Sigma)^*)^*$ is a finite *set of productions.*

The symbols : and $\rightarrow$ are used to separate the three components of a production: the *predecessor*, the *condition*, and the *successor*. For example,

$$\mathtt{A(s,t)} \;:\; \mathtt{s > t} \;\rightarrow\; \mathtt{B(s+1)CD(s \wedge 0.5, t-2)} \tag{1}$$

is a production with predecessor $\mathtt{A(s,t)}$, condition $\mathtt{s > t}$ and successor

$$\mathtt{B(s+1)CD(s \wedge 0.5, t-2)}.$$

A production *matches* a module in a parametric word if the following conditions are met:

- the letter in the module and the letter in the production predecessor are the same,
- the number of actual parameters in the module is equal to the number of formal parameters in the production predecessor, and
- the logical expression C evaluates to *true* if the actual parameter values are substituted for the formal parameters in the production.

A matching production can be *applied* to the module. The result is a string of modules specified by the production successor. During the evaluation of arithmetic expressions, the actual parameter values in the module are substituted for the formal parameters in the production. For example, production (1) above matches a module $\mathtt{A(9,2)}$ since the letter A in the module is the same as in the production predecessor, there are two actual parameters in the module $\mathtt{A(9,2)}$ and two formal parameter in the predecessor $\mathtt{A(s,t)}$, and the logical expression $\mathtt{s > t}$ is true for $\mathtt{s} = 9$ and $\mathtt{t} = 2$. The result of the production application is the parametric word $\mathtt{B(10)CD(3,0)}$.

If a module x produces a parametric word χ as the result of a production application in an L-system G, we write $x \mapsto \chi$. Given a parametric word $\mu = x_1 x_2 \ldots x_m$, we say that the word $\nu = \chi_1 \chi_2 \ldots \chi_m$ is *directly derived* from (or *generated* by) μ, and write $\mu \Longrightarrow \nu$ if and only if $x_i \mapsto \chi_i$ for $i = 1, 2, \ldots, m$. A parametric word ν is generated by G in a *derivation of length* n if there exists a sequence of words $\mu_0, \mu_1, \ldots, \mu_n$ such that $\mu_0 = \omega$, $\mu_n = \nu$ and $\mu_0 \Longrightarrow \mu_1 \Longrightarrow \ldots \Longrightarrow \mu_n$.

An example of a parametric OL-system and the words obtained in the first three derivation steps are shown below. If no matching production is found in the set P, a module is replaced by itself.

$$\begin{array}{llll} \omega: & \mathtt{B(1)A(4,4)} & & \\ p_1: & \mathtt{A(a,b)} & : \mathtt{b} < 2 & \rightarrow \mathtt{A(a*3,a+b)} \\ p_2: & \mathtt{A(a,b)} & : \mathtt{b} >= 2 & \rightarrow \mathtt{B(a)A(a/b,1)} \\ p_3: & \mathtt{B(a)} & : \mathtt{a} < 1 & \rightarrow \mathtt{C} \\ p_4: & \mathtt{B(a)} & : \mathtt{a} >= 1 & \rightarrow \mathtt{B(a-1)} \end{array}$$

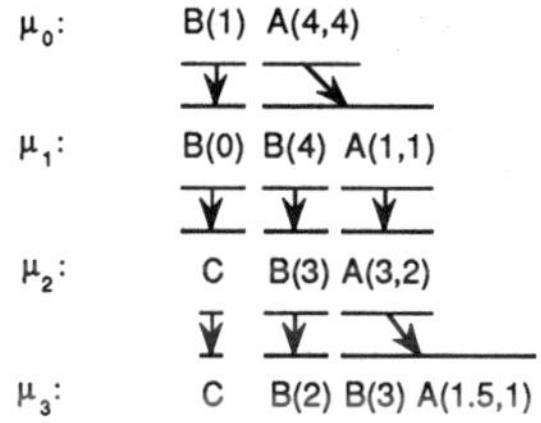

Productions in OL-systems are *context-free*, i.e., applicable regardless of the context in which the predecessor appears. They are sufficient to capture information transfer from the parent module to its descendants. However, *context-sensitive* extensions of OL-systems are necessary to model information exchange between neighboring modules. For example, in a *2L-system* the predecessor consists of three components, called the *left context*, the *strict predecessor*, and the *right context*. By convention, they are separated using the symbols $<$ and $>$. Formal parameters from all components may appear in the condition and the production successor.

A sample context-sensitive production is given below:

$$\mathtt{A(a)} < \mathtt{B(b)} > \mathtt{C(c)} : \mathtt{a}+\mathtt{b}+\mathtt{c} > 10 \rightarrow \mathtt{E((a+b)/2)F((b+c)/2)} \qquad (2)$$

Production (2) can be applied to the module B(5) in a string

$$\cdots \mathtt{A(4)B(5)C(6)} \cdots$$

since the sequence of letters A, B, C in the production and in the string is the same, the numbers of formal parameters and actual parameters coincide, and the condition $4+5+6>10$ is true. As a result of production application, the module B(5) will create the pair of modules E(4.5)F(5.5). Naturally, other productions will be applied to the modules A(4) and C(6) in the same derivation step.

Parametric L-systems are parallel counterparts of attribute grammars [13, 14]. In both cases, nodes of a derivation tree are labeled by symbols and variables. The variable values are calculated using rules associated with productions. Although attribute grammars allow for bi-directional information flow, up and down the derivation tree, parametric L-systems pass information only from the top down. Using the terminology of attribute grammars, they support *inherited* attributes, not *synthesized* ones.

Parametric L-systems were introduced to model plants [23, 24], but their domain of application includes standard computational problems as well. Whether L-systems could serve as a general model of computation, and how that model would relate to existing machine architectures, is addressed in the following section.

4 L-systems as a model of parallel computation

In order to capture the general features that characterize computation using parametric L-systems, let us examine a few examples.

Example 4.1. The L-system

$$\begin{array}{lll} \omega: & \mathtt{A(1,1)} & \\ p_1: & \mathtt{A(x,y)} & \rightarrow \mathtt{A(x*y,y+1)} \end{array}$$

computes the value of n factorial in n derivation steps:

$$A(1,1) \Rightarrow A(1,2) \Rightarrow A(2,3) \Rightarrow A(6,4) \Rightarrow \ldots \Rightarrow A(n!, n+1).$$

Example 4.2. The L-system

$$\begin{array}{ll} \omega: & \mathtt{A(1,1)} \\ p_1: & \mathtt{A(x,y)} \rightarrow \mathtt{A(y,x+y)} \end{array}$$

computes consecutive terms of the Fibonacci series:

$$A(1,1) \Rightarrow A(1,2) \Rightarrow A(2,3) \Rightarrow A(3,5) \Rightarrow \ldots$$

Example 4.3. The following L-system implements the *odd-even-transposition* parallel algorithm for sorting numbers $a_1 a_2 \ldots a_n$ [2, pages 89-91]. It is assumed that for each predecessor module the first matching production in the list is used.

$$\begin{array}{lllllll} \omega: & \multicolumn{6}{l}{\mathtt{A(a_1)B(a_2)A(a_3)B(a_4)\ldots}} \\ p_1: & & \mathtt{A(x)} & > \mathtt{B(y)} & : x > y & \rightarrow & \mathtt{B(y)} \\ p_2: & & \mathtt{A(x)} & & & \rightarrow & \mathtt{B(x)} \\ p_3: & \mathtt{A(x)} < & \mathtt{B(y)} & & : x > y & \rightarrow & \mathtt{A(x)} \\ p_4: & & \mathtt{B(y)} & & & \rightarrow & \mathtt{A(y)} \end{array}$$

This L-system operates by exchanging parameter values a_i and a_{i+1} if $a_i > a_{i+1}$. Pairs whose first element has an odd-numbered index i are considered in odd-numbered derivation steps; pairs whose first element has an even-numbered index i are considered in even-numbered steps. This principle is illustrated by the following sample derivation.

$$\begin{array}{ll} \omega: & \mathtt{A(5)B(4)A(3)B(2)A(1)} \\ \mu_1: & \mathtt{B(4)A(5)B(2)A(3)B(1)} \\ \mu_2: & \mathtt{A(4)B(2)A(5)B(1)A(3)} \\ \mu_3: & \mathtt{B(2)A(4)B(1)A(5)B(3)} \\ \mu_4: & \mathtt{A(2)B(1)A(4)B(3)A(5)} \\ \mu_5: & \mathtt{B(1)A(2)B(3)A(4)B(5)} \end{array}$$

In spite of its simplicity, the parallel odd-even-transposition algorithm sorts n numbers in $O(n)$ time, which is better than the $O(n \log n)$ lower bound for sequential sorting.

Example 4.4. The L-system given below generates prime numbers using a parallel version of the *sieve of Eratosthenes*.

$$\begin{array}{llllll} \omega: & \multicolumn{5}{l}{\mathtt{N(1)E}} \\ p_1: & & \mathtt{N(k)} & & \rightarrow & \mathtt{N(k+1)A(k+1)} \\ p_2: & & \mathtt{A(k)} & & \rightarrow & \epsilon \\ p_3: & \mathtt{A(k)} < & \mathtt{E} & & \rightarrow & \mathtt{P(k)E} \\ p_4: & \mathtt{A(k)} < & \mathtt{P(n)} & : !(\mathtt{n}\%\mathtt{k} = 0) & \rightarrow & \mathtt{P(n)A(k)} \end{array}$$

As specified by production p_1, module $\mathtt{N(k)}$ generates a series of modules $\mathtt{A(k)}$ representing consecutive integers. Each module $\mathtt{A(k)}$ propagates to the right through a sequence of modules $\mathtt{P(n)}$ representing prime numbers found so far. If $\mathtt{A(k)}$ immediately precedes $\mathtt{P(n)}$ in a string, and $\mathtt{n}$ does not divide $\mathtt{k}$, $\mathtt{A(k)}$ moves to the right of $\mathtt{P(n)}$. Specifically, production p_4 copies A(k) to the right side of P(n.), while p_2 removes its previous occurrence. If n divides k, production p_2 acts alone and removes A(k) from the string. Thus, A(k) reaches the end of the string marked by module E if and only if k is prime. A(k) is then transformed to P(k) by production p_3. These operations are illustrated by the following derivation steps.

μ_8 : N(9)A(9)P(2)P(3)A(7)P(5)E
μ_9 : N(10)A(10)P(2)A(9)P(3)P(5)A(7)E
μ_{10} : N(11)A(11)P(2)P(3)P(5)P(7)E

As shown by the above examples, the most obvious characteristic of computation using parametric L-systems is the *monadic* nature [28] of production application. Information is passed *unilaterally* from the modules specified by the production predecessor to the modules in the successor. This contrasts L-systems with applicative languages, such as C or Lisp, in which computation occurs in *dyads*: a function is called, then returns a value to the calling environment.

The relationship between L-systems and the *SIMD* model of parallel computation [2, Chapter 1] is illustrated by Example 4.3. The string of modules corresponds to a collection of processors which act according to the same production set (*S*ingle *I*nstruction), but hold different parameters (*M*ultiple *D*ata). In this case, the processors are arranged in a linear array in which the neighbors can communicate. Since each production creates one module, the array contains a constant number of processors connected in a fixed way.

An extension to the SIMD model is suggested by Example 4.4, which allows productions to create new modules and destroy existing ones. We can envision the corresponding parallel architecture as a linear array of processors which may *self-replicate* and *die* in the course of computation. While current technology does not offer such processors, a model of parallel computation assuming their existence should not be dismissed as unrealistic; living organisms that form strings of cells (filaments) provide an example implemented by nature. De Koster and Lindenmayer describe the development of a filamentous blue-green bacteria using differential equations [5], but we may also view this filament as a parallel computing device solving differential equations using self-replicating processors.[2] The design and analysis of algorithms for such devices is an open problem.

In a more mundane setting, a computational model with self-replicating and self-destroying processors can be simulated using existing computers. An implementation of L-systems on a Connection Machine, proposed by R. and S. Pinter [19], is the closest approximation of this model to date. A separate processor is allocated to each module of the generated string. After a derivation step, the processors are reallocated to accommodate productions that create or delete modules. As reallocation may dominate the computation time, conclusions drawn from this implementation cannot be automatically extrapolated to the original model.

L-systems can also be implemented on coarse-grained parallel and single-processor machines. Below we analyze the use of L-systems as a programming language for these architectures. Extensions to parametric L-systems are introduced in order to make programs more efficient. Some of these extensions will have an effect on the way parallelism in L-systems is understood and implemented.

[2]This role reversal, pushed to the extreme, leads to the view that the universe is a computer [29, Part1: Edward Fredkin].

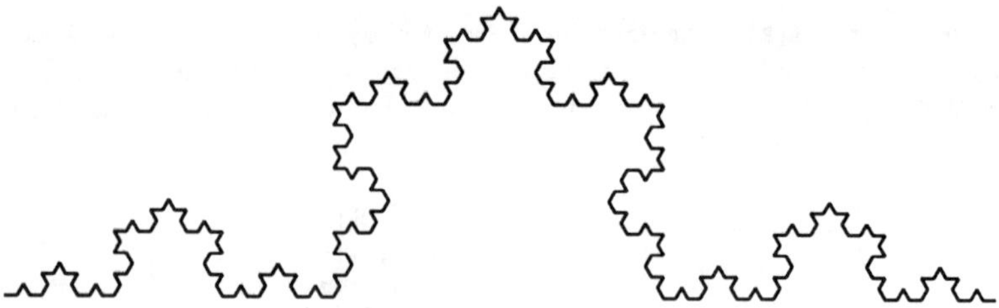

Figure 3: The Koch snowflake curve

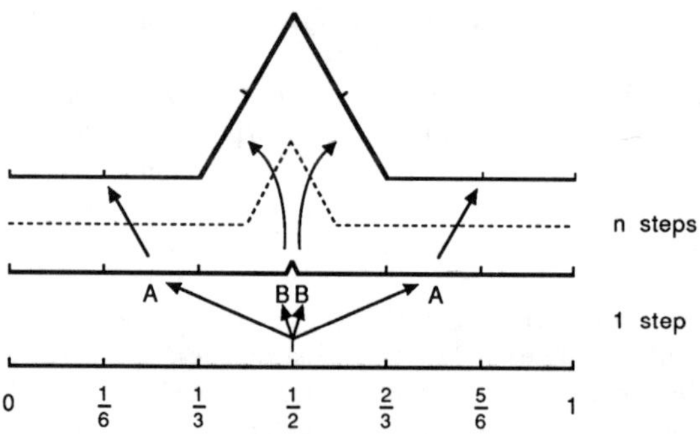

Figure 4: Construction of the snowflake curve

5 L-systems as a concurrent programming language

As a running example in this section, let us consider the generation of the Koch snowflake curve (Figure 3). Unlike the original construction which replaces edges by polygons in each derivation step [17, page 39], the L-system given below develops the curve gradually.

L-system 5.1

```
#define n 5            /* there are n+1 steps between line divisions */
#define e 0.01         /* error margin */
```

ω : F(1,1)
p_1 : F(a,d) : $a > d*(1+e) \rightarrow$ F(a − d/(2 ∗ n), d)
p_2 : F(a,d) : $a < d*(1-e) \rightarrow$ F(a + d/n, d)
p_3 : F(a,d) : $(a > d*(1-e))\&(a <= d*(1+e))$
$\rightarrow$ F(a/2, d/3) + F(0, d/3) − −F(0, d/3) + F(a/2, d/3)

Module F(a,d) draws a line of length a, and modules + and − rotate the turtle by $\pm 60°$. The number e is a small positive constant ($0 < e << 1$) introduced to avoid exact

Figure 5: Developmental sequence of the Koch snowflake curve

comparisons of floating-point numbers. In the first derivation step, production p_3 splits the unit-length line F(1,1), specified by the axiom, into four segments labeled **A**, **B**, **B**, **A** in Figure 4. The initial length of segments **A** equals 1/2, and the initial length of segments **B** equals 0. During the subsequent n steps, production p_1 decreases the length of each segment **A** by 1/2n per step, while production p_2 increases the length of each segment **B** by 1/n. Thus, all segments reach a length of 1/3 simultaneously in step n+1. At that time, the segments subdivide again, and the whole process repeats recursively. The sequence of curves obtained for n = 5 during the first 24 derivation steps is shown in Figure 5.

L-system 5.1 does not implement the underlying algorithm efficiently. The inefficiencies are imposed by the definition of parametric L-systems, which lacks useful constructs such as local and global variables. In order to introduce them without inventing a special notation, the syntax of the C programming language [12] is used. Subsets of the production set are defined as functions, with productions sharing the same predecessor grouped into a single definition. For example, the three productions of L-system 5.1 are represented by function F(a,d) below.

L-system 5.2 — productions only

```
F(a,d)
float a,d;
{
    if (a<d*(1-e))
        produce(F(a-d/(2*n),d));
    else if (a>d*(1+e))
        produce(F(a+d/n,d));
    else
        produce(F(a/2,d/3)+F(0,d/3)--F(0,d/3)+F(a/2,d/3));
}
```

Depending on the conditions, the successor module or string of modules is formed using a `produce()` statement, which *creates* the specified module or sequence of modules, *evaluates* all expressions such as `a/2` or `a-d/(2*n)`, and *passes* the results to the modules. The implementation of `produce()` depends on the architecture of the machine used to execute the L-system. A sample implementation for a sequential machine is discussed in Section 8.

The first improvement aimed at increasing the efficiency of L-system 5.2 is achieved by introducing *local variables* to avoid repetitive evaluation of the same expression within a module. The last `produce()` statement is replaced by the following block:

```
{
    float x, y;
    x=a/2; y=d/3;
    produce(F(x,y)+F(0,y)--F(0,y)+F(x,y));
}
```

The amount of computation can be further reduced if we notice that in any derivation step all segments **A** have the same length, and so do all segments **B**. To take advantage of this observation, values required by productions are precomputed and passed as *global variables.* This idea is illustrated by L-system 5.3 which consists of two functions, `F(a)` and `main()`. The function `F(a)` is a modification of `F(a,d)` from L-system 5.2, taking advantage of the global variables. The function `main()` updates the global variables and calls function `apply_productions()` which performs a derivation step. The implementation of `apply_productions()` is machine-dependent, as was the case with `produce()`. A sequential implementation is described in Section 8.

L-system 5.3

```
#define n 5          /* there are n+1 steps between line divisions */
#define e 0.01       /* error margin */
float aA=1, aB=0;    /* lengths of segments A and B */
float d=1;           /* length of A and B before next division */
float dA, dB;        /* upper and lower approximations of d */

F(a)
float a;
{
    if (a>dA)
        produce(F(aA));
    else if (a<dB)
        produce(F(aB));
    else
        produce(F(aA)+F(aB)--F(aB)+F(aA));
}
```

a) b)

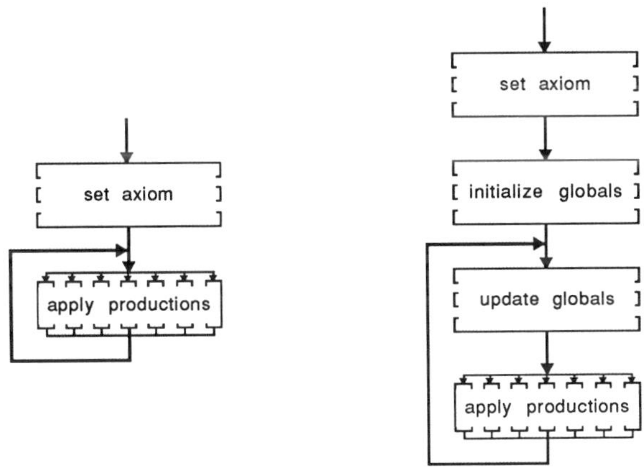

Figure 6: Operation of an L-system without global variables (a) and with global variables (b)

```
main()
{
    int i, N=100;          /* variables controlling the derivation */
    produce(F(aA));        /* set the axiom */
    for (i=1; i<N; i++) {
        /* update global variables */
        dA=d*(1+e); dB=d*(1-e);
        if (aA>dA) {       /* equivalent to aB<dB */
            aA=aA-d/(2*n); aB=aB+d/n;
        }
        else {
            aA=aA/2; aB=0; d=d/3;
        }
        /* perform one derivation step */
        apply_productions();
    }
}
```

The updating of global variables, illustrated by the above example, modifies the notion of derivation in L-systems; it is no longer simply a sequence of parallel production applications (Figure 6a), but has become a cycle of alternating sequential and parallel computations (Figure 6b).

In each step, productions can read global variables. Should they be able to write them as well? The answer to this question depends on the way parallelism in L-systems is understood and implemented. Strictly speaking, the definition of L-systems requires productions to be applied *simultaneously* in each derivation step. In this context, allowing productions to modify and write global variables leads to inconsistencies when the productions attempt to set the same variable to different values. The situation changes if productions are viewed as *asynchronous concurrent processes* [7, Chapter 4]. Various synchronization techniques are available to resolve the write conflict by allowing only one process at a time to modify a shared variable. For example, its value may be read, changed, and written back inside a *critical section* in each production [7, page 82]. Of course, the mutual exclusion of critical sections implies that productions are no longer applied simultaneously. Even with no order of production application imposed within a derivation step, the result of computation may be deterministic, as is the case when a shared variable is used to count the number of times a particular production has been applied.

Concurrent application of productions is important for two reasons. First, it is conceptually close to the definition of L-systems as a class of parallel rewriting systems. Second, it is needed for an efficient implementation of L-systems on multiprocessor machines. This is true not only for the hypothetical model with self-replicating processors, discussed in Section 4, but also for more traditional parallel architectures with a fixed, possibly small, number of processors. In the latter case, each processor carries out the derivation for a segment of the string.

In summary, programs expressed using L-systems can be implemented on machines with various architectures, as long as functions `produce()` and `apply_productions()` are available. The semantics of these functions is machine independent and results directly from the definition of L-systems. Due to their firm theoretical background, L-systems are attractive as a parallel programming language.

6 L-systems as a sequential programming language

Sequential implementations of L-systems are important in practice, since most computers currently in use have a traditional single-processor architecture. The usual technique for performing the derivation step $\mu_i \Longrightarrow \mu_{i+1}$ is to scan the string μ_i from left to right, apply productions sequentially to consecutive modules, and append the resulting successors to the string μ_{i+1} [22, pages 107-108]. Since all modules of μ_i are considered before the next derivation step, the result is the same as if productions were applied in parallel. In addition, programs can take advantage of the sequential application of productions within each derivation step, and use constructs that rely on it. An example is given by L-system 6.1, in which function `main()` from L-system 5.3 has been modified in order to move all calculations of global variables to a module `U`. The updating of variables at the beginning of each derivation step has been achieved by specifying module `U` as the first symbol of the axiom and, in consequence, as the first symbol of each string derived. The division of computational tasks between function `main()` and the productions is no longer necessary. The productions carry out all computations; `main()` is confined to controlling the sequence of derivation steps.

L-system 6.1 — replacement for function main() of L-system 5.3

```
main()
{
    int i, N=100;              /* variables controlling the derivation */
    produce(UF(aA));           /* set the axiom */
    for (i=1; i<N; i++) {
        apply_productions(); /* perform one derivation step */
    }
}


U()                            /* update global variables */
{
    dA=d*(1+e); dB=d*(1-e);
    if (aA>dA) {
        aA=aA-d/(2*n); aB=aB+d/n;
    }
    else {
        aA=aA/2; aB=0; d=d/3;
    }
    produce(U);
}
```

The above example illustrates the impact of L-systems on the concept of a sequential program. Its execution is viewed as a sequence of production applications. Typically, a production performs some computation and creates new tasks (represented as modules of the successor string) using the function produce(). The sequential order of the derivation steps, and of production applications within each step, implies that a new task will be performed after the completion of all tasks created earlier. Consequently, computation can be viewed as a *queue of tasks*, with the current task taken from the beginning of the queue, and newly created tasks added to its end. The statement produce() can be compared to a function call in that it delegates some computational tasks to another function. However, due to the monadic character of production application (Section 4), the results are never passed back. The differences between a function call and the execution of a produce() statement are further explained in Table 1.

Function call	Produce() statement execution
Puts the actual parameters and the return address at the top of a *stack*. Transfers control to the called function.	Puts the actual parameters and the address of the function specified as an argument at the end of a *queue*. Control is not transferred to the queued function.
Once a function is completed, it returns control to the calling function at the top of the stack. The stack organizes computation in a first-in, last-out fashion.	Once a function is completed, control is transferred to the next function in the queue. The queue organizes computation in a first-in, first-out fashion.
The called function may return a value to the calling function.	There is no concept of "returning a value". The effect of computation is to create new modules and supply parameters for them.

Table 1: Differences between a function call and the execution of a produce() statement

To further illustrate these differences, let us refer to another example. A binary tree is searched for occurrences of nodes with a given value of a numeric search key. The structure of a node is defined as follows:

```
struct node {
    int key;
    char data[100];
    struct node *left;
    struct node *right;
};
```

The recursive function given below searches the externally defined tree in depth-first order and prints data from the nodes that match the search key.

```
/* Depth-first search of a binary tree. */
/* The tree and the search key are defined as external variables */

search(current)
struct node *current;
{
    if (current->key == search_key)
        printf("%s\n", current->data);
    if (current->left != NULL)
        search(current->left);
    if (current->right != NULL)
        search(current->right);
}
```

By changing all function calls to produce() statements, an L-system production that searches the same tree in breadth-first order is obtained.

L-system 6.2

```
search(current)
struct node *current;
{
    if (current->key == search_key)
        produce(printf("%s\n", current->data));
    if (current->left != NULL)
        produce(search(current->left));
    if (current->right != NULL)
        produce(search(current->right));
}
```

The statement produce() provides an alternative to recursion as a programming construct by which a function can refer to itself. The general methodology for designing and analyzing programs relying on this construct remains open for further research.

7 Summary of extensions

The following extensions of parametric L-systems were introduced and discussed in Sections 4 to 6.

- **Local variables.** To increase the efficiency of production application, repetitive calculation of the same expression can be prevented by assigning its value to a temporary variable.
- **Global variables available for reading.** Variables that can be read by all productions may be used to represent the environment in which derivation takes place. Since they are updated between derivation steps, sequencing of events during derivation is required.
- **Global variables available for writing.** These variables provide a means for communicating between distant modules in a string. In addition, they can be used to collect global data, such as the number of times a particular production has been applied in a given derivation step. The use of global variables may create write conflicts which can be solved by proper synchronization of production applications within a derivation step.

Two more constructs were tacitly assumed.

- **Typed parameters.** In programming practice it is inconvenient to assume that all parameters are of the same type. For example, the L-systems in Section 5 require real-valued parameters, but L-systems 4.1, 4.2, and 4.4 operate on integer-valued parameters. The introduction of typed parameters and variables allows for proper specification of the operations involved; it may also reduce the amount of memory needed to store a string.

- **Built-in functions.** In addition to logical and arithmetic operators, productions may include function calls. For example, print functions are needed to produce output (L-system 6.2). Standard mathematical functions (trigonometric, logarithmic, random) are useful in arithmetic and logical expressions.

The above extensions augment the expressive power of L-systems and may be employed to speed up the derivation process on computers with various parallel and sequential architectures.

8 An implementation of L-systems

In Sections 5 and 6, a series of extensions to L-systems was introduced using the syntax of C. This suggests that L-systems could be treated as C programs, compiled, and executed. A simple programming environment consisting of two externally defined strings and a library of two functions is sufficient for this purpose. The first function, `append()`, implements the statement `produce()`, found in the examples, in a manner consistent with C. For instance,

```
produce(A(x,y)BC(2));
```

is expressed as a sequence of function calls:

```
append(2, A, x, y);
append(0, B);
append(1, C, 2);
```

Function `append()` adds a single module to a predefined string. The first argument indicates the number of actual parameters of the module, the second is the module's symbol, and the remaining arguments are parameter values.

In addition to `append()`, the library contains the function `apply_productions()`, which performs a derivation step by manipulating two predefined strings of modules. String 1 is scanned, the matching productions are applied, and the resulting successors are appended to string 2. When a derivation step is completed, the content of string 2 is moved to string 1 and string 2 is cleared in anticipation of the next derivation step. At this stage, another call to `apply_productions()` may take place.

In the case of context-free L-systems, the two strings can be replaced by a single queue. The predecessor modules are obtained from the beginning of the queue and the successors put at the end. A special module producing itself, such as `U` in L-system 6.1, may be used to indicate the start of a derivation step.

The expression of productions in a high-level language opens the possibility for experimenting with L-systems without an extensive, specialized programing environment. Unfortunately, the syntax of C does not provide any construct for specifying context-sensitive productions.

9 Conclusions

We have shown that the ramifications of L-systems extend beyond the generally recognized areas of formal languages, biological models, and computer graphics. Specifically, parametric L-systems provide a model for parallel computation, and may serve as a basis for a class of programming languages.

As a model of computation, L-systems represent an architecture consisting of self-replicating processors. Although this architecture is not available at the present time, it is feasible in principle. The design and analysis of algorithms for self-replicating parallel architectures is open for further research.

As a paradigm for programming languages, L-systems are related to two notions: monadic computation and concurrency. Monadic computation consists of tasks that create new tasks and pass results to them. This corresponds to the execution of productions, which create new modules and supply them with actual parameter values. There is no concept of returning a value to the predecessor module. The parallel application of productions within a derivation step imposes constraints on the set of operations that the productions may perform; specifically, they must not write shared variables. This limitation can be removed if productions are viewed as concurrent processes that write global variables inside critical sections.

Parametric L-systems with the extensions introduced in this paper can be expressed in a high-level programming language complemented with two functions, `produce()` and `apply_productions()`. Their implementation depends on the underlying machine architecture, whether it is massively parallel, coarse-grain parallel, or single processor, but the semantics is machine independent; `apply_productions()` performs a derivation step by invoking productions that create new modules using `produce()`. These two functions provide a vehicle for creating programs that are machine independent, but can take advantage of parallel machine architecture.

In the special case of single-processor machines, productions are applied sequentially, and programs may take advantage of this. Specifically, all computation can be performed by productions. The sequential execution of productions leads to a view of a program as a queue of tasks. This view, as opposed to the traditional view of a structured program as a sequence of function calls, may lead to more efficient implementations of sequential programs, since no return of values and no resumption of the calling function is required.

Introduced more than twenty years ago, L-systems have proved to be a fruitful concept. It is exciting that they continue to unify notions and inspire research in diverse areas.

10 Acknowledgments

We are deeply indebted to Professor Lindenmayer for having guided us through the world of L-systems. We would like to thank Lynn Mercer for her comments and suggestions on this paper. This research was sponsored in part by an operating grant and graduate scholarship from the Natural Sciences and Engineering Research Council of Canada.

References

[1] H. Abelson and A. A. diSessa. *Turtle geometry.* M.I.T. Press, Cambridge, 1982.

[2] S. G. Akl. *The design and analysis of parallel algorithms.* Prentice Hall, Englewood Cliffs, 1989.

[3] R. Baker and G. T. Herman. Simulation of organisms using a developmental model, parts I and II. *Int. J. of Bio-Medical Computing*, 3:201–215 and 251–267, 1972.

[4] T. W. Chien. Graphical interpretation of the biological development modelled by VDOL-systems. Master's thesis, University of Western Ontario, 1989.

[5] C. G. de Koster and A. Lindenmayer. Discrete and continuous models for heterocyst differentiation in growing filaments of blue-green bacteria. In *Acta Biotheoretica*, volume 36, pages 249–273. Kluwer Academic, the Netherlands, 1987.

[6] F. J. de Vries, G. Havenith, and J. M. Janssen. *CELIA manual.* Theoretical Biology Group, University of Utrecht, The Netherlands, 1985.

[7] H. M. Dietel. *An introduction to operating systems.* Addison-Wesley, Reading, 1983.

[8] D. Frijters and A. Lindenmayer. A model for the growth and flowering of *Aster novae-angliae* on the basis of table (1,0)L-systems. In G. Rozenberg and A. Salomaa, editors, *L Systems*, Lecture Notes in Computer Science 15, pages 24–52. Springer-Verlag, Berlin, 1974.

[9] G. T. Herman and W. H. Liu. The daughter of CELIA, the French flag, and the firing squad. *Simulation*, 21:33–41, 1973.

[10] P. Hogeweg and B. Hesper. A model study on biomorphological description. *Pattern Recognition*, 6:165–179, 1974.

[11] J. M. Janssen and A. Lindenmayer. Models for the control of branch positions and flowering sequences of capitula in *Mycelis muralis* (L.) Dumont (Compositae). *New Phytologist*, 105:191–220, 1987.

[12] B. W. Kernighan and D. M. Ritchie. *The C programming language.* Prentice-Hall, Englewood Cliffs, 1978.

[13] D. E. Knuth. Semantics of context-free languages. *Mathematical Systems Theory*, 2(2):127–145, 1968.

[14] P. M. Lewis II, D. J. Rosenkrantz, and R. E. Stearns. *Compiler design theory.* Addison-Wesley, Reading, 1978.

[15] A. Lindenmayer. Mathematical models for cellular interaction in development, Parts I and II. *Journal of Theoretical Biology*, 18:280–315, 1968.

[16] A. Lindenmayer. Adding continuous components to L-systems. In G. Rozenberg and A. Salomaa, editors, *L Systems*, Lecture Notes in Computer Science 15, pages 53–68. Springer-Verlag, Berlin, 1974.

[17] B. B. Mandelbrot. *The fractal geometry of nature.* W. H. Freeman, San Francisco, 1982.

[18] S. Papert. *Mindstorms: Children, computers and powerful ideas.* Basic Books, New York, 1980.

[19] R. Y. Pinter and S. S. Pinter. Efficient processing of L-systems on the Connection Machine, or parallel breadth-first expansion, 1989. Preprint, Computer Science Department, Yale University, New Haven 1989.

[20] P. Prusinkiewicz. Graphical applications of L-systems. In *Proceedings of Graphics Interface '86 — Vision Interface '86*, pages 247–253. CIPS, 1986.

[21] P. Prusinkiewicz. Applications of L-systems to computer imagery. In H. Ehrig, M. Nagl, A. Rosenfeld, and G. Rozenberg, editors, *Graph grammars and their application to computer science; Third International Workshop*, pages 534–548. Springer-Verlag, Berlin, 1987. Lecture Notes in Computer Science 291.

[22] P. Prusinkiewicz and J. Hanan. *Lindenmayer systems, fractals, and plants*, volume 79 of *Lecture Notes in Biomathematics.* Springer-Verlag, Berlin, 1989.

[23] P. Prusinkiewicz and J. Hanan. Visualization of botanical structures and processes. In D. Thalmann, editor, *Scientific Visualization and Graphics Simulation*, pages 183–201. J. Wiley & Sons, Chichester, 1990.

[24] P. Prusinkiewicz and A. Lindenmayer. *The algorithmic beauty of plants.* Springer-Verlag, New York, 1990. With J. Hanan, F. D. Fracchia, D. R. Fowler, M. J. M. de Boer and L. Mercer.

[25] G. Rozenberg and A. Salomaa. When L was young. In G. Rozenberg and A. Salomaa, editors, *The Book of L.*, pages 383–392. Springer-Verlag, Berlin, 1986.

[26] M. F. Shebell. Modeling branching plants using attribute L-systems. Master's thesis, Worcester Polytechnic Institute, 1986.

[27] A. L. Szilard and R. E. Quinton. An interpretation for DOL systems by computer graphics. *The Science Terrapin*, 4:8–13, 1979.

[28] S. A. Ward and R. H. Halstead. A syntactic theory of message passing. *Journal of the ACM*, 27(2):365–383, April 1980.

[29] R. Wright. *Three scientists and their gods.* Times Books, New York, 1988.

Parameterized L Systems for Modelling: Potential and Limitations[1]

T. W. Chien and H. Jürgensen

Department of Computer Science, The University of Western Ontario, London, Ontario, Canada, N6A 5B7.

Abstract: L systems have been used as a tool for computer graphics and computer animation. Occasionally their limitations in this context have been criticized. In order to evaluate the potential and limitations of L systems in this context rigorously, however, one requires a sufficiently general formal notion of graphical interpretation of L systems.

In this paper, we provide a definition of parameterized D0L systems and parameterized D0L systems with branching—this concept has been used for quite some time with rather ad hoc definitions—and prove that the recursive parameterized D0L systems are capable of simulating all deterministic L systems with interaction.

We then define the notion of graphical interpretation of parameterized D0L systems in very general terms—certainly general enough to include all existing applications of L systems in computer graphics. We indicate the characteristics of a prototype implementation of this general concept and show examples.

On the basis of the general model we formulate a few preliminary conclusions about the limitations of L systems as far as their application to graphical modelling is concerned.

Keywords: L system, simulation, plant modelling, computer graphics.

1. L Systems as Modelling Tools

It was (probably) at the 1975 conference on "Formal Languages, Automata, and Development" held at Nordwijkerhout in The Netherlands, that one of us (H. J.) first saw an L system based movie [20]. It had been produced by H. Maurer and his group, then of Karlsruhe University. What it showed? Trees—various L systems for trees and some more or less arbitrary graphical renditions of the trees on a computer screen. These trees grew as the rewriting steps of the L systems were performed. Their growth was rather jumpy—as only very large developmental steps were simulated. The end of that movie was as follows: Some rather awkward L system was shown followed by a jungle of messy lines across the screen; this was followed by pictures of beautiful real trees—what a difference!

---

[1] This work was supported by the Natural Science and Engineering Research Council of Canada, Grant OGP0000243. It is in part based on the first author's M. Sc. thesis [4]. An extended version of this paper containing a few more illustrations, their specifications, and some proofs omitted below is available as a technical report [5].

To be just, that movie, despite its shortcomings, was probably the first to point out a potential of L systems, that is, to serve as a formal tool for computer animation. Since then, L systems have been considered and used for computer graphics on several occasions. Especially, we mention the work by Aono and Kunii [1, 2], by Prusinkiewicz and others [22–27], by de Reffye and others [28], by Smith [33], and by Szilard and others [34]. A recent book by Prusinkiewicz and Lindenmayer [27] and a bibliography by the same authors [14] contain additional important references.

L systems were introduced to model certain "logical" aspects of the development of organisms [17]. Essentially, an L system consists of an alphabet Σ, some kind of system δ of rewriting rules, and a start word ω. The symbols in the alphabet represent differentiated parts of the organism. The rules describe the "developmental program." The start word or axiom represents some kind of initial situation. The state of the organism at time i, $i \geq 0$, is expressed by a word w_i over Σ, where $w_0 = \omega$. From w_i a description of a possible next state w_{i+1} is obtained using the rules of δ. Note that δ may allow for non-deterministic or probabilistic behaviour of some kind. Usually, the type of an L system is described by an abbreviation consisting of some letters and digits followed by the letter 'L.' In this paper we mainly consider a variant of D0L systems, that is, L systems whose rewriting rules are deterministic and require no interaction between symbols. Since the early papers [17], a large amount of research has been conducted on L systems. The books by Herman and Rozenberg [9] and by Rozenberg and Salomaa [29] are standard references.

What can and cannot be done with L systems in terms of graphics depends strongly on the way how strings over Σ are interpreted as pictures. Hence, without a rigorous graphical semantics of L systems no real statement about their potential for specifying pictures can be made.

In this paper we lay the foundations for a mathematically rigorous evaluation of the potential and limitations of the application of L systems in graphical modelling. We propose a quite general notion of graphical interpretation of L systems which incorporates all existing applications of L systems in graphics and which, indeed, could be argued to provide the formal framework for all conveivable "natural" applications of L systems in graphics. While more specialized notions could suffice to point out certain aspects of the potential of L systems, this more general approach is required for a rigorous discussion of their limitations.

We base our poposal on new definitions of parameterized D0L systems and of parameterized branching D0L systems. Such systems have been in use for quite some time (see [19, 6, 8, 27]) with mostly ad hoc definitions. In [26, 27] and [4] formal definitions are provided which, however, seem to lack algebraic elegance or are, intuitively, too restrictive. We then prove that the recursive parameterized D0L systems are sufficiently powerful to simulate all DIL systems. This result places parameterized D0L systems fairly high in the L system hierarchy. Moreover, while not very surprising, it can be interpreted as an indicator of the modelling potential of parameterized D0L systems.

The feasibility of our general notion of graphical interpretation of L systems is demonstrated by a prototype implmentation. A few characteristics of the implementation as well as some examples of its application are provided in Section 6 below.

With this paper we also hope to contribute to the discussion on where to place the boundary between syntactic description and graphical semantics. More specifically, we argue that, in keeping with the general philosophy of physical modelling, the issues of object simulation and of graphical rendering should be kept apart. This is in contrast to most of the existing applications of L systems to graphics where the graphical information itself is encoded in the L system, for instance in terms of branching directions or the turtle geometry. Our main argument in favour of separating these issues is that of oimplicity.

Consider for example the task of modelling the development of a plant. Depending on the level abstraction, one can distinguish several aspects of the developmental process. There is a *logical* component[2] which concerns the development of plant organs, inflorescence patterns, and similar issues. There is a *phyico-chemical* component which concerns for instance the concentration levels of chemical substances, the rigidity or the weight of parts. There is also a *geometrical component* which concerns the actual shape of the plant.

L systems have been proven to be adequate tools for modelling the logical component of plant development at a fairly high level of abstraction—as witnessed by the biological literature. The main problem with modelling at lower levels of abstraction—modelling a complete plant at the cell level, for instance—or with including the physico-chemical and geometrical aspects at an acceptable resolution is the huge number of symbols and rules required for the L system. For example, we do not believe that the so-called time-delayed L systems ([35, 23]) are a proper model for time delays—because they use excessively large sets of rules and symbols to describe a very simply describable phenomenon—while, at the same time, we quite agree that time delays need to be modelled. In a similar sense one could argue that the stochastic component of probabilistic L systems (first introduced in [10, 11]) should mainly affect the physical, chemical, and geometrical aspects of the developmental model.

The physico-chemical and geometrical aspects of development are essentially non-discrete. It seems that discretization renders their description far more complex than necessary. Moreover, it seems natural to assume that the geometrical component is highly dependent on the physical and chemical details.

These considerations lead us to consider the following paradigm for the application of L systems in the context of graphical modelling:

(1) L systems will model the logical aspects of devlopment only.

(2) For the logical component, different levels of resolution are realized using hierarchies of L systems.

(3) To take the physico-chemical component and part of the geometrical aspects into account, one uses parameterized L systems in which parameters have been added to the symbols of the alphabets.

(4) For the graphical interpretation the symbols are considered as functions of their parameters with connectivity information obtained from left-to-right neighbourhoods in the string descriptions.

In this paper, we mainly address items 1, 3, and 4. Some ideas for hierarchies of L systems (item 2) could result from generalizing the considerations of [12].

The paper is structured follows: In Section 2 we briefly introduce the required notation; for all definitions in the area of D0L systems we refer to [9] and [29]. In

[2] Other terms like *genetic, structural,* or *morphological* were considered by us, but rejected to avoid their special connotations. However, at a certain level of abstraction, the logical component may indeed be precisely the genetic component, for instance.

Chapter 3 we establish a precise formalism for L systems with branching.[3] In Section 4 we add possibly continuous parameters to D0L systems in a way similar to [19, 6, 8, 27]. The resulting systems are called VD0L systems.[4] While the intention is to use them like D0L systems they are powerful enough to simulate even systems which would otherwise require interaction. In Section 5, we introduce a graphical semantics for VD0L systems. A prototype implementation and some examples are described in Section 6. In Section 7, we summarize some preliminary conclusions concerning the potential and limitations of L systems with respect to graphical modelling.

2. Basic Notions and Notation

As usual, $\mathbb{N}$ denotes the set of positive integers, $\mathbb{N}_0 = \mathbb{N} \cup \{0\}$, and $\mathbb{R}$ is the set of real numbers. An *alphabet* is a finite, non-empty set. Let Σ be an alphabet. Then Σ^* denotes the set of *words* over Σ including the *empty word* ε. Let $\Sigma^+ = \Sigma^* \setminus \{\varepsilon\}$. For a word $w \in \Sigma^*$, $|w|$ denotes the length of w. For i and j with $1 \leq i \leq j \leq |w|$, let $w^{(i,j)}$ be the word $\overline{w}$ such that $w = u\overline{w}v$ with $|u| = i - 1$ and $|\overline{w}| = j - i + 1$. Moreover, let $w^{(i)} = w^{(i,i)}$.

A *D0L system* is a triple $G = (\Sigma, \delta, \omega)$ such that Σ is an alphabet, δ is an endomorphism of Σ^* and ω is a non-empty word over Σ, the *axiom* of G. The *derivation sequence* of G is the sequence $\delta^0(\omega) = \omega$, $\delta(\omega)$, $\delta^2(\omega)$, ... Occasionally we use the symbol $\vdash$ to denote derivation steps. For all further L system notions we refer to the books [9] and [29].

3. Branching D0L Systems

In this section we define the notions of a D0L system with branching—essentially a well-known concept—and of logical neighbourhoods.

First we introduce some auxiliary notation. Let Σ be an alphabet with $|\Sigma| \geq 3$, let $\langle, \rangle$ be two distinct elements of Σ, and let ν be a mapping of Σ into $\mathbb{N}_0$ such that $\nu(\langle) = \nu(\rangle) = 0$. The quadruple $(\Sigma, \langle, \rangle, \nu)$ is called a *branching alphabet with branch delimiters* $\langle$ *and* $\rangle$. For $i = 0, 1, \ldots$ let $\Sigma_i = \{a \mid a \in \Sigma \setminus \{\langle, \rangle\}, \nu(a) = i\}$. The set $D(\Sigma, \langle, \rangle, \nu)$ is the language consisting of precisely the following words over Σ:

- If $a \in \Sigma \setminus \{\langle, \rangle\}$ and $w_1, w_2, \ldots, w_{\nu(a)} \in D(\Sigma, \langle, \rangle, \nu)$ then $a\langle w_1\rangle\langle w_2\rangle \cdots \langle w_{\nu(a)}\rangle \in D(\Sigma, \langle, \rangle, \nu)$.
- If $w_1, w_2 \in D(\Sigma, \langle, \rangle, \nu)$ then $w_1 w_2 \in D(\Sigma, \langle, \rangle, \nu)$.

Intuitively, the symbols $\langle$ and $\rangle$ indicate the beginning and the end of a branch. For a symbol a different from these, the number $\nu(a)$ is the number of branches originating from it. In $a\langle w_1\rangle\langle w_2\rangle \cdots \langle w_{\nu(a)}\rangle$ the subword $\langle w_i\rangle$ with $1 \leq i \leq \nu(a)$ is called the ith branch of a. Note that the homomorphism which maps $\langle$ onto $($, $\rangle$ onto $)$, and everything else onto the empty word transforms $D(\Sigma, \langle, \rangle, \nu)$ into a subset of the Dyck language D_1, that is, the language of strings of nested brackets.[5] The language $D(\Sigma, \langle, \rangle, \nu)$ does not contain the empty word. Moreover, $D(\Sigma, \langle, \rangle, \nu)$ is non-empty if and only if Σ contains a symbol a different from $\langle$ and $\rangle$ such that $\nu(a) = 0$.

---

[3] Of course, such L systems have been considered informally since the very first days of L systems.

[4] V=valued; the letter 'V' is also in use for something else; unfortunatly, there do not seem to be any free letters left.

[5] $D(\Sigma, \langle, \rangle, \nu)$ can be considered as the free semigroup over the free algebra generated by $\emptyset$ with operation symbols in $\Sigma \setminus \{\langle, \rangle\}$ and arities given by ν.

Definition 3.1 A *D0L system with branching (a BD0L system)* is a sixtuple $G = (\Sigma, \delta, \omega, \langle, \rangle, \nu)$ with the following properties:

(a) (Σ, δ, ω) is a D0L system.
(b) $(\Sigma, \langle, \rangle, \nu)$ is a branching alphabet.
(c) $\delta(\langle) = \langle$ and $\delta(\rangle) = \rangle$.
(d) $\omega \in D(\Sigma, \langle, \rangle, \nu)$.
(e) For every $a \in \Sigma_0$ one has $\delta(a) \in D(\Sigma, \langle, \rangle, \nu)$.
(f) For every $a \in \Sigma_i$ with $i > 0$ one has $\delta(a) \in D(\Sigma, \langle, \rangle, \nu)^* \Sigma_i$.

Note that a BD0L system is always *propagating,* that is, δ does not yield the empty word nor even a branch with only the empty word in it. The generating mechanism of a BD0L system is that of the underlying D0L system. The conditions on δ guarantee that branching introduced in a step is maintained in all future steps.

Remark 3.2 *Let $G = (\Sigma, \delta, \omega, \langle, \rangle, \nu)$ be a BD0L system. Then $L(G) \subseteq D(\Sigma, \langle, \rangle, \nu)$.*

Let $(\Sigma, \langle, \rangle, \nu)$ be a branching alphabet. On the words $w \in D(\Sigma, \langle, \rangle, \nu)$ we define *logical neighbourhoods.* Suppose that $w = a_1 a_2 \cdots a_n$ with $a_i \in \Sigma$ for $i = 1, \ldots, n$. We define a relation Λ_w on the set $\{1, \ldots, n\}$. One has $(i, j) \in \Lambda_w$ if and only if $1 \le i < j \le n$ and one of the following conditions is satisfied:

- $i = j - 1$ and $a_i \neq \rangle$.
- $a_{j-1} = \rangle$ and $(i, k) \in \Lambda_w$ for some k with $i < k < j - 1$ such that $a_k = \langle$ and $a_{k+1} \cdots a_{j-2} \in D(\Sigma, \langle, \rangle, \nu)$.

We say that a_j is a *logical right neighbour* of a_i in w if and only if $(i, j) \in \Lambda_w$. Moreover, a_j is the kth logical right neighbour of a_i if there exist indices $j_1, \ldots, j_k$ such that $j_1 < \cdots < j_k = j$, $(i, j_1), \ldots, (i, j_k) \in \Lambda_w$, and $(i, j') \notin \Lambda_w$ for $i < j' < j_k$, $j' \notin \{j_1, \ldots, j_k\}$. Let $\Lambda_{k,w}$ be the set of all $(i, j) \in \Lambda_w$ with a_j the kth logical right neighbour of a_i.

For example, suppose that $\Sigma = \{a, b, \langle, \rangle\}$ with $\nu(a) = 0$ and $\nu(b) = 2$. For the string[6] $w = a_1 b_2 \langle_3 b_4 \langle_5 a_6 \rangle_7 \langle_8 a_9 \rangle_{10} \rangle_{11} \langle_{12} a_{13} a_{14} \rangle_{15} a_{16}$ the set Λ_w is computed as

$$\Lambda_w = \{(1,2), (2,3), (2,12), (2,16), (3,4), (4,5), (4,8), (5,6), (6,7), (8,9), (9,10), (12,13), (13,14), (14,15)\}.$$

Intuitively, the strings $\langle b \langle a \rangle \langle a \rangle \rangle$ and $\langle aa \rangle$ denote branches originating at the first b in w while the last a in w is the direct continuation of the main string.

For the graphical interpretation of strings to be defined later, the notion of logical right neighbourhood will serve the following purpose: The representation of the symbols a_i and a_j will share information if and only if $(i, j) \in \Lambda_w$.

Lemma 3.3 *Let $(\Sigma, \langle, \rangle, \nu)$ be a branching alphabet, and let $w \in D(\Sigma, \langle, \rangle, \nu)$. Suppose that $w = ua\langle w_1 \rangle \langle w_2 \rangle \cdots \langle w_{\nu(a)} \rangle bv$ with $u, v \in \Sigma^*$, $a \in \Sigma \setminus \{\langle, \rangle\}$, $b \in \Sigma \setminus \{\langle\}$, and $w_1, w_2, \ldots, w_{\nu(a)} \in D(\Sigma, \langle, \rangle, \nu)$. Let $n = |u|$, $m = |v|$, and $l_i = |w_i|$ for $i = 1, \ldots, \nu(a)$. Then Λ_w has the following properties:*

(1) $(n + 1, j) \in \Lambda_w$ *if and only if* $j \in \{n + 2(k + 1) + \sum_{i=0}^{k} l_i \mid 0 \le k \le \nu(a)\}$.
(2) For $j \in \{n + 2(k + 1) + \sum_{i=0}^{k} l_i \mid 0 \le k < \nu(a)\}$ *one has* $(j, j') \in \Lambda_w$ *if and only if* $j' = j + 1$.
(3) For every i *with* $1 < j \le |w|$ *there is a unique* j *with* $1 \le j < |w|$ *such that* $(j, i) \in \Lambda_w$.

[6] We indicate symbol positions by subscripts in this particular case.

One uses Lemma 3.3 to derive an algorithm for determining the logical neighbourhoods as required for the graphical rendering process. The algorithm reads the input word $w \in D(\Sigma, \langle, \rangle, \nu)$ once from left to right, uses a stack, and, while scanning a symbol in the input word, indicates the symbol in the string of which the current symbol is a right neighbour.

4. VD0L Systems

In this section we provide a definition of D0L systems with continuous parameters, VD0L systems, and we prove that a restricted type of VD0L systems is powerful enough to simulate DIL systems.

Let Σ be an alphabet. With every $a \in \Sigma$ a number $\pi(a) \in \mathbb{N}_0$ of real-valued parameters $v_1, \ldots, v_{\pi(a)} \in \mathbf{R}$ is associated resulting in a *V-symbol* denoted by $a(v_1, \ldots, v_{\pi(a)})$ or by $a\mathfrak{v}$. The pair (Σ, π) is called a *V-alphabet.* Words over such a V-alphabet have the general form

$$w = a_1(v_{1,1}, \ldots, v_{1,\pi(a_1)}) \cdots a_r(v_{r,1}, \ldots, v_{r,\pi(a_r)}) = a_1\mathfrak{v}_1 \cdots a_r\mathfrak{v}_r$$

where $a_i \in \Sigma$ and $v_{i,j} \in \mathbf{R}$ for all i, j. The *underlying structure of* w is the word $\sigma(w) = a_1 \cdots a_r$. Note that the round parentheses are used for readability only—they are redundant (on this abstract level). Let $\mathfrak{S}$ denote the set of all V-words over the V-alphabet (Σ, π). Obviously, $\mathfrak{S}$ is a free semigroup with respect to concatenation and σ is a semigroup morphism of $\mathfrak{S}$ onto Σ^+.

For the purpose of defining VD0L systems we consider only a restricted class of mappings as morphisms[7] of $\mathfrak{S}$. In particular, an *endomorphism* ϑ of $\mathfrak{S}$ is a pair $\vartheta = (\vartheta_1, \vartheta_2)$ with the following properties:

(1) For $a \in \Sigma$, $\vartheta_1(a)$ is a mapping of $\mathbf{R}^{\pi(a)}$ into Σ^* such that the image $\vartheta_1(a)(\mathbf{R}^{\pi(a)})$ is finite.[8]

(2) For $a \in \Sigma$ and $a_1 \cdots a_l \in \vartheta_1(a)(\mathbf{R}^{\pi(a)})$ with $a, a_1, \ldots, a_l \in \Sigma$, $\vartheta_2(a, a_1 \cdots a_l)$ is an l-tuple $(f_1, \ldots, f_l)$ of mappings[9] $f_i : \mathbf{R}^{\pi(a)} \to \mathbf{R}^{\pi(a_i)}$.

For a V-symbol $a\mathfrak{v}$, its image $\vartheta(a\mathfrak{v})$ is the V-word $a_1 f_1(\mathfrak{v}) \cdots a_l f_l(\mathfrak{v})$ where $\vartheta_1(a)(\mathfrak{v}) = a_1 \cdots a_l$ and $\vartheta_2(a, a_1 \cdots a_l) = (f_1, \ldots, f_l)$. The image of a V-word $w = a_1\mathfrak{v}_1 \cdots a_r\mathfrak{v}_r$ is then defined as

$$\vartheta(w) = \vartheta(a_1\mathfrak{v}_1) \cdots \vartheta(a_r\mathfrak{v}_r).$$

Note that in the definition of an endomorphism ϑ of $\mathfrak{S}$, the property (1) results in a finite case distinction for what the symbol a is mapped onto depending on the value $\mathfrak{v}$ of its parameters. After the appropriate case has been chosen, ϑ_2 then determines a transformation of the parameter values.

Definition 4.1 A *VD0L system* is a quadruple $G = (\Sigma, \vartheta, \omega, \pi)$ with the following properties:

(a) (Σ, π) is a V-alphabet.

(b) ϑ is an endomorphism of the set $\mathfrak{S}$ of all V-words over (Σ, π).

(c) $\omega \in \mathfrak{S}$ is the *axiom.*

---

[7] We need and define only endomorphisms. It is, however, obvious how to generalize this definition.

[8] As usual, $\mathbf{R}^0$ is defined to be a singleton set. Hence, $\vartheta_1(a)(\mathbf{R}^{\pi(a)})$ will consist of a single word if $\pi(a) = 0$.

[9] In certain practical cases some randomiztion of the f_i may be required for realistic modelling. To include this, some obvious changes would be needed in these definitions. However, this issue is not essential for our considerations.

In a VD0L system the decision which string to replace a given symbol by will depend on the value of its parameters; this is achieved using ϑ_1. By ϑ_2 it is determined how the real-valued parameters of a symbol are transformed into real-valued parameters of its off-spring. In the definition of VD0L systems no assumptions are made about the kind of functions used by ϑ_2. From a practical point of view it is obvious that significant restrictions will be warranted. A rather elementary restriction used in the sequel is formulated in the following definition.

Definition 4.2 Let $R \subseteq \mathsf{R}$ be a recursive domain. A VD0L system $G = (\Sigma, \vartheta, \omega, \pi)$ is said to be *recursive over* R if it has the following properties:
(a) If $a\mathfrak{v}$ is a V-symbol occurring in ω then $\mathfrak{v} \in R^{\pi(a)}$.
(b) For all $a \in \Sigma$, the restriction of $\vartheta_1(a)$ to $R^{\pi(a)}$ is recursive.
(c) For all $a \in \Sigma$ and all $v \in \vartheta_1(a)(R^{\pi(a)})$ assume that $\vartheta_2(a, v) = (f_1, \ldots, f_{|v|})$. Then, for $i = 1, \ldots |v|$, the restriction of f_i to $R^{\pi(a)}$ is a recursive mapping of $R^{\pi(a)}$ into $R^{\pi(a_i)}$.

A VD0L system is, essentially, a D0L system[10] with the infinite alphabet $\{a\mathfrak{v} \mid a \in \Sigma, \mathfrak{v} \in \mathsf{R}^{\pi(a)}\}$. The usual notions for D0L systems—like that of *derivation sequence*—can be generalized to VD0L systems in a natural way. Especially, we need the notion of a *branching VD0L system*, or *VBD0L system*.

Definition 4.3 A *VBD0L system* is an septuple $G = (\Sigma, \vartheta, \omega, \langle, \rangle, \nu, \pi)$ with the following properties:
(a) $(\Sigma, \langle, \rangle, \nu)$ is a branching alphabet.
(b) $(\Sigma, \vartheta, \omega, \pi)$ is a VD0L system.
(c) $\sigma(\omega) \in D(\Sigma, \langle, \rangle, \nu)$.
(d) $\vartheta_1(\langle\mathfrak{v}) = \langle$ and $\vartheta_1(\rangle\mathfrak{u}) = \rangle$ for all $\mathfrak{v} \in \mathsf{R}^{\pi(\langle)}$ and $\mathfrak{u} \in \mathsf{R}^{\pi(\rangle)}$.
(e) For $a \in \Sigma_0$ and $\mathfrak{v} \in \mathsf{R}^{\pi(a)}$ one has $\vartheta_1(a\mathfrak{v}) \in D(\Sigma, \langle, \rangle, \nu)$.
(f) For $a \in \Sigma_i$ with $i > 0$ and $\mathfrak{v} \in \mathsf{R}^{\pi(a)}$ one has $\vartheta_1(a\mathfrak{v}) \in D(\Sigma, \langle, \rangle, \nu)^*\Sigma_i$.

In the context of plant modelling, in the V-symbol $a\mathfrak{v}$ the symbol a denotes the abstract unit like apex while $\mathfrak{v}$ could describe physico-chemical and geometrical properties like age or length. In a VBD0L system $G = (\Sigma, \vartheta, \omega, \langle, \rangle, \nu, \pi)$ there is an underlying non-deterministic branching L system which models the underlying branching structure and the logical components of the development.

Our definition of VD0L and VBD0L systems captures some of the ideas discussed in [19]. It contains the definition of a parametric D0L system given in [26, 27] as a very special case.[11] At present it is an open question to which extent the greater degree of generality in our definition is actually useful for realistic modelling.

As an example of a VDB0L system, we show how the system proposed in [19] for the description of the branching growth of barley roots can be expressed in our formalism.

Example 4.4 We distinguish two types of root segments, basal and internodal segments denoted by b and apical segments denoted by a. For each of these, a parameter l will indicate the length of the segment. Apical segments will grow by an amount g_i in each step where g_i applies to i^{th} order branches. Branching will occur once the present length

---

[10] Note that every D0L system is a VD0L system where $\pi(a) = 0$ for all a.

[11] As a subtle point, we mention that the way by which case distinctions are made according to the definitions of [26] and [27] may lead to partially undefined transitions. This is probably not intended as it would render the general theory of D0L systems inapplicable to parametric D0L systems.

exceeds a threshold t_i. Formally, let $\Sigma = \{a, b, \langle, \rangle\}$, $\pi(a) = 2$, $\pi(b) = 1$, $\pi(\langle) = \pi(\rangle) = 0$, $\nu(a) = 0$, $\nu(b) = 1$, $\nu(\langle) = \nu(\rangle) = 0$, and $\omega = a(0, g_0)$. The mapping $\vartheta = (\vartheta_1, \vartheta_2)$ is given by

$$\vartheta_1(a)(i,l) = \begin{cases} a, & \text{if } l < t_i, \\ b\langle a\rangle a, & \text{if } t_i \le l, \end{cases}$$

$$\vartheta_2(a,a) = (f_1), \qquad \vartheta_2(a, b\langle a\rangle a) = (f_2, f_3, f_4, f_5, f_6),$$

and

$$\vartheta_1(b)(l) = b, \qquad \vartheta_2(b,b) = (f_7),$$

where $f_1(i,l) = (i, l + g_i)$, $f_2(i,l) = t_i$, $f_4(i,l) = (i+1, 0)$, $f_6(i,l) = (i, l + g_i - t_i)$, $f_7(l) = l$, and where f_3 and f_4 are trivial.

With a VD0L system one can simulate other kinds of L systems. The correspondence between the derivational sequences of the two L systems is established by a *coding* in the usual sense of L system theory. The following example illustrates this idea.

Example 4.5 Recall that a time-delayed 0L has an alphabet in which some letters may exist in several numbered copies to facilitate time counting [35]. They are used extensively in L system based graphics (see [23]). For instance, consider the D0L system $G = (\Sigma, \delta, \omega)$ with $\Sigma = \{a_0, a_1, a_2, b\}$, $\delta(a_0) = a_1$, $\delta(a_1) = a_2$, $\delta(a_2) = a_0 b$, $\delta(b) = b$, and $\omega = a_2$. The sequence produced by this system is

$$a_2 \vdash a_0 b \vdash a_1 b \vdash a_2 b \vdash a_0 b^2 \vdash \ldots,$$

that is,

$$\delta^i(\omega) = a_{i+2 \bmod 3} b^{\lceil i/3 \rceil}.$$

For the simulating VD0L system, consider the alphabet $\Sigma' = \{a, b\}$ and let $\pi(a) = 1$, $\pi(b) = 0$. $\vartheta = (\vartheta_1, \vartheta_2)$ is given by

$$\vartheta_1(a)(v) = \begin{cases} a, & \text{if } v < 2, \\ ab, & \text{if } v \ge 2, \end{cases} \qquad \vartheta_2(a,a) = (f_1), \qquad \vartheta_2(a, ab) = (f_1', f_2),$$

and

$$\vartheta_1(b)() = b, \qquad \vartheta_2(b,b) = (),$$

where $f_1(v) = v + 1$, $f_2'(v) = v - v$, and f_2 is arbitrary. As axiom we use $\omega' = a(2)$. A coding which establishes the simulation of the D0L system by this VD0L system is given by

$$\varphi(a(v)) = \begin{cases} a_0, & \text{if } v \le 0, \\ a_1, & \text{if } 0 < v \le 1, \\ a_2, & \text{if } 1 < v, \end{cases} \qquad \varphi(b()) = b.$$

This simulation indicates why, in general, our definition of VD0L systems makes the choice of the replacement strings for a symbol depend on the parameter values (as in ϑ_1).

We now demonstrate the power of VD0L systems by showing that they can be used to simulate even DIL systems, that is, deterministic L systems, in which neighbouring symbols interact.

Theorem 4.6 *For every DIL system G there effectively exists a VD0L system which simulates G and is recursive over* $\mathbb{N}_0$. *Moreover, the VD0L system can be chosen in such a way that every symbol has exactly one parameter.*

Proof: Let $G = (\Sigma, \delta, \omega)$ be the given DIL system where δ is a finite deterministic set of production rules with interaction. A VD0L system $G' = (\Sigma, \vartheta, \omega', \pi)$ which simulates G can be constructed using the following ideas: For $a \in \Sigma$ let $\pi(a) = 1$. For integers n of the form $n = 2^i 3^j$ we interpret $a(n)$ as indicating that a is the jth symbol in $\delta^i(\omega)$. Accordingly, if $\omega = a_1 \cdots a_r$ with $a_1, \ldots, a_r \in \Sigma$, one defines $\omega' = a_1(3) \cdots a_r(3^r)$.

Assume now that the symbol a indeed occurs as the jth symbol in $\delta^i(\omega)$. This fact determines the applicable rule of G uniquely and this defines $\vartheta(a)(n)$ where $n = 2^i 3^j$; for all those values of n which do not describe occurrences of a let $\vartheta(a)(n)$ be an arbitrary, but fixed word—we do not care what it is because it will never be used.

Given $n = 2^i 3^j$ describing an occurence of a one can compute the positions of the symbols of $\vartheta_1(a)(n)$ in $\delta^{i+1}(\omega)$ by simulating $i+1$ steps of G and keeping track of where symbols are. This defines $\vartheta_2(a, \vartheta_1(a)(n))$ as vector of functions which multiply n by 2 to indicate the new step and by appropriate powers of 3 to indicate the positions in the string.

One verifies that the intended interpretation of $a(n)$ is correct, hence $\sigma(\vartheta^i(\omega')) = \delta^i(\omega)$ for all i. Clearly, σ is a coding. □

The simulation used in the proof of Theorem 4.6 can be modified in such a way that the VD0L system has only a singleton alphabet.

Corollary 4.7 *For every DIL system G there effectively exists a VUD0L system which simulates G, is recursive over* $\mathbb{N}_0$, *and in which the only symbol in the alphabet has just one parameter.*

Proof: Let Σ be the alphabet of the given DIL system and assume the letters in Σ are numbered in an arbitrary, but fixed way as $a_1, a_2, \ldots, a_m$. The VUD0L system to be constructed has the single symbol a with a single parameter. If $n = 2^i 3^j 5^k$ with $1 \le k \le m$ then $a(n)$ is intended to denote the fact that a_k occurs in $\delta^i(\omega)$ as the jth symbol. The rest of the construction follows the ideas of the proof of Theorem 4.6. □

The results of Theorem 4.6 and Corollary 4.7 are not surprising. They can be viewed as special cases of the general phenomenon that an L system, in which every symbol can obtain complete global information about the derivation sequence, will be very high up in the hierarchy of L systems. Further examples of this type can be found in [31, 32, 12].

In a similar fashion one can prove that every VD0L system can be simulated by a VUD0L system in which the information about the symbols is carried in an additional parameter value. Both this and the preceding simulations of DIL systems seem to be unnatural from the point of view of biological modelling. It is an open problem which restrictions should be imposed in addition to or instead of recursivity. The trivial simulation of a time-delayed D0L system as shown in Example 4.5 is certainly closer to the spirit of D0L system modelling, but may be too simple. Indeed, it seems that the ability of a VD0L system to maintain some global information or some context information locally in the parameters could be a useful and quite adequate feature. On the one hand, it permits one to model consistency—for instance in terms of the shape of connecting surfaces—of neighbouring parts; it could also allow one to simplify the rather complicated context rules for branching IL systems.

It is an open question in which way the class of functions used in the definition of a specific recursive VD0L system determines its simulation power. In the proof of

Theorem 4.6 we use functions in whose computation one actually simulates the developmental sequence of the DIL system and extracts the relevant information about symbol occurrences from it. We conjecture that the restricted classes of functions permitted in [27] do not suffice for this purpose.

Generalizing the definition of VD0L systems to include non-determinism or interaction—resulting in V0L, VDIL, or VIL systems—is obviously possible. Using a natural definition of V0L systems[12] and proof ideas from above and from [12], it seems that recursive V0L systems would suffice to simulate all EIL systems, hence that the languages generated by recursive V0L systems are precisely the recursively enumerable languages. This again raises the question as to which effect complexity bounds imposed on the functions defining the V0L systems would have.

5. The Graphical Interpretation of VD0L Systems

So far we have concentrated on the syntactical aspects of parameterized D0L systems. We now turn to their graphical semantics. Unlike most other applications of L systems to graphical modelling our proposal does not include the drawing instructions in the L system itself. Instead, we consider the graphical representation as a completely separate issue with information about shape, texture, and so on to be computed from the parameter values.

Let $G = (\Sigma, \vartheta, \omega, \langle,\rangle, \nu, \pi)$ be a VBD0L system and let $w = a_1\mathfrak{v}_1 a_2\mathfrak{v}_2 \cdots a_r\mathfrak{v}_r$ be a V-word. For the graphical interpretation w we need a *status vector* e. Typical information provided by components of e could include the co-ordinates of reference points, the present orientation, accumulated weight of a branch, and so on. Let $e^{(0)}$ denote the *initial status vector*.

The graphical rendition of w will be obtained in two steps: first one generates a set of points in an appropriate attribute space $\mathfrak{A}$—for instance in $\mathfrak{A} = \mathsf{R}^4$ to accomodate 3D positions and colour—which is then mapped onto the actual output representation. In this paper we focus on the first step, or more precisely on a general framework for this step. Initially, let $P^{(0)} \subseteq \mathfrak{A}$ be given to describe the *environment* of the word w to be rendered. Then the set $\varrho(w, e^{(0)}, P^{(0)}) \subseteq \mathfrak{A}$ constructed below is called the *rendering* of w.

With each symbol $a \in \Sigma$ we associate its *rendering function*[13] ϱ_a. Suppose that $\mathfrak{v} \in \mathsf{R}^{\pi(a)}$, that e is a status vector, and that $P \subseteq \mathfrak{A}$. Then $\varrho_a(e, \mathfrak{v}, P)$ is a pair $(\mathfrak{e}, Q)$ to be interpreted as follows:

- $\mathfrak{e} = (e_1, \ldots, e_{\nu(a)+1})$ where each e_i is a status description. Intuitively, for $i \leq \nu(a)$, we use e_i when rendering the ith branch starting at a, while $e_{\nu(a)+1}$ is used for the continuation of the main word.
- $Q \subseteq \mathfrak{A}$ is the set of points representing a. Note that Q may depend on all three parameters e, $\mathfrak{v}$, and P and may, therefore, incorporate information resulting, for instance, from physical or temporal properties of the object as well as of the rendering environment.

Intuitively, a set of points representing a is generated in such a way as to take into account the present status vector e, the parameters $\mathfrak{v}$, and the currently existing points

[12] Essentially, in a finite substitution $\vartheta = (\vartheta_1, \vartheta_2)$, $\vartheta_1(a)$ would be a finite set of mappings of $\mathsf{R}^{\pi(a)}$ into Σ^* for every $a \in \Sigma$. Moreover, ϑ_2 would depend on a, on $\psi \in \vartheta_1(a)$, and on $a_1 \cdots a_l \in \psi(\mathsf{R}^{\pi(a)})$, that is, $\vartheta_2(a, \psi, a_1 \cdots a_l)$ would be an l-tuple $(f_1, \ldots, f_l)$ of mappings $f_i : \mathsf{R}^{\pi(a)} \to \mathsf{R}^{\pi(a_i)}$.

[13] Also here some randomization may be required; this issue is beyond the scope of this paper.

P. Incorporating P is important because in modelling "solid" objects like plants, parts must avoid each other.[14]

The rendering $\varrho(w, e^{(0)}, P^{(0)})$ is defined recursively as follows: For $j = 1, \ldots, |w|$ let

$$(\mathfrak{e}^{(j)}, Q^{(j)}) = \begin{cases} \varrho_{a_1}(e^{(0)}, \mathfrak{v}_1, P^{(0)}), & \text{if } j = 1, \\ \varrho_{a_j}(e_k^{(i)}, \mathfrak{v}_j, P^{(j-1)}), & \text{if } j > 1 \text{ with } (i,j) \in \Lambda_{k,w}, \end{cases}$$

and

$$P^{(j)} = P^{(j-1)} \cup Q^{(j)},$$

where

$$\mathfrak{e}^{(i)} = (e_1^{(i)}, \ldots, e_{\nu(a_i)+1}^{(i)}).$$

Then

$$\varrho(w, e^{(0)}, P^{(0)}) = P^{(|w|)}$$

is the *graphical interpretation* of w.

This rather general definition of graphical interpretation of VD0L systems captures the essential features of all graphical interpretation schemes for L systems, which we have seen in the literature. A prototype implementation of a slightly more restricted model to test the feasibility of this general proposal formed part of the thesis [4].

6. The Prototype Implementation and Examples

In this section we briefly summarize the characteristics of the prototype implementation and show two examples of graphical plant simulation using our model. The prototype was written in C as part of the thesis [4]. Its main features are the following:

- The input, the VBD0L system $G = (\Sigma, \vartheta, \omega, \langle,\rangle, \nu, \pi)$, is specified in a simple extension of C. As a consequence, all C constructs are available for the definition of ϑ. Hence, precisely the recursive VBD0L systems can be defined in this notation.
- As far as the graphical interpretation is concerned we impose a few restrictions:
  - The status vector consists of the co-ordinates of a reference point and information about orientation.
  - The functions ϱ_a for $a \in \Sigma$ only depend on the first two parameters, that is, on e and $\mathfrak{v}$ and not on the environment P. Moreover, for every $a\mathfrak{v}$ there is a protype description $(\mathfrak{e}(a\mathfrak{v}), Q(a\mathfrak{v}))$ such that $\varrho_a(e, \mathfrak{v})$ can be determined from it using only translations and rotations as determined by e and implicit reference point information contained in $(\mathfrak{e}(a\mathfrak{v}), Q(a\mathfrak{v}))$.

  These restrictions of our definition of graphical interpretation were introduced mainly to simplify the implementation of the prototype program. On the other hand, they still permit us to demonstrate the feasibility of our general proposal by modelling reasonably complicated features.
- We did not spend much effort on the quality of the actual printed rendition of the graphical interpretation. This was done deliberately as, on the one hand, the implementation was only performed as proof of feasibility in principle and as, on the other hand, the result $\varrho(w, e^{(0)}, P^{(0)})$ could always be post-processed by some of the more sophisticated graphics systems already available. As a consequence of this decision the graphical quality of the images shown below is quite low.

[14] For example, if an old branch is in the way of a new one, the new one might bend. This should be treated as an entirely mechanical problem and certainly neither as an L system problem nor as a graphics problem. Hence, the inclusion of P as an argument of ϱ_a.

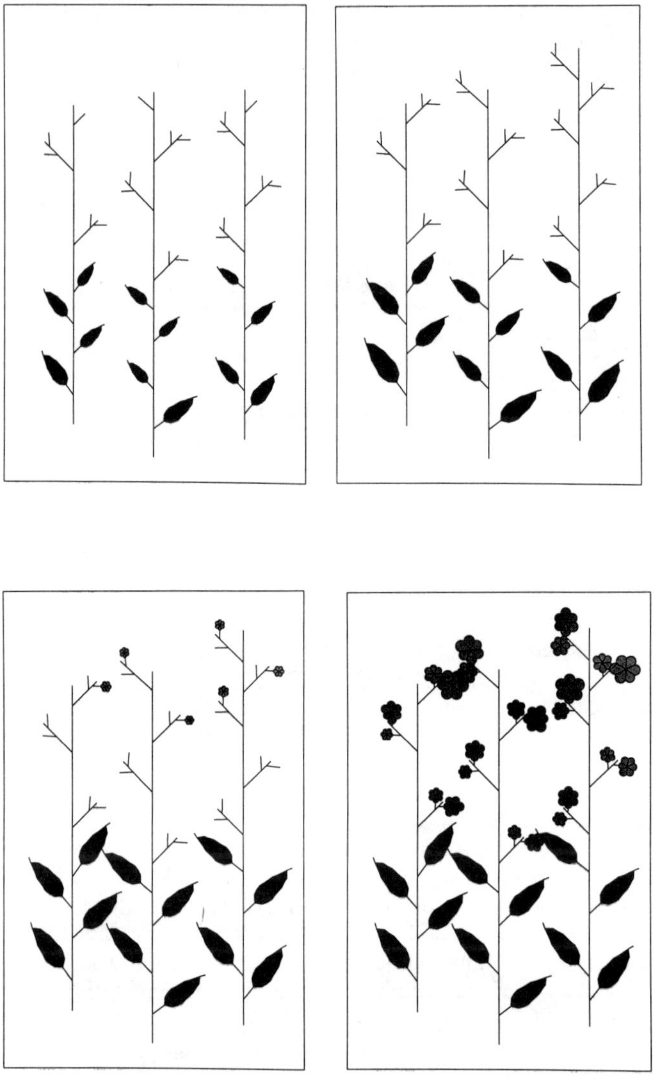

Figure 1. Simulation of an artificial flower: steps 12, 15, 18, and 21.

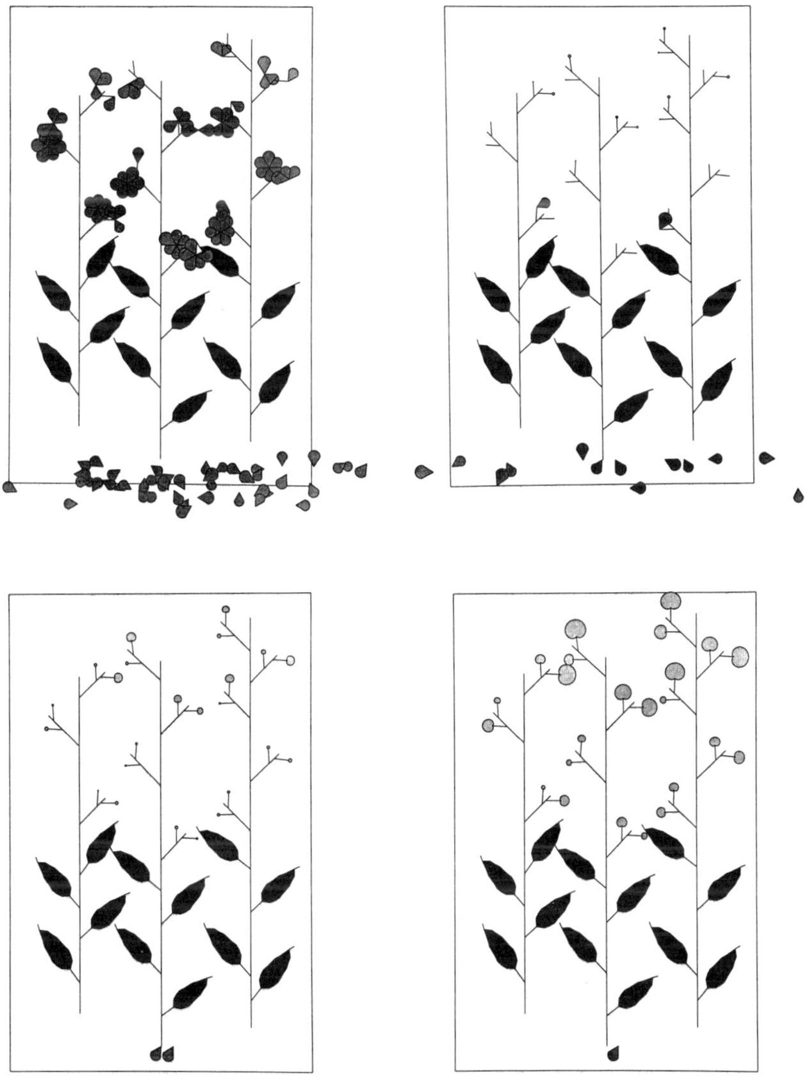

Figure 2. Simulation of an artificial flower: steps 24, 27, 30, and 33.

In Figures 1, 2, and 3 we illustrate some of the elementary capabilities of our prototype. In both cases the plants are artificial in the sense that they are not intended to simulate any specific natural plants. Their main purpose is to demonstrate certain features of our system.

The inflorescence of the plant in Figures 1 and 2 has a pattern which is known to be outside the scope of D0L systems. It can be modelled using interaction or parameterized symbols (see [27]). The falling petals are obtained by placing reference points appropriately. Petals are blown away by the wind—their interpretation results in an empty set of points.

The VBD0L system used to model this development has the following alphabet

symbol	meaning	symbol	meaning
a	apex	m	leaf
d	first order branch	f	second order branch
e	bud	g	young petal
i	adult petal	j	fallen petal
k	vanished petal	p	fruit
h	stamen	l	other

and the following substitution as defined by ϑ_1:

$$a \to \{a, l\langle m\rangle a, l\langle d\rangle a\} \qquad d \to \{d, l\langle e\rangle l\langle e\rangle d\}$$
$$e \to \{f\langle g\rangle\langle g\rangle\langle g\rangle\langle g\rangle\langle g\rangle\langle g\rangle\langle g\rangle h\} \qquad g \to \{g, i\}$$
$$i \to \{i, j\} \qquad j \to \{j, k\}$$
$$h \to \{h, p\}.$$

All other symbols rewrite as themselves. Some randomization is used in the definition of ϑ_2. For the graphical interpretation only very simple shapes and straight lines are used.

In Figure 3 we show the simulation of another artificial plant. This example illustrates (in an admittedly oversimplified fashion) how gravity and rigidity can be taken into account. In the graphical interpretation quadrilaterals are used to represent stem segments. All components grow in size; however, their growth rates are age dependent. Matching of connecting parts is established through the parameters. Again, some randomization is used in ϑ_2.

7. Conclusions: Potential and Limitations

To provide a basis for a rigorous evaluation of the potential and the limitations of L systems in the context of graphical modelling we have proposed definitions of parameterized D0L systems with and without branching and of their graphical interpretation. These seem general enough to permit meaningful statements about the limitations of L systems. On the other hand, they also suffice to demonstrate the potential of L systems both theoretically—in terms of the L system hierarchy and practically—using a prototype implementation to show feasibility in principle. We summarize a few rather preliminary conclusions:

- That L systems are most adequate for certain tasks in graphical modelling—of plants in particular—is well-known from the literature. Our general paradigm not

Figure 3. Simulation of an artificial plant: steps 15, 18, 21, and 24 are shown.

only corroborates this, but also shows that even scenes with some motion could be represented well using the derivation sequence of an L system.

- The limitations of L systems as tools for graphical modelling essentially result from the fact that the structure of words or words with branching implies that on the object or scene to be modelled a fixed tree structure has to be assumed. This seems to be acceptable for scene sequences in which a part can always be rendered once its parent with respect to the tree structure has been rendered. Thus, motion of an object would be limited to motion relative to its "parent" object. This seems to exclude, for instance, modelling a caterpillar crawling along a plant.
- In some cases, it may be possible in principle to use an L system for modelling a sequence scenes. However, the L system may look quite "unnatural." It turns out that our preceding statement concerning motion is not quite true. For instance, the crawling caterpillar could be modelled by attaching appropriate caterpillar parts to every single part of the plant and—using the fact that rendering of a letter can result in an empty set of points—by showing these or some of them only during certain time steps. This certainly does not look like a very natural way of modelling this sequences of scenes.[15]

These considerations indicate that the crucial question in assessing the potential and limitations of L systems is what one considers as a "natural" model. Our definitions provide the formal framework in which to ask this question. An acceptable answer may involve restrictions on the computational complexity of the functions used in VD0L systems and their renderings. Moreover, it would help clarifying the paradigm for L system based modelling as stated in Section 1 of this paper.

8. References

1. M. Aono, T. L. Kunii: Botanical Tree Image Generation. *IEEE, Computer Graphics and Applications* **4** (1984), 10–34.
2. M. Aono, T. L. Kunii: *Botanical Tree Generation*. Video Tape, approximately 1983.
3. J. Bloomenthal: Modelling the Mighty Maple. *SIGGRAPH '85, Computer Graphics* **19**(3) (1985), 305–311.
4. T. W. Chien: *Graphical Interpretation of the Biological Development Modelled by VD0L-Systems*. M. Sc. Thesis, The University of Western Ontario, London, Canada, 1989.
5. T. W. Chien, H. Jürgensen: *Parameterized L Systems for Modelling: Potential and Limitations*. Report 283, Department of Computer Science, The University of Western Ontario, 1991.
6. D. Frijters: Principles of Simulation of Inflorescence Development. *Ann. Bot.* **42** (1978), 549–560.
7. D. Frijters: Mechanisms of Developmental Integration of *Aster novae-angliae* L. and *Hieracium murorum* L. *Ann. Bot.* **42** (1978), 561–575.
8. D. Frijters, A. Lindenmayer: A Model for the Growth and Flowering of *Aster novae-angliae* on the Basis of Table $\langle 1,0 \rangle$-L-Systems. [30], 24–52.
9. G. T. Herman, G. Rozenberg: *Developmental Systems and Languages*. North-Holland, Amsterdam, 1975.
10. H. Jürgensen: *Substochastische L-Systeme*. Report 7501, Institut f. Informatik u. Prakt. Math., Univ. Kiel, 1975.
11. H. Jürgensen: Probabilistic L Systems. [18], 211–225.
12. H. Jürgensen, D. Wood: Self-Regulating L Systems. *EIK, J. Information Proc. Cybern.* **18** (1982), 375–387.
13. C. G. Langton (editor): *Artificial Life*. Addison-Wesley Publ. Co., Redwood City, 1989.

[15] The fact that a letter can be rendered as an empty set of points is not really important in this example. The really crucial fact is, that the timing of when certain parts are shown can be controlled using the simulation of interaction between symbols.

14. A. Lindenmayer, P. Prusinkiewicz: An Annotated Bibliography of Plant Modelling and Growth Simulation. [13], 625–643.
15. A. Lindenmayer, P. Prusinkiewicz: Developmental Models of Multicellular Organisms: A Computer Graphics Perspective. [13], 221–250.
16. A. Lindenmayer: Developmental Algorithms: Lineage versus Interactive Control Mechanisms. In: S. Subtelny, P. B. Green (editors), *Developmental Order, Its Origin and Regulation. Proc. 40th Ann. Sympos. of the Society for Developmental Biology.* Alan Liss, New York, 1982, 219–245.
17. A. Lindenmayer: Mathematical Models for Cellular Interaction in Development, Parts I and II. *J. Theor. Biol.* **18** (1968), 280–315.
18. A. Lindenmayer, G. Rozenberg (editors): *Automata, Languages, Development.* North-Holland, Amsterdam, 1976.
19. A. Lindenmayer: Adding Continuous Components to L Systems. [30], 53–68.
20. *Growth of Plants Using D0L Systems.* Super-8 movie, unpublished, Institut für Angewandte Informatik und Formale Beschreibungsverfahren, Universität Karlsruhe, approximately 1975.
21. P. E. Oppenheimer: Real Time Design and Animation of Fractal Plants and Trees. *SIGGRAPH '86, Computer Graphics* **20**(4) (1986), 55–64.
22. P. Prusinkiewicz: Applications of L-Systems to Computer Imagery. In: H. Ehrig, M. Nagl, A. Rosenfeld, G. Rozenberg (editors), *Graph Grammars and Their Application to Computer Science.* Lecture Notes in Computer Science **291**, Springer-Verlag, Berlin, 1987, 534–548
23. P. Prusinkiewicz, J. Hanan: *Lindenmayer Systems, Fractals, and Plants. Lecture Notes in Biomathematics* **79**. Springer-Verlag, New York, 1989.
24. P. Prusinkiewicz, A. Lindenmayer, J. Hanan: Developmental Models of Herbaceous Plants for Computer Imagery Purposes. *SIGGRAPH '88, Computer Graphics* **22**(4) (1988), 141–150.
25. P. Prusinkiewicz: Graphical Applications of L Systems. *Proc. Graphics Interface '86—Vision Interface '86.* CIPS, 1986, 247–353.
26. P. Prusinkiewicz, J. Hanan: Visualization of Botanical Structures and Processes Using Parametric L System. In: D. Thalmann (editor), *Scientific Visualization and Graphics Simulation.* J. Wiley & Sons, New York, 1990, 183–201.
27. P. Prusinkiewicz, A. Lindenmayer: *The Algorithmic Beauty of Plants.* Springer-Verlag, New York, 1990.
28. P. de Reffye, C. Edilin, J. Françon, M. Jaeger, C. Puech: Plant Models Faithful to Botanical Structure and Development. *SIGGRAPH '88, Computer Graphics* **22**(4) (1988), 151–158.
29. G. Rozenberg, A. Salomaa: *The Mathematical Theory of L Systems.* Academic Press, New York, 1980.
30. G. Rozenberg, A. Lindenmayer (editors): *L Systems.* Lecture Notes in Computer Science **15**, Springer-Verlag, Berlin, 1974.
31. W. Rülling: *Informationsübertragung in L-Systemen,* Dissertation, TH Darmstadt, 1983.
32. W. Rülling: Real Time Simulation of Concentration Dependent L Systems by PD$\langle 1,1 \rangle$L Systems with Fast Cell Communication. In: I. Peák (editor), *Papers on Automata and Languages* **VI**, 17–45, Department of Mathematics, Karl Marx University of Economics, Report 1984-3, 1984.
33. A. R. Smith: Plants, Fractals, and Formal Languages. *SIGGRAPH '84, Computer Graphics* **18**(3) (1984), 1–10.
34. A. L. Szilard, R. E. Quinton: An Interpretation for D0L Systems by Computer Graphics. *The Science Terrapin* **4**(2) (1979), 8–13.
35. D. Wood: Generalized Time-Delayed 0L Languages. *Information Sciences* **12** (1977), 151–155.

A High-Level Language for L-systems and Its Applications

Narendra S. Goel and Ivan Rozehnal

Department of Systems Science, State University of New York, Binghamton, NY 13902-6000, USA

Abstract: A high-level language which incorporates several extensions of the L-system's grammar is described. A computer program ELSYS, consisting of a L-system compiler for this language, a graphic module and a user interface for Sun workstations, has been developed. Some applications of the ELSYS to animation, architecture, optical scattering and remote sensing are given.

Keywords: Lindenmayer systems, L-systems, animation, architecture, scattering, radiosity, remote sensing, fractals.

1. **Introduction**

Lindenmayer [10] introduced the original parallel rewriting systems for describing the development and growth of living systems. As noted in Goel and Rozehnal [6], L-systems have the following attractive features:

(1) Simplicity: Because they use iterative algorithms, simple rules can generate complex structures and behaviors.

(2) Variety: By changing some of the parameters in the iterative algorithm, one could get a large variety of structures and behaviors.

(3) Modularity: By combining several iterative algorithms or production rules, one could get a variety of complex structures and behaviors with components from several systems.

(4) Universality: By interpreting the variables involved, with the same production rules one could model systems from different disciplines of arts, science and engineering.

The most basic system, the 0L-system is an ordered triplet (V,s,P); V is a collection of symbols, letters, or characters, called the **alphabet**, s is a sequence of letters from V, called the **axiom** (initial word), and P is a finite set of **productions** or production rules. A pair (a,x) of letter a and word x is called a production rule, and is written as a --> x. The letter a and the word x are called the predecessor and the successor of this production, respectively. An 0L-system is deterministic (D0L-system) if and only if for each letter a of alphabet V there is only one word x such that a --> x.

Application of the production rules expands the starting word into a long string of characters from the alphabet. All this is somewhat abstract. L-systems take on a "beautiful life" when one interprets the strings graphically [12,13]. It is the graphical interpretation which is the basis for generating botanical structures using L-systems. In this interpretation, one uses turtle graphics [1] for mapping strings of symbols into graphic objects on a computer display screen, by maneuvering a little imaginary creature, called a turtle, that exists on the screen. This turtle can respond to simple commands, like move forward, turn right or left by a specified angle, make a U-turn, etc. The turtle's movement in response to a sequence of commands traces a line on the screen, leading to interesting patterns or pictures.

With the turtle graphic interpretation, L-systems have found their applications beyond the biological regime. In [6] we have shown that L-systems can be used to (1) generate a variety of interesting and complex structures and designs by graphically interpreting on an IBM PC compatible computer the strings of simple L-systems, (2) to rapidly create designs and written directions for weaving a given piece of cloth, (3) to configure electronic components in a standard rack for the purpose of manufacturing, and (4) to design contemporary and colonial houses and lake cottages, and to generate their architectural plans.

A mathematical interpretation of the symbols allows L-systems to be used in mathematics. In [6] we have shown that L-systems can be used to carry out many mathematical computations like calculating factorial, summing series, finding the values of various functions, etc. Goel and Goodwin [4] have developed a language for carrying out symbolic computation using L-systems (e.g., to show that $(x^2-1)/(x+1)$ is equal to $(x-1)$).

By interpreting the symbols as musical notes, tempo, rests, etc., one [6] can also use L-systems for writing musical compositions and playing it on a microcomputer. By interpreting symbols as the alphabets of the English language, L-systems can be used for developing a public encryption system for cryptology (for coding and encoding secret messages) [7,14].

Since their introduction, L-systems have been modified in many ways.

In the context-dependent L-systems, the production rule depends upon the (left, right or both) context of the predecessor. In the probabilistic L-systems, there are more than one production rules for the same predecessor which are applied with pre-specified probabilities. (For these systems, the structures generated by the L-systems tend to be less regular and more akin to natural objects.) In the table L-systems, a particular set of productions are applied for a specific number of iterations (parallel rewriting), and then another set of rules for another specified number of iterations, and so on. Such systems are used to describe two or more successive phases of growth by means of a change in the production rules.

In this chapter we describe a few extensions of the L-systems' grammar and implementation of the extended L-systems in a higher level language. Through a few applications, we show that the resulting software makes L-systems even a more powerful, universal and useful tool.

2. Key Features of the Extended L-system

In this section, we will describe the key features of the extended L-system, referred to as the EL-system. Although EL-system is a general purpose language, great many of the features of this language were built with respect to the graphics applications of the system.

We have developed a integrated computer program, ELSYS, all written in C, consisting of a "compiler" for this language, a graphic module, and a user interface for Sunview window environment for Sun workstations (primarily to view and manipulate the geometrical structure generated by the L-system code). ELSYS is not a true compiler; the input source code is compiled into a sequence of instructions which are executed in software, independent of a particular machine instruction set. The instruction set is actually a set of subroutines, written in C, which manipulate global data, stack, and jumps, perform basic arithmetic operations etc. The resulting target code produced by the compiler consists of an array of pointers to these procedures which are then executed sequentially. We chose this approach because it retains many advantages of a compiler system without the complexity of writing a real compiler for a particular hardware architecture and without sacrificing portability of the code.

The major extensions in ELSYS include (a) use of identifiers (not just single letters) for production rules, functions and variables; (b) use of variables (integer, real, and vector); (3) passing of the parameters to a production rule; (4) full execution control within a

Table 1

Set of symbols and their 3-D turtle graphic interpretation.

F   Move forward by a pre-defined step Fsize in the direction of heading and draw a line between initial and final points.

f   As F except without drawing a line.

+   Turn right by an angle Delta.

-   Turn left by an angle Delta.

&   Pitch down by an angle Delta.

^   Pitch up by an angle Delta.

\   Roll left by an angle Delta.

/   Roll right by an angle Delta.

| Turn away or make a U-turn.

[Push current state of the turtle into a push-down stack.

]   Pop a state from the stack, and make it the current state of the turtle.

(The turtle will carry out the instructions within the pair of symbols [ ] and then return to its initial position.)

{   start saving the subsequent turtle positions as the vertices of a polygon to be filled.

}   End the polygon and fill it using the current color c.

=   As f except that the new turtle position is not saved as a vertex (if used within polygon braces { and })

rule (if, goto and loop statements); and (5) built-in functions and operators.

To make our discussion of ELSYS more pedagogical, it is useful for us to summarize an existing L-system [5,6], together with the 3-D turtle graphic interpretation of the symbols.

The alphabet V for the L-system used consists of the Ascii character set from ! to ~ (decimal values from 33 to 126). Out of these characters, some are interpreted graphically, using 3D turtle graphics [1]. In 3D, a state of the turtle is defined by its position and its orientation. A vector **P** represents the turtle's **position** in Cartesian coordinates. The **orientation** of the turtle is represented by three unit vectors **H, L, U**, indicating the turtle's **heading**, the direction to the **left**, and the **up** direction, respectively. These vectors are perpendicular to each other, with HxL = U. In addition, we include the current drawing color c in the state of the turtle.

Thus, a turtle state is represented by a quintuplet (**P**, **H**, **L**, **U**, c). The turtle can move from one location to another, and can also rotate around any of the three vectors defining its orientation. The set of characters (letters) of the alphabet V which are interpreted graphically (and their interpretations) are given In Table 1.

We now describe basic language features in more details.

(1) Case sensitivity: All variables, functions, production rules etc. are case sensitive. Thus, names "var1" and "Var1" refer to different variables.

(2) Comments: Comments can appear anywhere in the source code. The format is: /* comment */.

(3) Variables: There are two classes of variables: (i) User-defined and (ii) standard (built-in). User-defined variables are (a) global, if defined in "global" statement, with their scope being the entire source file and (b) local, defined by appearance (as in FORTRAN), with the scope limited to the production rule within which they appear.

There are three types of variables or data types: integer, real and three-dimensional (3-D) vector of reals. Vectors are triples of real numbers and are used to represent points or directions in 3D space as well as colors described as triplets (R,G,B). Vector variable u can be indexed in order to access its real components: u = (u(1), u(2), u(3)).

The type of user-defined variable is not explicitly defined in the source code. It is determined in all cases by its first letter. Names which start with i,j,k,l,m,n,I,J,K,L,M,N refer to integers; those which start with u,v,w,U,V,W refer to vectors; and all other names represent reals. Standard (built-in) variables do not conform to this convention. Their type is explicitly defined in Table 2. The conversions between different data types are performed automatically, whenever possible. Parameters to production rules are always passed by address which enable rules to modify parameters passed. As with arithmetic operations, the conversions between formal and actual types of parameters are performed automatically.

(4) Data structures: There are two kinds of user-defined data structures: global (multidimensional) arrays of integers and reals, allocated at compile time, and tree-like dynamically created data structures (see the description of mountain in Section 3) which can be used as linked lists or trees with arbitrary number of leaves.

Table 2

Standard Built-in Variables

Name	Type	Initial value	Meaning
pi	Real	3.14159265	
TPos	Vector	(0,0,0)	Turtle position
Head	Vector	(1,0,0)	Turtle heading
Left	Vector	(0,1,0)	Direction left for the turtle
Up	Vector	(0,0,1)	Direction up for the turtle
Delta	Real	1	Value (in degrees) for + - / \ & ^ turtle operators
Fsize	Real	1	Current distance for F and f operators
FileCounter	Real		Counter for animation file index
Color	Vector	(1,1,1)	Current (reflected) color
TransColor	Vector	(0,0,0)	Current transmitted color
black	Vector	(0,0,0)	
blue	Vector	(0,0,1)	
green	Vector	(0,1,0)	
cyan	Vector	(0,1,1)	
red	Vector	(1,0,0)	
magenta	Vector	(1,0,1)	
brown	Vector	(.5,.5,0)	
yellow	Vector	(1,1,0)	
white	Vector	(1,1,1)	

(5) <u>L-system source file:</u> It consists of four basic parts in the order given below:

1. global variable definitions (optional)
2. user-defined functions definitions (optional)
3. axiom for the L-system
4. definitions of production rules for the L-system

The source file must adhere to the following conventions and syntaxes:

(a) Numbers and identifiers or two identifiers must be separated by blanks. For example, K --> 2F is a syntax error; the correct form is K --> 2 F.

(b) Whenever two language elements are separated by a blank, an arbitrary number of blanks or empty lines can be used to improve readability of the source code.

(c) All identifiers (user-defined variables, functions and rules) must start with a letter or an underscore and other characters must be either letters, digits or an underscore, with no limitation on the identifier length.

(d) Global variables are defined by using a "global" statement, consisting of a list of identifiers, separated by commas. The type of the variable is determined by the first letter convention described above. If a variable name is followed by a list of digits (e.g., (3,3,5)) it is assumed to be a definition of a global array. Arrays of vectors are not allowed. An example of declaration of global variables is

```
global (a,b,i,vecl,S(10,3),I0(2,2,6),x,p,Q(3))
```

where a, b, x, and p are real variables, i is a integer, vecl is a vector, S is a 10x3 real matrix, I0 is a three dimensional array of integers, and Q is an array of 3 real elements.

(e) Function Definitions. Aside from built-in functions (see (i) below), one can define one's own functions using the following format:

```
def ( function_name ( parameter_list ) := expression )
```

or

```
def ( function_name := expression )
```

for function with no parameters. The expression can contain only variables from parameter list, constants, global variables and calls to other functions. The type returned by the function must observe the first-letter convention described earlier. The type of arguments in the parameter list is also determined by the first-letter convention. Two examples of function definition are:

```
def (A(x1,y1) := sqrt(x1) + sin(x1+y1) + 1)
def (ufun3 (x,y,z,t) := (x,y,z)/t )
```

Here function ufun3 returns vector composed from parameters x,y,z divided by t. Functions are used in expressions as per following format:

```
function_name ( expression, expression, expression, ... )
```

e.g., ... x + A(a+b*d,6) * 3 ....

(f) Assignments of variables have the following form:

```
let ( variable := expression, variable := expression, ... )
```

For example,

```
let (x:=3., z(n):=2, i:=j+k, u:=blue, v:=(1,1,x), w:=v+u+(x+y,0,1) )
```

The variable on the left side of := must be indexed if it is declared as an array in global declaration. As noted above, conversions between different types of variables are performed automatically whenever possible.

(g) The following operators are available for mathematical computations, with their interpretation dependent upon their use.

+ - * /	Add, subtract, multiply and divide real numbers and integers. For vectors "*" is the cross-product, "." is the dot product, "+" and "-" are vector addition and subtraction. Operators "*" and "/" can be also used between vector and a number to produce a multiplied/divided vector.
** ^	Power function.
mod	Modulo function for integers.
and or not	Logical operators on integers according to the rules of boolean algebra. This enables one to use integers as boolean variables.
< > <= >=	For comparing numbers. These operators produce integer 0 or 1 depending whether the test resulted in truth or not.
=	For the equality test, both for numbers and vectors.
:=	For setting the value of a variable.

All the operators behave according to the standard priority rules. Conversions between different types are performed automatically whenever possible. Note that the operator "/" for 2 integers performs integer division with integer result. For example, 2/3 is 0, but 2./3, 2/3. and 2./3. all are 0.666666... A few examples showing the use of various operators follow.

```
let (i := x < y and u = v,  j := i and k = 0)
.... if(x >= y+3 or y < -1)then .... endif
.... if(i)then .... endif
```

Note that the operators / * + - ^ are also used as L-system turtle operators (see Table 1). The meaning of these symbols depends on the context in which they are used. Generally, these operators are arithmetic, if they are used in "let" (assignment) statement, in function definition, and in a parameter list for rule/function/built-in operator (e.g., R(x+y) H(a+b,c/d,t) setcolor(blue+yellow)). Otherwise, these symbols are interpreted as L-system turtle operators. For example, in

```
P -->  x+y/z F P
```

both + and / are turtle operators preceded by repeat counts of x and y respectively (see also the discussion of turtle graphics operators later in the section).

Table 3

Standard Built-in Operators

Name	Arguments	Meaning
run	List of integers	Specify number of iterations for the Table L-system
exit	None	Terminate execution
val	Real/Integer	Evaluate expression
inc	Real/Integer	Increment real or integer variable
dec	Real/Integer	Decrease real or integer variable
let	List of assignments	Performs assignment(s)
if	Integer	If statement
goto	Label	Jump within a rule
initrand	Real	Initialize random number generator with a given real seed
pt	Vector	Save point as a vertex of a polygon
write	List of any values	Write on the terminal (no newline)
writeln	List of any values	Write on the terminal (including newline)
save		Save the polygons generated so far into the animation file and increment the file index
reset		Discard all polygons generated so far
setname	String	Set animation file name
flush		Same as "save" followed by "reset"
setviewer	Real Real	Sets viewer's zenith and azimuth angles
setcolor	Vector, [Vector]	Sets current color (second optional arg = transmitted color)
Tree	Real	Create tree and return handle to it
Tset	Real, Real/Integer	Assign value to a tree
TGet	Real, Real/Integer	Get a previously assigned value from a tree
TChild	Real, Real	Create a child of a tree and return handle to it (first arg = parent, second arg = child)
TGetChild	Real, Real	Return first child of a tree (first arg = parent, second arg = first child)
TGetNext	Real, Real	Return next child of a tree (first arg = last child obtained, second arg = next child)

(h) Besides mathematical operators, there are many standard built-in operators given in Table 3. The meanings of most of the operators should be obvious, with the possible exception of operators Tree, Tset, TGet, TChild, TGetChild and TGetNext. These operators were introduced to generate a fractal mountain described in Section 3 where we have explained them in more detail. The writeln statement can use strings like

```
writeln (' result is ',x+y,' y is ',y, ' vector u is ',u)
```

(i) ELSYS has many built-in standard mathematical functions. They are: ln,exp,sqrt,asin,acos,atan,sin,cos,tan,abs,max,min,dist,rnd,srnd. Their meanings are standard except for a few which we now define. dist(u,v) gives the distance between two vectors (=abs(u-v)), rnd returns a real number in [0,1), rnd(a,b) a real number in [a,b) and srnd a real number in [-0.5,0.5), all from a random uniform distribution. All trigonometric functions use angle in degrees for their arguments; min and max can have an arbitrary number of arguments (integer and real); and abs returns absolute value for integers and reals but the vector norm for vectors.

(j) There are three types of control statements, which can be used within the axiom or a production rule. They are: (1) if statement, e.g., if (a > 0) then setcolor(blue) P(x) endif ; (2) goto statement, e.g.,

```
label1:    /* define a label */
   P(x) Q(y)
   if(x = 0)then
       goto label1  /* jump */
   endif
```

and (3) do loops, e.g.,

```
let (n := 20)
n*(   P(x)  Q(y)   )       /* repeat P(x) Q(y) 20 times */.
```

Both loops and if statements can be nested.

(6) L-System Implementation: It consists of two parts: expansion of the axiom into the final string using the production rules and the implementation of the turtle graphics operators, described below.

ELSYS allows Table L-systems where each symbol can have multiple (a maximum of 50) production rules (replacement strings). This feature is implemented as follows:

```
rule_heading ---> repl1 ; repl2 ; repl3 ....
```

or

```
rule_heading ---> repl1     /* single replacement */
```

The number of iterations to be executed for each replacement is specified in "run" command. By default, the number of iterations is set to the maximum value of 10,000 for all the tables so that all the replacements are always performed (as specified in the replacements rules). The "run" command sets explicitly the number of iterations for some of the tables;

for example, the statement

run (3,5,4)

will use "repl1" part of all rules in the system for the first 3 iterations, then use "repl2" part for the next 5 iterations, then "repl3" for 4 iterations and the rest of the replacements for the maximum number of iterations. If a rule does not have replacements of this higher order (as for the case of single replacement), no replacement is done. We should note that the regular mechanism of L-system execution can be altered by using "goto" or "if" statement. The recursive calls are handled in a standard way using global stack which holds the return addresses as well as the locations of local variables.

Each rule heading can be followed by a list of parameters. The scope of these parameters is local (as for local variables). Parameters can be used within the expressions in a production rule, and can be passed to other rules etc. If an assignment is made to a parameter, the value will be assigned to the actual variable in the actual call (FORTRAN style parameter passing). For example, in

```
....
....
R(x)  /* rule usage */
....
....
                        /* rule definition: */
R(a) --> T(a)           /* call another rule */
         let(a := 5)    /* assign 5 into the variable passed with R */
```

x gets assigned a value 5 if the number of iterations warrants the replacement of R(x) by the string defined in the rule definition. Otherwise, R(x) is ignored.

A rule usage can be preceded by a repeat factor. Thus

n R(x)

has the same meaning as

n*(R(x))

(see also control statements above).

The turtle graphics operators + - / \ & ^ | f F = { } (see Table 1) are implemented in a standard fashion. The translation and rotation operators + - / \ & ^ f F = can be preceded by a repeat count, which specifies the numerical value used by the operator. For example, n + means turn right by n degrees while n F means go forward by n times the current value of global variable Fsize, which controls the advancement of the turtle by F f and = operators. Without a repeat count the operators + -/ \ & ^ cause the turtle rotation by an amount equal to the global variable Delta.

If an expression is to be used for a repeat count, use function "val":

val(x+y) +

Otherwise, statement

```
x+y +
```

would be interpreted as "x times turn right then y times turn right". Val function can be also used for loops:

```
val(3*m+1) * ( P Q )
```

A polygon can be created without maneuvering the turtle by using the built-in operator "pt". Thus

```
{ pt(u)  pt(v) pt(w) }
```

creates a triangle with vertices given by vectors u, v, and w.

Colors for the turtle trail are defined as vectors containing 3 components R,G,B (see Table 2 for built-in variables for some standard colors). These colors can be set using the operator "setcolor". Thus

```
setcolor ( (a,b,c) )
```

sets R,G,B components of the current color equal to a, b, and c, respectively. This operator can also be used to specify the red, green and blue hemispherical reflectances and hemispherical transmittances of a polygon created by the turtle graphics. For example,

```
setcolor ( reflected_color, transmitted_color)
```

sets reflected color and transmitted color. These quantities are needed in calculating scattering in the optical region [9] from objects created by the L-system and for displaying an object according to the intensity of light scattered from its various parts (see also Section 3).

3. Some Applications of the Extended L-system

In this section, we will present some applications of ELSYS in two categories: (a) animation and (b) generation of objects and calculation of optical scattering from them. In the first category, we choose an example of a flowering plant growth, while in the second category we choose a snow-covered mountain, a castle and a cottage.

3.1 Animation

The animation feature of ELSYS is illustrated by the source code given in Fig. 1 which creates an animation sequence, shown in Fig. 2, describing various stages of a flowering plant growth.

The rule facet(h,delta,beta) creates a symmetric, four sided polygon, which is used for the construction of both the stem and the flower of the plant. The base of the polygon is drawn in the current direction of the turtle and has length 1; angles delta and beta instruct the turtle how to change the direction of its heading in order to draw the side of the polygon. The length of the side is specified by the parameter h.

```
animation(1,1, 7,20, 1,7, 60,170, 3)  /* axiom */
animation(kx,ky, n,m, h1,h2, delta1,delta2, nsteps) -->
let(i:= 0) setname ('an')
nsteps*( ky*( [kx*( [object(n,m, h1+(h2-h1)*real(i)/(nsteps-1),
                     delta1+(delta2-delta1)*real(i)/(nsteps-1))]
90+ val(2*n) f 90-)]  val(2*n) f) flush  inc(i))
object(n,m,h,delta) -->  trunk(n,m) layer(n,h,delta)
trunk(n,m)          -->  m*(layer(n,1,90) 90^ f 90&)
layer(n,h,delta)    --> [n*(facet(h,delta,90-180./n)f val(360./n)+)]
facet(h,delta,beta) --> [[beta+ delta^
                        { h f | h f ] f beta- | delta^ h f }]
```

Fig. 1. The source code for animation.

The rule layer(n,h,delta) generates a sequence of n polygons (by repeated calls to rule "facet") so that the bases of these polygons are inscribed into a circle. This rule is used to construct a "polygon-approximated" cone or cylinder.

The rule trunk(n,m) constructs a polygonal approximation of a cylinder which is m "stories" high and has its round base approximated by symmetric, n-sided polygons.

The rule object(n,m,h,delta) creates a flowering plant whose stem is constructed using rule trunk(n,m). The flower (constructed using the rule "layer") has the length h and grows out of the trunk at an angle delta. By changing the parameters h and delta, one can obtain various stages of the flowering process, as the flower grows bigger and spreads out of the stem at increasing angle.

This is accomplished in the rule animation(kx,ky,n,m,h1,h2,delta1,delta2,nsteps) which repeatedly builds a field of kx times ky flowers (kx flowers in x direction and ky in y direction) and saves various stages of growth into files an.001, an.002, an.003 etc. The animation file name is set using built-in operator "setname". Aside from kx and ky, the rule has parameters n and m describing the stem (as in the rule trunk), parameters h1 and delta1 describing the initial growth stage of the flower, parameters h2 and delta2 describing the final growth stage, and integer nsteps which specifies the number of steps for the animation to proceed from the initial stage to the final one. The rule animation generates the polygons describing the flower field in a loop where the parameters h and delta for the rule "object" are linearly interpolated between the values for the initial and final stages of growth. As soon as a particular growth stage is completed, polygons generated are saved into a file, file counter is

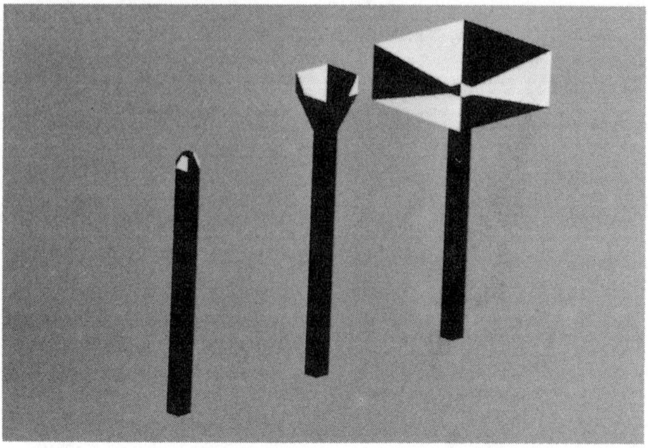

Fig. 2. Three growth stages of a flowering plant generated by using the source code of Fig. 1, with commands for shading added to show the flower petals.

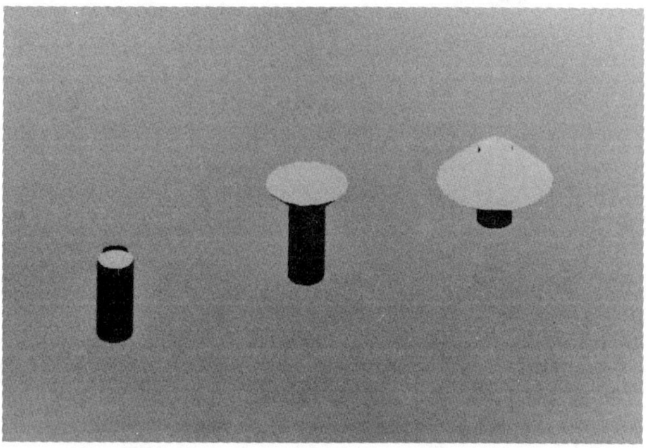

Fig. 3. Three growth stages of a mushroom generated by using the source code of Fig. 1 but the axiom modified (see text).

incremented, and number of polygons is reset to 0. These three functions are accomplished by the built-in operator "flush".

To illustrate the flexibility of this short sample code, and hence of the extended L-system, we present another animation given in Fig. 3, which shows three stages of growth generated by the ELSYS with the rules of Fig. 2, but with the axiom changed to

animation(1,1, 15,10, 1,7, 60,220, 3)

As one can see, this simple change of parameters generates a growth of a rather different object, resembling more a mushroom than a flower.

3.2 Optical Scattering and Remote Sensing

When solar radiation is incident on a terrestrial object, it is scattered and reflected, and its direction and spectral composition are altered in a complex manner by the object. Remote sensing of the terrestrial objects is based on the interception and measurement of this altered and reflected radiation by a satellite-borne sensor. A correct analysis of this data, hence success of the remote sensing, depends upon being able to relate reflectance measurements to the characteristics of the object. This can be done once the nature of the alteration in the radiation by the object and environmental factors (e.g., atmospheric conditions) is understood. Models of scattering and reflectance from objects of arbitrary shapes, arranged in an arbitrary configuration, play a key role in achieving this objective. Such models provide the logical connection between the physical, biological, and structural features of the object, the geometry of the radiometric interaction and the resulting alterations to the reflected radiation. Hence, the remotely sensed measured alterations can be used to infer the aforementioned features of the object [2,3].

Goel, Rozehnal and Thompson [9] have developed a novel approach, based on computer graphics, to model the scattering from a collection (canopy) of objects of arbitrary shapes in an arbitrary configuration. In this model, named DIANA, L-systems play an important role. The model basically has four components: (1) Computer generation of the object or the canopy (divided into 3000 to 30,000 3- and 4- sided polygons) using an L-system; (2) determination of visibility of a facet (2 per polygon) by sun (for determining sunlit facets) or by an observer (for reflectance calculations) using computer graphic techniques for generating realistic (to a human eye) display of objects; (3) determination of whether a facet is visible from another facet (for calculating multiple scattering of incident radiation), and if so, what fraction of energy flux leaving a facet reaches another facet, using the concept of radiosity, the amount of flux leaving per unit area per unit time from a facet; and (4) rendering of the canopy, with a 3D perspective, with brightness of a facet indicating the amount of radiation reaching part of the object or an observer for the calculation of measurable quantities like bidirectional reflectance factor (BRF) distribution which basically is a 3D plot of reflectance from the object as a function of view zenith and azimuth angle, for a fixed solar direction.

We used ELSYS to generate various objects, used the model DIANA to calculate the radiosity of each facet, and then rendered the object. We now give two examples.

(a) Fractal mountain: In Fig. 4 are given the optical scattering from a simulated snow-covered fractal mountain as seen by a viewer at two different viewing azimuthal angles (essentially from two different sides of the mountain). The mountain is generated by using ELSYS in conjunction with the source code given in Fig. 5. The source code is based on a well-known fractal surface algorithm which generates the surface of the mountain as a set of three dimensional triangles [11]. The coordinates of these triangles are generated by a recursive algorithm which replaces a given triangle by four smaller triangles using the midpoints of the original triangle. In addition, z-coordinates of these midpoints are modified by a random number scaled in proportion to the size of the original triangle. For our fractal surface, we used formula

z = (z1+z2)/2 + rand * ruggedness * abs(z1-z2)

where z is the modified z-coordinate of the midpoint, z1 and z2 are the z-coordinates of the two vertices, rand is a random number in the interval [-0.5, 0.5], and ruggedness is a value between 0 and 1 which determines the "roughness" of the mountainous surface.

In most implementations of this algorithm, a global two dimensional array is used to store the z-coordinates for the triangular grid which describes the surface. We have chosen another approach which utilizes dynamically built tree-like data structures which are part of our extended L-system language (see Table 3).

Each side of a triangle is associated with a tree node, which stores the midpoint computed using equation above. Each node has two children, corresponding to segments created by the subdivision of the triangle's side in half by its midpoint. In the source code presented in Fig.5, we use

Fig. 4. Optical scattering from a snow-covered fractal mountain, generated by using the source code of Fig. 5. Two different views are given.

```
global(rug)
/* function assigning randomly displaced midpoint : */
def (VecRand(u,w) := (u+w)/2 + (0,0,rug * srnd * abs(u(3)-w(3))) )
/* axiom : */
initrand(1)
let( rug:=0.5,  /* set ruggedness */
     a:=0.2,  b:=0.3, c:=0.15, d:=0.35,
     top:=0.4 ) /* set mountain peaks */
run(5) Mountain(a,b,c,d,top) /* create mountain */
/* rules: */
Mountain (a,b,c,d,top) --->
 let( u1:=(0,0,0), u2:=(1,0,0), u3:=(1,1,0), u4:=(0,1,0),
      u5:=(0.5,0.5,top) )
 Tree(T21)  Tree(T23)  Tree(T43) Tree(T41)
 Tree(T15)  Tree(T52)  Tree(T35) Tree(T54)
 DivLine(u2,u1,T21)  DivLine(u2,u3,T23)
 DivLine(u4,u3,T43)  DivLine(u4,u1,T41)
 DivLine1(u1,u5,T15,d)  DivLine1(u5,u2,T52,c)
 DivLine1(u3,u5,T35,b)  DivLine1(u5,u4,T54,a)
 Frac(u2,u1,u5,T21,T15,T52)  Frac(u2,u3,u5,T23,T35,T52)
 Frac(u3,u5,u4,T35,T54,T43)  Frac(u4,u1,u5,T41,T15,T54)
 writeln('done.')
DivLine(u,w,tree) -->
  let( v := VecRand(u,w) ) TSet(tree,v)
  TChild(tree,tfirst)  TChild(tree,tsec)
  DivLine(u,v,tfirst)  DivLine(v,w,tsec)
; let( v := VecRand(u,w) ) TSet(tree,v)
DivLine1(u,w,tree,z) -->
  let( v := (u+w)/2, v(3) := z ) TSet(tree,v)
  TChild(tree,tfirst)  TChild(tree,tsec)
  DivLine(u,v,tfirst)  DivLine(v,w,tsec)
; let( v := VecRand(u,w) ) TSet(tree,v)
Frac(u,v,w, Tuv,Tvw,Twu) -->
  TGet(Tuv, uv)  TGet(Tvw, vw)  TGet(Twu, wu)
  Tree(Tuv_wu)   Tree(Twu_vw)   Tree(Tvw_uv)
  DivLine(uv,wu, Tuv_wu)
  DivLine(wu,vw, Twu_vw)
  DivLine(vw,uv, Tvw_uv)
  TGetChild(Tuv,        Tuv_First)
  TGetNext (Tuv_First, Tuv_Sec)
  TGetChild(Tvw,        Tvw_First)
  TGetNext (Tvw_First, Tvw_Sec)
  TGetChild(Twu,        Twu_First)
  TGetNext (Twu_First, Twu_Sec)
  Frac(u,  uv, wu,   Tuv_First,  Tuv_wu,    Twu_Sec)
  Frac(uv,  v, vw,   Tuv_Sec,    Tvw_First, Tvw_uv )
  Frac(vw,  w, wu,   Tvw_Sec,    Twu_First, Twu_vw )
  Frac(uv, wu, vw,   Tuv_wu,     Twu_vw,    Tvw_uv )
 ;  { pt(u)        pt((u+v)/2)  pt((u+w)/2) }
    { pt(v)        pt((u+v)/2)  pt((v+w)/2) }
    { pt(w)        pt((u+w)/2)  pt((v+w)/2) }
    { pt((u+v)/2) pt((v+w)/2)  pt((w+u)/2) }
/* End of the code */
```

Fig. 5. The source code for the mountain.

the following built-in functions for creation and management of trees:

Tree(t) create new tree node and assign its unique reference value (handle) into t; value in t is used for any subsequent reference to this node.

TChild(t,ch) .... create new tree node as a child of existing tree node t; assign child's handle into ch. Additional children of a tree t are created by repeated calls to TChild.

TGetChild(t,ch) .... for a tree node t, assign its first child's handle into ch.

TGetNext(ch_old,ch) ... using the handle of an existing child ch_old, assign next child's handle into ch

TSet(t,x) .... assign a value x to an existing tree node t (x can be of any type).

TGet(t,x) .... retrieve value of tree node t into variable x (x should be the same type as the one assigned into t by TSet).

Rule DivLine (u,w,tree) generates complete binary tree associated with the segment [u,w] using its midpoint and z-coordinate displacement as described above. The depth of the subdivision is controlled by the number of iterations of the L-system, as specified in the axiom by the "run" statement. Rule Frac(u,v,w, Tuv,Tvw,Twu) generates fractal surface for a given triangle with vertices u,v,w. Together with these three vectors, it accepts as an input three trees, one for each triangle side. Each of these trees is assumed to contain complete subdivisions of the line segment to which it corresponds (this subdivision is accomplished using rule DivLine). The depth of the triangle subdivisions in Frac is again controlled by the number of iterations of the L-system.

The described functionality of the rules DivLine and Frac can be conveniently achieved using the Tl L-system rules, which provide each rule with two alternative replacements, applied consecutively, each for a specific number of iterations. For radiosity calculations, the last part of the Frac rule subdivides the final triangle into four additional triangles (using again midpoints of the triangle sides, this time without random displacement of the z-coordinate).

The rule Mountain(a,b,c,d,top) "designs" fractal mountain with user-prescribed values of height a,b,c,d,top at various points on the mountainous surface (see figure below). To accomplish this purpose, rule DivLine1 assigns directly the z-coordinate of the node value of the given

Fig. 6. Optical scattering from a castle generated by using ELSYS.

Fig. 7. Optical scattering from a cottage generated by using ELSYS.

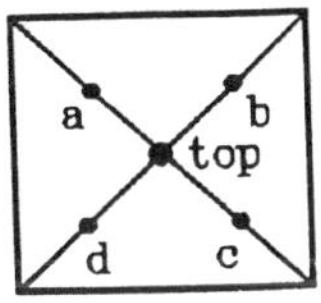

line segment as specified by the parameter z, instead of computing it using random number generator.

(b) Castle and cottage: ELSYS, in conjunction with DIANA, has been used for architectural applications [8]. In Figs. 6 and 7 are given optical scattering from a castle and a cottage using these programs. The details of the source codes for these objects are given in [8].

4. Acknowledgments:

This research was in part supported by NASA.

5. Literature

1. Abelson, H. and diSessa, A. A.: Turtle Geometry. Cambridge. Mass.: MIT Press 1982
2. Goel, N. S.: Models of vegetation canopy reflectance and their use in estimation of biophysical parameters from reflectance data. Remote Sens. Reviews, 4, 1-222 (1988)
3. Goel, N. S.: Inversion of canopy reflectance models for estimation of biophysical parameters from reflectance data, In Theory and Applications of Optical Remote Sensing, (G. Asrar, ed.). pp.205-251. New York: Wiley-Interscience 1989
4. Goel, N. S. and Goodwin, M. D.: Symbolic computations using L-systems. Appl. Math. Computation **42**, 223-253 (1991)
5. Goel, N. S., Knox, L. B., and Norman, J. M.: From artificial life to real life: computer simulation of plant growth. Int. J. General Systems, **18**, 00-00 (1991)
6. Goel, N. S. and Rozehnal, I.:Some non-biological applications of L-systems. Int. J. General Systems, **18**, 00-00 (1991)
7. Goel, N. S. and Rozehnal, I.:Explorations with L-systems. (monograph in preparation).
8. Goel, N. S., Regan, N. and Rozehnal, I.: Some architectural applications of L-systems with radiosity based computer display. Int. J. General Systems (submitted)
9. Goel, N. S., Rozehnal, I. and Thompson, R. L.: A computer graphics based model for scattering from objects of arbitrary shapes in the optical region. Remote Sens. Environ. (in press)
10. Lindenmayer, A.: Mathematical models for cellular interaction in development, I and II. J. Theor. Biol., **18**, 280-315 (1968)
11. Pokorny, C. K. and Gerald, C. F.: Computer Graphics. Irvine, CA.: Franklin, Beedle and Associates.
12. Prusinkiewicz, P. and Hanan, J.: Lindenmayer Systems, Fractals, and Plants. New York: Springer-Verlag 1989
13. Prusinkiewicz, P. and Lindenmayer, A.: The Algorithmic Beauty of Plants. New York: Springer-Verlag 1990
14. Salomaa, A.: **Computation and Automata**. Cambridge: Cambridge Univ. Press 1985

On Synchronized Lindenmayer Picture Languages

Jürgen Dassow[1] and Juraj Hromkovic[2]

[1]Technische Universität Magdeburg, Fakultät für Mathematik, O-3080 Magdeburg, Germany
[2]Universität-Gesamthochschule Paderborn, Fachbereich Mathematik/Informatik, W-4790 Paderborn, Germany (present address)
Comenius University, Institute of Computer Science, CS-84215 Bratislava, Czechoslovakia (permanent address)

Abstract: Picture languages considered in this paper are obtained by a chain code interpretation of words. We determine the hierarchy of languages families generated by different types of Lindenmayer systems and discuss some decision problems.

Keywords: Lindenmayer systems, chain code pictures

1. Introduction

In order to draw a picture by a plotter or on the screen of a computer the following basic commands can be used:
go (and draw) one unit line up from the current point,
go (and draw) one unit line down from the current point,
go (and draw) one unit line to the right of the current point,
go (and draw) one unit line to the left of the current point.
Then the work of the plotter and the resulting picture can be described as a sequence of these elementary commands. If we abbreviate the commands by u, d, r, l, respectively, each picture can be described by a word over the alphabet $\{u, d, r, l\}$, and conversely, each word over this alphabet describes a picture. Moreover, languages over $\{u, d, r, l\}$ correspond to sets of pictures and vice versa. These connections between words/languages and pictures/sets of pictures is first studied by Freeman in 1968 (using diagonal directions in addition). An intensive investigation of such picture languages (called chain code picture languages) started with a paper by Maurer, Rozenberg and Welzl in 1983. However, the theoretical studies concentrate on the case of sequential grammars. A summary of results in this direction can be found in [1] .
On the other hand, Lindenmayer systems are used to generate fractals and pictures of plants (see [8]). However, the type of the Lindenmayer system varies often in order to desribe the desired picture. In this paper we present some theoretical investigations on the concept of chain code picture languages generated by Lindenmayer systems. We discuss only the case that each letter of the basic alphabet $\{u, d, r, l\}$ is substituted by a word which describes an enlarged version of the line that corresponds to the letter. Therefore

any substitution of the L system acts like a microscope.
Thus this approach is also related to the concept of hierarchical design used in engineering and design of VLSI structures, which is reflected in some studies on graph grammars where nodes/sets of nodes are sequentially rewritten (see e.g. [5]). Moreover, most generations of fractals in this area are done by the use of Lindenmayer systems with this restriction.

2. Definitions

Throughout the paper we assume that the reader is familiar with the basic concepts of formal language theory (see [3] and [9]).
We specify a grammar $G = (V_N, V_T, P, S)$ by the set V_N of nonterminals, the set V_T of terminals, the set P of productions, and the axiom S. A TOL system $G = (V, \mathbf{P}, w)$ is specified by an alphabet V, a set $\mathbf{P}$ of finite substitutions, and an axiom w; the special cases of a DTOL and OL system are obtained by considering homomorphisms as substitutions or systems with only one substitution, respectively; moreover, a DOL system appears if we combine both these restrictions.
For a point $n = (x, y)$ in the grid given by the two-fold cartesian product Z^2 of the set Z of integers, we set $x = x(n)$ and $y = y(n)$. For $b \in \{u, d, r, l\}$ and a point $n = (x, y)$, we set

$$b(x,y) = \begin{cases} (x, y+1) & \text{for } b = u \\ (x, y-1) & \text{for } b = d \\ (x+1, y) & \text{for } b = r \\ (x-1, y) & \text{for } b = l \end{cases}$$

For $n \in Z^2$ and $b \in \{u, d, r, l\}$, we denote by $< n, b(n) >$ the (undirected) unit line connecting n and $b(n)$.
By a *picture* we mean a finite set of such unit lines in the grid. The *size* $|p|$ of a picture p is defined as the number of its unit lines. The *area size* $s(p)$ of a picture p (which is of interest in VLSI considerations) is (x, y) iff p fits in a rectangle with length x and heigth y and x and y are minimal with this property. We say that $s(p) \leq (x', y')$ if $s(p) = (x, y), x \leq x'$, and $y' \leq y$.
With a word w over $\{u, d, r, l\}$ we now associate a picture p with distinguished point n in the following way:
If $w = \lambda$, then $p = \emptyset$ and $n = (0, 0)$.
If $w = w'x$ and p' and n' are associated with w', then $p = p' \cup \{< n', b(n') >\}$ and $n = b(n')$.
(Intuitively, we start in the point $(0, 0)$ and draw the picture unit line by unit line following the sequence of letters in w). By $shift(w)$ we denote the distinguished node n and by $p(w)$ the picture associated with w, and, for a language $L \subseteq \{u, d, r, l\}^*$ we set

$$p(L) = \{p(w) : w \in L\}.$$

Now let G be a phrase structure grammar or a Lindenmayer system such that the generated language $L(G)$ is contained in $\{u, d, r, l\}^*$. Then we set

$$B(G) = \{p(w) : w \in L(G)\}.$$

Such languages are called (chain code) *picture languages*, and the generating grammar or system is called a *picture grammar* or *picture system*. By $\mathbf{B}(CS), \mathbf{B}(CF)$, and $\mathbf{B}(REG)$ we denote the families of all picture languages generated by context-sensitive, context-free and regular grammars, respectively.

Definition. *A (chain code) picture TOL system*

$$G = (\{u, d, r, l\}, \{\tau_1, \tau_2, \dots, \tau_r\}, w)$$

is called synchronized if, for any i, $1 \leq i \leq r$, there exist numbers $k_i \in N$ and $d_i \in N$ such that for any $w_a \in \tau_i(a)$ the following conditions are satisfied:

$$shift(w_a) = \begin{cases} (k_i, 0) & for\ a{=}r \\ (-k_i, 0) & for\ a{=}l \\ (0, k_i) & for\ a{=}u \\ (0, -k_i) & for\ a{=}d \end{cases}$$

and for any point n of $p(w_a)$,

$$\begin{array}{c} 0 \leq x(n) \leq k_i \text{ and } -d_i \leq y(n) \leq d_i \text{ for } a = r, \\ -k_i \leq x(n) \leq 0 \text{ and } -d_i \leq y(n) \leq d_i \text{ for } a = l, \\ -d_i \leq x(n) \leq d_i \text{ and } 0 \leq y(n) \leq k_i \text{ for } a = u, \\ -d_i \leq x(n) \leq d_i \text{ and } -k_i \leq y(n) \leq 0 \text{ for } a = d. \end{array}$$

We refer to these systems as $sDOL$, sOL, $sDTOL$, and $sTOL$ systems. By $\mathbf{B}(sTOL)$, $\mathbf{B}(sDTOL)$, $\mathbf{B}(sOL)$, and $\mathbf{B}(sDOL)$ we denote the families of all chain code picture languages generated by synchronized chain code picture TOL, DTOL, OL, and DOL systems, respectively.

We say that $G = (V, \mathbf{P}, w)$ is a (k, d)-sTOL system if

$$k = max\{k_i : 1 \leq i \leq r\} \text{ and } d = max\{d_i : 1 \leq i \leq r\}.$$

3. Hierarchies

The following theorem presents our hierarchy results as a diagram. An arrow from X to Y denotes the proper inclusion of X in Y, and sets which are not connected are incomparable.

Theorem 1.

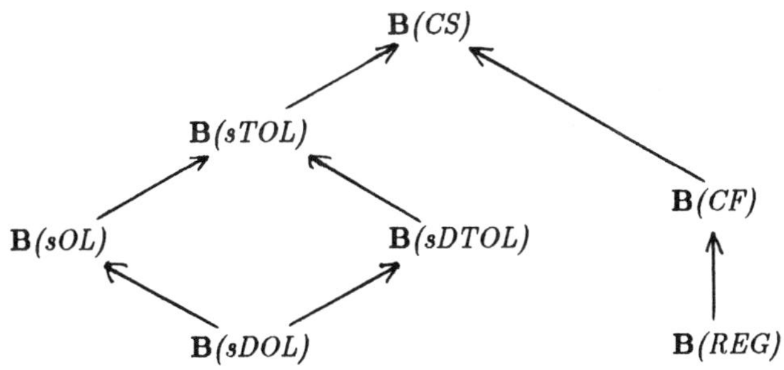

Proof. The inclusions follow by definition and by the known relations for the families of word languages. The properness of the inclusions $\mathbf{B}(REG) \subseteq \mathbf{B}(CF)$ and $\mathbf{B}(CF) \subseteq \mathbf{B}(CS)$ is shown in [7] . Thus it is sufficient to show the following statements:

i) $\mathbf{B}(sDTOL)$ *is not contained in* $\mathbf{B}(sOL)$.
ii) $\mathbf{B}(sOL)$ *is not contained in* $\mathbf{B}(sDTOL)$.
iii) $\mathbf{B}(sDOL)$ *is not contained in* $\mathbf{B}(CF)$.
iv) $\mathbf{B}(REG)$ *is not contained in* $\mathbf{B}(sTOL)$.

We now prove these statements.

i) We consider the sDTOL system

$$G = (\{u, d, r, l\}, \{\tau_1, \tau_2\}, rdlu)$$

with

$$\tau_1(u) = uluru, \tau_1(d) = drdld, \tau_1(r) = rurdr, \tau_1(l) = ldlul$$

and

$$\tau_2(u) = uluru, \tau_2(d) = drdld, \tau_2(r) = rdrur, \tau_2(l) = luldl.$$

Then the smallest generated pictures corresponding to the axiom $rdlu, \tau_1(rdlu)$, and $\tau_2(rdlu)$ are

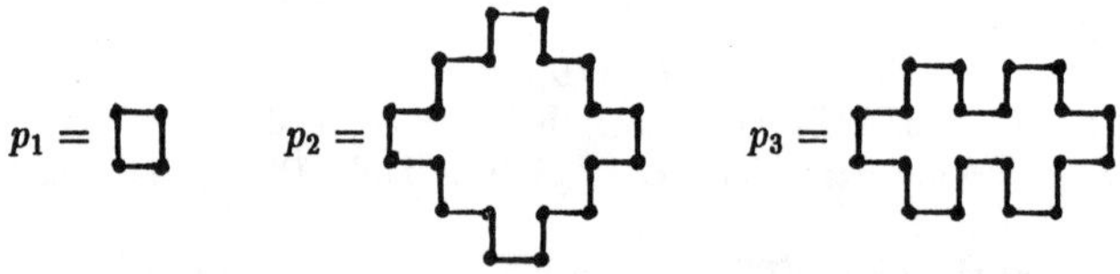

Assume that $B(G) = B(G')$ for some sOL system $G = (\{u, d, r, l\}, \tau, w)$. Obviously, the axiom w of G' gives a description of p_1. Without loss of generality we assume that $w = rw'$ (the modifications for the other cases are left to the reader). Then, for the number k associated with τ we obtain $k = 3$ and in order to generate p_2 the following cases are possible for $w_r \in \tau(r)$:
Case 1. $p(w_r) = p(rurdr)$,
Case 2. $p(w_r) = p(rdrur)$.

We discuss only Case 1 (the argumentation in the other case is similar). Then the drawing of p_2 starts in the point marked by a square. If $w = rlw''$ holds, then $p(w_l) = p(luldl)$ for some $w_l \in \tau(l)$. But then also the lower horizontal of p_1 is transformed in this picture and we obtain the picture

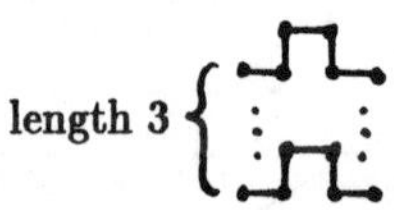

which is not contained in $B(G)$. Hence we get $w = rdw''$ and $p(w_d) = p(drdld)$ for some $w_d \in \tau(d)$. Continuing in this way we obtain $w = rdluz$ and $p(w_l) = p(ldlul)$ and $p(w_u) = p(uluru)$ for some $w_l \in \tau(l)$ and $w_u \in \tau(u)$. In order to generate p_3 by using the same word w_r we have to start in the point marked by a circle, and as above we then obtain $p(w'_d) = p(urulu)$ which contradicts $k \geq o$. If we start the drawing of p_3 with some $w'_r \in \tau(r)$ corresponding to Case 2, then $v = w'_r w_d w_l w_u \tau(z)$ belongs to $L(G')$, however,

$p(v) \neq p_3$. Thus we obtain a contradiction, again. Hence $B(G)$ is not in $\mathbf{B}(sOL)$.
ii) We consider the sOL system

$$H = (\{u, d, r, l\}, \tau, rdlu)$$

with

$$\tau(u) = \{uluru\}, \tau(d) = \{drdld\}, \tau(r) = \{rurdr, rdrur\}, \tau(l) = \{luldl, ldlul\}.$$

We mention two facts on the generated language.

Fact 1. *If $v \in \tau^2(a)$ for $a \in \{u, d, r, l\}$, then v contains any of the words ur, ul, dr, dl, ru, rd, lu, ld at least once and no other words of length 2.*

This fact can be seen easily by going through all possible v.

Fact 2. *Let h be the homomorphism given by*

$$h(u) = \tau(u), h(d) = \tau(d), h(r) = rurdr, h(l) = ldlul.$$

Then, for any $n \geq 0$ and any $a \in \{u, d, r, l\}$,

$$h^n(rdlu) \in L(H) \text{ and } \#_a(h^n(rdlu)) = 5^n.$$

This follows by induction using the growth matrix

$$\begin{pmatrix} 3 & 0 & 1 & 1 \\ 0 & 3 & 1 & 1 \\ 1 & 1 & 3 & 0 \\ 1 & 1 & 0 & 3 \end{pmatrix}$$

of the DOL system $(\{u, d, r, l\}, h, rdlu)$ from the theory of growth of DOL systems.
Now assume that $B(H) = B(H')$ for some sDTOL system

$$H' = (\{u, d, r, l\}, \{h_1, h_2, \ldots, h_r\}, w).$$

Then, for sufficiently large n, we consider the picture p_n which is constructed in the following way from $h^{n-1}(rdlu)$: We substitute the 2^n-th, (2^n+1)-th,..., 2^{n+1}-th occurrence of r or l by $rurdr \in \tau(r)$ and $luldl \in \tau(l)$, respectively, and therefore by the picture

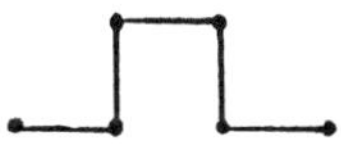

Figure 1.

and the remaining occurrences of r and l by $rdrur$ and $ldlul$, respectively, corresponding to the picture

Figure 2.

and the occurrences of u and d by $h(u)$ and $h(d)$. By Fact 1 and Fact 2, p_n contains a subpicture q of size 2^{n+1} which contains no subpicture shown in Figure 2. q is described by some subword v of $z \in L(H')$ with $p(z) = p_n$. Since n is sufficiently large, $z = h_i(z')$ for some $i, 1 \leq i \leq r$, and $z' \in L(H')$. Let $v = h_i(v')$ for some subword v' of z'. By Fact 1, v' contains any letter r, l, u, d, and thus none of these letters can produce the picture of Figure 2, since it does not occur in $p(h_i(v'))$. Because h_i is a homomorphism this subpicture cannot occur in $p_n = p(z)$. This contradicts the construction of p_n. Therefore $B(H) \notin \mathbf{B}(sDTOL)$.

iii) We consider the sDOL system $G = (\{u,d,r,l\}, h, drul)$ with $h(a) = a^2$ for $a \in \{u,d,r,l\}$. Then $B(G)$ consists of all squares with a basic line of length 2^n for some n. In order to prove that $B(G)$ is not a context-free picture language we use the following iteration lemma due to Kim:

Let H be a context-free grammar. Then there is a constant k (depending on H) such that, for every picture $p \in B(H)$ with $|p| \geq k$, there is a word w and a decomposition $w = w_1w_2w_3w_4w_5$ such that
i) $|p(w_2w_3w_4)| \leq k$,
ii) $shift(w_2) \neq 0$ *or* $shift(w_4 \neq 0$,
iii) $p(w_1w_2^iw_3w_4^iw_5) \in B(H)$ *for all* $i \geq 0$.
(Note that $w \in L(H)$ is not required.)

Let now n be a sufficiently large number and consider the decomposition $w_1w_2w_3w_4w_5$ of the word which describes the square with a basic line of length 2^n. Without loss of generality let $shift(w_2) \neq 0$. Then w_2 cannot describe a part of the square which contains a corner, since $p(w_2^i)$ would contain at least i corners. Thus w_2 generates a part of a straight line and $|p(w_2)| < |p(w_2^i)|$ for $i \geq 2$.
Case 1. $shift(w_4) \neq 0$. Then, by the same arguments, w_4 generates a part of a straight line, too. Thus at least one of the pictures corresponding to the words w_1, w_3, w_5, say w_1, contains two corners and the line between them. Hence $p(w_1w_2^2w_3w_4^2w_5)$ is not a rectangle or the line of the rectangle containing $p(w_2^2)$ is longer than that contained in $p(w_1)$.
Case 2. $shift(w_4) = 0$. Then $p(w_4^i) = p(w_4)$ for all $i \geq 0$. Therefore $p(w_1w_2^iw_3w_4w_5) = p(w_1w_2^iw_3w_4^iw_5)$ is not a rectangle since the side containing $p(w_2^i)$ is longer than all other straight parts of the picture.

iv) $L = \{u^2, u^4\}$ is regular. Assume that $B(L)$ is a sTOL picture language generated by some system G. Obviously, $L(G) \subseteq \{u,d\}^*$. In order to describe $p(u^2)$ there has to be a word v in $L(G)$ which contains u^2 or d^2 as a subword. Without loss of generality we assume that v contains u^2. Moreover, for any substitution τ of G, $\tau(u)$ contains only words of the form u^r (generally, $\tau(u)$ contains only words describing $p(u^r)$ for some r, but we can restrict to the given special case). Then any word in $\tau(d)$ has the form d^r (if d occurs in some word of $L(G)$). If $r > 2$, then $p(\tau(v))$ contains $p(u^6)$ as a subpicture which is not the case for any picture of $B(L)$. If $r = 1$, then τ is the identity and can be omitted without a change of the generated language and picture language. If $r = 2$, then we consider $v' \in L(G)$ with $p(v') = u^4$ and prove by analogous arguments that $p(\tau(v'))$ contains the subpicture $p(u^8)$ which shows that $p(\tau(v')) \notin B(L)$.

Finally in this section we give a result on the growth of the area size of the pictures.

Theorem 2. *For the* (k,d)*-sTOL system* $G = (V, \mathbf{P}, w)$*, let*

$$s(p(w)) = (x, y) \quad \text{and} \quad g(n) = 2d\frac{k^n - 1}{k - 1}.$$

Then for any z *generated by* n *iterated applications of elements of* $\mathbf{P}$ *to* w*,*

$$s(p(z)) \leq (k^n x + g(n), k^n y + g(n)).$$

Proof. For any word $v \in V^*$ with $s(p(v)) = (x, y)$, $\tau \in \mathbf{P}$, and $u \in \tau(v)$, we obtain

$$s(p(u)) \leq (xk + 2d, yk + 2d).$$

Now the statement follows by induction.

4. Decision problems

We start with two basic problems of formal language theory which are decidable for context-free picture languages.

Theorem 3. *Given a picture* p *and an sTOL system* G*, it is decidable whether or not* $p \in B(G)$.

Proof. By [7], the set L_p of all descriptions of p is regular. Thus $L(G) \subseteq L_p$ is an ETOL language for which an ETOL system G' can be constructed effectively. Moreover, $p \in B(G)$ if and only if $L(G') \neq \emptyset$. Since the latter problem is decidable, $p \in B(G)$ is decidable.

Theorem 4. *Given an sTOL system* G*, it is decidable whether or not* $B(G)$ *is finite.*

Proof. Let $G = (\{u, d, r, l\}, \{\tau_1, \tau_2, \ldots, \tau_r\}, w)$. If $max_{1 \leq i \leq r} k_i = k \geq 1$, say without loss of generality $k = k_1$, then the sequence $|p(\tau_1^i(w))|$ contains an infinite monotone increasing subsequence and therefore $B(G)$ is infinite. If $k_1 = k_2 = \ldots = k_r = 1$ and $d_i \geq 1$ for some $i, 1 \leq i \leq r$, then we obtain the infinity of $B(G)$ by an analogous argument. If $k_i = 1$ and $d_i = 0$ for $1 \leq i \leq r$, then τ_i is the identity mapping (on pictures) for $1 \leq i \leq r$ and thus $B(G) = \{w\}$, i.e. $B(G)$ is finite.
By these considerations it is obvious how to decide the finiteness of $B(G)$.

The emptiness problem is trivial since $p(w) \in B(H)$ holds. We leave the equivalence problems and the complexity of the problems which are decidable by the above theorems as open problems. (The above algorithm reduces the membership problem to the PSpace-complete emptiness problem for ETOL systems, which seems to be much more hard than necessary.)

We now turn to some geometrical and graph-theoretical decision problems which are generalizations of the membership problem. If we consider a picture p as a graph, then the set of nodes is the set of grid points belonging to p and the set of edges is the set of unit lines of p. A graph is called a circuit if any node of the graph has the degree 2. A tree is a graph which has no circuit as a subgraph.

Theorem 5. *Given an sTOL system* G*, it is decidable in linear time whether or not* $B(G)$ *contains a tree.*

Proof. Assume that the axiom w of $G = (\{u,d,r,l\},\mathbf{P},w)$ does not describe a tree. Then $p(w)$ contains a subgraph q which is a circuit. By the definition of the synchronization it is easy to see that $p(\tau(w))$ contains a circuit as a subgraph which is formed by the images of the letters corresponding to edges of q under the substitution $\tau \in \mathbf{P}$. Iterating this considerations we obtain that $B(G)$ contains a tree if and only if $p(w)$ is a tree. In order to check whether or nor $p(w)$ is a tree we construct a spanning tree T of $p(w)$ in linear time by depth-first-search (see e.g. [6]) and compare T and $p(w)$ in linear time.

Theorem 6. *Given a sTOL system G, it is decidable whether or not B(G) contains a circuit.*

Proof. Assume that $v \in L(G), \tau_i \in \mathbf{P}, p(v)$ is not a circuit, and $p(\tau_i(v))$ is a circuit. Then $p(v)$ contains a node of degree ≥ 3 or degree 1. First we discuss the former case. Then $p(v)$ has without loss of generality a subgraph of the form  . It is easy to see that d_i has to be a multiple of k_i because otherwise we would obtain again a node of degree ≥ 3 in $p(\tau_i(v))$ which contradicts our assumption. Moreover, if a,b,c are the letters corresponding to the horizontal, the upper vertical, the lower vertical edge, then $p(\tau(a))$ is the same picture as generated from the considered subgraph, it has the "shape of a rectangle" and its left "side" consists of $p(\tau(b))$ and $p(\tau(c))$. Now we try to add further edges to the subgraph such that new connected subgraphs appear. In order to ensure that $p(\tau(v))$ is a circuit we obtain in this way only the possiblities

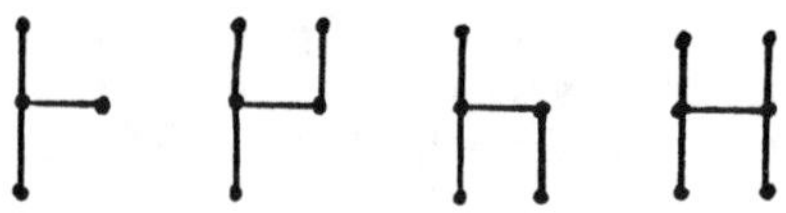

Figure 3.

Therefore we have the following decision procedure:

Step 1. Check whether or not $p(w)$ is a circuit.

If the answer is positive, we know that B(G) contains a circuit.

If the answer is negative, do the following:

Step 2. Using the algorithm for the membership problem (see Theorem 2) decide whether or not one of the pictures of Figure 3 or the corresponding pictures obtained by rotation by a multiple of 90^0 is contained in $B(G)$ and whether or not the above mentioned conditions for a,b,c are satisfied with respect to some substitution τ.

If the answer is positive then $B(G)$ contains a circuit.

If the answer is negative then $B(G)$ contains no circuit.

If no node of degree ≥ 3 but a node n of degree 1 exists in $p(v)$, we can give a similar algorithm starting with the edge containing n (we obtain then the critical pictures , 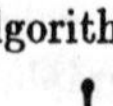 and 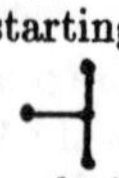). The details are left to the reader.

We note that the decision problems considered in the last two theorems are undecidable for regular chain code picture languages (see [2]).

References

[1] Dassow,J., Hinz,F., Kettenkode-Bildsprachen. Theorie und Anwendungen. Wiss. Zeitschr. d. Techn.Univ. Magdeburg 33 (1989) 1-12.

[2] Dassow,J., Hinz,F., Decision problems and regular chain code picture languages. Submitted for publication.

[3] Hopcroft,J.E., Ullman,J.D., Introduction to Automata Theory, Languages, and Computation. Addison-Wesley Publ. Co., 1979.

[4] Kim,C., Picture iteration and picture ambiguity. J.Comp. Syst. Sci. In press.

[5] Lengauer,T., Wagner,K.W., The correlation between the complexities of the non-hierarchical and hierarchical version of graph problems. In: Proc. STACS'87, LNCS 247, 100-113, Springer-Verlag, 1987.

[6] Manber,U., Introduction to Algorithms. Addison-Wesley Publ. Co., Reading, 1989.

[7] Maurer,H.W., Rozenberg,G., Welzl,E., Using string languages to describe picture languages. Inform. Control 54 (1982) 155-185.

[8] Pruzinkiewicz,P., Hanan,J., Lindenmayer systems, fractals and plants. Univ. of Regina (Canada), Dept. Computer Science, Techn. report.

[9] Rozenberg,G., Salomaa,A., The Mathematical Theory of L Systems. Academic Press, 1980.

Generation of Random Planar Tessellations by Hierarchical Decimation

Peter Meer and Azriel Rosenfeld

Center for Automation Research, University of Maryland, College Park, MD 20742-3411

Abstract: We describe a method of generating a class of irregular tessellations of a digital image using a random decimation process. The decimation is hierarchical and can generate regions of varying sizes. The tessellation can be generated in $O(\log \text{image size})$ time if the process is implemented in parallel on a cellular pyramid computer.

Keywords: Irregular tessellations, random decimation, hierarchical decimation, pyramid computers

Introduction

There are several standard approaches to generating random planar patterns. Once class of approaches makes use of Markov random fields; it randomly chooses an initial gray level for each pixel, and then repeatedly modifies the gray levels so that their joint frequencies of occurrence conform to the joint probabilities of the desired random field (e.g., [3]). This method is characterized by slow convergence, though it can be speeded up to some degree by parallel implementation [4]. It is also difficult to use the method for random fields that have large (or varying) correlation distances.

Another class of pattern generation methods makes use of random geometric processes [1]. For example, a set of randomly distributed points (e.g., a Poisson point process) defines a Voronoi tessellation of the plane into convex polygons; gray levels can then be assigned to the polygons so as to satisfy given probabilistic constraints. This approach can be used to generate arbitrarily coarse tessellations, but it gives little control over the nature of these tessellations (though this could be somewhat improved by using point patterns that have clustering or anticlustering tendencies [2]).

This paper describes a method of generating tessellations using a random decimation process. The decimation is hierarchical and can generate regions of varying sizes. The

The support of the Air Force Office of Scientific Research under Grant AFOSR-86-0092 is gratefully acknowledged, as is the help of Sandy German in preparing this paper.

tessellation can be generated in $O(\log \text{image size})$ time if the process is implemented in parallel on a cellular pyramid computer. For a more detailed description of the method and additional examples see [5].

The method

Assume we are given a planar point pattern and a weighted neighbor relationship defined on the points. Suppose a point P has r neighbors with weights $\lambda_1, \ldots, \lambda_r$. We associate outcomes of independent identically distributed random variables with all the points in the pattern. Let the outcomes for P and its neighbors be $x_0, x_1, \ldots, x_r$. P survives the decimation process iff $x_0 = \max(\lambda_1 x_1, \ldots, \lambda_r x_r)$. Note that the probability of survival depends on the number of neighbors and on the values of the λ's; note also that we can decimate nonisotropically by introducing directional biases into the λ's.

The process just described is performed in parallel. If we perform it just once, it yields a somewhat sparse pattern of survivors, since only points whose x values are weighted local maxima can survive. We can increase the density of the survivors by repeating the process; at each repetition, nonsurvivors are reconsidered for survival if they are not neighbors of survivors. (Survivors are not reconsidered.) Repeating this process yields, after a few iterations, a stable set of survivors that tends to have a more regular structure.

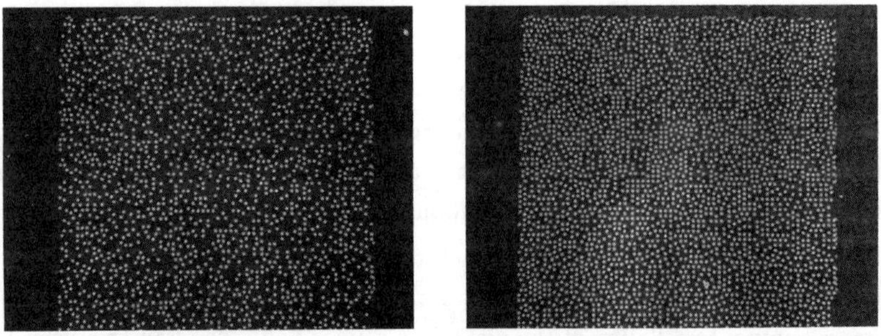

(a) (b)

Figure 1: Iterated decimation. (a) One iteration. (b) Four iterations.

Figure 1 illustrates how the iterated decimation process works. We begin with the set of all digital lattice points under 8-adjacency. The random variables are uniformly distributed in [0,1], and in this example the λ's are all equal to 1. A single iteration of the process yields the patterns shown in Figure 1a, which is somewhat random, but after four iterations the pattern is much more regular, as shown in Figure 1b.

To define a new neighbor relation on the survivors, we first associate each nonsurvivor with the nearest survivor(s), where "nearest" is defined in terms of distance in the graph defined by the previous neighbor relation. Two survivors are called neighbors if the same nonsurvivor was assigned to both of them, or if two neighboring nonsurvivors were assigned to the two of them. We can now repeat the entire decimation process using the set of survivors and the new neighbor relation. Each repetition yields a sparser point pattern.

When we decimate the set of digital lattice points, the assignment of nonsurvivors defines a tessellation of the digital plane; each survivor, together with the set of nonsurvivors assigned to it (with ties resolved arbitrarily), defines a cell of this tessellation. When we repeat the decimation process, and again assign nonsurvivors to survivors, we obtain a coarser tessellation whose cells are unions of the cells constructed at the previous decimation stage. Thus the repeated decimation process defines a hierarchy of tessellations. An example of such a hierarchy, for five stages of decimation of a 128×128 digital array, is shown in Figure 2, where gray levels have been assigned randomly to the cells obtained at each stage.

We can obtain "mixed" tessellations of an image by combining cells obtained at different stages. This is done by specifying the (approximate) fraction of the image area that should be covered by cells of each size. We randomly choose cells of the largest size (the last stage) whose total area is close to the desired fraction. In the remainder of the image, we randomly choose cells of the next largest size that have approximately the desired total area; and so on. Figure 3 shows a mixed tessellation constructed in this way from the five tessellations of Figure 2.

Finally, we can obtain anisotropic patterns by using directionally dependent λ's. (Some degree of anisotropy can also be achieved by directionally biasing the resolution of ties when the cells are defined.) Figure 4 shows a set of tessellations constructed in this way using $\lambda = 0.05$ for directions in the range 122.5°–157.5°, and $\lambda = 1$ otherwise.

Concluding remarks

We have described a general method of constructing hierarchical tessellations of the plane into irregular regions. It would be of interest to investigate the appropriateness of our method as a model for patterns generated by biological growth/competition processes—for example, the sampling structure found in the central part of the retina.

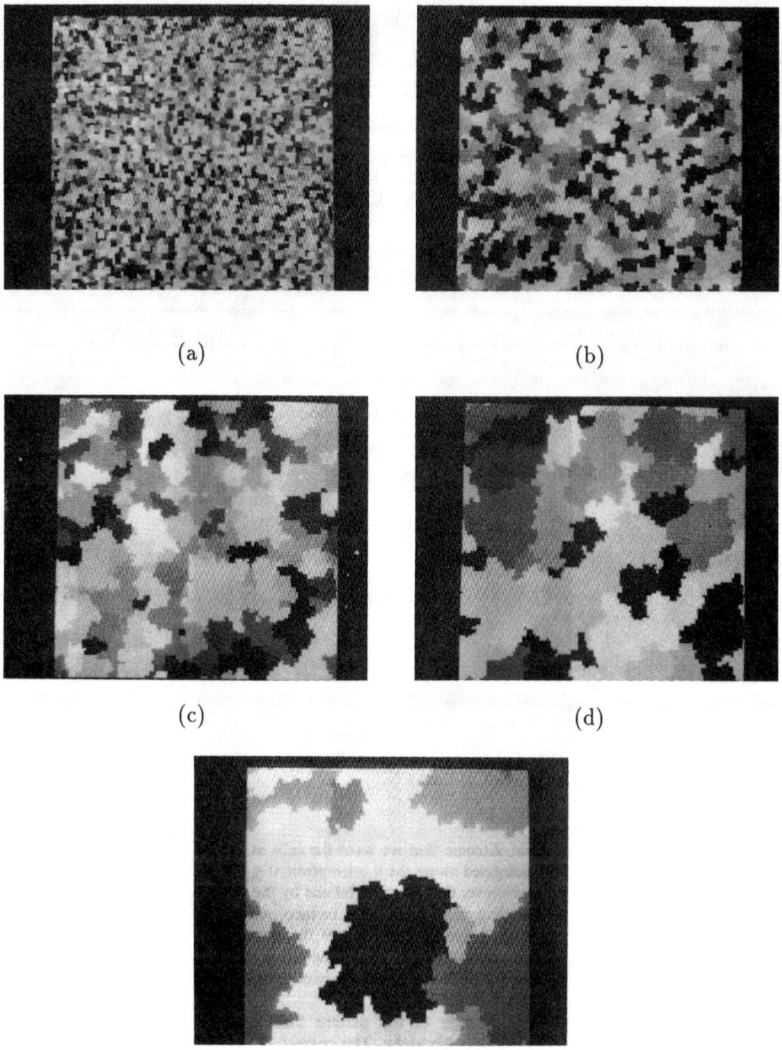

Figure 2: Five-level hierarchy of tessellations defined by repeated decimation.

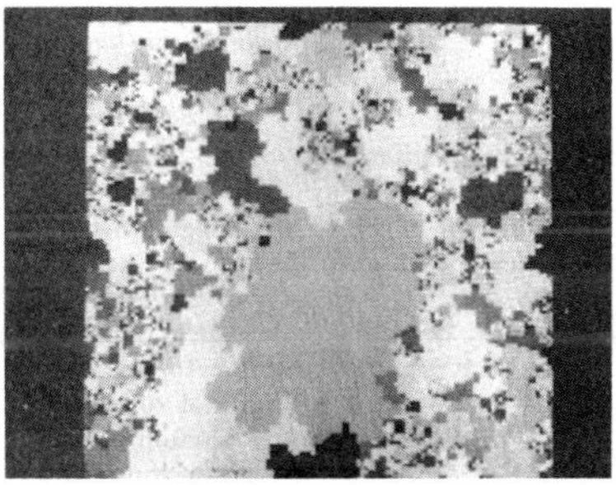

Figure 3: "Mixed" tessellation.

References

[1] N. Ahuja and B.J. Schachter, *Pattern Models.* Wiley-Interscience, New York (1983).

[2] S.K. Bhaskar, A. Rosenfeld, and A.Y. Wu, Models for neighbor dependency in planar dot patterns, *Pattern Recognition* **22**, 533–559 (1989).

[3] G.R. Cross and A.K. Jain, Markov random field texture models, *IEEE Trans. Pattern Analysis Mach. Intell.* **PAMI-5**, 25–39 (1983).

[4] A. Margalit, A parallel algorithm to generate a Markov random field image on a SIMD hypercube machine, *Pattern Recognition Letters* **9**, 263–278 (1989).

[5] P. Meer and S. Connelly, A fast parallel method for synthesis of random patterns, *Pattern Recognition* **22**, 189–204 (1989).

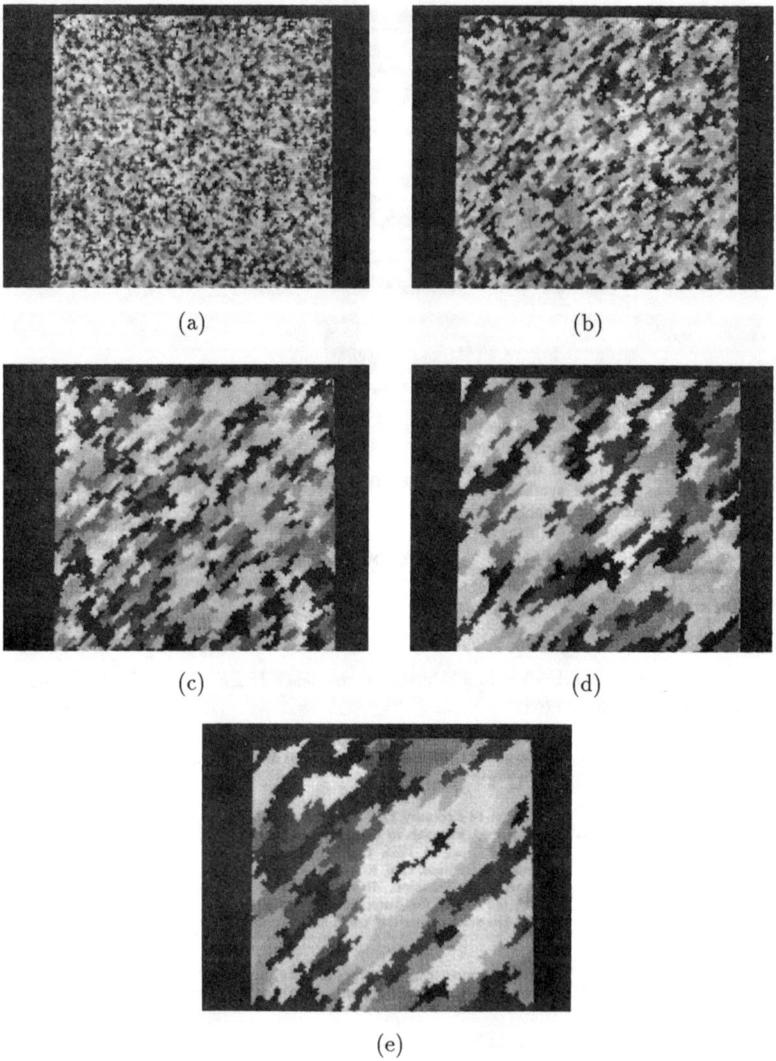

Figure 4: Anisotropic tessellations.

4. Walking in the Plane:

Graph Grammars and Map L Systems

Parallel Hyperedge Replacement

Hans–Jörg Kreowski

Universität Bremen, FB Mathematik und Informatik, Postfach 33 04 40, D–2800 BREMEN 33

Abstract: This paper presents the very first steps of a systematic study of hyperedge replacement grammars with a parallel mode of rewriting as known from L systems.

Keywords: hyperedge replacement, OL systems, graph grammars.

0 Introduction

The aim of this paper is to present the very first steps of a systematic study of hyperedge replacement grammars with a parallel mode of rewriting as known from L systems (see, e.g., Herman and Rozenberg [12], Rozenberg and Salomaa [17]).

Hyperedge replacement is one of the most elementary and frequently used concepts of graph transformation with the characteristics of context–free rewriting. Introduced in the early seventies by Feder [5] and Pavlidis [16] (under various pseudonyms), it has been intensively and systematically studied in the eighties (starting with the special case of edge replacement) by Bauderon and Courcelle [1, 2, 3], Habel, Kreowski and Vogler [13, 7, 8, 10, 11], Lengauer and Wanke [14] and others. But nearly all work done on hyperedge replacement concerns the sequential mode of rewriting (where the simultaneous replacement of hyperedges is not demanded). Readers who want to learn more about hyperedge replacement are refered to the literature and, in particular, Habel's thesis [6], which presents the area of hyperedge replacement in a neat, systematic and nearly complete way.

A hyperedge is an atomic item with a label and a fixed number of ordered tentacles. It can be attached to any kind of structure coming with a set of nodes in such a way that each tentacle is attached to a node. So the hyperedge controls a sequence of nodes, called the attachment, and can play the role of a placeholder, which may be replaced by some other structure eventually. In such a replacement step the hyperedge disappears, and the replacing structure is added and suitably connected to the original structure. For this purpose, the replacing structure is assumed to have a sequence of external nodes which are fused with the attachment of the replaced hyperedge. The replacement of a hyperedge by a structure can be iterated if the original structure or the replacing one are decorated by other hyperedges. In this paper, ordinary undirected and unlabelled graphs are employed as underlying structures of interest because they form a well–known class of graphs.

If the replacement is controlled by productions of a grammar and all hyperedges are replaced simultaneously, one gets a derivation process of the type of OL systems, which is stydied in this paper. Actually, the grammar is assumed to provide a finite set of tables each of them being a finite set of productions, and the productions applied in a single derivation steps must belong to one of the tables. Moreover, the structures being processed are of a hybrid nature consisting of nodes and edges of a graph that are not rewritten and a set of decorating hyperedges subject to rewriting. The acronym HTOL will indicate this situation.

The paper is organized in the following way. The model of HTOL hyperedge replacement grammars is introduced in section 3 together with some results concerning the composition and decomposition of derivations. In section 4, the decidability of the emptiness problem is shown. In section 5, the generative power of the HTOL model and of two special cases is compared with the generative power of hyperedge replacement grammars with a sequential mode of rewriting.

1 Hyperedge–decorated graphs

In this section, we recall the basic notions hyperedge replacement and hyperedge–replacement grammars are built upon. We choose ordinary unlabelled undirected graphs without loops and multiple edges as basic structures of interest. To generate sets of graphs, these graphs become decorated with *hyperedges* each of which has got a label and is attached to several nodes of the graph. Further, every such *(hyperedge–)decorated* graph has got a sequence of so–called *external* nodes. Every hyperedge of a decorated graph serves as a placeholder for a graph or — recursively — for another decorated graph, which can replace the hyperedge by fusing its external nodes with those the hyperedge is attached to.

1.1 Definition (graph)
A(n unlabeled, undirected) *graph* is a pair (V, E) where V is a set of *nodes* (or *vertices*) and E is a set of 2–elements subsets of V, called *edges*. The set of all graphs is denoted by $\mathcal{G}_0$. □

1.2 Definition (decorated graph)
Let N be a set of *nonterminal labels.* A (hyperedge–)*decorated graph* (over N) is a system $H = (V, E, Y, lab, att, ext)$ where $(V, E) \in \mathcal{G}_0$, called the *underlying graph* of H and denoted by $U(H), Y$ is a set of *hyperedges*, $lab : Y \to N$ is a mapping, called the *labelling*, $att : Y \to V^*$ is a mapping, called the *attachment*, and $ext \in V^*$, called the (sequence of) *external nodes.*

Remarks

1. The components V, E, Y, lab, att, and ext of H are also denoted by V_H,E_H,Y_H,lab_H, att_H, and ext_H, respectively.
2. The length of ext_H is called the *type* of H and denoted by $type(H)$. If $type(H) = m$, H is called an *m–graph.*
3. The length of $att_H(y)$ for $y \in Y_H$ is called the *type* of y, denoted by $type(y)$. If $type(y) = n$, y is called an *n–edge.*

4. Note that *ext* as well as $att(y)$, for $y \in Y_H$, may contain nodes twice or more.

5. The class of all decorated graphs over N is denoted by $\mathcal{G}(N)$. The class of all m–graphs over N is denoted by $\mathcal{G}_m(N)$.

6. A graph can be seen as a 0–graph without hyperedges. In this sense, $\mathcal{G}_0 \subseteq \mathcal{G}_0(N) \subseteq \mathcal{G}(N)$. Moreover, the subclass of $\mathcal{G}_m(N)$, for $m \geq 0$, consisting of all m–graphs without hyperedges is denoted by $\mathcal{G}_m$ while $\mathcal{G}$ denotes the union of all $\mathcal{G}_m$, for $m \geq 0$. In the description of m–graphs without hyperedges we will drop the components Y, *lab* and *att*.

7. A particular case is a decorated graph with a single hyperedge where external nodes and attachment coincide, i.e., $H = (V, \emptyset, \{y\}, lab, att, ext)$ with $ext = att(y)$. Such an H is called a *handle*. If $V = \{1, \ldots, n\}$, $lab(y) = A$, and $att(y) = ext = 1 \ldots n$, H is denoted by $(A, n)^{\bullet}$. □

Two decorated graphs are said to be isomorphic if there is an isomorphism between the underlying graphs preserving external nodes, and there is a bijective mapping between their hyperedges which is consistent with the graph isomorphism. If a decorated graph is restricted to subsets, one gets a subgraph. It should be noticed that the external nodes of subgraph and supergraph may be different.

1.3 Definition (isomorphic decorated graphs, decorated subgraph)
Let $H, H' \in \mathcal{G}(N)$.

1. H and H' are *isomorphic*, denoted by $H \cong H'$, if there are bijective mappings $f : V_H \to V_{H'}$ and $g : Y_H \to Y_{H'}$ such that $E_{H'} = \{\{f(x), f(y)\} \mid \{x, y\} \in E_H\}$, $lab_H(y) = lab_{H'}(g(y))$ and $f^*(att_H(y)) = att_{H'}(g(y))$ for all $y \in Y_H$, and $f^*(ext_H) = ext_{H'}$, where f^* is the natural extension of f to sequences.

2. H is a *subgraph* of H', denoted by $H \subseteq H'$, if $V_H \subseteq V_{H'}, E_H \subseteq E_{H'}, Y_H \subseteq Y_{H'}$, $att_H(y) = att_{H'}(y)$ and $lab_H(y) = lab_{H'}(y)$ for all $y \in Y_H$. □

1.4 Example (hyperedge–decorated graph)
The picture below shows a decorated graph consisting of a graph on six vertices having six edges together with four hyperedges. Its external nodes are drawn as filled circles (whereas all others are unfilled). The hyperedges are labelled with A, B, and C, and the order on their tentacles is indicated by their numbers. Similarly, we provide numbers for the external nodes indicating the ordering on *ext*. The type of this decorated graph as well as that of the A–labelled hyperedge is 3, while the hyperedges labelled with B are of type 2 and the one labelled with C is of type 1.

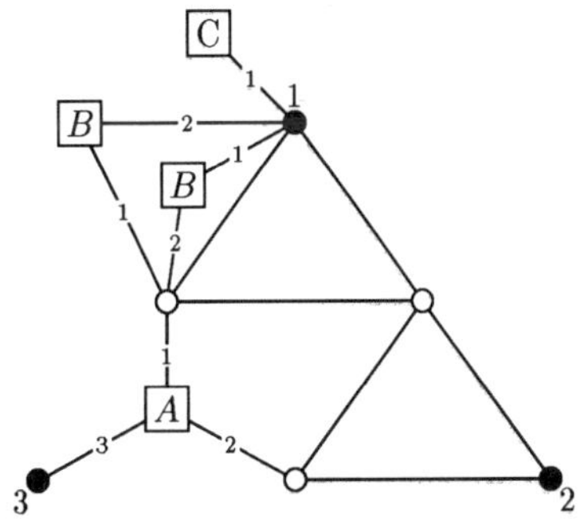

2 Hyperedge replacement

In this section, we will make precise what it means to replace a hyperedge e of a decorated graph by another decorated graph G. As we mentioned above, this procedure can intuitively be seen as substituting G for e in such a way that the external nodes of G become fused with the nodes e was attached to. In fact, we will give a definition which is a bit more general, because it allows to replace several hyperedges in one step. In order to be able to give a concise definition for this, three simple operations hyperedge replacement is based on are needed: Hyperedge removal, disjoint union of decorated graphs, and nodes fusion (as a special case of nodes–set exchange).

2.1 Definition (operations on decorated graphs)
Let $H, H' \in \mathcal{G}(N)$.

1. Let $B \subseteq Y_H$. Then the *removal* of B from H yields the decorated graph
$$H - B = (V_H, E_H, Y_H - B, lab, att, ext_H)$$
with $lab(y) = lab_H(y)$ and $att(y) = att_H(y)$ for all $y \in Y_H - B$.
2. The *disjoint union* of H and H' is the decorated graph
$$H + H' = (V_H \mathring{\cup} V_{H'}, E_H \mathring{\cup} E_{H'}, Y_H \mathring{\cup} Y_{H'}, lab, att, ext_H)$$
with $lab(y) = lab_H(y)$ and $att(y) = att_H(y)$ for all $y \in Y_H$ and $lab(y) = lab_{H'}(y)$ and $att(y) = att_{H'}(y)$ for all $y \in Y_{H'}$. Here $\mathring{\cup}$ denotes the disjoint union of sets.
For $H_1, \ldots, H_n \in \mathcal{G}(N)$ $\sum_{i=1}^{i} H_i$ means $H_1 + H_2 + \ldots + H_n$.
3. Let $f : V_H \to V$ be a mapping for some set V. Then the *nodes–set exchange* in H through f yields the decorated graph
$$H/f = (V, E, Y_H, lab_H, att, ext)$$
with $E = \{\{f(x), f(y)\} \mid \{x, y\} \in E_H, f(x) \neq f(y)\}$, $att(y) = f^*(att_H(y))$ for all $y \in Y_H$, and $ext = f^*(ext_H)$, where f^* is the natural extension of f to sequences.
4. Let δ be an equivalence relation on V_H, let V be the corresponding quotient set (of V_H through δ) and $nat : V_H \to V$ the natural mapping. Then H/nat is called the *nodes fusion* according to δ, and is also denoted by H/δ. If, $u = x_1 \cdots x_n$ and $v = y_1 \cdots y_n$ ($x_i, y_i \in V_H$ for $i = 1, \ldots, n$, $n \geq 0$, where $n = 0$ means $u = \lambda = v$) and $\equiv$ is the equivalence relation on V_H generated by $x_i \equiv y_i$ for $i = 1, \ldots, n$, we also write $H/(u = v)$ instead of $h/\equiv$.

Remarks

1. Removal removes some hyperedges without changing anything else. In particular, we have $type(H - B) = type(H)$.
2. It should be noted that the disjoint union is asymmetric with respect to the choice of the external nodes, which are borrowed from the first component only. Hence, $type(H + H') = type(H)$.
3. Nodes–set exchange may add some new nodes or rename old ones (including identifications of originally different nodes). Anything else is not changed, but adapted to the exchange. In particular, $type(H/f) = type(H)$. □

2.2 Definition (hyperedge replacement)
Let $H \in \mathcal{G}(N)$ and $B = \{y_1, \ldots, y_n\} \subseteq Y_H$. Let $repl : B \rightarrow \mathcal{G}(N)$ be a mapping with $type(y_i) = type(repl(y_i))$ for $i = 1, \ldots, n$. Then the *replacement* of B in H through *repl* yields the decorated graph $\text{REPLACE}(H, repl) = ((H - B) + \sum_{i=1}^{n} repl(y_i)) / (att = ext)$ with $att = att(y_1) \cdots att(y_n)$ and $ext = ext_{repl(y_1)} \cdots ext_{repl(y_n)}$

Remarks

1. Hyperedge replacement is a simple construction where some hyperedges are removed, the associated decorated graphs are added disjointly and their external nodes are fused with the corresponding nodes formerly attached to the replaced hyperedges.
2. Note that the component graphs replacing hyperedges are fully embedded into the resulting graph where their external nodes loose this status and may additionally be fused with other nodes.
3. If a mapping $repl : B \rightarrow \mathcal{G}(N)$ meets the assumption of the definition, it will be referred to as *well–typed.* □

2.3 Example (hyperedge replacement)
Consider again the decorated graph G given in Example 1.4. If we replace the hyperedge labelled with A by G itself, we get

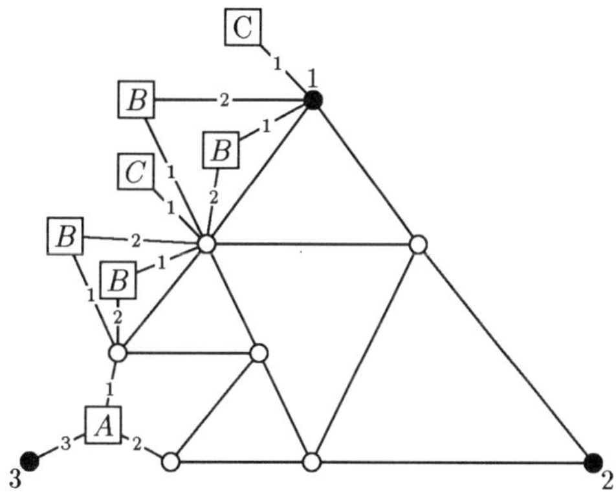

Observe that G cannot replace the other two hyperedges, because this would not be well–typed. □

In the following sections, some fundamental properties of hyperedge replacement will be used.

2.4 Lemma
Let $H \in \mathcal{G}(N), B \subseteq Y_H$, and $repl : B \longrightarrow \mathcal{G}(N)$ be a well-typed mapping.

Then the following properties hold:

1. $U(H) \subseteq U(REPLACE(H, repl))$

2. Let y_0 be the hyperedge of the handle $(A,m)^\bullet$ for some $A \in N$ and $m \geq 0$. Then $REPLACE((A,m)^\bullet, repl) = repl(y_0)$.

3. Let $H' \in \mathcal{G}(N)$ with $H \subseteq H'$. Then $REPLACE(H, repl) \subseteq REPLACE(H', repl)$.

4. Let $B' \subseteq Y_H$ with $B \subseteq B'$, and $repl' : B' \longrightarrow \mathcal{G}(N)$ be a well–typed mapping with $repl(y) = repl'(y)$ for all $y \in B$. Then $REPLACE(H, repl) \subseteq REPLACE(H, repl')$.

5. Let $B_0 \subseteq B$ with $repl(y) = (lab_H(y), type(y))^\bullet$ for each $y \in B - B_0$, and $repl_0 : B_0 \longrightarrow \mathcal{G}(N)$ be the restriction of $repl$ to B_0. Then $REPLACE(H, repl) = REPLACE(H, repl_0)$.

6. Let $B_1 \subseteq B$, and $B_2 = B - B_1$. Let, for $i = 1, 2, repl_i : B_i \longrightarrow \mathcal{G}(N)$ be the restriction of $repl$ to B_i. Then $REPLACE(REPLACE(H, repl_1), repl_2) = REPLACE(H, repl)$.

7. Let $K \in \mathcal{G}(N), X \subseteq Y_K$, and $subst : X \longrightarrow \mathcal{G}(N)$ be a well–typed mapping with $H = REPLACE(K, subst)$. Let, for $e \in X, B(e) = B \cap Y_{subst(e)}$, and $repl(e) : B(e) \longrightarrow \mathcal{G}(N)$ be given by $repl(e)(y) = repl(y)$ for all $y \in B(e) \subseteq B$. Let $subst' : X \longrightarrow \mathcal{G}(N)$ be given by $subst'(e) = REPLACE(subst(e), repl(e))$ for all $e \in X$. Then $REPLACE(H, repl) = REPLACE(K, subst')$. □

3 Parallel hyperedge replacement grammars

A hyperedge replacement is considered as a derivation step of a grammar with parallel mode of rewriting if (1) all hyperedges are replaced simultaneously, and (2) the label of each of the replaced hyperedges and the replacing decorated graph form a production of the grammar. If the first condition is omitted, one gets the derivation process of the usual hyperedge replacement grammars.

The actual model studied in this paper is HTOL hyperedge replacement. The letter H indicates that the underlying structures are of a hybrid nature because they consist of hyperedges to be processed and nodes and edges of graphs that are never rewritten. The letter T indicates that the considered grammars provide sets of tables where each table is a set of productions. Finally, OL refers to the type of rewriting of OL systems which is adapted to hyperedge replacement here.

3.1 Definition (production, base, direct derivation, derivation)

1. A *production* (over N) is a pair $p = (A, R)$ with $A \in N$ and $R \in \mathcal{G}(N)$. A is called the *left–hand side* of p and denoted by $lhs(p)$. R is called the *right–hand side* and denoted by $rhs(p)$.

2. Let $H \in \mathcal{G}(N)$ and $B \subseteq Y_H$. Let $\mathcal{P}$ be a set of sets of productions and $P \in \mathcal{P}$, which is called a *table*. A mapping $b : B \to P$ is called a *base* (over $\mathcal{P}$) in H if $lab_H(y) = lhs(b(y))$ and $type(y) = type(rhs(b(y)))$ for all $y \in B$. If $B = Y_H, b$ is also called an L *base*.

3. Let $H, H' \in \mathcal{G}(N)$ and $b : B \to P$ be a base in H. Then H *directly derives* H' *through* b if $H' \cong$ REPLACE$(H, repl)$ with $repl : B \to \mathcal{G}(N)$, $repl(y) = rhs(b(y))$ for all

$y \in B$. A direct derivation is denoted by $H \underset{b}{\longrightarrow} H'$ or $H \underset{\mathcal{P}}{\longrightarrow} H'$. If b is an L base, it is also called a *direct L derivation* and denoted by $H \underset{b}{\Longrightarrow} H'$ or $H \underset{\mathcal{P}}{\Longrightarrow} H'$.

4. A sequence of direct derivations of the form $H_0 \underset{b_1}{\longrightarrow} H_1 \underset{b_2}{\longrightarrow} \ldots \underset{b_k}{\longrightarrow} H_k$ where, for $i = 1, \ldots, k, b_i : B_i \to P_i$ is a base (over $\mathcal{P}$) in H_{i-1} is called a *derivation* (of *length* k) from H_0 to H_k and is denoted by $H_0 \overset{k}{\underset{\mathcal{P}}{\longrightarrow}} H_k$. The sequence of tables $P_1 \ldots P_k$ is called the *route* of the derivation. If all bases are L bases, the derivation is also called an *L derivation* and denoted by $H_0 \overset{k}{\underset{\mathcal{P}}{\Longrightarrow}} H_k$. Moreover, if $H \cong H'$, this may be denoted by $H \overset{0}{\underset{\mathcal{P}}{\Longrightarrow}} H'$ or $H \overset{0}{\underset{\mathcal{P}}{\longrightarrow}} H'$. If k in $H \overset{k}{\underset{\mathcal{P}}{\Longrightarrow}} H'$ or $H \overset{k}{\underset{\mathcal{P}}{\longrightarrow}} H'$ does not matter, it may be replaced by a star symbol. □

3.2 Definition (parallel hyperedge replacement grammars and languages)

1. A *hybrid table–controlled OL hyperedge replacement grammar* (an *HTOL HR grammar* for short) is a system $LG = (N, \mathcal{P}, Z)$ where N is a set of *nonterminals*, $\mathcal{P}$ is a finite set of *tables* each of which being a finite set of productions (over N), and $Z \in \mathcal{G}_0(N)$, called the *axiom*. If $Z = (S, 0)^\bullet$ for some $S \in N$ we may denote LG by $(N, \mathcal{P}, S)$.
2. Given such a grammar, the *generated graph language* consists of all graphs derivable from $Z : L(LG) = \{G \in \mathcal{G}_0 \mid Z \overset{*}{\underset{\mathcal{P}}{\Longrightarrow}} G\}$.
3. The class of all HTOL HR grammars is denoted by $\mathcal{HTOL}$.

□

3.3 Example

The HTOL HR grammar *Sierpinski* $= (\{S,T\}, \{\{(S,T_0),(T,T_1)\}, \{(S,T_0),(T,T_2)\}\}, S)$ with

$T_0 =$ 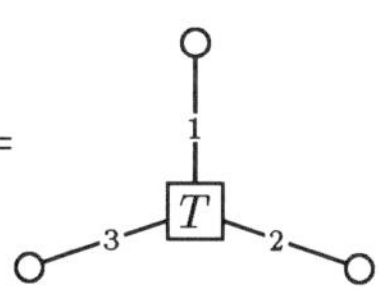

$T_1 =$ 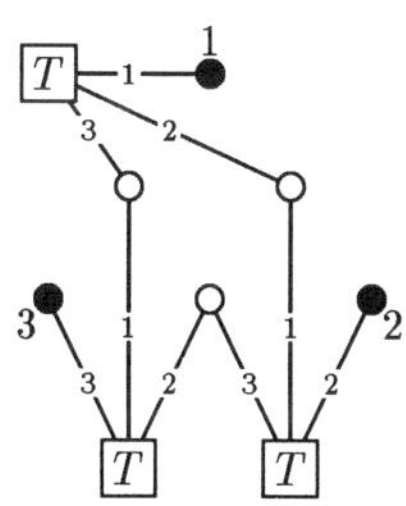

and $T_2 =$ 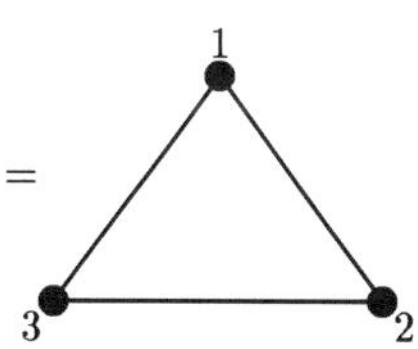

generates the set of all Sierpinski triangles

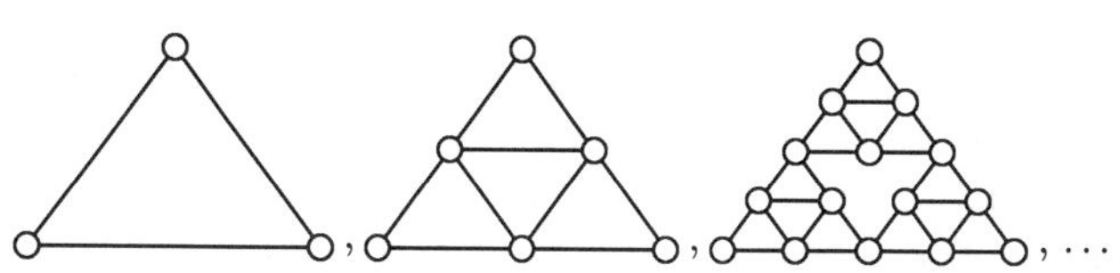

One of the fundamental properties of hyperedge replacement is its context–free character. Given a hyperedge in a decorated graph and a decorated subgraph, the hyperedge is either outside the subgraph or it belongs to the subgraph together with its attachment. In the latter case, all replacements of the hyperedge in the supergraph remain possible in the subgraph and conversely. This allows one the parallel compositions and decompositions of derivations. If L derivations are composed in parallel, an additional synchronization is required because all hyperedges must be replaced through productions from the same table.

3.4 Theorem (decomposition)
Let $H \underset{\mathcal{P}}{\overset{*}{\Longrightarrow}} H'$ be an L derivation with route $P_1 \ldots P_k$, and $H_0 \subseteq H$.

Then there is an L derivation $H_0 \underset{\mathcal{P}}{\overset{*}{\Longrightarrow}} H_0'$ with route $P_1 \ldots P_k$ and $H_0' \subseteq H'$.

Proof: Consider first a direct L derivation $H \underset{b}{\Longrightarrow} H'$ through $b : Y_H \longrightarrow P (P \in \mathcal{P})$ with $H' = REPLACE(H, repl)$ and $repl(y) = rhs(b(y))$ for all $y \in Y_H$. Then $b_0 : Y_{H_0} \longrightarrow P$ given by $b_0(y) = b(y)$ for all $y \in Y_{H_0}$ is an L base in H_0 defining the direct L derivation $H_0 \underset{b_0}{\Longrightarrow} H_0'$. Due to Lemma 2.4.3 and 4, the following holds: $H_0' = REPLACE(H_0, repl_0) \subseteq REPLACE(H, repl_0) \subseteq REPLACE(H, repl) = H'$ with $repl_0(y) = rhs(b_0(y)) = rhs(b(y))$ for all $y \in Y_{H_0}$. The rest follows by induction on the length of derivations. □

3.5 Theorem (parallel composition)
Let $K \in \mathcal{G}(N)$, and $subst : Y_K \longrightarrow \mathcal{G}(N)$ be a well–formed mapping. Let $\mathcal{P}$ be a set of tables, and $r \in \mathcal{P}^*$. Let, for each $e \in Y_K, s(e)$ be an L derivation $subst(e) \underset{\mathcal{P}}{\overset{*}{\Longrightarrow}} G(e)$ with route r. Then there is an L derivation $REPLACE(K, subst) \underset{\mathcal{P}}{\overset{*}{\Longrightarrow}} REPLACE(K, subst')$ with route r where $subst' : Y_K \longrightarrow \mathcal{G}(N)$ is given by $subst'(e) = G(e)$ for all $e \in Y_K$.

Proof: The case of direct L derivations is considered first, i.e. $s(e) = (subst(e) \underset{b(e)}{\Longrightarrow} G(e))$ with $b(e) : Y_{subst(e)} \longrightarrow P$ and $r = P$. Let $H = REPLACE(K, subst)$. By definition, Y_H is the disjoint union of the $Y_{subst(e)}$ for all $e \in Y_K$. Hence, one can define an L base $b : Y_H \longrightarrow P$ by $b(y) = b(e)(y)$ for all $y \in Y_{subst(e)}$, $e \in Y_K$ establishing a direct L derivation $H \underset{b}{\Longrightarrow} H'$ with route $r = P$. By definition and Lemma 2.4.7, the following holds: $H' = REPLACE(H, repl) = REPLACE(REPLACE(K, subst), repl) = REPLACE(K, subst')$ with $repl : Y_H \longrightarrow \mathcal{G}(N)$ given by $repl(y) = rhs(b(y))$ for all $y \in Y_H$. The rest of the statement follows by induction on the length of derivations. □

4 Decidability of the emptiness problem

HTOL HR grammars are specifications of graph languages based on parallel hyperedge replacement. Do the syntactic descriptions allow one to reveal further interesting information on the generated languages and their members? A first example of this type of questions is the decidability of the emptiness problem.

To show this, the notion of active pairs is used. An active pair consists of the label and the type of each hyperedge that may occur in the derivation process of a grammar.

4.1 Definition
Let $LG = (N, \mathcal{P}, Z) \in \mathcal{HTOL}$. Let $H = Z$ or $H = rhs(p)$ for some $p \in P \in \mathcal{P}$. Let $y \in Y_H$. Then $(lab_H(y), type(y))$ is called an *active pair* of LG. $AP(LG)$ denotes the set of all active pairs of LG. □

4.2 Theorem
For $\mathcal{HTOL}$, it is decidable whether, given $LG \in \mathcal{HTOL}$ as input, $L(LG) = \emptyset$ or not.

Proof: For each $LG = (N, \mathcal{P}, Z) \in \mathcal{HTOL}$, a finite, labelled, directed graph TEST is constructed in such a way that TEST contains a certain path if and only if $L(LG) \neq \emptyset$.

The set of nodes of TEST is the powerset of $AP(LG)$. Given $\alpha, \beta \subseteq AP(LG)$, TEST contains an edge from α to β with label $P \in \mathcal{P}$ iff, for every $(A, m) \in \alpha$, there exists $p(A, m) = (A, R) \in P$ such that $(lab_R(y), type(y)) \in \beta$ for all $y \in Y_R$.

An edge in TEST has the following significance where the notation $AP(H) = \{(lab_H(y), type(y) \mid y \in Y_H\}$ for $H \in \mathcal{G}(N)$ is used: If $AP(H) \subseteq \beta$, then there is $H \underset{b}{\Longrightarrow} H'$ through the base $b : Y_H \longrightarrow P$ given by $b(y) = p(lab_H(y), type(y))$ for all $y \in Y_H$, and $AP(H') \subseteq \beta$.

Hence, if TEST contains a path from $AP(Z)$ to $\emptyset$, then this induces $Z \underset{\mathcal{P}}{\overset{*}{\Longrightarrow}} G$ for some G with $AP(G) = \emptyset$ which means $Y_G = \emptyset$ and, therefore, $G \in L(LG)$.

Conversely, let $L(LG) \neq \emptyset$. Then there is $Z \underset{\mathcal{P}}{\overset{*}{\Longrightarrow}} G$ for some $G \in \mathcal{G}_0$. In particular, $AP(G) = \emptyset$. Let $H_0 \underset{b_0}{\Longrightarrow} H_0'$ through $b_0 : Y_{H_0} \longrightarrow P_0$ be one of the steps of the given derivation. Let $(A, m) \in AP(H_0)$. Choose $p(A, m) = b_0(y_0)$ for some $y_0 \in Y_{H_0}$ with $lab_H(y_0) = A$ and $type(y_0) = m$ which exists by definition of $AP(H_0)$. For $y \in Y_R$ with $R = rhs(p(A, m))$ gets $(lab_R(y), type(y)) \in AP(H_0')$. In other words, TEST contains an edge from $AP(H_0)$ to $AP(H_0')$. Following the derivation step by step, one finds a path from $AP(Z)$ to $AP(G) = \emptyset$ in TEST.

This decides the emptiness problem because the graph TEST can be constructed effectively. □

5 Special cases and their generative power

In this section, special cases of HTOL HR grammars are introduced, and their generative power is studied. It turns out that one looses generative power if only one table is available or if, in addition, the table contains productions that guarantee the replacement of each hyperedge by itself. In both cases, the generative power coincides with the one of (sequential) hyperedge replacement.

5.1 Definition
Let $LG = (N, \mathcal{P}, Z) \in \mathcal{HTOL}$.

1. If $\mathcal{P} = \{P\}$ for some set P of productions, then LG is called a *hybrid OL HR grammar* (a *HOL HR grammar* for short). The class of all such grammars is denoted by $\mathcal{HOL}$.
2. LG is called *quasi-sequential* if $LG \in \mathcal{HOL}$ and the only table contains the production $(A, (A, m)^\bullet)$ for each active pair (A, m) of LG. The class of all such grammars is denoted by $\mathcal{QHOL}$.

3. A *hyperedge replacement grammar* (a *HR grammar* for short) is a system $HRG = (N, P, Z)$ such that $(N, \{P\}, Z) \in \mathcal{HOL}$. The *generated language* of HRG consists of all graphs derivable from Z:

$$L(HRG) = \{G \in \mathcal{G}_0 \mid Z \xrightarrow[\mathcal{P}]{*} G\}$$

Remarks

1. The main difference between an HR grammar and an HOL HR grammar is the kind of bases used in generating a language. While arbitrary bases are allowed in the first case, L bases are required in the second case.
2. Let $\mathcal{C} \subseteq \mathcal{HTOL}$. Then the class of languages generated by grammars from $\mathcal{C}$ is denoted by $\mathcal{L}(\mathcal{C})$. Using this notation, one gets by the definition above the following inclusions:
$$\mathcal{L}(\mathcal{QHOL}) \subseteq \mathcal{L}(\mathcal{HOL}) \subseteq \mathcal{L}(\mathcal{HTOL}).$$
3. The class of all HR grammars is denoted by $\mathcal{HRG}$, the class of their generated languages by $\mathcal{L}(\mathcal{HRG})$. □

It will turn out that the second inclusion is proper whereas the first one is not. Both results are consequences of the following comparisons of $\mathcal{QHOL}, \mathcal{HOL}$, and $\mathcal{HTOL}$ with $\mathcal{HRG}$ with respect to the generative power.

5.2 Theorem

$$\mathcal{L}(\mathcal{HTOL}) - \mathcal{L}(\mathcal{HRG}) \neq \emptyset.$$

Proof: In Habel [6], it is shown that the language generated by an HR grammar is sublinear in the following sense: If the generated graphs are ordered by size, the difference in size between each two successive graphs is bounded by a constant. If the Sierpinski triangles of Example 3.3 are ordered by size, the difference in size between each two successive ones is growing beyond any bound. Hence, the set of Sierpinski triangles in not in $\mathcal{L}(\mathcal{HRG})$, but in $\mathcal{L}(\mathcal{HTOL})$ (as shown in 3.3). □

5.3 Theorem

$$\mathcal{L}(\mathcal{HOL}) \subseteq \mathcal{L}(\mathcal{HRG}).$$

Proof: Let $LG = (N, \{P\}, Z) \in \mathcal{HOL}$ and $HRG = (N, P, Z)$. Consider $G \in L(LG)$ i.e. $Z \overset{*}{\underset{\{P\}}{\Longrightarrow}} G$ and $G \in \mathcal{G}_0$. By definition, this implies $Z \xrightarrow[\{P\}]{*} G$ such that $G \in L(HRG)$. Conversely, let $G \in L(HRG)$, i.e. $Z \xrightarrow[\{P\}]{*} G$ and $G \in \mathcal{G}_0$. Without loss of generality, one can assume that this is a canonical derivation which means the following for HR grammars: Each production is either applied in the first step, or it is applied to a hyperedge that is produced in the preceeding step. This means in other words that all hyperedges are replaced in each step because $G \in \mathcal{G}_0$ does not have any hyperedge left. Hence the used bases are L bases, and one gets $Z \overset{*}{\underset{\{P\}}{\Longrightarrow}} G$ such that $G \in L(LG)$. Altogether, $L(LG) = L(HRG)$ proving the statement. □

5.4 Theorem

$$\mathcal{L}(\mathcal{HRG}) \subseteq \mathcal{L}(\mathcal{QHOL})$$

Proof: Let $HRG = (N,P,Z) \in \mathcal{HRG}$ and $LG = (N,\{P'\},Z)$ with $P' = P \cup \{(A,(A,m)^\bullet) \mid (A,m) \in AP((N,\{P\},Z))$. By construction $LG \in \mathcal{QHOL}$.

Consider $G \in L(HRG)$, i.e. $Z \xrightarrow[\{P\}]{*} G$ and $G \in \mathcal{G}_0$. By Claim 1 and induction on the length of derivations, the given derivation can be transformed into an L derivation $Z \overset{*}{\underset{\{P\}}{\Longrightarrow}} G$ such that $G \in L(LG)$. Conversely, let $G \in L(LG)$, i.e. $Z \overset{*}{\underset{\{P'\}}{\Longrightarrow}} G$ and $G \in \mathcal{G}_0$. By Claim 2 and induction on the length of derivations, this L derivation can be transformed into a derivation $Z \xrightarrow[\{P\}]{*} G$ such that $G \in L(HRG)$. Altogether, one gets $L(LG) = L(HRG)$ proving the statement.

It remains to show the following two claims.

Claim 1: $H \xrightarrow[b]{} H'$ through $b : B \longrightarrow P$ for some $B \subseteq Y_H$ implies $H \underset{b'}{\Longrightarrow} H'$ through $b' : Y_H \longrightarrow P'$ given by $b'(y) = b(y)$ for $y \in B$ and $b'(y) = (lab_H(y), (lab_H(y), type(y))^\bullet)$ otherwise.

Claim 2: $H \underset{b'}{\Longrightarrow} H'$ through $b' : Y_H \longrightarrow P'$ implies $H \xrightarrow[b]{} H'$ through $b : B \longrightarrow P$ with $b(y) = b'(y)$ for all $y \in B$ where B is the largest subset of Y_H with $b'(B) \subseteq P$.

Proof of Claim 1: Let $repl : B \longrightarrow \mathcal{G}(N)$ be given by $repl(y) = rhs(b(y)), repl'' : Y_H \longrightarrow \mathcal{G}(N)$ by $repl''(y) = rhs(b'(y))$ and $repl' : Y_H - B \longrightarrow \mathcal{G}(N)$ by $repl'(y) = (lab_H(y), type(y))^\bullet$). Then one has by Lemma 2.4.5 and 6: $H' = REPLACE(H, repl) = REPLACE(REPLACE(H, repl'), repl) = REPLACE(H, repl'')$. Therefore, $H \underset{b'}{\Longrightarrow} H'$ if $H \xrightarrow[b]{} H'$.

Proof of Claim 2: Choose $repl, repl'$ and $repl''$ as in the proof of Claim 1 such that one gets the following by the same arguments in different order: $H' = REPLACE(H, repl'') = REPLACE(REPLACE(H, repl'), repl) = REPLACE(H, repl)$. Therefore, $H \xrightarrow[b]{} H'$ if $H \underset{b'}{\Longrightarrow} H'$. □

6 Conclusion

In this paper, the investigation of parallel hyperedge replacement grammars has been initiated by studying HTOL HR grammars. The paper demonstrates that hyperedge replacement is easily equipped with a parallel mode of rewriting as known from OL systems and that the framework offers the prospect of nice theoretical results.

Here are some further directions of research to continue the very first steps in this paper.

(1) Other decidability problems than the emptiness problem should be considered. For example, one may like to ask for certain graph–theoretic properties of the generated graphs.

(2) Other subclasses of $\mathcal{HTOL}$ than $\mathcal{HOL}$ and $\mathcal{QHOL}$ may be investigated.

(3) In the literature, one encounters quite a variety of graph grammar approaches with a parallel mode of rewriting (see, e.g., the contributions by Ehrig, Kreowski, Mayoh, Nagl, Paz, Rosenfeld and Rozenberg in Lindenmayer and Rozenberg [15] and by de Boer and Lindenmayer as well as J. Lück and H.B. Lück in Ehrig, Nagl, Rozenberg and Rosenfeld [4]). They all seem to have a much more complicate derivation mechanism than parallel hyperedge replacement. Hence, it may be worthwhile to compare the other models with HTOL HR grammars.

References

1. M. Bauderon, B. Courcelle: Graph expressions and graph rewriting. Mathematical Systems Theory 20, 83-127, 1987.
2. B. Courcelle: On context-free sets of graphs and their monadic second-order theory. Lecture Notes in Computer Science 291, 133-146, 1987.
3. B. Courcelle: A representation of graphs by algebraic expressions and its use for graph rewriting systems. Lecture Notes in Computer Science 291, 112-132, 1987.
4. H. Ehrig, M. Nagl, G. Rozenberg, A. Rosenfeld: Graph Grammars and Their Application to Computer Science. Lecture Notes in Computer Science 291, 1987.
5. J. Feder: Plex languages. Information Science 3, 225-241, 1971.
6. A. Habel: Hyperedge Replacement: Grammars and Languages. Dissertation, Univ. Bremen, 1989.
7. A. Habel, H.-J. Kreowski: Characteristics of graph languages generated by edge replacement. Theoretical Computer Science 51, 81-115, 1987.
8. A. Habel, H.-J. Kreowski: May we introduce to you: hyperedge replacement. Lecture Notes in Computer Science 291, 15-26, 1987.
9. A. Habel, H.-J. Kreowski: Some structural aspects of hypergraph languages generated by hyperedge replacement. Lecture Notes in Computer Science 247, 207-219, 1987.
10. A. Habel, H.-J. Kreowski, W. Vogler: Decidable boundedness problems for hyperedge replacement graph grammars. Lecture Notes in Computer Science 351, 275-289, 1989.
11. A. Habel, H.-J. Kreowski, W. Vogler: Metatheorems for decision problems on hyperedge replacement graph languages. Acta Informatica 26, 657-677, 1989.
12. G.T. Herman, G. Rozenberg: Developmental Systems and Languages. North-Holland, Amsterdam- Oxford 1975.
13. H.-J. Kreowski: A pumping lemma for context-free graph languages. Lecture Notes in Computer Science 73, 270-283, 1979.
14. T. Lengauer, E. Wanke: Efficient analysis of graph properties on context-free graph languages. Lecture Notes in Computer Science 317, 379-393, 1988.
15. A. Lindenmayer, G. Rozenberg: Automata, Languages, Development. North-Holland, Amsterdam-New York-Oxford 1976.
16. T. Pavlidis: Linear and context-free graph grammars. Journal of the ACM 19(1), 11-23, 1972.
17. Rozenberg, A. Salomaa: The Mathematical Theory of L Systems. Academic Press, New York-London-Toronto-Sydney-San Francisco 1980.

From Parallel Graph Grammars to Parallel High-Level Replacement Systems

Hartmut Ehrig, Gabriele Taentzer

Computer Science Department, TU Berlin, Franklinstr. 28/29, Sekr. FR 6-1, D-1000 Berlin 10

Abstract: Different concepts of parallel replacement for graphs are introduced extending the theory of parallel graph grammars. This framework contains concepts of parallel graph derivations with constant interfaces which are generalized by allowing dynamic interfaces. Moreover concepts of two–level derivations are introduced combinig parallel and sequential derivation concepts. In the Dynamic Parallel Derivation Theorem a bijective correspondence between amalgamated two–level derivations and dynamic parallel derivations with context completion is shown. In order to formalize parallel rewriting for different kinds of structures parallel high–level replacement systems are introduced. This allows to extend all concepts and results from parallel graph grammars to parallel high-level replacement systems.

Keywords: Parallel Graph Grammars, Parallel Derivations, Dynamic Interfaces, High–Level Replacement Systems

1 Introduction

The idea of parallel graph grammars is to formalize the concept of parallel replacement of graphs. They are a parallel version of sequential graph grammars on one hand, on the other hand parallel graph grammars generalize Lindenmayer–systems which were introduced to model the development of organisms. An extended bibliography on different kinds of sequential and parallel graph grammars can be found in [CER79].

L–systems can be considered as context–free parallel replacement systems on strings. Parallel derivations on graphs presented in [EK76] are not restricted to be context–free but arbitrary graphs for the left hand sides of productions are allowed. For parallel replacement the mother graph has to be covered by occurences of left hand sides of productions. The interfaces between different occurences are required to remain constant during the replacement process. If these interfaces are allowed to be changed by interface productions we obtain dynamic parallel derivations. Parallel derivations can be combined with sequential derivations which allow the application of only one production at a time. This leads to amalgamated two–level derivations where first a new production is amalgamated from a set of elementary productions and then this production is applied. Two–level derivations are useful if only one part of a graph has to be replaced in parallel and the other part is context.

In order to extend the field of applications the concepts of parallel derivations are introduced for high–level replacement systems. This framework generalizes the double–

pushout approach known from sequential graph grammars and other types of replacement systems to high-level structures like partial graphs, relational structures, graphs with partially ordered color sets, hypergraphs, place–transition nets,etc. Several results like Church–Rosser–Properties, Parallelism Theorem and Concurrency Theorem, which are known for the graph case are shown for HLR–systems in [EHLP91]. To formalize the parallel rewriting of these high–level structures parallel high–level replacement systems (PHLR–systems) are introduced. This concept generalizes parallel graph grammars by defining parallel replacement also for other structures than graphs with colored nodes and edges and color–preserving graph morphisms as described in [EK76].

For example, in [ST91] parallel transformations are defined for graphs with a particular node color set which is partially ordered to model object transformations in configuration spaces.

Map grammars as described for example in [Lin86] with maps defined as sets of vertices, edges and regions can be viewed as a replacement system of a special kind of topological spaces which is shown to be a parallel high–level replacement system.

The main concepts and results of this paper are summarized in section 2 on an informal level. In section 3 to 5 we present the formal theory and explicit examples. For the proofs of the new results in this paper we refer to [Ta91].

2 Informal Approach to Parallel Derivations

Before we start to describe different kinds of parallel derivations we want to give a short overview of the derivation concepts introduced for graph grammars (gragra), high–level replacement systems (HLRS) and parallel high–level replacement systems (PHLRS) shown in figure 1.

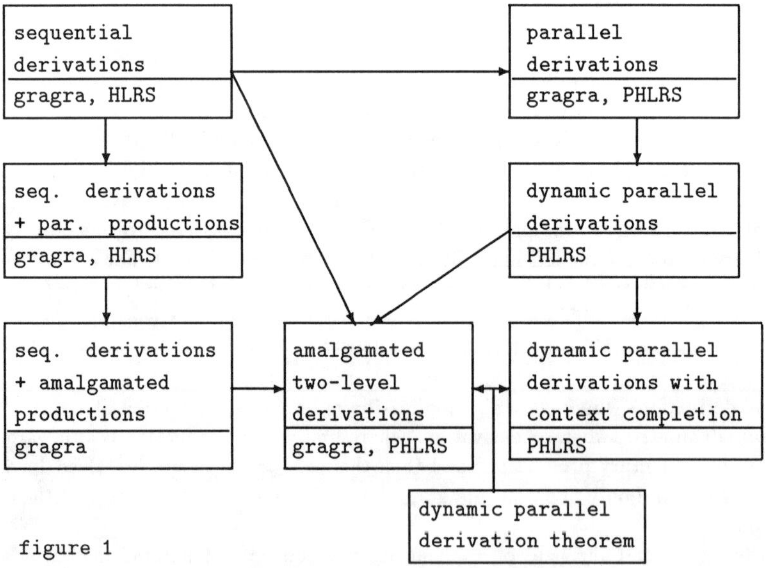

figure 1

The basic derivation is the **sequential derivation** where one production is applied to a structure. It is introduced in various approaches to graph grammars (like [Eh79], [JR80], [Na87],etc.) and in high–level replacement systems ([EHLP91]).

The sequential derivations can be extended by **derivations with parallel productions** which are parallel independent. They are also presented for graphs and high–level replacement systems (in [Eh79] and [EHLP91]).

Derivations with amalgamated productions which are parallel productions associated with each other are only studied in the graph case in [BFH87].

Parallel derivations are already known from parallel graph grammars ([EK76]). We want to define them also for high–level replacement systems.

They are defined for constant interfaces. If the interfaces can also be transformed we arrive at **dynamic parallel derivations.**

The **amalgamated two–level derivations** combine the concepts of dynamic parallel derivations and sequential derivations. On the first level an amalgamated production is constructed using a dynamic parallel derivation. On the second level this production is applied in a sequential way. (See also mixed rewriting in [Na87].) Two–level derivations are already examined for special graphs with partial ordered node labels (in [ST91]). We want to generalize them also to high–level replacement systems.

The amalgamated two–level derivations can be compared with a special kind of dynamic parallel derivations: the **dynamic parallel derivations with context completion.** This is a parallel replacement where the context remains constant by applying an identical context production. The equivalence of these last two kinds of derivations is shown in the **dynamic parallel derivation theorem.**

Now we explain informally the different kinds of parallel derivations for graphs. The explicit constructions are shown in subsequent sections.

2.1 Parallel Derivation

For deriving a daughter graph from a given mother graph the whole mother graph has to be covered completely by occurences of left hand sides of productions. In the case $n = 3$ we have the following diagram where L_i, $1 \leq i \leq n$, are occurences of productions and I_{ij}, $1 \leq i < j \leq n$, are interfaces.

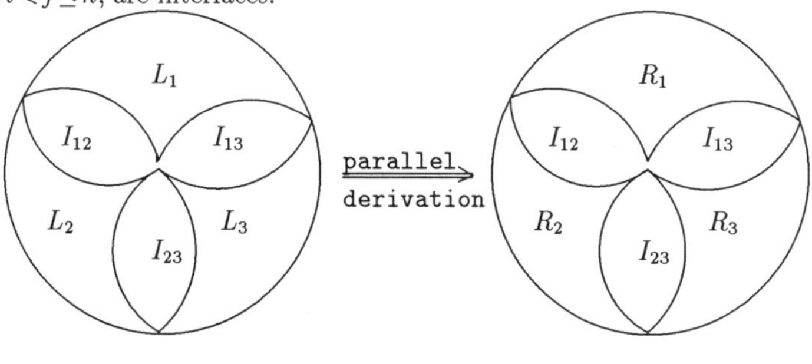

mother graph daughter graph

Transforming all these occurences L_i in parallel this yields the daughter covering consisting of all R_i, $1 \leq i \leq n$. All interfaces bewteen these occurences I_{ij} remain constant during the derivation.

2.2 Dynamic Parallel Derivation

The dynamic parallel derivation transforms also a given covering of a mother graph into a covering of a daughter graph. In contrast to usual parallel derivations also the interfaces of the mother covering, L_{ij}, are transformed into daughter interfaces R_{ij}.

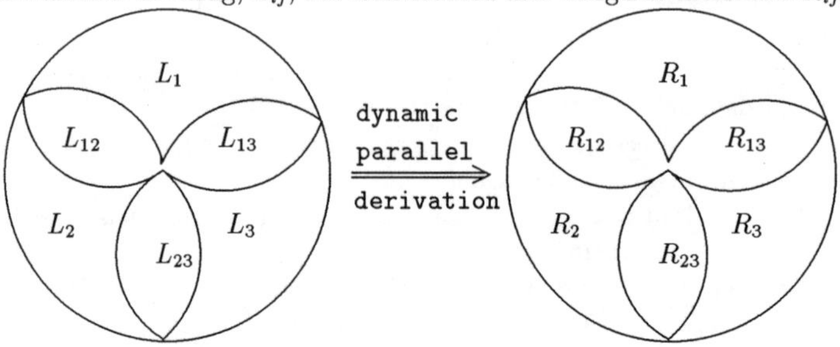

mother graph daughter graph

2.3 Dynamic Parallel Derivation with Context Completion

If only a part of the mother graph is covered by occurences of productions the context graph C has to be considered to complete the covering. Then a dynamic parallel derivation as described above can be executed over the completed covering using an identical production to shift the context graph unchanged from the mother to the daughter graph.

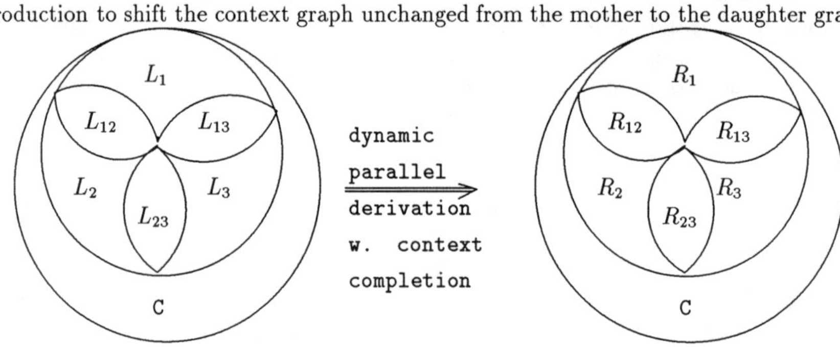

mother graph daughter graph

2.4 Amalgamated Two–Level Derivation

The amalgamated two–level derivation consists of the following two levels:

On the first level a new production is generated from a set of elementary productions by using a dynamic parallel derivation. Here that part of the mother graph which is covered by occurences of elementary productions is transformed into a daughter covering part.

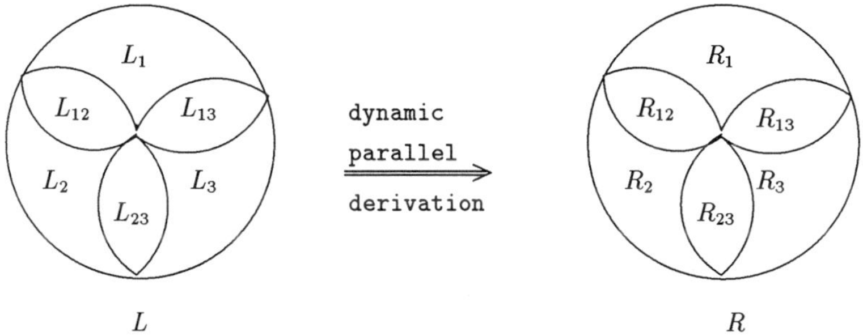

The mother covering part forms the left hand side of the new production L and the daughter covering part forms the right hand side R.

On the second level the new generated production is applied to the mother graph by a sequential derivation.

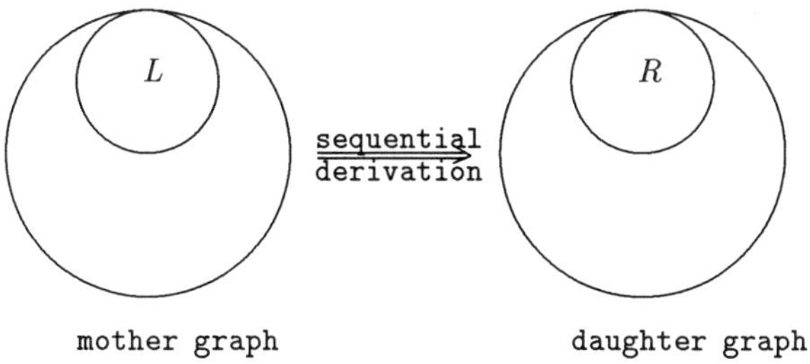

2.5 Example

The following example shows a context dependent graph transformation defined by an amalgamated two–level derivation.

A node x is to be replaced by a handle (y,z) (which is an edge with a source and a target node) in an arbitrary graph.

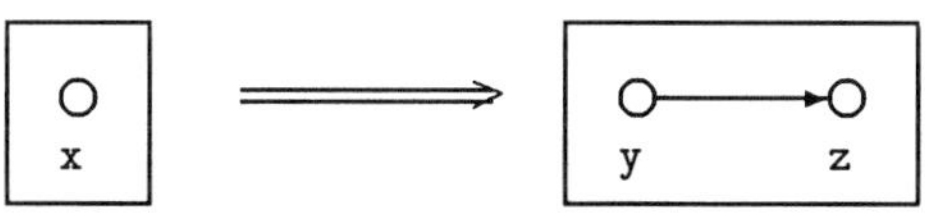

(1)

For adding the handle (y,z) to a graph an embedding specification is given regulating the relations between the handle and the context:

1. If there is an edge from node a to node x then there is an edge from a to node y.

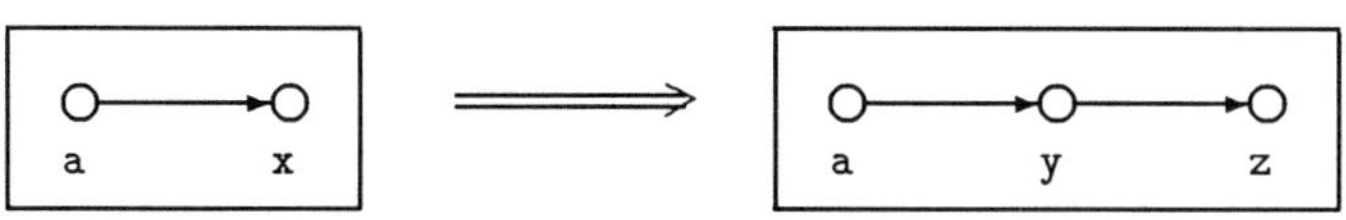

(2)

2. If there is an edge from node x to node b then there is an edge from node z to b.

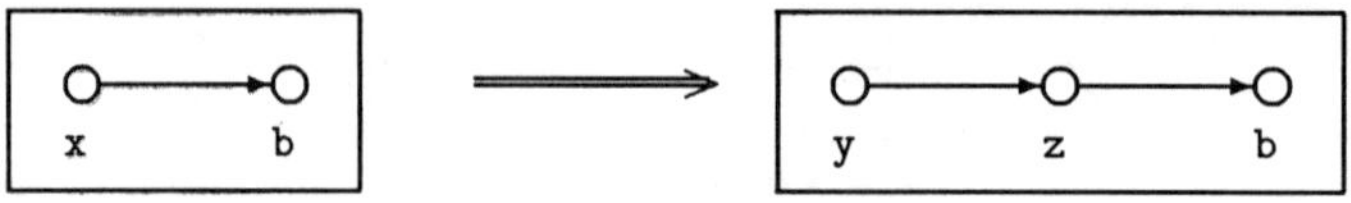

(3)

(See also the NLC–approach to graph grammars for embedding relations. [JR80])

For explicit graphs such a replacement with the above specified embedding relation would yield the following derivations:

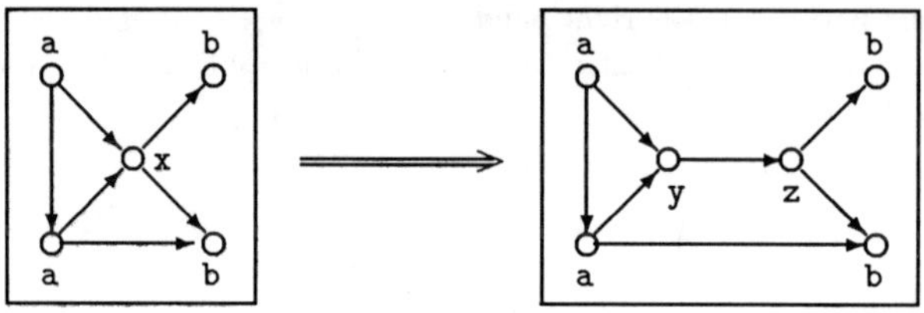

A *A′*

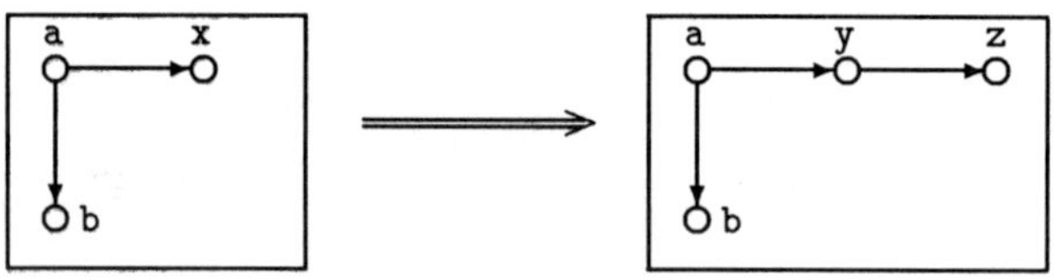

B *B′*

If we use the embedding productions (2) and (3) as elementary productions describing the relation of node x to the context, node x and its particular context in graph A can be covered by occurences of these elementary productions. All four occurences overlap in node x so this node can be viewed as an interface. The actual part of the graph which is transformed is shown in L.

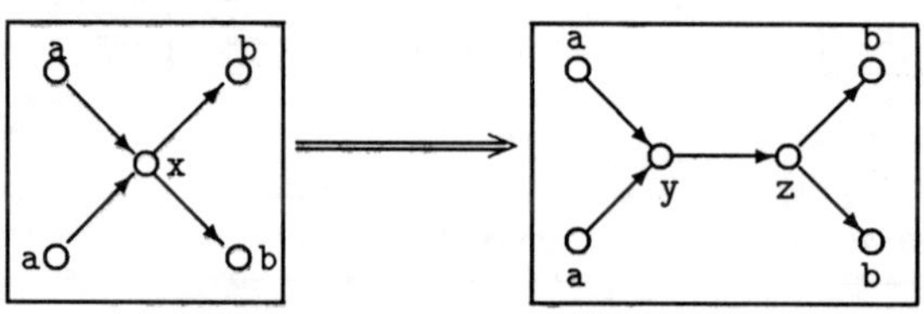

L *R*

If we would replace all handles of *L* in parallel using the productions (2) and (3) with interface node x only we would obtain four copies of the handle (y,z). But if we apply production (1) to interface node x we obtain handle (y,z) as derived interface. And parallel gluing using the derived interface on the right hand side yields graph R which contains handle (y,z) only once.

Now the actual part of the graph which is transformed, the graph L, and this graph R form a new production $L \Longrightarrow R$ which fits to the particular context of graph A.

The generated production $L \Longrightarrow R$ can now be applied to graph A leading to a sequential derivation $A \Longrightarrow A'$. Similarly a new production can be generated from the same elementary productions (1) - (3) for the special context given in graph B leading to a sequential derivation $B \Longrightarrow B'$. (See replacements for explicit graphs above.)

3 Parallel Graph Grammars

In this section starting with definitions for graphs and graph morphisms, we review the covering approach and gluing constructions for graphs, given in [EK76] and [Eh79], leading to parallel derivations, parallel graph grammars and corresponding graph languages.

A **graph** $G = (G_A, G_N, s_G, t_G, m_{GA}, m_{GN})$ consists of the sets G_A of arcs and G_N of nodes, the mappings $s_G, t_G : G_A \longrightarrow G_N$, called source resp. target map, and the coloring maps for arcs and nodes, $m_{GA} : G_A \longrightarrow C_A$ and $m_{GN} : G_N \longrightarrow C_N$ where C_A and C_N are the color alphabets for arcs and nodes.

If we don't want to distinguish between arcs and nodes we use the notation $x \in G$ which means $x \in G_A$ or $x \in G_N$.

Given two graphs G and G' a **graph morphism** $f : G \longrightarrow G'$ is a pair of maps $(f_A : G_A \longrightarrow G'_A, f_N : G_N \longrightarrow G'_N)$ such that $f_N \circ s_G = s_{G'} \circ f_A$, $f_N \circ t_G = t_{G'} \circ f_A$, $m_{G'A} \circ f_A = m_{GA}$ and $m_{G'N} \circ f_N = m_{GN}$.

A graph morphism f is said to be injective (surjective) if both f_A and f_N are injective (surjective) mappings.

A graph morphism between two graphs is represented by additional numbers that indicate the nodes or arcs which are mapped to each other.

The basic concept in the algebraic approach to graph grammars is a **gluing construction** for graphs which is a pushout in the sequential case and a pushout–star for parallel gluing of graphs. Pushout–stars which comprise pushouts as a special case are colimits in the sense of the category theory. The universal definitions are given in section 5.

Given graph morphisms $I \longrightarrow G_1$ and $I \longrightarrow G_2$ the gluing construction yields a graph G which can be obtained from the disjoint union of G_1 and G_2 where all items from the interface graph I are identified. (See [Eh79] for details of the gluing construction.) The gluing construction is a pushout in the category of graphs. (See definition 5.1.)

For parallel replacement we need the gluing of more than two graphs in one step. Here the **star gluing contruction** is used where each pair of graphs G_i and G_j are glued together via a separate interface graph I_{ij} with graph morphisms $g_{ij} : I_{ij} \longrightarrow G_i$ and $g_{ji} : I_{ij} \longrightarrow G_j$.

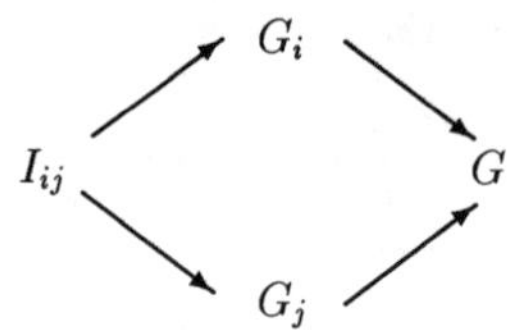

Let $\bar{\mathcal{I}}$ be the equivalence relation generated by the relation $\mathcal{I} = \{(g_{ij}(x), g_{ji}(x)) \mid x \in I_{ij},\ 1 \leq i < j \leq n\}$. Then the star gluing graph G is defined by $G = (\coprod_{1 \leq i \leq n} G_i)_{/\bar{\mathcal{I}}}$, the quotient set of the disjoint union of all G_i. The functions $G_i \longrightarrow G$ for all $1 \leq i \leq n$ in the diagram send each element of G_i to its equivalence class in G. (See [EK76] for details of the star gluing construction.) The star gluing construction is a pushout–star in the category of graphs. (See 5.2.) In the case $n = 2$ this is equal to the gluing construction introduced above.

A **production** p is is a pair of graph morphisms $p = (L \longleftarrow I \longrightarrow R)$ where the graphs L,R and I are called left hand side, right hand side and interface graph.

First let us review the notion of sequential derivations as given in [Eh79].

3.1 Definition *Given a production $p = (L \longleftarrow I \longrightarrow R)$ and a graph C, the context graph, together with a graph morphism $I \longrightarrow C$ a* **sequential derivation** $G \Longrightarrow G'$ *consists of the following two pushout constructions:*

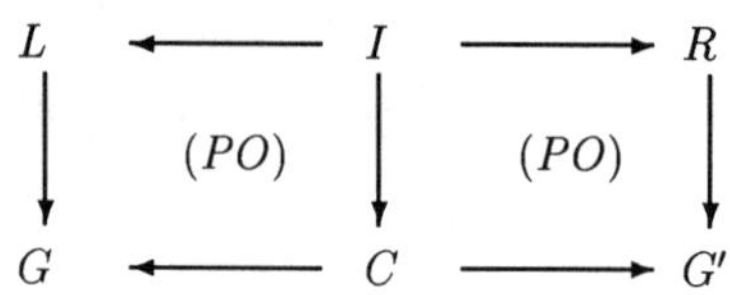

In the case that production $p = (L \overset{l}{\longleftarrow} I \overset{r}{\longrightarrow} R)$ and graph G together with an occurence $g : L \longrightarrow G$ are given a context graph C has to be constructed such that G is the gluing graph of L and C via interface I and G' is the gluing graph of R and C via I. For the construction of a context graph C we need a certain **gluing condition** consisting of an identification condition and a dangling condition which are based on the set $GP(p)$ of gluing points of p with $GP(p) = l(I)$, identification points

$$ID(g) = \{x \in L \mid \forall y \in L : g(x) = g(y) \Rightarrow x = y\}$$

and dangling points

$$DANG(g) = \{x \in L_N \mid \forall a \in L_A \backslash g_A(L_A) : g(x) = s_G(a) \vee g(x) = t_G(a)\}$$

The conditions are the following:

Identification Condition:	$ID(g) \subseteq GP(p)$
Dangling Condition:	$DANG(g) \subseteq GP(p)$
Gluing Condition:	$ID(g) \cup DANG(g) \subseteq GP(p)$

If the gluing condition is satisfied we can construct a graph $C = G - g(L) + g(l(I))$ as subgraph of G such that G is gluing graph of $(L \longleftarrow I \longrightarrow C)$. (See [Eh79].)

In the case of parallel replacement the graph G has to be covered completely by occurences of left hand sides of productions $p_i = (L_i \longleftarrow I_i \longrightarrow R_i)$. Thus a family $(g_i : L_i \longrightarrow G)_{1 \leq i \leq n}$ of graph morphisms is called covering of G if the induced graph morphism $\coprod_{1 \leq i \leq n} L_i \longrightarrow G$ from the disjoint union $\coprod_{1 \leq i \leq n} L_i$ of all L_i is surjective. Let all graph morphisms g_i be injective and $S = (L_i \longleftarrow I_{ij} \longrightarrow L_j)_{1 \leq i < j \leq n}$ the family of intersections of $(L_i \longrightarrow G \longleftarrow L_j)$ $\forall\, 1 \leq i < j \leq n$. Then the star gluing construction of S yields graph G and graph morphisms g_i. (See theorem 3 in [EK76].)

3.2 Definition *Given a set of productions $\mathcal{P} = \{p_i | p_i = (L_i \longleftarrow I_i \longrightarrow R_i),\ 1 \leq i \leq n\}$ and a graph G a* **splitting** *of G by $\mathcal{P}$, called $SPLIT_{\mathcal{P}}(G)$, is a graph morphism star $(L_i \longleftarrow I_{ij} \longrightarrow L_j)_{1 \leq i < j \leq n}$ if graphs I_{ij} and graph morphisms $I_{ij} \longrightarrow L_i$, $\forall\, 1 \leq i < j \leq n$, exist such that the following conditions are satisfied:*

1. *The graph G is pushout-star object of $(L_i \longleftarrow I_{ij} \longrightarrow L_j)_{1 \leq i < j \leq n}$.*
2. *There exist graph morphisms $I_{ij} \longrightarrow I_i$ and $I_{ij} \longrightarrow I_j$ respectively, satisfying $I_{ij} \longrightarrow I_i \longrightarrow L_i = I_{ij} \longrightarrow L_i$ and $I_{ij} \longrightarrow I_j \longrightarrow L_j = I_{ij} \longrightarrow L_j$, respectively. $(\forall\, 1 \leq i < j \leq n)$*

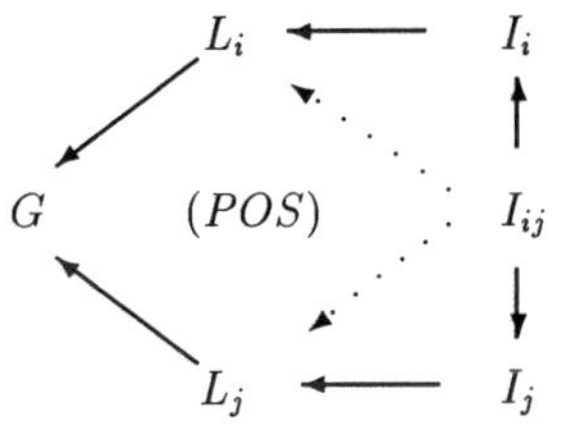

Now a parallel derivation can be defined using such a splitting:

3.3 Definition *Given $SPLIT_{\mathcal{P}}(G) = (L_i \longleftarrow I_{ij} \longrightarrow L_j)_{1 \leq i < j \leq n}$ with $\mathcal{P} = \{p_i | p_i = (L_i \longleftarrow I_i \longrightarrow R_i),\ 1 \leq i \leq n\}$ a* **parallel derivation** *$G \Longrightarrow G'$ is given by the following two pushout-stars if we define $I_{ij} \longrightarrow R_i = I_{ij} \longrightarrow I_i \longrightarrow R_i$ and $I_{ij} \longrightarrow R_j = I_{ij} \longrightarrow I_i \longrightarrow R_j$ $\forall\, 1 \leq i < j \leq n$:*

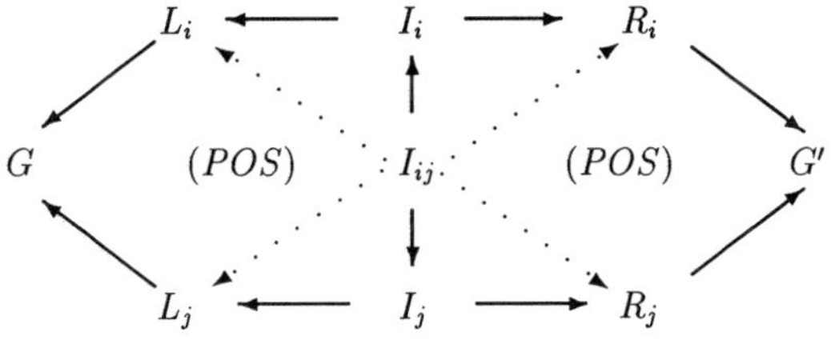

3.4 Definition *A parallel graph grammar* $GG = (S, \mathcal{P}, \Longrightarrow, T)$ *is given by a start graph* S*, a set of productions* $\mathcal{P}$*, a parallel derivation relation* $\Longrightarrow$ *and a class* T *of graphs, called terminal graphs.*

The language $L(GG)$ *of a parallel graph grammar* $GG = (S, \mathcal{P}, T)$ *is given by a set of those terminal graphs* G *which are derivable from* S *using parallel derivations.*

This notion of parallel graph grammars was defined in [EK76]. Replacing parallel by sequential derivations we obtain sequential graph grammars in the sense of [Eh79].

4 Dynamic Parallel Derivations

If the interfaces between different occurences of productions do not have to be constant during the derivation, i.e. they can be dissolved or extended, we obtain the concept of dynamic parallel derivations. In this case the interfaces are modelled not only by a graph but by a graph production which has to be a subproduction of all productions that use it as interface. Using dynamic parallel derivations parts of the left or right hand sides of the productions are amalgamated in order to avoid deletion or addition of subgraphs inside the whole graph several times.

4.1 Definition

Given a production $p = (L \longleftarrow I \longrightarrow R)$ *a production* $s = (L_s \longleftarrow I_s \longrightarrow R_s)$ *is called* **subproduction** *of* p *if there are graph morphisms* $L_s \longrightarrow L$, $I_s \longrightarrow I$ *and* $R_s \longrightarrow R$ *such that* $I_s \longrightarrow L_s \longrightarrow L = I_s \longrightarrow I \longrightarrow L$ *and* $I_s \longrightarrow R_s \longrightarrow R = I_s \longrightarrow I \longrightarrow R$*, i.e. the following diagrams commute.*

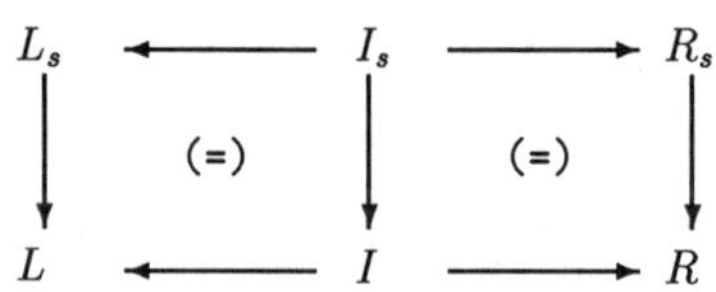

Now the splitting defined in 3.2 has to be redefined using interface productions instead of interface graphs.

4.2 Definition *Given a set of productions* $\mathcal{P} = \{p_i | p_i = (L_i \longleftarrow I_i \longrightarrow R_i),\ 1 \leq i \leq n\}$*, a set of interface productions* $\mathcal{S} = \{s_{ij} | s_{ij} = (L_{ij} \longleftarrow I_{ij} \longrightarrow R_{ij}),\ 1 \leq i < j \leq n\}$ *and a graph* G *a* **dynamic splitting** *of* G *by* $\mathcal{P}$ *and* $\mathcal{S}$*, called* $DSPLIT_{\mathcal{P},\mathcal{S}}(G)$*, is a graph morphism star* $(L_i \longleftarrow L_{ij} \longrightarrow L_j)_{1 \leq i < j \leq n}$ *if the following conditions are satisfied:*

1. *The graph* G *is pushout–star object of* $(L_i \longleftarrow L_{ij} \longrightarrow L_j)_{1 \leq i < j \leq n}$.
2. *Each production* s_{ij} *is a subproduction of* p_i *and* p_j $\forall\, 1 \leq i < j \leq n$.

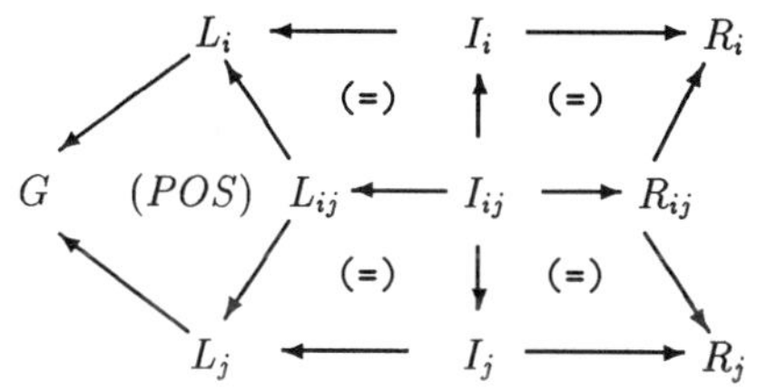

4.3 Definition *Given a dynamic splitting* $DSPLIT_{\mathcal{P},\mathcal{S}}(G) = (L_i \longleftarrow L_{ij} \longrightarrow L_j)_{1 \leq i < j \leq n}$ *as above the* **dynamic parallel derivation** $G \Longrightarrow G'$ *over this splitting is given by the following diagram*

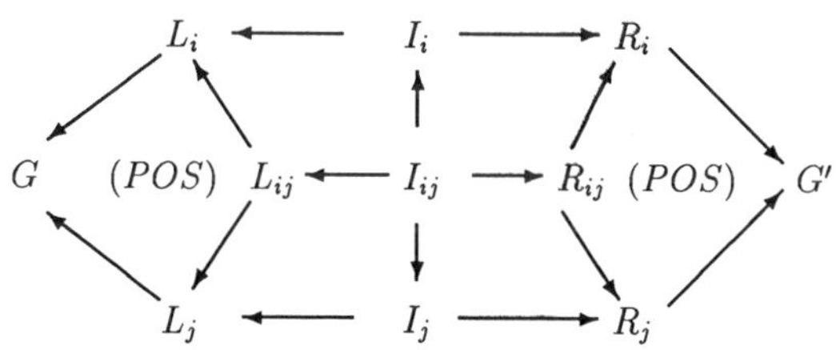

where the pushout–star construction of $(R_i \longleftarrow R_{ij} \longrightarrow R_j)_{1 \leq i < j \leq n}$ *yields graph* G' *and graph morphisms* $R_i \longrightarrow G'$ $1 \leq i \leq n$.

Remark: In the case that every subproduction s_{ij} is an identical production the dynamic parallel derivation is equal to the parallel derivation with constant interfaces as defined in 3.3. Similarily to definition 3.4 we obtain dynamic parallel graph grammars if $\Longrightarrow$ is the dynamic parallel derivation relation defined above.

If only a part of graph G is covered by occurences of productions we have a partial dynamic splitting:

4.4 Definition

Given sets of productions $\mathcal{P}$ *and* $\mathcal{S}$*, two graphs* G *and* L *and a graph morphism* $L \longrightarrow G$ *a* **partial dynamic splitting** *of* G *by* $\mathcal{P}$ *and* $\mathcal{S}$*, called* $PDSPLIT_{\mathcal{P},\mathcal{S},L}(G)$*, is a complete dynamic splitting of* L*,* $DSPLIT_{\mathcal{P},\mathcal{S}}(L)$*.*

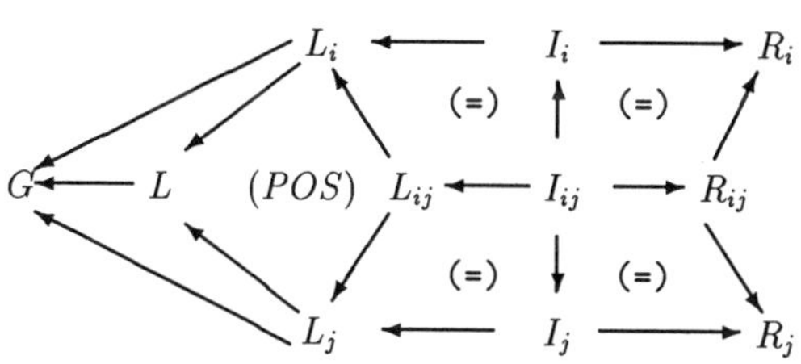

4.5 Definition *Given a partial dynamic splitting* $PDSPLIT_{\mathcal{P},\mathcal{S},L}(G) = (L_i \longleftarrow L_{ij} \longrightarrow L_j)_{1 \leq i < j \leq n}$ *with sets of productions* $\mathcal{P} = \{p_i | p_i = (L_i \longleftarrow I_i \longrightarrow R_i),\ 1 \leq i \leq n\}$ *and* $\mathcal{S} = \{s_{ij} | s_{ij} = (L_{ij} \longleftarrow I_{ij} \longrightarrow R_{ij}),\ 1 \leq i < j \leq n\}$ *and a graph morphism* $L \longrightarrow G$ *an* **amalgamated two–level derivation** $G \Longrightarrow G'$ *over this splitting consists of the following two levels:*

1. *Construction of an amalgamated production* $(L \longleftarrow I \longrightarrow R)$ *by the dynamic parallel derivation* $L \Longrightarrow R$ *with the above splitting. The pushout–star construction of the interface graph morphism star* $(I_i \longleftarrow I_{ij} \longrightarrow I_j)_{1 \leq i < j \leq n}$ *yields the interface graph* I. *The graph morphism* $I \longrightarrow L$ $(I \longrightarrow R)$ *is the universal morphism such that* $I_i \longrightarrow I \longrightarrow L = I_i \longrightarrow L_i \longrightarrow L$ $(I_i \longrightarrow I \longrightarrow R = I_i \longrightarrow R_i \longrightarrow R)$ $\forall\, 1 \leq i \leq n$.

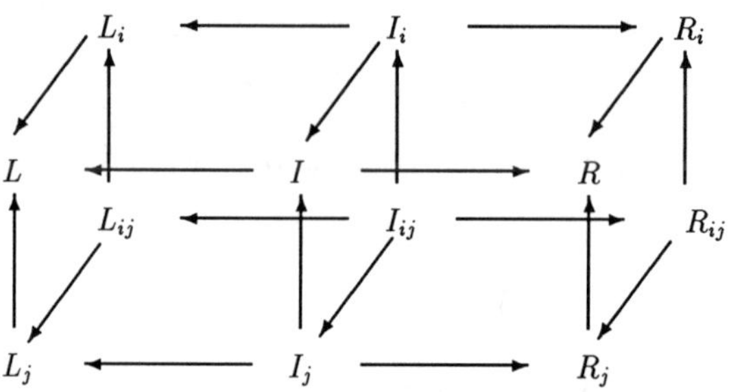

2. *Application of the amalgamated production* $(L \longleftarrow I \longrightarrow R)$ *to the graph* G *with the occurence* $L \longrightarrow G$ *by a sequential derivation* $G \Longrightarrow G'$. *(See definition 3.1.)*

Remark: The r–amalgamated derivation defined with additional application conditions in [BFH87] is a special case of the amalgamated two–level derivation if $n = 2$.

4.6 Examples

1. The context dependent graph transformation given informally in example 2.5 can now be constructed explicitly using an amalgamated two–level derivation.

   Let all items that are persistent in productions (1) - (3) form the interface graphs of those productions. Then production (1) which has an empty interface graph is subproduction of (2) and (3) where node a and b, respectively, persist.

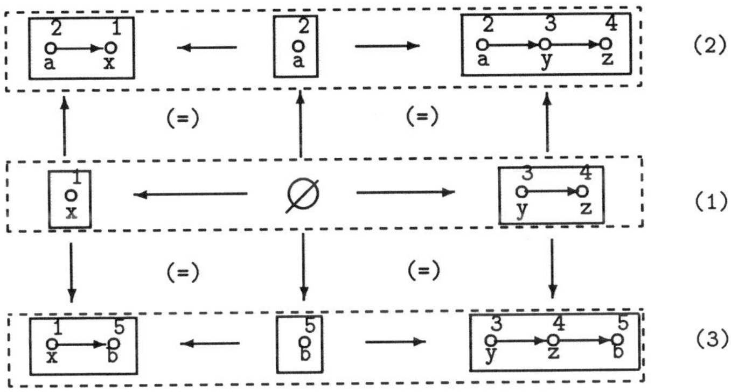

The partial dynamic splitting of A by productions (1) - (3) comprising graph L is given by the following diagram:

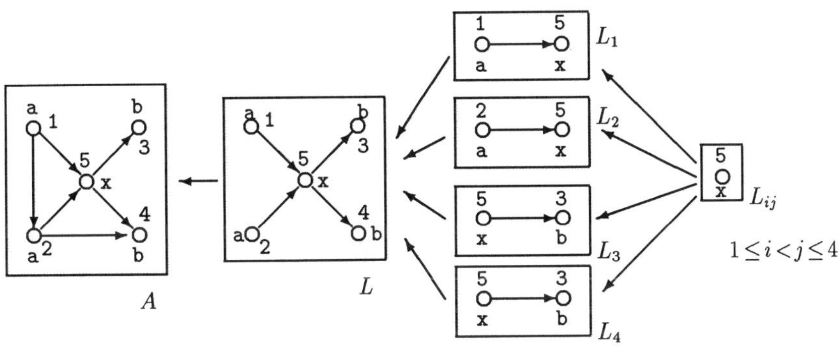

The interface graph I is constructed as follows:

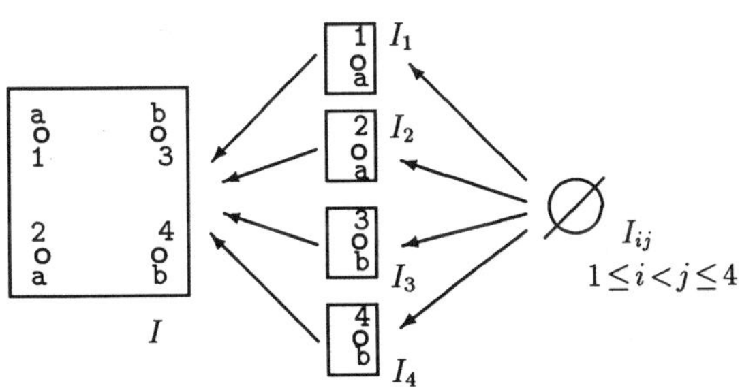

Executing a dynamic parallel derivation the right hand side R of the new generated production ($L \longleftarrow I \longrightarrow R$) is obtained:

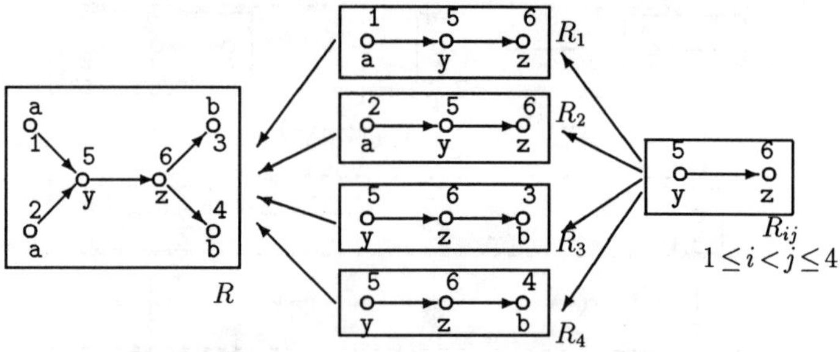

Applying the production $(L \longleftarrow I \longrightarrow R)$ to the graph A by a sequential derivation we get exactly the derivation introduced in example 2.5:

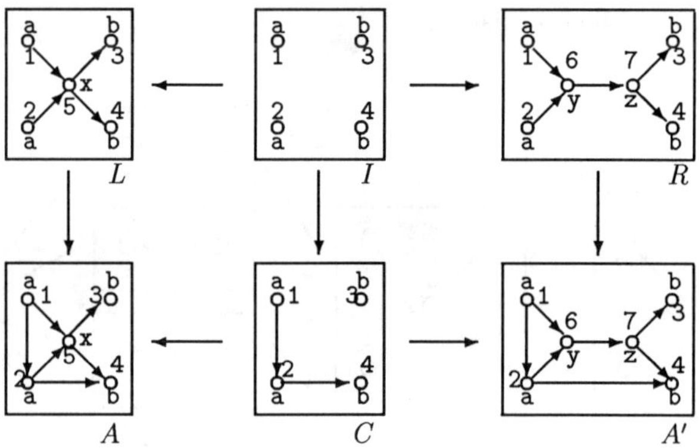

2. The following example shows the deletion of arbitrary subtrees indicated by a black root using amalgamated two–level derivations. We assume that the trees are unlabelled except the label black. After the deletion the black root is to be reserved so the following deletion productions seem to be suitable:

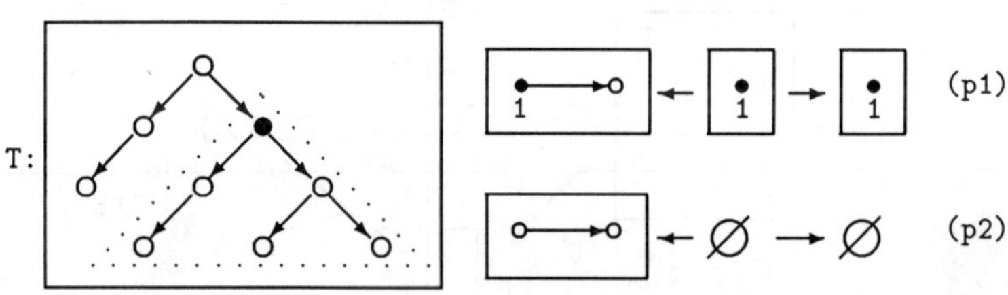

The interface productions are the empty production $\emptyset \longleftarrow \emptyset \longrightarrow \emptyset$ and the following:

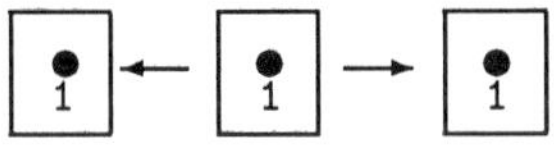 (s1) 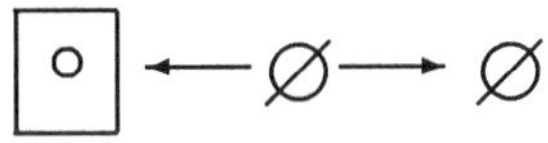 (s2)

Nodes of productions which are persistent, i.e. which are interface nodes are indicated by numbers.

Now we have to look for the largest occurence in the tree T such that the dangling condition (see section 3) for the amalgamated production to be constructed below is not violated, i.e. the context of T is a tree. The result is exactly the indicated subtree:

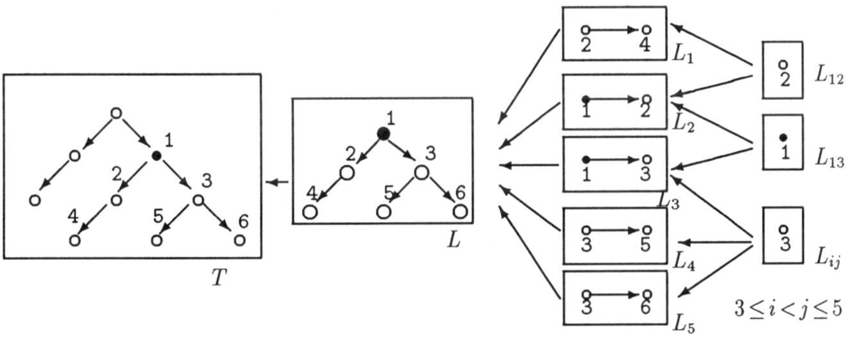

All L_{ij} $1 \leq i < j \leq 5$ not occuring in this diagram are empty. The interface graph I consists of the black root, only. All interface graphs I_i $1 \leq i \leq 5$ and I_{ij} $1 \leq i < j \leq 5$ except I_2, I_3 and I_{23} are empty:

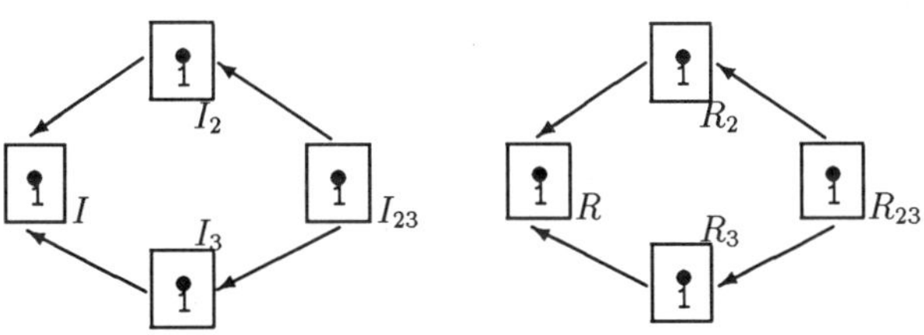

On the right hand side R the tree is deleted except the black root. All other right hand sides are empty.

Now the particular subtree L in the tree T is deleted by the following sequential derivation:

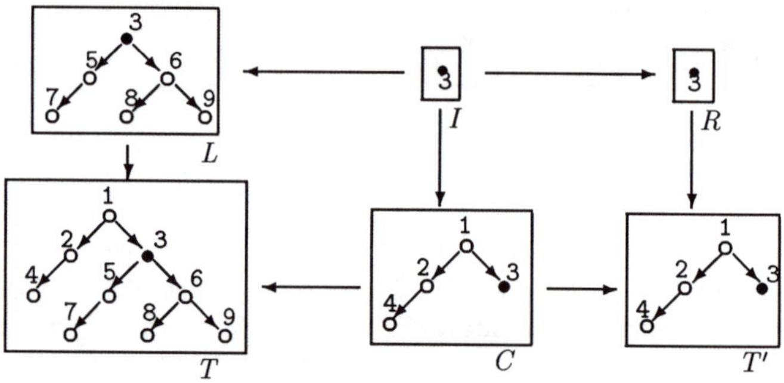

This example makes extensive use of the dangling condition because if the tree T would by covered by occurences more than L_{ij} $1 \leq i < j \leq 5$ this would violate the dangling condition: Adding handle (2,4) for example the edge (1,2) would dangle. Adding moreover the handle (1,2) the same would happen with edge (1,3). Handle (1,3) cannot be added because there is no suitable production.

The amalgamated two–level derivation $G \Longrightarrow G'$ can be compared with a special kind of dynamic parallel derivation. To complete the given partial splitting the context graph has to be determined and joined to the derived graph part.

4.7 Proposition *Given a partial dynamic splitting* $PDSPLIT_{\mathcal{P},\mathcal{S},L}(G) = (L_i \longleftarrow L_{ij} \longrightarrow L_j)_{1 \leq i < j \leq n}$ *the graph* C *with graph morphisms* $I_i \longrightarrow C \ \forall\, 1 \leq i \leq n$ *and* $C \longrightarrow G$ *is the context graph for the sequential derivation* $G \Longrightarrow G'$ *in the second level of an amalgamated two–level derivation (see 4.5), if the pushout–star construction of* $S = (L_i \longleftarrow L_{ij} \longrightarrow L_j)_{1 \leq i < j \leq n+1}$ *with* $L_{n+1} = C$ *and* $L_{i(n+1)} \longrightarrow L_i = I_i \longrightarrow L_i \ \forall\, 1 \leq i \leq n$ *yields graph* G*. (*C *is called pushout–star complement.) The dynamic splitting completed in such a way, called* $CDSPLIT_{\mathcal{P},\mathcal{S}}(G)$*, consists of the graph morphism star* S*. The* **completed splitting** $CDSPLIT_{\mathcal{P},\mathcal{S}}(G)$ *is a dynamic splitting in the sense of definition 4.2 if the set of productions* $\mathcal{P}$ *is extended by the identical production* $p_{n+1} = (C \longleftarrow C \longrightarrow C)$ *and the set of interface productions* $\mathcal{S}$ *by* $s_{i(n+1)} = (I_i \longleftarrow I_i \longrightarrow I_i) \ \forall\, 1 \leq i \leq n$*.*

4.8 Definition *Given a partial dynamic splitting* $PDSPLIT_{\mathcal{P},\mathcal{S},L}(G) = (L_i \longleftarrow L_{ij} \longrightarrow L_j)_{1 \leq i < j \leq n}$ *with sets of productions* $\mathcal{P} = \{p_i | p_i = (L_i \longleftarrow I_i \longrightarrow R_i),\ 1 \leq i \leq n\}$ *and* $\mathcal{S} = \{s_{ij} | s_{ij} = (L_{ij} \longleftarrow I_{ij} \longrightarrow R_{ij}),\ 1 \leq i < j \leq n\}$ *a* **dynamic parallel derivation with context completion** $G \Longrightarrow G'$ *is given by the dynamic parallel derivation* $G \Longrightarrow G'$ *over the completed dynamic splitting* $CDSPLIT_{\mathcal{P},\mathcal{S}}(G)$*.*

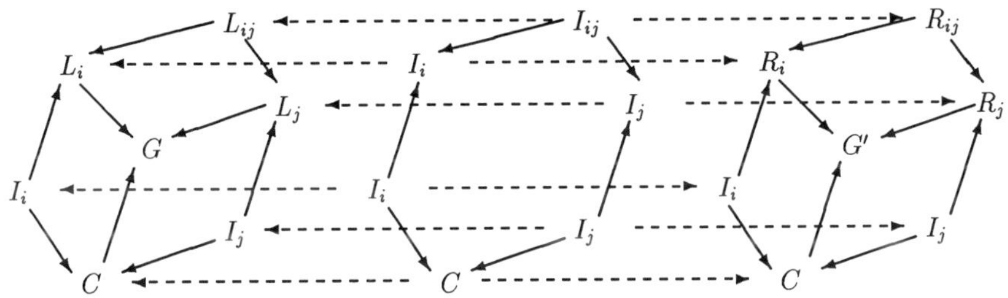

Remark: A special case of the dynamic parallel derivation with context completion is the sequential derivation. For $n = 1$ the partial dynamic splitting consists of only one object L_1 and the context completion is the context graph C together with the morphisms $I_1 \longrightarrow C$ and $C \longrightarrow G$ such that the pushout construction over $L_1 \longleftarrow I_1 \longrightarrow C$ yields G. The sets of productions are extended by the identical rules $(C \longleftarrow C \longrightarrow C)$ and $(I_1 \longleftarrow I_1 \longrightarrow I_1)$ and the result of the derivation G' is obtained by a pushout construction over $R_1 \longleftarrow I_1 \longrightarrow C$.

4.9 Dynamic Parallel Derivation Theorem *Given sets of productions* $\mathcal{P} = \{p_i | p_i = (L_i \longleftarrow I_i \longrightarrow R_i),\ 1 \le i \le n\}$ *and* $\mathcal{S} = \{s_{ij} | s_{ij} = (L_{ij} \longleftarrow I_{ij} \longrightarrow R_{ij}),\ 1 \le i < j \le n\}$ *the following statements are equivalent:*

1. *There is a dynamic parallel derivation with context completion* $G \Longrightarrow G'$ *over a partial dynamic splitting* $PDSPLIT_{\mathcal{P},\mathcal{S},L}(G) = (L_i \longleftarrow L_{ij} \longrightarrow L_j)_{1 \le i < j \le n}$.
2. *There is an amalgamated two-level derivation* $G \Longrightarrow G'$ *over the same partial dynamic splitting* $PDSPLIT_{\mathcal{P},\mathcal{S},L}(G) = (L_i \longleftarrow L_{ij} \longrightarrow L_j)_{1 \le i < j \le n}$.

5 Parallel High–Level Replacement Systems

In the last sections we presented different derivation concepts for graphs, now these concepts are generalized also for other types of high–level structures. Basic concepts of high–level replacement systems like productions and sequential derivations are presented. In order to extend results like Church–Rosser–Properties, the Parallelism Theorem and the Concurrency Theorem from graph grammars to high–level replacement systems suitable conditions have to be satisfied which are studied in [EHLP91].

Extending the framework of high–level replacement systems by parallel derivation concepts parallel high–level replacement systems (PHLR–systems) are formulated satisfying certain conditions, the PHLR–conditions.

Finally we give some examples of parallel high–level replacement systems.

Let **CAT** be a category where the **objects** can be regarded as high-level structures and the **morphisms** as structure preserving functions between high-level structures.

This kind of categories corresponds exactly to that introduced in [EHLP91] where, however, a special class M of morphisms is distinguished. If the results presented in [EHLP91] are applied to two–level derivations such a class has to be introduced to satisfy the corresponding conditions.

A **production** p is a pair of morphisms $(L \longleftarrow I \longrightarrow R)$ between the objects L, I and R, called left hand side, interface and right hand side respectively.

Remark: The morphisms $I \longrightarrow L$ and $I \longrightarrow R$ in [EHLP91] are required to belong to the special class M of morphisms.

The **sequential derivation** in the category **CAT** is defined like the sequential derivation of graphs replacing the special gluing construction for graphs by a general pushout construction in **CAT**. (See definition 3.1.)

5.1 Definition *Given the objects I,G_1 and G_2 and two morphisms $I \longrightarrow G_1$ and $I \longrightarrow G_2$ the* **pushout** *of these morphisms consists of an object G together with two morphisms $G_1 \longrightarrow G$ and $G_2 \longrightarrow G$ such that:*

1. *commutativity:* $I \longrightarrow G_1 \longrightarrow G = I \longrightarrow G_2 \longrightarrow G$
2. *universal property: For all objects G' and morphisms $G_1 \longrightarrow G'$ and $G_2 \longrightarrow G'$ satisfying $I \longrightarrow G_1 \longrightarrow G' = I \longrightarrow G_2 \longrightarrow G'$ there is a unique morphism $G \longrightarrow G'$ such that $G_i \longrightarrow G \longrightarrow G' = G_i \longrightarrow G'$ for $1 \leq i \leq 2$.*

In this case the following diagram is called pushout diagram, or short pushout (PO).

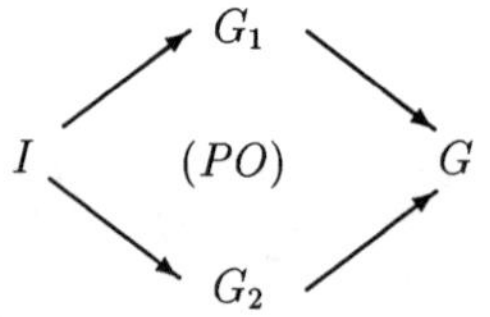

The star gluing construction used for gluing more than two graphs is generalized to the pushout–star construction.

5.2 Definition *A morphism star of degree $n \geq 1$ is a diagram $MS = (S_i \longleftarrow I_{ij} \longrightarrow S_j)_{1 \leq i < j \leq n}$ consisting of objects S_i,S_j,I_{ij} and morphisms $I_{ij} \longrightarrow S_i$ resp. $I_{ij} \longrightarrow S_j$. In the case $n = 1$ we only have the single object S_1. An object S together with the morphisms $S_i \longrightarrow S$ for $1 \leq i \leq n$ is called* **pushout–star** *of MS (POS) if we have:*

1. *commutativity:* $I_{ij} \longrightarrow S_i \longrightarrow S = I_{ij} \longrightarrow S_j \longrightarrow S$ *for all* $1 \leq i < j \leq n$
2. *universal property: For all objects S' and morphisms $S_i \longrightarrow S'$ satisfying $I_{ij} \longrightarrow S_i \longrightarrow S' = I_{ij} \longrightarrow S_j \longrightarrow S'$ for all $i < j$ there is a unique morphism $S \longrightarrow S'$ such that $S_i \longrightarrow S \longrightarrow S' = S_i \longrightarrow S'$ for all $1 \leq i \leq n$.*

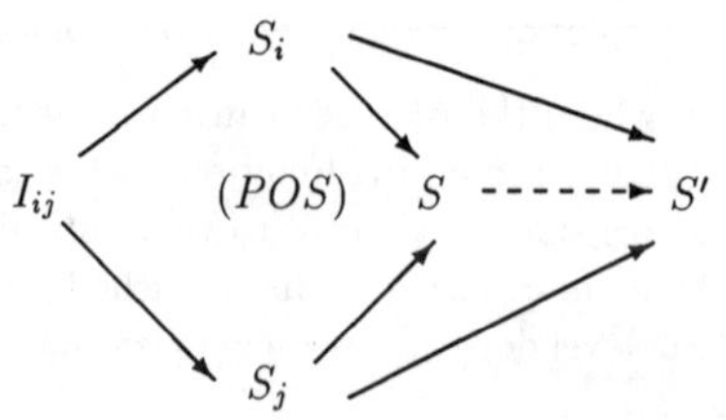

In this case the morphism star MS together with S and $S_i \longrightarrow S$ is called pushout-star diagram, or short pushout-star (POS).

In the case $n = 2$ the pushout-star is called pushout. (See definition 5.1.)

Now the different **parallel derivation** concepts for graphs can be generalized to parallel derivation concepts in the category **CAT** using objects and morphisms instead of graphs and graph morphisms and general pushout-star constructions instead of star gluing constructions for graphs.

5.3 Definition *A* **parallel high-level replacement system** $PHLR = (S, \mathcal{P}, \Longrightarrow, T)$ *based on the category* **CAT** *is given by a start object S in* **CAT**, *a set $\mathcal{P}$ of productions, a parallel derivation relation $\Longrightarrow$,and a class T of objects in* **CAT**, *called terminal colored objects.*

Different kinds of splittings, subproductions, parallel derivations, dynamic parallel derivations and amalgamated two-level derivations are defined similarily to the definitions in section 3 and 4.

The equivalence between amalgamated two-level derivations and dynamic parallel derivation with context completion stated in the dynamic parallel derivation theorem (4.9) can be proofed in any category which satisfies the following conditions.

5.4 Definition *The following two conditions are called PHLR-conditions and each category* **CAT** *which satisfies these conditions is called PHLR-category.*

1. *There is an initial object in* **CAT**.
2. *For each morphism star $(S_i \longleftarrow I_{ij} \longrightarrow S_j)_{1 \leq i < j \leq n}$ in* **CAT** *there is a pushout-star diagram as in definition 5.2.*

The proof of theorem 4.9 in the framework of PHLR-systems in [Ta91] makes use of commutativity properties of colimits [ML72] which are also used in the framework of module specifications (see commutativity of colimits in [EM90]). In our case we have to apply it to pushouts and pushout-stars which are special colimits.

5.5 Examples

1. In section 3 and 4 all different derivation concepts introduced for parallel high-level replacement systems are presented for directed graphs with colored nodes and edges and color-preserving morphisms. Taking these graphs as objects and these graph morphisms as arrows the category GRAPH is obtained. (Definitions of graphs and graph morphisms are given in section 3.) This category has the empty graph as initial object and for each graph morphism star there exists a pushout-star. ([EK76]) So the category GRAPH is a PHLR-category.

2. Another example of a PHLR-system is given in [ST91] where the movement of objects in configuration spaces is modelled by graph grammars where the morphisms are not color-preserving. Configuration spaces are modelled by grid structured graphs where the node colors indicate the position of objects. Movement of objects is modelled via two-level derivations changing node colors if corresponding objects

are moved. These two–level derivations move elementary parts of objects in parallel and let their context unchanged. The derivations are built on graph morphisms which allow the changing of node colors along a given partial order. The underlying category C–GRAPH is the category of graphs with color alphabet $C = \{w, b\}$ where the morphisms are allowed to map w to w, b to b and w to b, but not b to w. C–GRAPH satisfies also the PHLR–conditions ([ST91]) and is thus another PHLR–category.

3. Each category with colimits is a PHLR–category and can be used to build up PHLR–systems. So it would be interesting to define the category for map grammars as used in [Lin86] where a map is a special kind of topological space. At least the category TOP of topological spaces has arbitrary colimits ([ML72]). It has an initial object, the empty topological space, and pushout–stars and is therefore a PHLR–category.

6 Conclusions

In this paper we have introduced two–level derivations combining concepts of parallel and sequential derivations. Here, first a new production is generated by a parallel derivation and then this production is applied using a sequential derivation. So on one hand we have the advantage of using parallel replacements speeding up the derivation process, on the other hand all results which are known for sequential derivations (like Church–Rosser–Properties, Parallelism Theorem and Concurrency Theorem, etc.) can be used also for this type of derivation if the corresponding conditions for HLR–systems (see [EHLP91]) are satisfied.

A second–level production is generated from a set of elementary productions. So the two–level derivations can also be used to describe derivations over production families. Dependent on the actual structure a suitable second–level production is generated from a given set of elementary productions describing an operation independent of particular structures.

The different kinds of parallel derivations introduced in this paper can also be seen to model some concepts of distribution. A global state structure can be split into local structures which are transformed in parallel, each independent of the others. Each pair of local structures is combined by an interface which can be transformed, too. After the local transformations all structures obtained are joined to a global state according to the interfaces obtained. All local transformations run concurrently without any synchronization. So the only way to synchronize the local transformations is to switch to the global state. (See also [BEHL87], [EL91])

Using two–level derivations for safe deletion of objects in arbitrary context a kind of garbage collection can be modelled. An object is only deleted if there is only garbage which points on it. (See example 4.6.2 where arbitrary subtrees can be deleted if there are not any other references to the subtree than the specified one.)

A further class of application is indicated in example 4.6.1 where the relation of the added object to its context has to be specified before. Another way to model such an operation is given by the NLC–approach to graph grammars using embedding relations ([JR80],[Roz87]).

References

[ArMa75] M.A. Arbib, E.G. Manes: Arrows, Structures and Functors, Academic Press, New York - San Francisco - London,1975

[BEHL87] P. Boehm, H. Ehrig, U. Hummert, M. Loewe: Towards distributed graph-grammars in Proc. 3rd Int. Workshop on Graph Grammars, Springer LNCS 291(1987),p.86–98

[BFH87] P. Boehm, H.-R. Fonio, A. Habel: Amalgamation of Graph Transformations: A synchronization Mechanism in JCSS,34:377-408 (1987)

[CER79] V. Claus, H. Ehrig, G. Rozenberg: Graph Grammars and Their Applications to Computer Science and Biology in LNCS 73, Springer Verlag, 1979

[Eh79] H. Ehrig: Introduction to the Algebraic Theory of Graph Grammars in LNCS 73, Springer Verlag, Berlin, 1979

[EHLP91] H. Ehrig, A. Habel, M. Loewe, F. Parisi-Presicce: From Graph Grammars to High-Level Replacement Systems in 4rd Int. Workshop on Graph Grammars and Their Applications to Computer Science, to appear in LNCS 1991

[EK76] H. Ehrig, H.-J. Kreowski: Parallel Graph Grammars in A. Lindenmayer, G. Rozenberg: Automata, Languages, Development, 425-447, Amsterdam: North Holland 1976

[EL91] H. Ehrig, M. Loewe: From Parallel to Distributed Derivations of Graphs in the Single Pushout Approach, to appear in Springer Informatik Fachberichte (1991)

[EM90] H. Ehrig, B. Mahr: Fundamentals of Algebraic Specification 2: Module Specifications and Constraints, Springer New York - Heidelberg - Berlin 1990

[Lin86] A. Lindenmayer: An Introduction to Parallel Map Generating Systems in Proc. 3rd Int. Workshop on Graph Grammars, Springer LNCS 291(1987),p.27–40

[ML72] S. MacLane: Categories for the working mathematician, Springer New York - Heidelberg - Berlin 1972

[JR80] D. Jansens, G. Rozenberg: On the structure of node–label controlled graph grammars in Information Science, 20:191 - 216,1980

[Na87] M. Nagl: Set Theoretic Approaches to Graph Grammars in Proc. 3rd Int. Workshop on Graph Grammars, Springer LNCS 291(1987),p.41–54

[Roz87] G. Rozenberg: An introduction to the NLC way of rewriting graphs in 3rd Int. Workshop on Graph Grammars and Their Applications to Computer Science, LNCS 291, 1987

[ST91] H. Schween, G. Taentzer: Movement of Objects in Configuration Spaces modelled by Graph Grammars in 4rd Int. Workshop on Graph Grammars and Their Applications to Computer Science, to appear in LNCS, 1991

[Ta91] G. Taentzer: Parallel High–Level Replacement Systems, Technical Report, TU Berlin (1991), to appear

On Relationships between Graph L-Systems and Picture Languages: Part II

Kunio Aizawa[1] and Akira Nakamura[2]

[1]Department of Applied Mathematics, Hiroshima University, Higashi-Hiroshima, 724 Japan

[2]Department of Information Science, Meiji University, Kawasaki, Kanagawa, 214 Japan

Abstract: In this paper, a path controlled embedding graph grammar (PCE graph grammar) having the confluent property is proposed. Then it is extended to a parallel case, i.e., graph L-system. The generative powers of these systems are compared with that of context-free array grammars.

Keywords: graph L-systems, graph grammars, partial path groups, confluent grammars, array languages

1. Introduction

Many models for graph grammars have been proposed as a kind of hyper-dimensional generating systems. The basic primitive in a graph grammar is a directed (or non-directed) graph which is a set of labelled vertices and labelled edges. Thus, most of the graph grammars have some kind of the "embedding" rules as well as the "production" rules to embed the newly replaced graph to its host graph. One of the reasons why graph generating systems were proposed is to generalize the (one-dimensional or string) formal grammars to two-dimensional or more complex structures. However, few has known about the relationships between graph grammars and two-dimensional formal grammars. It is obvious that the graph grammars with no restrictions are more powerful than or incomparable with the two-dimensional formal grammars since there exist many graphs which cannot be embedded in any two-dimensional structures. Thus, in order to investigate such relationships, the graph grammars are restricted such that they can generate planer graphs only. In the two-dimensional formal grammars, the "array" is used as an implicit structure of the generated patterns (i.e.,two-dimensional words). Even if restricted to planer graphs, almost all structures, on the other hand, can be formed in the graph grammars by making use of various "embedding" rules. Therefore, it is necessary to restrict the array grammars within the graphs having the two-dimensional array structure. It has been shown in Nakamura and Aizawa [6], however, that such restriction may bring sufficient generative power as a isometric array grammar into a graph L-system having production rules of the context-free style.

In the recent years, many models for context-free graph grammars have been proposed (see e.g., Ehrig, et al. [2]). Some grammars are node rewriting and others are edge (or hyper-edge) rewriting. These grammars are context-free in the sense that one node is replaced without considering any other part of the rewrited graph. However, they may still be context-sensitive in the sense that generated graph depends on the order in which the production rules are applied. A graph grammar that does not suffer from this context-sensitivity is said to be confluent (see, e.g., Engelfriet [3]).

In general, graph grammars are less powerful to describe structures in contrast of their generative power. In Aizawa and Nakamura [1], we introduced a graph grammar called node-replacement graph grammar with path controlled embedding (nPCE grammars) which use a sequence of edges instead of the single edge to embedding a newly replaced graph into the host graph. It has been shown that there exists a subclass of nPCE grammars generating context-free array languages. In this paper, a path controlled embedding graph grammar (PCE graph grammar) having the confluent property is proposed. Then it is extended to a parallel case, i.e., graph L-system. The generative powers of these systems are compared with that of context-free array grammars.

We assume for readers to be familiar with the theories of two-dimensional grammars and graph grammars (see e.g., Rosenfeld [8] and Nagl [5]). The following notations will be used in the rest of this paper.

(1) Let X be set. By 2^X we denote the set of subsets of X and if X is finite, then #X denotes the cardinality of X.

(2) Let $\pi=\{c_1c_2...c_i\}$ be a string. By π^R we denote the reverse string of π, i.e., $\pi^R=\{c_i...c_2c_1\}$, and $|\pi|$ denotes the length of π, i.e., i.

(3) A *graph* is a system $H = (V, E, \Sigma_V, \Sigma_E, \phi_V, \phi_E)$, where V is a finite nonempty set of natural numbers called the *set of nodes*, E is a set of pairs of two elements from V called the *set of edges*, Σ_V is a finite nonempty set called the *set of node labels*, Σ_E is a finite nonempty set called the *set of edge labels*, ϕ_V is a mapping from V into Σ_V called the *nodes labelling function*, and ϕ_E is a mapping from E into Σ_E called the *edges labelling function*. H is called a graph over (Σ_V, Σ_E). Throughout of this paper, V(H) and E(H) denote the set of nodes and the set of edges of H, respectively.

(4) Let $H = (V, E, \Sigma_V, \Sigma_E, \phi_V, \phi_E)$ be a graph and let (x,y) is an edge of H. We say that the edge (x,y) is *incident with* the nodes x and y, and the nodes x, y are *neighbors*.

(5) Let $H = (V, E, \Sigma_V, \Sigma_E, \phi_V, \phi_E)$ be a graph and let x be a node of H. Then the *degree of x*, denoted as deg(x), is the number of edges incident with x.

(6) Let $A = (V, E, \Sigma_V, \Sigma_E, \phi_V, \phi_E)$ and $B = (V', E', \Sigma_V', \Sigma_E', \phi_V', \phi_E')$ be graphs. A is a *subgraph of B* if $V'\supseteq V$, $E'\cap\{(x, y)|\ x, y\in V\}\supseteq E$, $\Sigma_V'\supseteq\Sigma_V$, $\Sigma_E'\supseteq\Sigma_E$, $\phi_V'(x)= \phi_V(x)$ for

$\forall x \in V$, and $\phi_E'((x, y)) = \phi_E((x, y))$ for $\forall(x, y) \in E$. In this case, we call A the subgraph *spanned by V in B*. By B-A we denote the subgraph spanned by V'\V in B.

(7) Let $A = (V, E, \Sigma_V, \Sigma_E, \phi_V, \phi_E)$ and $B = (V', E', \Sigma_V, \Sigma_E, \phi_V', \phi_E')$ over (Σ_V, Σ_E). An *isomorphism from A into B* is a bijective mapping h from V into V' such that $\phi_V' h = \phi_V$ and $E' = \{(h(x), h(y)) \mid (x, y) \in E\}$. We say that *A is isomorphic to B*.

(8) A graph $A = (V, E, \Sigma_V, \Sigma_E, \phi_V, \phi_E)$ is connected if for every x, y in V, there exists a sequence $x_1, x_2, \ldots, x_n$ of nodes in V such that $x_1 = x$, $x_n = y$ and for $1 \le i \le n-1$, x_i is a neighbor of x_{i+1}.

2. Path controlled embedding

In this section, we define a kind of graph grammars called U-nPCE graph grammars and languages. They are restricted version of nPCE graph grammars proposed in Aizawa and Nakamura [1].

Definition 2.1. For any given graph H, its node P, and a string $\pi = \{c_1 c_2 \ldots c_i\}$ of its edge labels, $P\pi$ is *realizable on H* if and only if there exists a set of nodes $\{P_0, P_1, \ldots, P_i\}$ such that $P_0 = P$ and P_j is a neighbor of P_{j-1} joined by an edge labelled with c_j ($1 \le j \le i$).

Definition 2.2. *A node-replacement graph grammar with path controlled embedding*, denoted as nPCE grammar, is a construction

$G = < \Sigma_N, \Sigma_E, P, Z, \Delta_N, \Delta_E >$, where

Σ_N is a finite nonempty set of node labels,

Σ_E is a finite nonempty set of edge labels,

Δ_N is a finite nonempty subset of Σ_N, called terminal node labels,

Δ_E is a finite nonempty subset of Σ_E, called terminal edge labels,

P is a finite set of productions of form (a, β, ψ), where a is a node, $\beta = (V, E, \Sigma_V, \Sigma_E, \phi_V, \phi_E)$ is a connected graph, ψ is a mapping from Σ_E^+ into $V(\beta) \times \Sigma_E$. ψ is called *embedding function.*

Z is a connected graph over (Σ_N, Σ_E) called *the axiom.*

A direct derivation step in a nPCE grammar is performed as follows:

Let $H = (V, E, \Sigma_V, \Sigma_E, \phi_V, \phi_E)$ be a graph. Let $p = (a, \beta, \psi)$ be a production in P. Let β' be isomorphic to β (with h an isomorphism from β' into β), where β' and H–a have no common nodes. Then the result of the application of p to H (by using h) is obtained by first removing a from H, then replacing a with β' and finally adding edges (u,v) between every nodes u in β' and every v in H-a such that there exists a path $\pi \in$ domain(ψ) with v = $a\pi$ in H. The production p is applicable to the graph H if following conditions hold.

(1) If a has at least one neighbor in H, there exist at least one realizable path π, and

(2) if a node v of H-a is adjacent to the node labelled with a in H, v must be adjacent to the node 1 of β'.

Note here that the embedding function ψ bring no significant context into the generation procedure of an nPCE grammar since no node labels are referred in any place of embedding steps except the one of the newly replaced graph.

Formally the notion of a direct derivation step is defined as follows:

Definition 2.3. Let G = < Σ_N, Σ_E, P, Z, Δ_N, Δ_E > be a nPCE grammar and let H, H' be graphs over (Σ_V, Σ_E).

(1) H *directly derives* H' in G, denoted as H $\Rightarrow_G$ H', if there exists a production p = (a, β, ψ) in P, a graph β' with V(β') $\cap$ V(H-a) = $\emptyset$ and an isomorphism h from β' into β such that H' is isomorphic to the graph X constructed as follows:

X = (V, E, Σ_V, Σ_E, ϕ_V, ϕ_E), where

V = V(H-a) $\cup$ V(β'),

E = { (x,y) | x,y $\in$ V(H-a) and (x,y) $\in$ E(H) }

$\cup$ { (x,y) | x,y $\in$ V(β') and (h(x),h(y)) $\in$ E(β)}

$\cup$ { (x,y) | x $\in$ V(β'), y $\in$ V(H-a), there exists a path $\pi \in$ domain(ψ) with y = $a\pi$ and <h(x), e> $\in$ $\psi(\pi)$ for some e in Σ_E },

ϕ_V is equal to the node labelling function of H for nodes in V(H-a), equal to the node labelling function of β' for nodes in V(β'),

ϕ_E is equal to the edge labelling function of H for edges between the nodes of H, is equal to the edge labelling function of β' for nodes in V(β'),

We also say that H' is derived from H by replacing a using the production p.

(2) We will denote the reflexive and the transitive closure of $\Rightarrow_G$ by $\Rightarrow_G^*$ and the transitive closure of $\Rightarrow_G$ by $\Rightarrow_G^+$.

(3) The language of G, denoted as L(G), is defined by L(G) = { H | H is a graph over (Σ_V, Σ_E) and Z $\Rightarrow_G^*$ H }.

We here present an example of the derivations of nPCE grammars.

Example 2.1. The application of the production in Fig. 1a to the graph in Fig. 1b results the graph in Fig. 1c.

By making use of the path controlled embedding mechanism defined above, it is possible to construct nPCE grammars generating array languages under some way to regard a graph as a two-dimensional array as in Fig. 2. In Aizawa and Nakamura [1], it is shown that, for any given context-free array grammar G, there exists an nPCE grammar G' such that L(G') is regarded as a set of two-dimensional array, L(G). However, describing the nPCE grammars generating array languages needs somewhat clumsy production rules especially in path descriptions. So, nPCE grammars using partial path group to describe paths in the embedding functions seems to be more suitable for describing patters having some geometrical structures. We review here the definitions of the partial path groups. For more precise definitions of the partial path groups, see Rosenfeld [9].

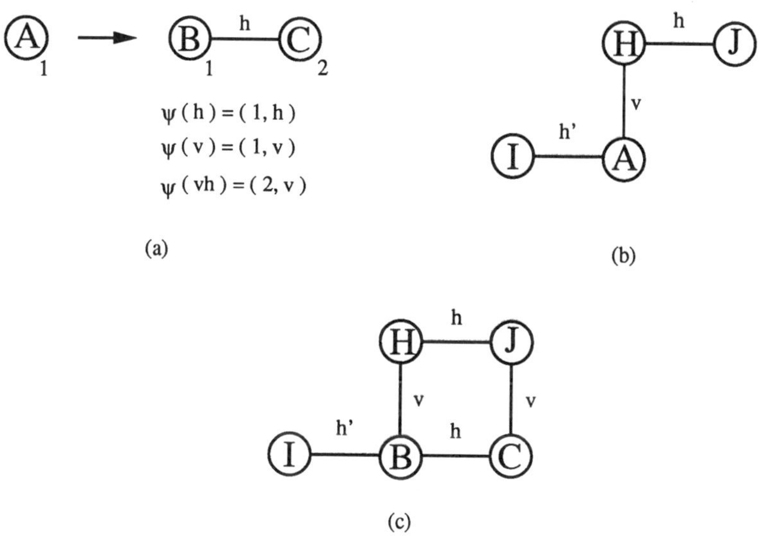

Fig. 1. An example of derivations of nPCE grammar.

Definition 2.4. Let H be a connected graph of degree d, i.e., no more than d edges emanate from any node. By an *edge coloring* of H we mean an assignment of colors to the edges of H such that the edges emanating from any given node all have different colors.

Definition 2.5. Let p be a node of a graph H and let π be a string of colors. Then $p\pi$ is defined as the terminal node of the path defined by π starting from p, provided this path is realizable. For convenience, let us define a fictitious "blank" color representing "no move". It is obvious that the structure defined above resembles a group structure called "partial group".

From now on we shall refer to this partial group as the *partial path group* of H defined by given coloring, and denote it by $\Pi(H)$.

Note that the partial path groups can be defined on a graph generated by a graph grammar by regarding the edge labels of the generated graph to the colors mentioned in above definitions.

Definition 2.6. Let $\alpha = c_{i_1} c_{i_2} \ldots c_{i_k}$ be a string of colors. The string α' is called an *elementary reduction* of α if one of following statements is true:

a) $k>1$; for some $1 \leq j \leq k$ we have $c_{i_j} = c_0$; and $\alpha' = c_{i_1} \ldots c_{i_{j-1}} c_{i_{j+1}} \ldots c_{i_k}$.

b) $k>2$; for some $1 \leq j < k$ we have $c_{i_j} = c_{i_{j+1}}$; and $\alpha' = c_{i_1} \ldots c_{i_{j-1}} c_{i_{j+2}} \ldots c_{i_k}$.

c) $k=2$; $c_{i_1} = c_{i_2}$; and $\alpha' = c_0$.

The string α' is called a *reduction* of α if there exists strings $\alpha=\alpha_0, \alpha_1, \ldots, \alpha_m=\alpha'$ such that α_i is an elementary reduction of α_{i-1}, $1 \leq i \leq m$. α is called *fully reducible* if $\alpha=c_0$, or if c_0 is a reduction of α. $\Pi(G)$ is called *free* if $P\alpha=P$ implies that α is fully reducible.

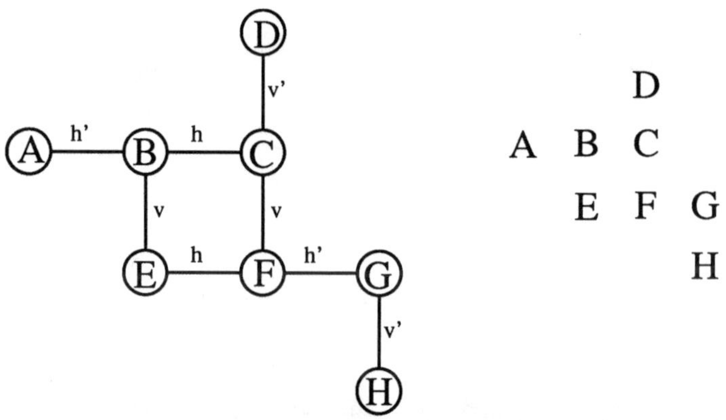

Fig. 2. A graph and its corresponding array.

Definition 2.7. A partial path group $\Pi(H)$ is called *near-abelian* if each color except blank color commutes with all but one of the other colors, and does not commute with the remaining one.

It is shown in Rosenfeld [9] that the free near-abelian partial path groups with four colors correspond to a subgraph of a two-dimensional array.

We define nPCE grammars using embedding functions with near-abelian partial path groups to describe paths.

Definition 2.8. A *nPCE grammar with 4 colors free near-abelian partial path groups*, denoted as nPCE$_{\Pi 4}$ grammar, is a construction

G = <Σ_N, Σ_E, P, Z, Δ_N, Δ_E>, where Σ_N, Σ_E, Z, Δ_N, Δ_E are same as in the definition of nPCE grammars provided that $\Sigma_E=\Delta_E$ and $|\Delta_E|=4$. P is a finite set of production rules of form (a, β, $\psi_{\Pi 4}$), where $\psi_{\Pi 4}$ is a mapping from $\Sigma_E{}^+$ into $(V(\beta)\times\Sigma_E)$ provided that if $\psi_{\Pi 4}$ maps π into (i, c) for some $c\in\Delta_E$, then c is the reduction of σπ, where σ is the path between node 1 to i in β; $\psi_{\Pi 4}$ is called embedding function with 4 colors near-abelian partial path groups. For a path π, $\psi_{\Pi 4}(\pi)$ implies the shortest realizable path equivalent to π under near-abelian partial path groups instead of π itself.

In the nPCE$_{\Pi 4}$ grammars, we use $\psi_{\Pi 4}$ instead of the embedding function ψ of nPCE grammars. In this case, arbitrarily long paths can be used to embed the newly replaced graphs. Since the derivation steps of our nPCE$_{\Pi 4}$ grammars are proceeded in the context-free node-replacement style, there are no insurance that the degree of each node is always at most 4 and the edges emanating from any given node have always different colors (see, e.g., Fig. 3). In such cases, even if α is fully reducible, Pα may not be P. However, the embedding mechanism of nPCE grammars still works correctly for such cases as far as using the shortest realizable paths.

One of the advantages to introduce the partial path groups into the path controlled embedding mechanism is flexibility of describing various structures. As mentioned above, two-dimensional square arrays correspond to free near-abelian groups with 4 colors, and other types of arrays (triangular, hexagonal) correspond to other types of abelian groups. The partial path groups allow us to treat other classes of graphs (such as trees and hypercubes etc.) in addition to arrays, and they have very simple characterizations. For detail discussions of those structures, see Rosenfeld [9]. The partial path groups for more complicated structures are discussed in Melter [4].

Definition 2.9. An nPCE$_{\Pi 4}$ grammar is called having *uniform embedding function*, denoted as U-nPCE$_{\Pi 4}$ grammar, if the following conditions for embedding functions hold.

(1) For each pair of production rules $p_i=(a_i, \beta_i, \psi_{\Pi 4_i})$, $p_j=(a_j, \beta_j, \psi_{\Pi 4_j})$ and a path $\pi\in\Sigma_E{}^+$, if both of $\psi_{\Pi 4_i}(\pi)$ and $\psi_{\Pi 4_j}(\pi)$ are defined, then $\psi_{\Pi 4_i}(\pi)=\psi_{\Pi 4_j}(\pi)$.

(2) For each production rule p=(a, β, $\psi_{\Pi 4}$), if $\psi_{\Pi 4}(\pi)=(v, c)$ is defined for some $\pi\in\Sigma_E{}^+$ and v=1, then $\psi_{\Pi 4}(\pi^R)$ is also defined and $\psi_{\Pi 4}(\pi)=\psi_{\Pi 4}(\pi^R)$.

(3) For each production rule $p=(a, \beta, \psi_{\Pi 4})$, if $\psi_{\Pi 4}(\pi)=(v, c)$ and $v \neq 1$, then there exist ρ and σ such that $\pi=\rho\sigma^R$, $\psi_{\Pi 4}(\rho)=\psi_{\Pi 4}(\rho^R)=(1, c)$ and $\psi_{\Pi 4}(\rho^R\sigma)=\psi_{\Pi 4}(\pi)$ are defined, and σ is a shortest realizable path between node 1 and v in β.

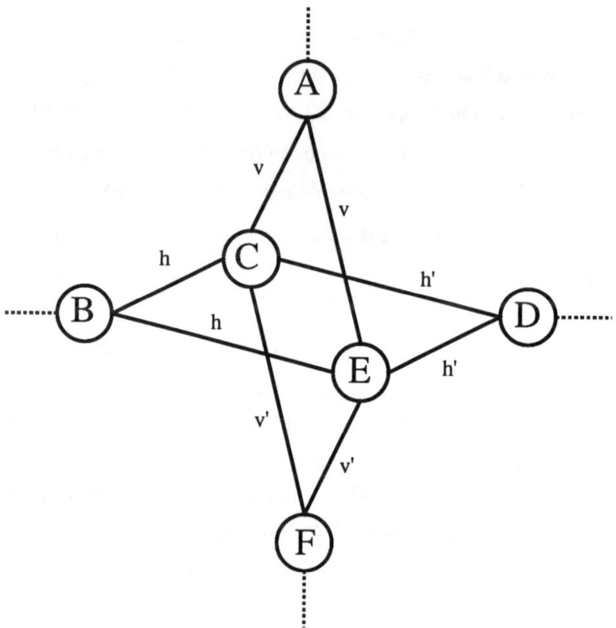

Fig. 3. Nodes with more than one edges having same label.

So we can describe a U-nPCE$_{\Pi 4}$ grammar G as a construction $G = <\Sigma_N, \Sigma_E, P, \psi_{\Pi 4}, Z, \Delta_N, \Delta_E>$, where $\psi_{\Pi 4}= \bigcup_{p_i \in P} \psi_{\Pi 4_i}$ and each production rule has no exclusive embedding function.

3. Confluent derivations on PCE embedding

As defined in the last section, nPCE grammars are context-free in the sense that one node is replaced without considering any other part of the rewrited graph. However, the grammar may still be context-sensitive in the sense that generated graph depends on the order in which the production rules are applied. A graph grammar is said to be *confluent*, if derivation steps on distinct vertices can be done in any order.

Definition 3.1. Let G be an nPCE grammar. G is *confluent*, denoted as C-nPCE grammar, if the following condition holds for every sentential form H of G: Let v_1 and v_2 be distinct nodes of H labelled with nonterminal labels, and let p_1 and p_2 be production rules applicable to v_1 and v_2, respectively. If H_{12} is derived from H by applying p_1 at first then p_2 and H_{21} is derived by applying p_2 at first then p_1, then $H_{12}=H_{21}$.

The class of all graph languages generated by U-nPCE grammars are included in the class of all graph languages generated by C-nPCE grammars.

Theorem 3.1. For any given $nPCE_{\Pi 4}$ grammar G, G is confluent if G has uniform embedding function.

Proof: Let $G=<\Sigma_N, \Sigma_E, P, \psi_{\Pi 4}, Z, \Delta_N, \Delta_E>$ be U-$nPCE_{\Pi 4}$ grammar. For any given sentential form H of G, let v_1 and v_2 be distinct nodes of H labelled with nonterminal labels, and let $p_1=(a_{v_1}, \beta_1)$ and $p_2=(a_{v_2}, \beta_2)$ be production rules applicable to v_1 and v_2, respectively. If v_1 and v_2 are not neighbors of H and any pair of nodes from β_1 and β_2 do not become neighbors by applying both rules, then obviously $H_{12}=H_{21}$. If v_1 and v_2 are neighbors of H and the node 1s of β_1 and β_2 are not neighbors, then at least one of the production rules is not applicable. This is a contradiction. Assume here that v_1 and v_2 are not neighbors in H and become neighbors after applying p_1 and p_2. There exist four possible cases.

Case 1: Node 1s of β_1 and β_2 become neighbors.
From the definition of uniform embedding, there exists a path π from v_1 to v_2 and $\psi_{\Pi 4}(\pi)=\psi_{\Pi 4}(\pi^R)=(1, c)$. Thus obviously $H_{12}=H_{21}$.

Case 2: Node $i\neq 1$ of β_1 and node 1 of β_2 become neighbors.
From the definition of uniform embedding, there exists a path π from v_1 to v_2 such that $\psi_{\Pi 4}(\pi)=(i, c)$ and $\pi=\rho\sigma$ where σ is the path between node 1 to i in β_1 and c is the reduction of ρ. $\psi_{\Pi 4}(\rho^R)=(1, c)$ is also defined. Since c is an element of Δ_E, c is also the reduction of ρ^R. Then $H_{12}=H_{21}$.

Case 3: Node 1 of β_1 and node $i\neq 1$ of β_2 become neighbors.
Same as in the Case 2.

Case 4: Node $i\neq 1$ of β_1 and node $j\neq 1$ of β_2 become neighbors.
In this case, the path σ_i from node 1 to i in β_1 is equal to the path σ_j from 1 to j in β_2 under $\Pi 4$ since c of $\psi_{\Pi 4}(\pi\sigma_j)=(i, c)$ is the reduction of π, and also the reduction of $\sigma_i^R\pi\sigma_j$. Then from the fact $\psi_{\Pi 4}(\pi\sigma_j)=\psi_{\Pi 4}(\pi^R\sigma_j^R)$, $H_{12}=H_{21}$.

The case in which v_1 and v_2 are neighbors in H and are also neighbors after applying p_1 and p_2 is proved in the almost same way as in the above case.

4. Array languages defined by U-nPCE grammars

In this section, we introduce two mappings which map graph languages generated by U-nPCE$_{\Pi 4}$ grammar into the set of array languages. Then, we investigate the generative powers of U-nPCE$_{\Pi 4}$ grammars as the array patterns generators.

Definition 4.1. A mapping k from a graph generated by U-nPCE$_{\Pi 4}$ grammars into two-dimensional arrays is such that

(1) The horizontal neighborhood on array is defined as the set of edges labelled with two colors, say h and h', which do not commute each other.

(2) The vertical neighborhood on array is defined as the set of edges labelled with remaining two colors, say v and v', which also do not commute each other.

(3) The label of each node is the symbol in the corresponding position of array.

(4) For a graph H which has tmore than one node mapped to a position, k(H) is undefined.

The mapping k can be extended to the set of graphs in two different ways.

Definition 4.2. For any given graph language L(G) generated by a U-nPCE$_{\Pi 4}$ grammar,

(1) $K(L(G)) = \begin{cases} K=\{k(g) \mid g \in L(G)\} & \text{if } k(g) \text{ is defined for all elements of } L(G) \\ \text{undefined} & \text{otherwise} \end{cases}$

(2) K'(L(G))={k(g) | g∈L(G) and k(g) is defined}.

Example 4.1. Let L(G) be a graph language consisting of two graphs in Fig. 4a. Then K(L(G)) is undefined and K'(L(G)) is array language consisting of only one array in Fig. 4b.

Both K and K' can be extended to the families of languages in the natural way, i.e., K(X)={K(L(G)) | G is a grammar in X} and K'={K'(L(G)) | G is a grammar in X}.

From the definitions of nPCE$_{\Pi 4}$ grammars and uniform embedding, it is not so difficult to see the following lemma holds:

Lemma 4.1. $\mathcal{F}$(K'(U-nPCE$_{\Pi 4}$)) = $\mathcal{F}$(CFAG).

The same result is obtained in Aizawa and Nakamura [1] but the concept of uniform embedding is not used in it. If all production rules of a U-nPCE$_{\Pi 4}$ grammar is restricted to be *strongly linear*, i.e., the right hand side of each rule has at most one nonterminals and there exists a single-stroke path covering the whole of the right hand side (see Yamamoto, et al. [10]

for more detail definition), denoted as U-nPCE$_{\Pi 4}$-SLAG grammar, then following corollary is obtained:

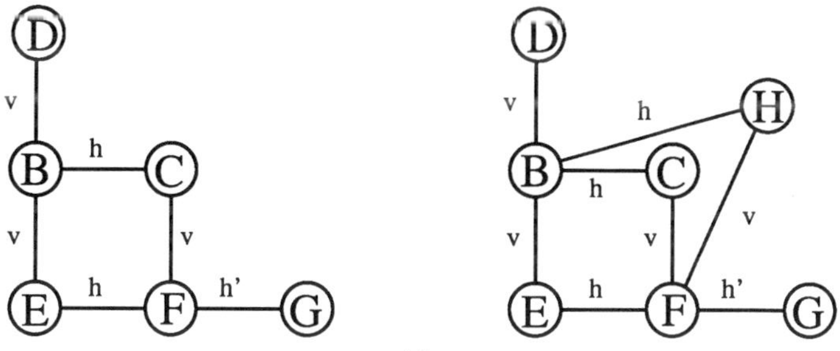

(a)

D
B C
E F G

(b)

Fig. 4. Graphs and their images on array.

Corollary 4.1. $\mathcal{F}$(K'(U-nPCE$_{\Pi 4}$-SLAG)) = $\mathcal{F}$(RAG).

Once we use the mapping K instead of K', the situation is entirely changed. In fact, $\mathcal{F}$(K(U-nPCE$_{\Pi 4}$) is no longer equal to $\mathcal{F}$(CFAG).

Lemma 4.2. $\mathcal{F}$(K(U-nPCE$_{\Pi 4}$)) $\subset$ $\mathcal{F}$(CFAG).
$\mathcal{F}$(K(U-nPCE$_{\Pi 4}$)) is incomparable with $\mathcal{F}$(RAG).

Proof: It is easy to see that $\mathcal{F}$(K(U-nPCE$_{\Pi 4}$) $\subseteq$ $\mathcal{F}$(CFAG). Thus, to prove $\mathcal{F}$(RAG) is not included in $\mathcal{F}$(K(U-nPCE$_{\Pi 4}$)), assume that there exists a U-nPCE$_{\Pi 4}$ grammar G such that K(G) is the set R of all rectangles. As shown in Yamamoto, et al. [10], R is in $\mathcal{F}$(RAG). If such grammar G exist, at least one production rule P is applied to a nonterminal node which is generated from a nonteminal node of the right hand side of P itself. Since otherwise arbitrary large pattern cannot be generated by application of rules whose right hand sides have constant size. If so, we can remove the array pattern generated from the first application of P and then connect the array pattern generated from the second application of P without any shearing effects since K is defined for G. Such a removed array pattern is finitely large unless there exists another repeated rule P'. So G is not the set of all rectangles. If such P' exists, then repeated application process of these rules proceed independently. Again G is not the set of all rectangles. This is a contradiction.

5. Parallelism on PCE embedding

In this section, we extend nPCE grammars to the parallel systems, i.e., graph L-systems or GL-systems. Then the relationships between nPCE GL-systems and two-dimensional array grammars are investigated.

Definition 5.1. An extended *nPCE graph L-system with 4 colors free near-abelian partial path groups*, denoted as EnPCE$_{\Pi 4}$ GL-system, is a construction
$G = <\Sigma_N, \Sigma_E, P, Z, \Delta_N, \Delta_E>$, where $\Sigma_N, \Sigma_E, Z, \Delta_N, \Delta_E$ are same as in the definition of nPCE$_{\Pi 4}$ grammars. P is a finite set of production rules of form $(a, \beta, \psi_{\Pi 4})$, where $\psi_{\Pi 4}$ is a mapping from $\Sigma_E{}^+ \times \Sigma_E{}^*$ into $(V(\beta) \times \Sigma_E)$ provided that if $\psi_{\Pi 4}$ maps $\pi \times \rho$ into (i, c) for some $c \in \Delta_E$, then c is the reduction of $\sigma\pi\rho$, where σ is the path between node 1 to i in β.

The derivation steps of EnPCE$_{\Pi 4}$ GL-system proceed in the almost same way as any other L-systems, i.e., all nodes are rewritten in parallel within a direct derivation step. A direct derivation step in a nPCE$_{\Pi 4}$ GL-system is performed as follows: Let $H = (V, E, \Sigma_V, \Sigma_E, \phi_V, \phi_E)$ be a graph. Let $\{v_1, v_2, ..., v_n\}$ be the set of nodes of H and $p_1, p_2, ..., p_n$ are production rules applicable to $v_1, v_2, ..., v_n$, respectively, where $p_i=(a_i, \beta_i, \psi_i)$, $1 \le i \le n$. Let β_i' be isomorphic to β_i (with h_i an isomorphism from β_i' into β_i), where β_i's have no common nodes. Then the result of the application of p_is to H (by using h_is) is obtained by first removing v_is from H, then replacing v_i with β_i' and finally adding edges (u,w) between every nodes u in β_i' and every w in β_j', $i \ne j$, such that either there exists a path $\pi \times \rho \in \text{domain}(\psi_i)$ with $w=v_i\pi\rho$ in H and $\psi_i(\pi \times \rho)=(u, c)$ for some c in Δ_E or there exist a path $\pi \in \text{domain}(\psi_i)$ and a path ρ from node 1 to w in β_j' with $v_j=v_i\pi$ in H, $w=1\rho$ in β_j', and $\psi_i(\pi \times \rho)=(u, c)$ for some c in Δ_E.

The uniform embedding functions for EnPCE$_{\Pi 4}$ GL-systems can be defined in the same way as nPCE$_{\Pi 4}$ grammars.

Lemma 5.1. $\mathcal{F}$(CFAG)$\subset\mathcal{F}$(K'(U-EnPCE$_{\Pi 4}$ GL))$\subset\mathcal{F}$(K'(U-ETnPCE$_{\Pi 4}$ GL))$\subset\mathcal{F}$(MAG).
Proof: It is obvious that U-EnPCE$_{\Pi 4}$ GL-systems can simulate the derivation steps of CFAGs by adding "do not rewrite" rules, e.g., (a, a). MAGs can simulate the derivation steps of U-ETnPCE$_{\Pi 4}$ GL-systems (and, of course, U-EnPCE$_{\Pi 4}$ GL-systems) by making use of

traversal capability of TBAA. As in the string case, the class of tableless systems is included properly in the class of tabled systems. It is not so difficult to see that $\mathcal{F}$(CFAG) ≠ $\mathcal{F}$(K'(U-EnPCE$_{\Pi 4}$ GL)) since the language $\{a^n b^n \mid n \geq 1\}$ is not an element of $\mathcal{F}$(CFAG), (see, e.g., Nirmal and Krithivasan [7]), on the other hand it is an element of $\mathcal{F}$(K'(U-EnPCE$_{\Pi 4}$ GL)). Finally, the set of all hollow rectangles of thickness 1 over one-letter alphabet is in the class $\mathcal{F}$(MAG)-$\mathcal{F}$(K'(U-ETnPCE$_{\Pi 4}$ GL)).

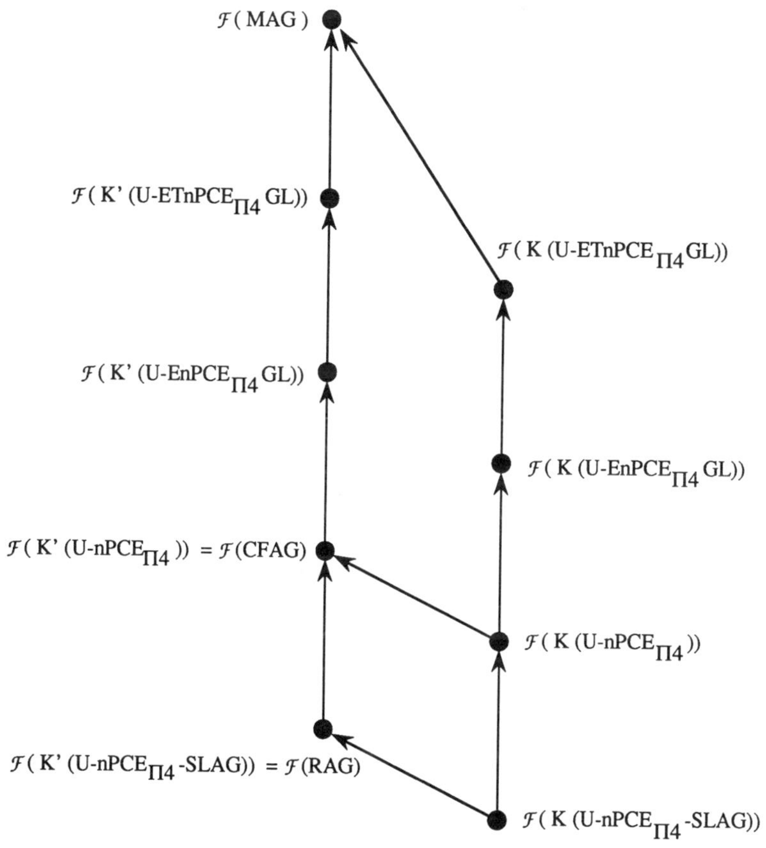

Fig. 5. Hierarchical results.

From the results of Lemmas 4.1, 4.2, and 5.1, it is not so difficult to see the following results.

Lemma 5.2.

$$\mathcal{F}(K(U\text{-}nPCE_{\Pi 4})) \subset \mathcal{F}(K(U\text{-}EnPCE_{\Pi 4}\ GL)) \subset \mathcal{F}(K(U\text{-}ETnPCE_{\Pi 4}\ GL)) \subset \mathcal{F}(MAG)$$

It is not clear that $\mathcal{F}(K(U\text{-}ETnPCE_{\Pi 4}\ GL))$ includes $\mathcal{F}(CFAG)$ or not (but we conjecture strongly that they are incomparable). Then the results of thisand last sections are summarized in the following theorem:

Theorem 5.1. The diagram in Fig. 5 holds.

References

[1] Aizawa, K. and A. Nakamura: Graph grammars with path controlled embedding, *Theoretical Computer Science* (in printing), 1991.

[2] Ehrig, H., M. Nagl, G. Rozenberg and A. Rosenfeld (ed.): *Graph-Grammars and Their Application to Computer Science*, Lecture Notes in Computer Science, 291, Springer-Verlag, Berlin, 1987

[3] Engelfriet, J.: Context-free NCE graph grammars, *Lecture Notes in Computer Science*, 380, pp. 148-161, 1989.

[4] Melter, R. A.: Tessellation graph characterization using rosettas, *Pattern Recognition Letters*, 4, pp. 79-85, 1986.

[5] Nagl, M.: A tutorial and bibliographical survey on graph grammars, *Lecture Notes in Computer Science*, 73, pp. 70-126, 1979.

[6] Nakamura, A. and K. Aizawa: On a relationship between graph L-systems and picture languages, *Theoretical Computer Science*, 24, pp. 161-177, 1983.

[7] Nirmal, N. and K. Krithivasan: Filamentous systems with apical growth, *International Journal of Computer Mathematics*, 12, pp. 203-215, 1983.

[8] Rosenfeld, A.: *Picture Languages*, Academic Press, New York, 1979.

[9] Rosenfeld, A.: Partial path groups and parallel group constructions, in G. Rozenberg and A. Salomaa (ed.), *The Book of L*, Springer-Verlag, Berlin, pp. 369-382, 1986.

[10] Yamamoto, Y., K. Morita and K. Sugata: Context-sensitivity of two-dimensional regular array grammars, *International Journal of Pattern Recognition and Artificial Intelligence*, 3, pp. 295-319, 1989.

MAP L SYSTEMS WITH MULTIPLE MARKERS

Rani Siromoney, K.G. Subramanian And T. Robinson

Department of Mathematics, Madras Christian College, Tambaram, Madras - 600 059, India.

Abstract : A new class of map OL systems with multiple markers is introduced. This class is shown to properly include cyclic controlled and marker controlled, binary, propagating OL systems and map systems with handles. This class will also include map OL systems with marker, if we allow erasing of edges. Thus the new system introduced here includes all the existing map system models. Applications of these systems to fractal images, slicible and non-slicible VLSI floorplan designs and kolam patterns are discussed.

Keywords: Map-generating system, Map L system

1. Introduction

Lindenmayer and Rozenberg [1] are the first to introduce map L systems motivated by biological considertions. Their model which is now referred to as cyclic controlled map OL system [2], allows binary cell division and consists of two phases. In the first phase each oriented edge in the wall or cell boundaries is replaced by a sequence of daughter edges, by edge productions. In the second phase, correct vertices are found for edge insertion in walls and new edges are inserted in existing walls in such a way that no more than two walls are produced from a single wall. This is controlled by the circular word which is the boundary word of a region. Subsequently markers were introduced by Nakamura et al in [4], to indicate edge insertion locations, resulting in marker controlled map systems. In the second phase of these systems, pairs of matching markers within the same region are connected together. A compact survey of map L systems is presented by Lindenmayer[1]

Siromoney et al [7,8] introduced a model using maps with handles for the generation of maps generalizing the ideas in [1], by allowing division of a wall or cell into finitely many cells instead of binary division. Patterns found on the wings of insects can be described by this model.

In this paper, we generalize the notion of markers and allow multiple markers at any location on the boundary of a wall. Again different pairs of matching markers in a region are connected together and this gives rise to a region being split into finitely many regions. The rules of this new system are compact and general enough to include existing types of rules.

All these map systems are propagating in the sense that no edges are erased. Nakamura et al [3] have considered map L systems with markers which are neither binary nor propagating. They have outlined the effect of erasing edge productions and have shown that membership problem is recursively unsolvable. But they have not used the feature of multiple markers.

In section 2, we describe the map OL systems with multiple markers proposed in this paper and compare them with other map systems. We prove that map systems introduced here strictly include the cyclic controlled map OL systems [1], the map OL systems with markers [4] and the map systems with handles [7,8]. If we allow erasing edge productions as in [3], our map systems with multiple markers will include map systems in [3]. In section 3, we discuss the application of the map systems to generation of fractal images, of slicible and non-slicible VLSI floorplan designs and of kolam patterns.

2. Definitions And Comparisons

In this section we recall certain necessary notions. We then define the new map systems and compare them with the existing map systems.

Maps consist of vertices, edges and walls or cells. Maps can be considered as connected planar graphs with directed edges embedded in the plane with all vertices and edges lying on the boundaries of walls.

Map derivations controlled by labels assigned to edges have been investigated in [1,3,4]. We deal with systems in which edges are single labelled. Single labelled, cyclic controlled binary, propagating map OL (BPMOL) systems [1] are referred to as sc- systems; single labeled, marker controlled BPMOL systems [4] are called sm-systems; propagating map systems with handles [7,8] are PMHOL systems.

An edge production rule may determine a transformation of an edge into one or several edges

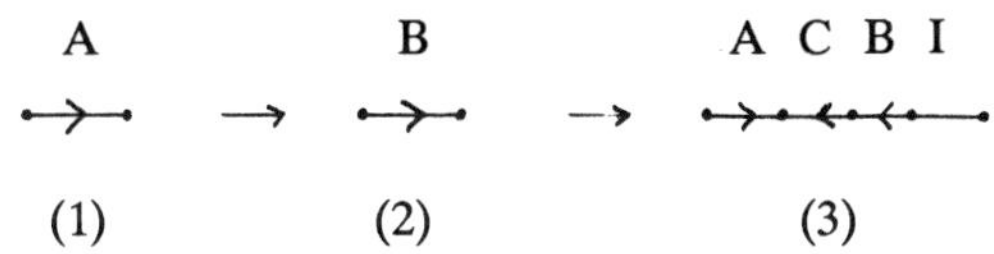

Figure (1) Transformation and subdivision of edges

For instance, transformation from (1) to (2) in figure 1, consists of a change of label from A to B. The corresponding edge production is $A \rightarrow B^{+}$. + indicates that the new edge has the same direction as the mother edge. We use - if the new edge has opposite direction, and no symbol is used if there is no direction at all. The transformation from (2) to (3) results in splitting the edge labelled B into 4 new edges with new lables and directions. The corresponding edge production is

$$B \rightarrow A^{+}\ C^{-}\ B^{-}\ I$$

Furthermore edge productions can contain markers \ and / (left and right slashes) which control insertion of edges into regions. For instance the edge production in figure 2

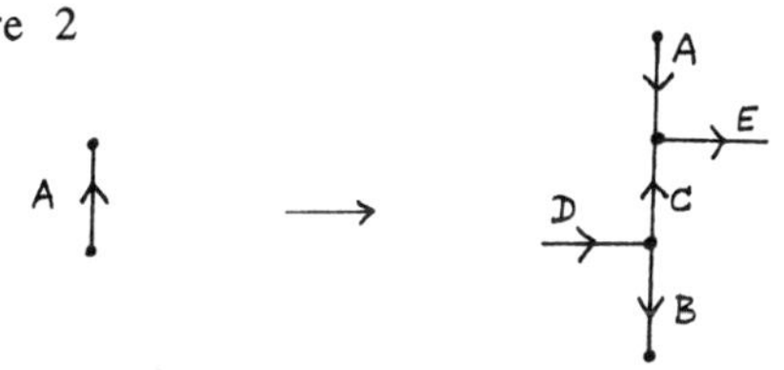

Figure 2 : An edge production with single markers

is written as $A \rightarrow B^{-} \setminus (D^{-})\ C^{+}\ /\ (E^{+})\ A^{-}$

The slashes \ and / indicate the insertion points of new edges to the left or right of the mother edge. Only one label appears within a pair of matching parentheses.

Here we extend the marker notation to the concept of multiple markers. We use multiple left and right slashes such as \\ , /// to indicate insertion of two or more markers to the left or right of the mother edge. The edge labels of multiple markers will be written as a string within matching paranthesis, ( ), with as many labels as the number of slashes preceding the parantheses. We also allow a new labeled edge to be created at a division point, which is then followed by multiple markers as mentioned earlier. The new edge is oriented away or towards the mother edge (indicated by + or - in its lable) (if the new edge has no orientation, no symbol is used) and will be to the left or right of the mother edge (indicated by \ or /). The new edge label for the new edge with markers at its "tip" is enclosed in rectangular parantheses [,]. For instance the edge production in figure 3

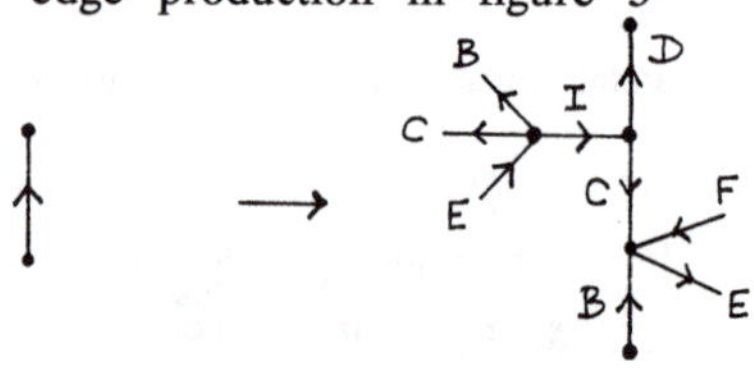

Figure 3 : An edge production with multiple markers

is written as $A \rightarrow B^+ \; // \; (E^+F^-) \; C^- \; \backslash \; [I \; \backslash\backslash\backslash \; (E^- \; C^+ \; B^+)] \; D^+$

We now define map OL system with multiple markers

Definition 1.1. A map OL system with multiple markers is G consisting of

i) a finite alphabet Σ of edge labels,

ii) a finite set P of edge productions

iii) an axiom map ω, and

iv) special symbols : signs + and -; markers \ and / ; opening and closing parentheses (,) and [,].

Each edge production is of the form $A \rightarrow \alpha$, where $A \in \Sigma$ and α is a sequence of elements of Σ, and signs +, - and multiple markers and matched parentheses [,] and (,). An example of an edge production is in figure 3.

In these systems, a derivation step consists of two phases. In the first phase the rewriting of all the edges surrounding a given wall and of finding the multiple markers pointing inwards into the wall are done. In the second phase inserting new edges connecting markers with matching labels is done.

The language generated by a map OL system with multiple markers consists of all maps generated from the axiom map by application of productions of the system in two phases as described earlier.

Example 1.1: Consider the map system G, with multiple markers, whose edge productions are

$A \rightarrow \backslash[I\backslash\backslash(C^{-}A^{+})]D^{+}$

$B \rightarrow \backslash[I\backslash\backslash(A^{-}B^{+})]D^{+}$

$C \rightarrow \backslash[I\backslash\backslash(B^{-}C^{+})]D^{+}$

$D \rightarrow D^{+}$

$I \rightarrow I$ and axiom map is as in figure 4.

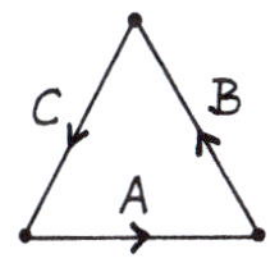

Figure 4.

The system G generates maps of the form shown in figure 5

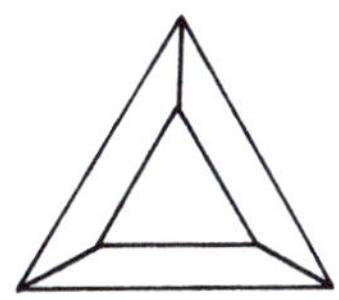

Figure 5.

A derivation step is shown in figure 6.

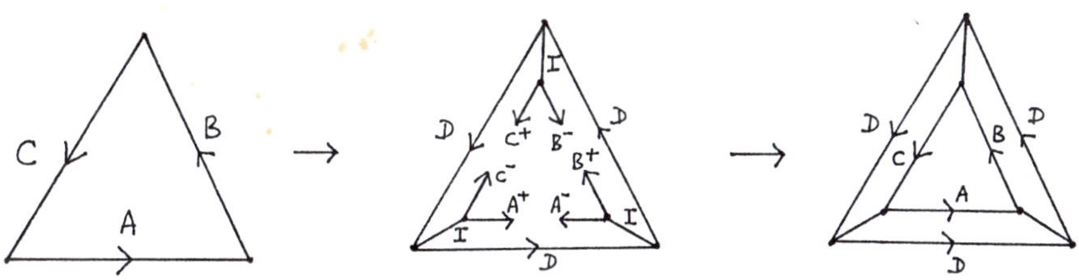

Figure 6 : A derivation step.

We now compare the new map system with the existing models.

Theorem: The class of map sequences generated by Map OL systems with multiple markers includes properly the classes of map sequences generated by sc-systems, sm-systems [2] and propagating map systems with handles [7,8] (when the edge directions are ignored) and includes map OL systems [3] when erasing of edges is allowed.

Proof: It has been proved in [2] that sm-systems properly include sc-systems. Other inclusions follow from definitions, and proper inclusions are seen from the following examples.

Example 2.1: The map OL system, G_1 with multiple markers having the following edge production rules, and axiom ω as in figure 7, generates maps (figure 8) which can not be generated by any sm-system.

The rules are

$$A \rightarrow //\ (A^{-}A^{+})\ B^{+}$$

$$B \rightarrow B^{+}$$

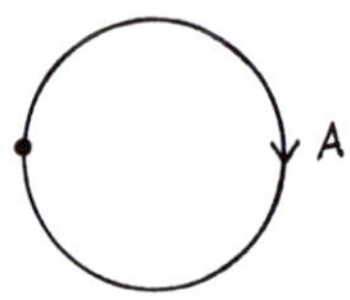

Figure 7 : axiom ω

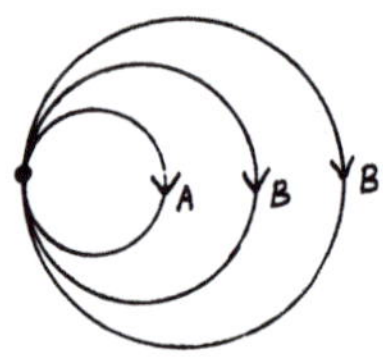

Figure 8 : A map of G_1

Example 2.2: The map OL system G_2 with multiple markers having the following edge production rules and axiom ω, generates maps (figure 10). These maps can not be described by any map system with handles.

The rules are

$$A \rightarrow \backslash\backslash\backslash\backslash(A^{+}A^{-}B^{+}B^{-})D^{+}$$

$$B \rightarrow \backslash\backslash\backslash\backslash(A^{+}A^{-}B^{+}B^{-})D^{+}$$

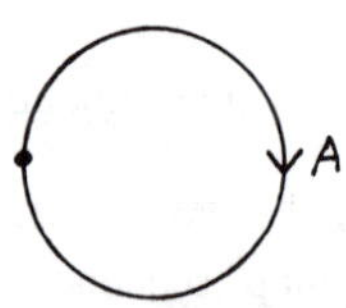

Figure 9: Axiom ω

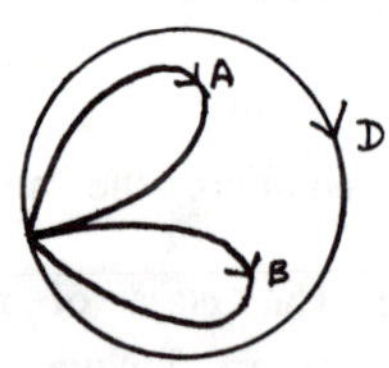

Figure 10 : A map of G_2

3. Applications

In this section we shall discuss the application of map systems to different areas.

Fractal images:

Recently, methods of generation of fractal images have been considered. The map system developed here is found suitable for generating images of fractals.

In the following example the fractals called "Sierpinski triangles" a member of which shown in figure 11 can be generated with edge productions.

$$A \rightarrow A^{+} \;//\; (B^{-}C^{-})\; A^{+}$$

$$B \rightarrow B^{+} \;\backslash\backslash\; (A^{-}C^{+})\; B^{+}$$

$$C \rightarrow C^{+} \;//\; (A^{+}B^{+})\; C^{+}$$

and axiom as in figure 11

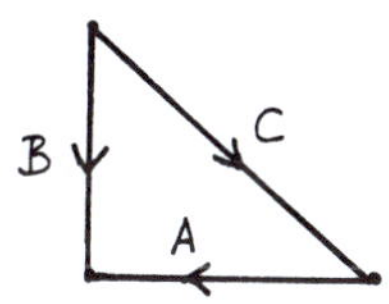

Figure 11

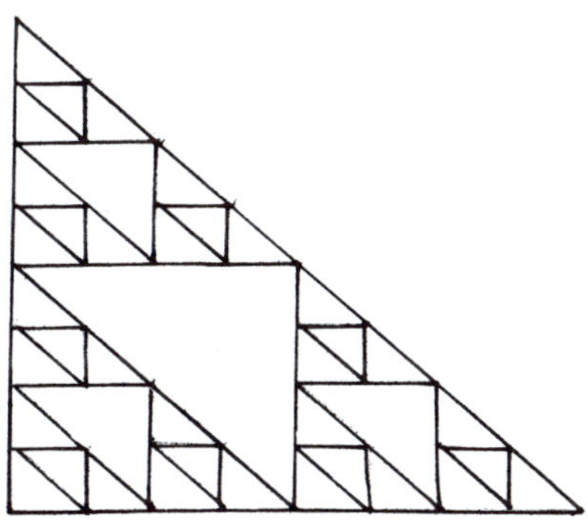

Figure 12 : Sierpinski triangle

Floorplan Designs:

A floorplan is essentially a rectanlge which is a subdivision of an overlapping rectangle by horizontal and vertical line segments into a finite number of indivisible non-overlapping rectangles. A rectangle that is not subdivided further, is called a basic rectangle and such indivisible non-overlapping rectangles correspond to the functional modules in the floorplan.

Floorplans are divided into two kinds; slicible and non-slicible. A floorplan is said to be slicible if it is a rectangular disssection obtained by recursively dividing rectangles into smaller rectangles, by horizontal or vertical lines (called cutlines) until each non-overlapping rectangle is indivisible and corresponds to a module.

Figure 13(a) is slicible whereas Figures 13 (b) and (c) are non- slicible. The standard convention is that two

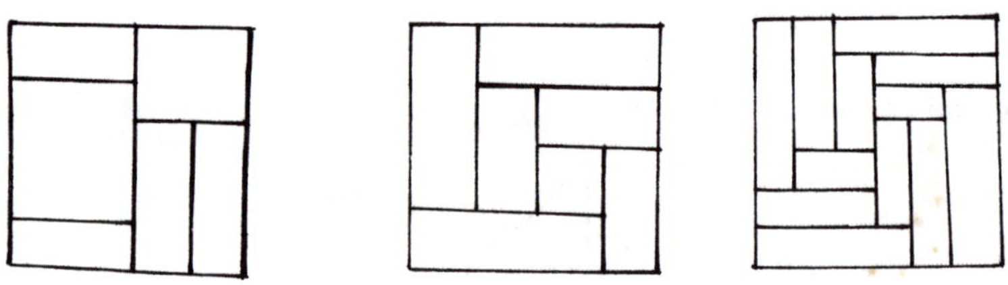

(a) (b) (c)

Figure 13 : Floor plans

perpendicular cutlines can meet but not cross. In otherwords, intersections of cutlines form T-junctions only. We give two examples of map systems that generate slicible and non-slicible floorplans respectively.

Example 3.2 : The set of rules

$$A \to F^{+} \; / \; (E) \; F^{+},$$

$$B \to F^{+} \; / \; (EA') \; B^{+},$$

$$C \to C^{+} \; / \; [D//(A^{+}E)] \; F^{+},$$

$$D \to D,$$

$$F \to F$$

with axiom map given in figure 14 generates the sequence of slicible floorplans shown in figure 15.

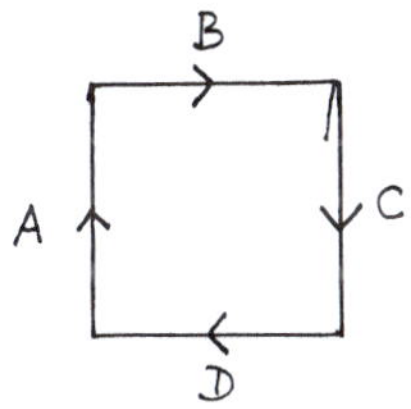

Figure 14

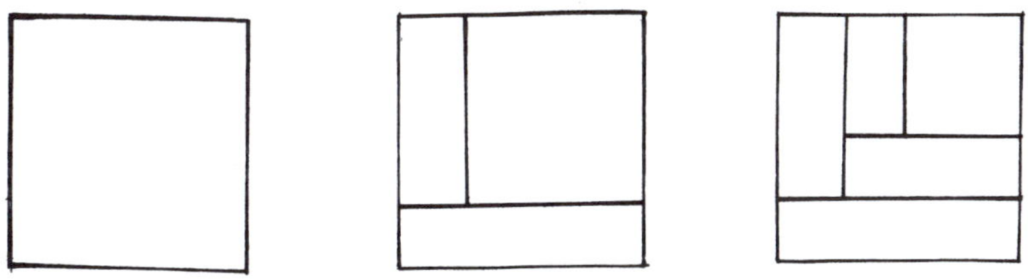

Figure 15 : Slicible floorplans

Example 3.3 : With same axiom map of example 3.2 and the following set of rules

$A \rightarrow E \ / \ [E \ // \ (A^{-}B^{+})]E,$

$B \rightarrow E \ / \ [E \ // \ (B^{-}C^{+})]E$

$C \rightarrow E \ / \ [E \ // \ (C^{-}D^{+})]E,$

$D \rightarrow E \ / \ [E \ // \ (D^{-}A^{+})]E$

$E \rightarrow E,$

the map system generates non-slicible floorplans shown in figure 16.

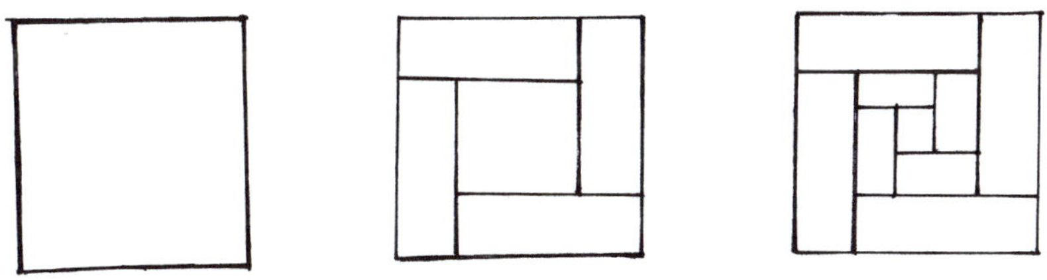

Figure 16 : Non-slicible floorplans

Kolam patterns

Kolam is a traditional art practised extensively in the southern part of India, for decorating courtyards of dwellings. Specific instructions for drawing certain kinds of kolam patterns graphically was given in [6]. A formal model for describing Kolam patterns was proposed in [5]. Generally, Kolam pattern can be divided into three classes : non-recursive designs, recursive designs with an exponential growth, and recursive designs with a polynomial growth. Certain families of Kolam patterns namely the Kambi Kolam are made up of closed curve with or without loops. Methods of generating Kambi Kolams, using turtle moves, chain code interpretation and Kolam moves are examined in [9].

We give an example to generate a Kambi Kolam of polynomial growth by Map OL system with multiple markers.

Example 3.3: The map system with production rules

$$A \rightarrow A,$$

$$B \rightarrow B,$$

$$C \rightarrow B^{+} \backslash\backslash (C^{+}[B\backslash\backslash\backslash(C^{-}A^{+}A^{-})])B^{+}$$

with axiom map shown in figure 17, generates a sequence of Kambi Kolams described in figure 18.

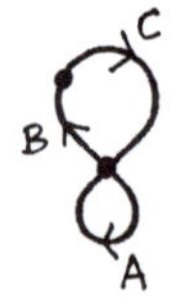

Figure 17 :

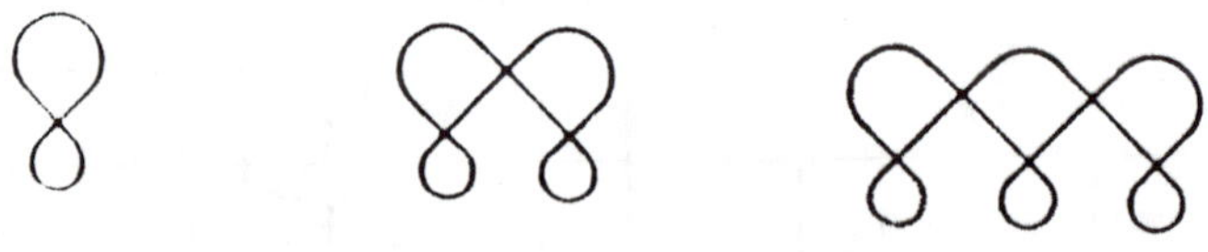

Figure 18 : Sequence of Kambi Kolam

References

1. Lindenmayer. A and Rozenberg. G, Parallel generation of map : Developmental systems for cell layers. In V. Claus, H. Ehrig, and G. Rozenberg, editors, Graph grammars and their application to computer sicence; First International Workshop, pp. 301-316, Lecture Notes in Computer Science vol.73, Springer-Verlag, 1979.

2. Martin J.M. de. Boer, Lindenmayer. A, Map OL-systems with edge label control : comparison of marker and cyclic systems. Editors : H. Ehrig, M. Nagl, G. Rozenberg, and A. Rosenfeld, Graph - Grammars and their application to computer sicence; Third International Workshop pp.378-392, Lecture Notes in Computer Science Vol.291, Springer-Verlag, 1986.

3. Nakamura A., Lindenmayer, A and Aizwa. K, Map OL systems with markers Editors : H. Ehrig, M. Nagl, G. Rozenberg, and A. Rosenfeld, Graph Grammars and their application to computer science, Third International Workshop pp.479-495. Lecture Notes in Computer Science Vo291, Springer-Verlag 1986.

4. Nakamura, A., Lindenmayer A and Aizwa, K. Some systems for map generation. In "The Book of L", Editors : G. Rozenberg and A. Salomaa, pp 323-332, Springer-verlag, 1986.

5. Prusinkiewicz, P., Graphical Applications of L-systems proceedings of Graphics Interface 86 - vision interface 86, pp.247-253, 1986.

6. Siromoney G., Siromomey, R and Krithivasan, K., Array grammars and Kolam, Computer Graphics and Image Processing, 3, pp 63-82, 1974.

7. Siromoney, G. Subramanian K.G and Siromoney, R. A map generating system, Proceedings of the Third conference on Foundations of Software Technology and Theoretical Computer Science, pp 145-170, 1983.

8. Siromoney, G. Subramanian, K.G. and Siromoney R, Map systems with handles, In "A perspective in Theoretical Computer Science"; commemorative volume for Gift Siromoney, Editor R. Narasimhan, pp. 324-352 Series in Computer Science - Vol.16. World Scientific, 1989.

9. Siromoney G, Siromoney R., and Robinson, T, Kambi Kolam and Cycle Grammars, In, "A perspective in Theoretical Computer Science"; commemorative volume for Gift Siromoney, Editor R. Narasimhan, pp 267-300 Series in Computer Science - vol 16. World Scientific, 1989.

5. Enjoying Flora Along the Path:
Biological Aspects and Models

Locally Generated Colourings of Hexagonal Cell Division Patterns: Application to Retinal Cell Differentiation

Zsolt Tuza[1] and Aristid Lindenmayer[2]

[1]Computer and Automation Institute, Hungarian Academy of Sciences, H–1111 Budapest, Kende u. 13-17, Hungary
[2]Theoretical Biology Group, State University of Utrecht, Utrecht 3584 CH, Padualaan 8, The Netherlands

Abstract. We investigate map D0L-systems which are models of developmental systems within the framework of formal language theory. The main result proves that D0L-systems are appropriate models not only for generating the division process of cells but also for governing their differentiation. For this purpose, we assign a "colour" to each cell type, and then generate colouring patterns for the clones obtained after any number of division steps. Our method provides an arbitrarily selected periodic colour arrangement in every step, hence it satisfies the requirement of size-independence that is essential when applications in biology are considered.

A particular case of the general model is then applied to retinal cell differentiation. With as few as nine cell labels (cellular states) and just one edge label we generate a pattern of cell types that agrees with the one in vertebrate retinal receptors, as observed experimentally by Morris [J. Comp. Neurol. 140 (1970) 359-387], within a few per cent of accuracy.

We also provide an infinite family of D0L-systems supporting Cheung's conjecture, and raise some related open problems.

Keywords: developmental system, parallel rewriting, map D0L-system, parallel graph grammar, colouring, cell division, cell differentiation, retinal receptor.

1. Introduction

Pattern formation in growing cell layers is a widely investigated problem of developmental biology. In some epidermal layers two differentiated cell types form the pattern in such a way that one of the types occurs in delimited areas (spots or stripes), surrounded by cells of the other type. In other cases, particularly in the receptor cell layer of the retina, several differentiated cell types are uniformly distributed in the cell layer, i.e., cells of the same type do not occur as neighbours. This latter kind of pattern formation is related to the problem of properly colouring the regions of a map but not necessarily with a minimum number of colours. Such a pattern generating process should be size invariant, i.e., it should be possible for the process to stop at an arbitrary size.

The control of cell division orientation has been considered previously but not how the distribution of differentiated cells is controlled. It has been shown that the regular orientation of newly inserted cell walls in consecutive cell generations can be programmed by the uniform application of local rules. These rules are based either on the temporary states of the cells, or on markers on the walls (edges) between them, or on both. The question we are asking in this paper is whether by such local rules it is also possible to program the proper positioning of differentiated cells in a dividing layer of cells. We present a mathematical framework (that of map D0L-systems) within which local rules can be formulated and applied. We demonstrate that with such constructs not only wall orientation patterns but also differentiation patterns can be produced. Models are considered for the distiribution of receptor cells in the retina of the vertebrate eye.

2. Map Generating Systems

Cellular layers in living tissues can be represented by a set of touching polygons in the plane. Division patterns in such layers arise by insertion of new edges in some of the polygons. In recent years we have attempted to describe cell division patterns in terms of the theory of automata and formal languages. Constructs such as parallel graph grammars and map grammars have been proposed for this purpose in several subsequent papers, including [1, 9, 11, 14, 3, 6]. Operations have been defined on planar graphs or maps such that, at each application of the grammar, an increasing number of nodes or regions (cells) is produced. For instance, patterns are known which produce increasingly larger arrays of hexagons from previous arrays of hexagons [8]. Cell division patterns in plant epidermal layers have been modelled by map systems with double edge labelling, too [10].

Below, first we describe marker-controlled map grammars and then in Section 2.2 we modify them in order to obtain a similar but more powerful type of grammars. Instead of giving a formal definition, we provide here an informal description of those map generating systems and give a demonstration of their operation with reference to two particular examples.

2.1. Marker Systems with Edge Labels

Map generating systems with markers were introduced by Nakamura, Lindenmayer, and Aizawa [13]. Maps are considered to be planar graphs embedded in the plane. To each edge of the graph we assign a label (from a finite set Σ of symbols) and orientation (one of the two directions along the edge). Edges are rewritten according to production rules. If all production rules in a system G are context-free (they depend only on the label of the edge itself), then G is called a *map 0L-system.* In such a system, production rules are of the form $a \rightarrow w$, where a is an element of Σ and w is composed of elements of Σ with orientation signs " + " and " – ", and of markers $\downarrow$, $\uparrow$ with elements of Σ and signs in parentheses. For example, $a \rightarrow b^+ c^+ \downarrow(d^-) e^-$ is such a rule.

The rewriting of an edge consists of the splitting of the edge into as many new edges as there are elements of Σ in w without markers (in the example, there are three such elements: b, c, and e). The elements of Σ are assigned to the new edges in the order as they appear in w, taking into account the orientation of the edge being rewritten. If an element has sign " + " then the orientation of the corresponding new edge is identical to that of the original edge, and if its sign is " – " then the new orientation is opposite.

Markers indicate the places where division edges are to be inserted. (Division edges are those edges which divide an existing cell into two.) Up- or down-pointing markers indicate whether a division edge is to be inserted to the left or to the right of the descendants of the edge being rewritten, as seen according to the orientation of this edge. Division edges can be spanned between two markers pointing into the same region and having the same element of Σ with opposite signs. This element becomes the label of the division edge, whose orientation is determined by the relative positions of the + and – signs of the element (the new edge is oriented from + to –).

In one generation step, first all edges are rewritten according to the production rules, and then all possible division edges are inserted according to matching markers. Map generation can be either deterministic or non-deterministic (for instance in the case where more than two matching markers are present in a cell). In our paper we consider deterministic systems only; they will be called *map D0L-systems.*

The following two examples illustrate map D0L-systems. Both of them are such that in every generation step all regions (cells) of the map divide, all regions are hexagons, and all vertices (except those on the environmental boundary) have degree 3. The first system has the following rules:

$$a \rightarrow c^- \downarrow(a^+)\, b^- \uparrow(a^-)\, c^-$$
$$b \rightarrow a^+$$
$$c \rightarrow b^+$$

Assuming that the starting map consists of a single cell with a sequence of consecutive edges $a^+ b^+ c^+ a^- b^- c^-$ (as seen in the clockwise direction), the first three generation steps are shown in Fig. 1.

This division pattern was discovered in plant epidermal layers by Korn and Spalding [4] and its production rules were found by Lindenmayer and Rozenberg [9]. We note that it can also be generated by other map D0L-systems with three letters, or using more than three letters.

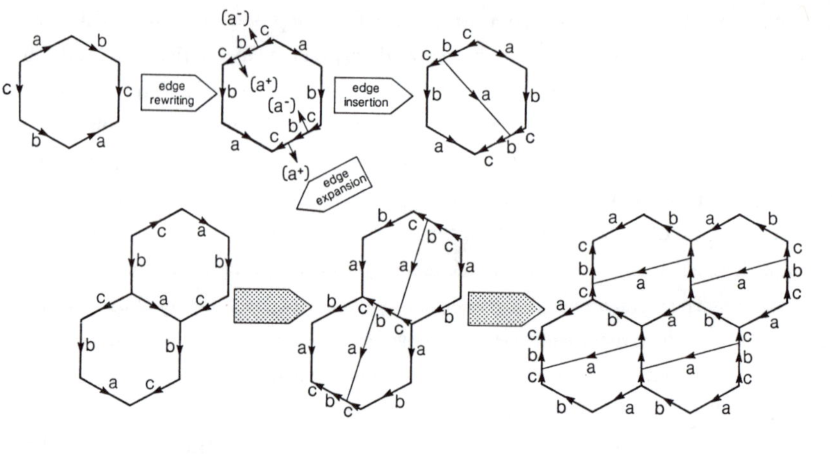

Figure 1. The first three generation steps of the Korn–Spalding pattern; the first step is shown in three phases.

Another system with hexagonal cells all of which divide in every step was originally proposed by Lewis [5] in the 1920s as a possible division pattern for the retina. Production rules for this system were first presented in [9]. One of the possibilities for rewriting is

$$
\begin{aligned}
a &\rightarrow b^{+} \\
b &\rightarrow c^{+} \uparrow (b^{+})\, a^{-} \\
c &\rightarrow e^{-} \downarrow (d^{-})\, f^{-} \\
d &\rightarrow a^{-} \uparrow (d^{+})\, e^{-} \\
e &\rightarrow f^{-} \downarrow (b^{-})\, c^{+} \\
f &\rightarrow d^{+}
\end{aligned}
$$

The starting map can be a region whose circular edge sequence is either $a^{+}\,b^{+}\,c^{+}\,a^{-}\,e^{-}\,d^{-}$ or $f^{+}\,e^{+}\,d^{+}\,f^{-}\,b^{-}\,c^{-}$.

From a biological point of view, the following difference between the Korn–Spalding and Lewis patterns is essential. Whichever generation is considered, the former system yields a cellular array (clone) with smooth boundaries, while the latter produces more and more branches which are attached to the rest of the clone by just one cell. We demonstrate this important fact in Fig. 2, where the structure of the sixth clone of the Lewis pattern is compared with the seventh clone of the Korn–Spalding pattern. Arrows point to a "bridge," i.e., an edge that completely separates two parts of the configuration. The existence of bridges shows that if a convex surface is covered by a cell layer then it is not very likely that those cells have been generated in accordance with the production rules of the Lewis pattern. Therefore, from now on we deal with the Korn–Spalding pattern. It has a more compact structure without any bridges, with a similar behaviour in every cell, and growing at the same rate in each direction.

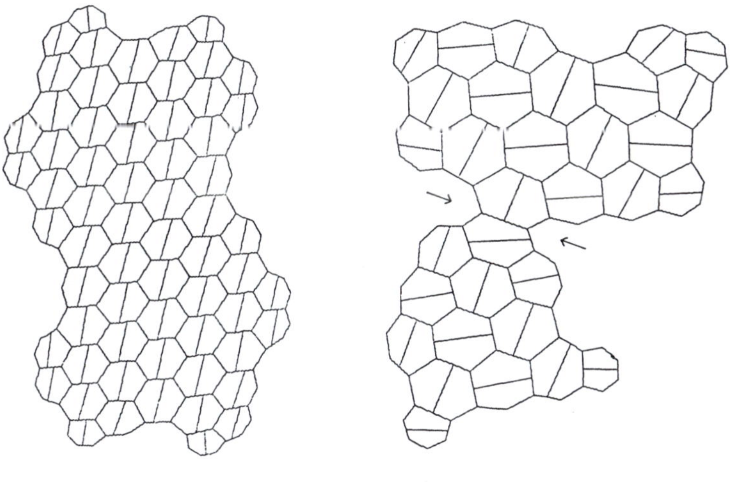

(a) (b)

Figure 2. Two cell clones with distinct division patterns. (a) The Korn–Spalding pattern. (b) The Lewis pattern with a "bridge." Computer generated grawings by Mark de Does.

2.2. Marker Systems with Region Labels

In this section we introduce a modified version of marker systems, in which distinct cell types are represented by *region labels*. We denote by Δ the (finite, non-empty) set of region labels, and by Σ the set of edge labels.

In edge- and region-labelled map D0L-systems the production rules determine the types of daughter cells after cell division. The left-hand side of each production rule is a triplet (A, a, B), where $a \in \Sigma$ is an edge label; A and B (elements of Δ) denote the region labels of the two cells incident to the edge labelled a. (Here A stands for the cell to the left when seen from the starting point of the corresponding edge.) For example, if a cell division is of the form

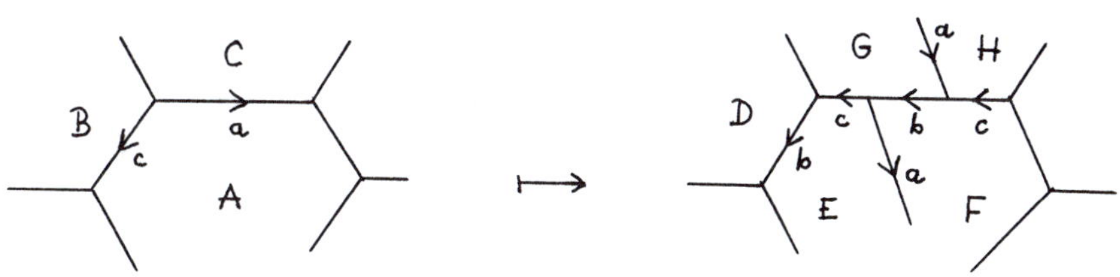

then the "non-splitting" edge c appears in the production rule

$(A, c, B) \ \rightarrow\ (E, b^+, D)$

while the "splitting" edge a is rewritten as

$(C, a, A) \ \rightarrow\ (G, c^-, E) \downarrow (F, a^+, E)\ (G, b^-, F) \uparrow (G, a^-, H)\ (H, c^-, F)$.

The daughter cells appear in these productions in the left-right order as they are seen from the points splitting the corresponding edge. Since the region labels are redundantly specified in a production rule of the above form, we shall write the same rule more concisely as

$(C, a, A) \ \rightarrow\ c^- \downarrow (F, a^+, E)\ b^- \uparrow (G, a^-, H)\ c^-$.

The convention of using markers and signs is the same as in marker systems described above.

It may be pointed out that these edge- and region-labelled map 0L-systems are duals of certain types of propagating graph 0L-systems (PG0L-systems) introduced in [1] and further investigated in [2, 7]. The duals of map regions are nodes in graphs, and the duals of map edges are edges in graphs. Then, for each of the maps produced by a region-labelled system there is a dual graph produced by the corresponding PG0L-system. The converse is true only if the graph system is planar and uniquely embeddable in the plane [6]. It is easy to see that for each production rule of a map 0L-system there can be found a graph rewriting rule of the corresponding PG0L-system, and vice versa. For instance, in the previous example of edge- and region-labelled map production rules the following graph rules can be constructed (by adopting a convention for the edge orientations of maps and graphs as shown in the figure below, and by indicating the source (s) and target (t) nodes with reference to the two nodes in the left-hand sides of the productions). The graph rewriting rules, in turn, can be reformulated to obtain the node productions and the edge productions of a PG0L-system.

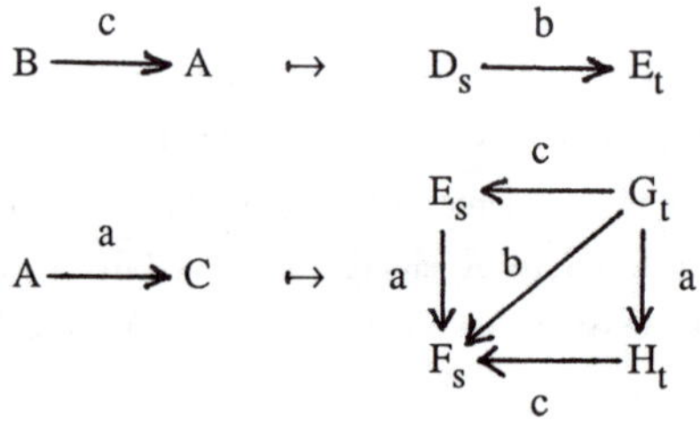

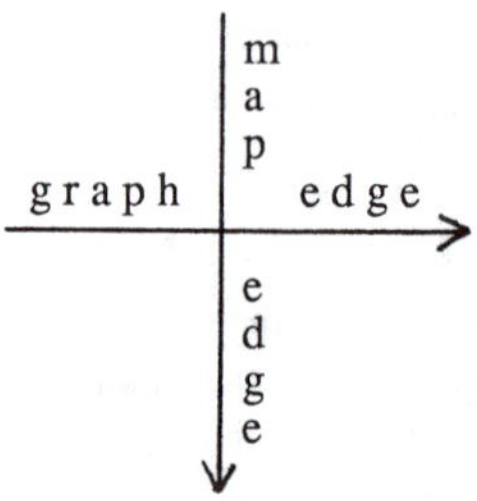

By virtue of the relationship between our map systems and the previously investigated graph systems, many of the results on the latter can now be transferred to the former.

Let us recall that in their original form, the production rules $a' \rightarrow w$ are composed of triplets (and markers), where each triplet consists of two region labels and one edge label. Thus, relabelling the edges by triplets, the system obtained is equvalent to a map system without region labels. As a consequence, the next statement follows.

Proposition 1. If M is an edge- and region-labelled map D0L-system with p edge labels and q region labels, then M is equivalent to an edge-labelled map D0L-system with at most pq^2 edge labels. []

As a matter of fact, the upper bound of pq^2 corresponds to a "worst case" and in the systems to be constructed it can be improved considerably. (Sometimes just pq edge labels will be needed.)

3. Colourings of the Korn–Spalding Pattern

3.1. Coordinate Systems for Hexagonal Patterns

Representing a hexagonal array of cells by regular hexagons (with equal sides and angles), we can associate coordinates to each hexagon in two standard ways, as described in [8]. We consider the geometric center p(C) of cell C and take either three coordinate axes x, y, z mutually forming an angle of 120° or two axes x, y of degree 120°, as shown in Fig. 3.

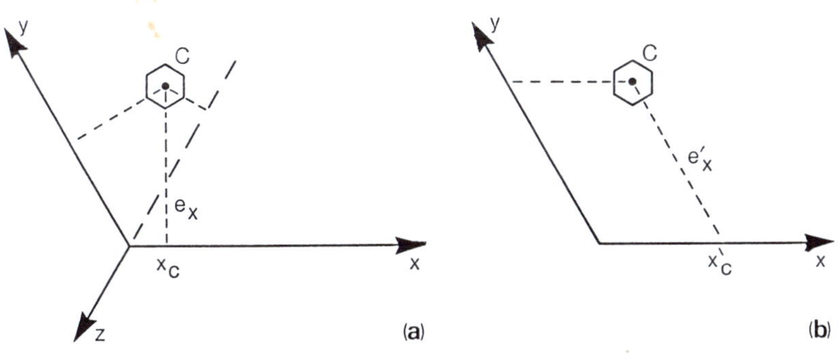

Figure 3. Two ways of defining coordinates of a cell C, by triplets (a) and pairs (b). Then $\langle C\rangle = \langle x_C, x'_C\rangle$.

In the first case, the coordinate x_C is determined by a straight line e_x passing through p(C) and being *perpendicular* to the *x*-axis, while in the second case, the line e'_x passes through p(C) and is *parallel* to the *y*-axis. Now we represent each cell C by the ordered pair

$\langle C \rangle := \langle x_C, x'_C \rangle$ (where the origin and the two axes x, y are in the same position when defining x_C and x'_C). In order to obtain integer coordinates, it is assumed that the centers p(C) form a triangular lattice of side length 2, and that the origin is the center of some cell C_0 (which, consequently, corresponds to the pair $\langle 0, 0 \rangle$). The coordinates $\langle C \rangle$ of some cells are exhibited in Fig. 4(a).

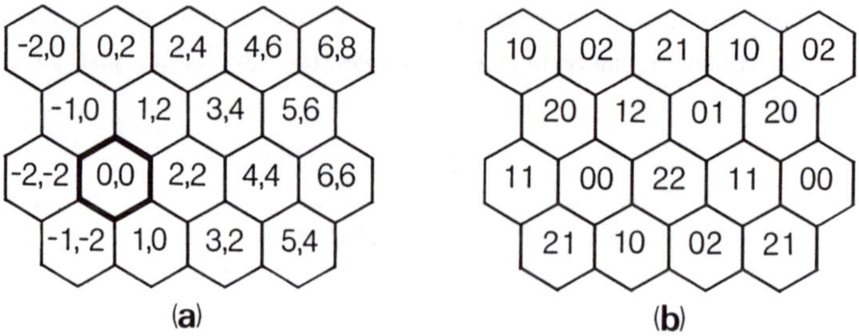

Figure 4. (a) The ordered pair $\langle C \rangle$ for some cells, the origin with bold edges. (b) The two coordinates of $\langle C \rangle$ modulo 3, as they appear in the 3^*-colouring f_3.

Clearly, the coordinate representation $\langle C \rangle$ of each cell C is unique, and $\langle C \rangle \neq \langle C' \rangle$ whenever the cells C and C′ are distinct. Moreover, a pair $\langle p, q \rangle$ can represent a cell if and only if p is an integer and q is an *even* number. Because of this one-to-one correspondence, we need not distinguish between C and $\langle C \rangle$. Thus, we can also handle cells as if they were 2-dimensional vectors. For example, defining $\mathbf{x} = \langle 2, 2 \rangle$ and $\mathbf{y} = \langle -1, 0 \rangle$, we have $\langle a, b \rangle + t\mathbf{x} = \langle a + 2t, b + 2t \rangle$ and $\langle a, b \rangle + t\mathbf{y} = \langle a - t, b \rangle$ for all integers a, b, t.

As we proved in [8], each self-generating hexagonal division pattern (i.e., a pattern producing hexagons from hexagons, in which all vertices have degree 3) defines a coordinate transformation corresponding to the images of the daughter cells. In other words, if daughter cells after a division step are again represented by regular hexagons, then their coordinates are uniquely determined (apart from rotations, translations, and reflections of the plane). One of the possible conventions is to assume that the bottom vertex of cell $\langle 0, 0 \rangle$ is the bottom vertex of the daughter cell with coordinates $\langle 0, 0 \rangle$, too. For any cell C in a hexagonal array, denote by $D_1(C)$ the daughter cell of C that contains the bottom vertex of C, and let $D_2(C)$ denote the other daughter cell of C.

Considering now the Korn–Spalding pattern, we say that we are in step 1 if splitting edges are perpendicular to the y-axis; step i + 1 (i = 1, 2) is obtained from step i by a rotation of 60° clockwise. We denote by **C** the array of regular hexagons covering the plane, and by $\mathbf{C}_i$ (i = 1, 2, 3) the hexagonal array obtained when step i is applied to **C**. From now on, all results are for the Korn–Spalding pattern, even if we do not mention it in the formulation of each statement. When applied with pairs of coordinates, the prime notation $\langle a, b\rangle'$ will always refer to daughter cells.

Lemma 2. If $D_1\langle 0, 0\rangle = \langle 0, 0\rangle'$ and the bottom vertex of $\langle 0, 0\rangle$ is identical to that of $\langle 0, 0\rangle'$, then

in step 1, $D_1\langle a, b\rangle = \langle a + b/2, 2b\rangle'$ and $D_2\langle a, b\rangle = \langle a + b/2 + 1, 2b + 2\rangle'$;

in step 2, $D_1\langle a, b\rangle = \langle 3a - b, 2a\rangle'$ and $D_2\langle a, b\rangle = \langle 3a - b - 2, 2a - 2\rangle'$;

in step 3, $D_1\langle a, b\rangle = \langle 2a - b, b\rangle'$ and $D_2\langle a, b\rangle = \langle 2a - b - 1, b\rangle'$.

Sketch of Proof. The statement is obvious for the cell $\langle 0, 0\rangle$. Moreover, it holds for $C + \mathbf{x}$ and $C + \mathbf{y}$ whenever it holds for a cell C. Since every cell can be written in the form $u\mathbf{x} + v\mathbf{y}$, an inductive argument can be applied to $|u| + |v|$, starting with $u = v = 0$. []

Note that in steps 1, 2, and 3 we have $D_2(C) = D_1(C) + \mathbf{x} + \mathbf{y}$, $D_2(C) = D_1(C) - \mathbf{x}$, and $D_2(C) = D_1(C) + \mathbf{y}$, respectively, for each cell C.

3.2. Periodic Colourings

Suppose that a colouring f is given on $\mathbf{C}$, i.e., a colour $f(C)$ is assigned to each mother cell $C \in \mathbf{C}$. We say that f is *t-periodic* $(t > 0)$ if $f(C + t\mathbf{x}) = f(C + t\mathbf{y}) = f(C)$ for every cell C. Let us view f as a function from the set of ordered pairs of coordinates to the set of colours. Then f assigns a colour to each daughter cell as well. We define the *division colouring* f' as $f'(C) = (f(D_1(C)), f(D_2(C)))$ (i.e., every cell is assigned to the pair of colours of its daughter cells). Note that steps 1, 2, and 3 define distinct division colourings (see Fig. 5). Each of them, however, satisfies the following property.

Proposition 3. If a colouring f in a Korn–Spalding pattern is t-periodic, then its division colouring f' is t-periodic, too.

Proof. Since the difference vector $D_2(C) - D_1(C)$ does not depend on the choice of the cell C, it is enough to prove that f' is t-periodic on the daughter cells $D_1(C)$, $C \in \mathbf{C}$. Moreover, the vectors $D_1(C + t\mathbf{x}) - D_1(C)$ and $D_1(C + \mathbf{y}) - D_1(C)$ are constant for $C \in \mathbf{C}$, i.e., the statement can be reduced to $f(D_1(t\mathbf{x})) = f(D_1(t\mathbf{y})) = f(\langle 0, 0\rangle)$. This equality follows from the t-periodicity of f by Lemma 2 in each step, since $t\mathbf{x} = \langle 2t, 2t\rangle$ and $t\mathbf{y} = \langle -t, 0\rangle$. []

A colouring f of $\mathbf{C}$ is called a *proper colouring* if $f(C) \neq f(C')$ holds for any two cells $C, C' \in \mathbf{C}$ sharing an edge.

Lemma 4. For every $t \geq 2$, in a Korn–Spalding pattern, there is a t-periodic proper colouring f of C with at most t^2 colours, with the property that for all $C, C' \in \mathbf{C}$, $f(C) = f(C')$ implies $f'(C) = f'(C')$.

Proof. Assign a distinct colour to each of the t^2 cells $u\mathbf{x} + v\mathbf{y}$, $0 \leq u, v \leq t-1$, and extend this partial colouring to a t-periodic colouring f of **C**. The extension is unique: $f(C) = f(C')$ for two cells $C = u\mathbf{x} + v\mathbf{y}$ and $C' = u'\mathbf{x} + v'\mathbf{y}$ if and only if $u \equiv u' \pmod t$ and $v \equiv v' \pmod t$. Since f' is t-periodic by Proposition 3 and $f(C) = f(C')$ implies $C' = C + ti\mathbf{x} + tj\mathbf{y}$ for some integers i and j, it follows that $f'(C) = f'(C')$ whenever $f(C) = f(C')$. []

The t-periodic colouring constructed in the previous proof will be used frequently, so that it will be worth giving a name to it.

Definition. For any natural number t, the t^*-*colouring* is the colouring pattern f_t in which two cells $C = u\mathbf{x} + v\mathbf{y}$ and $C' = u'\mathbf{x} + v'\mathbf{y}$ have the same colour if and only if $u \equiv u' \pmod t$ and $v \equiv v' \pmod t$. In particular, the 3^*-colouring f_3 is defined as $f_3(\langle a, b\rangle) = (a_0, b_0)$, where $a_0 \equiv a \pmod 3$, $b_0 \equiv b \pmod 3$, and $0 \leq a_0, b_0 \leq 2$ (see Fig. 4(b)). Then a_0 and b_0 are called the *components* of $f_3(\langle a, b\rangle)$.

Remark. Every t-periodic colouring can be obtained by defining it on the t^2 cells $u\mathbf{x} + v\mathbf{y}$, $0 \leq u, v \leq t-1$, because then the colour of every cell in **C** is uniquely determined. Thus, every t-periodic colouring can be obtained from the t^*-colouring, by taking the unions of some of its t^2 colour classes. Moreover, the colouring $f_{3,1} : \langle a, b\rangle \rightarrow a_0$, defined as the first component of f_3, is a 3-periodic *proper* colouring of **C** with three colours. (The division colouring of the latter, however, does not satisfy the requirements of Lemma 4.)

As we have mentioned above, each of the three possible directions of edges dividing mother cells into daughter cells (i.e., each of steps 1, 2, and 3) defines a distinct division colouring of f_3 (shown in Fig. 5). An essential simplification can be achieved by modifying the coordinate system in each step, as described in the next section.

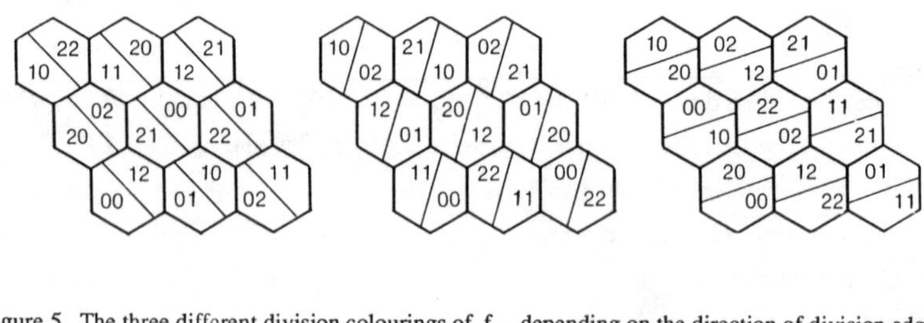

Figure 5. The three different division colourings of f_3, depending on the direction of division edges.

3.3. Rotated Cooridnate System

It is not necessary to use the same coordinate system in every step. Since in the Korn–Spalding system the division edges are rotated by 60° clockwise in each step, we might find it convenient to rotate the coordinate system in the same way. Clearly, this would result in a change of the division colouring f'.

According to the remark at the end of Section 3.2, the problem of generating a given periodic colouring pattern is reduced to that of the t^*-colouring. Therefore, we have to find $f_t(C)$, i.e., we need to know the " mod t " position of cell C relatively to ⟨0, 0⟩. For this purpose, our aim is to modify the production rules of the Korn–Spalding pattern in such a way that edge labels or region labels provide sufficient information concerning colours of cells. In other words, the colour of a cell C should be uniquely determined, either by the labels of edges surrounding C, or by the region label of C itself.

Let f^+ be the colouring obtained from f_t by a clockwise rotation of 60° around the geometric center of ⟨0, 0⟩. Now we introduce f'^+, the *rotated division colouring* of f :

$f'^+(C) = (f^+(D_1(C)), f^+(D_2(C))$.

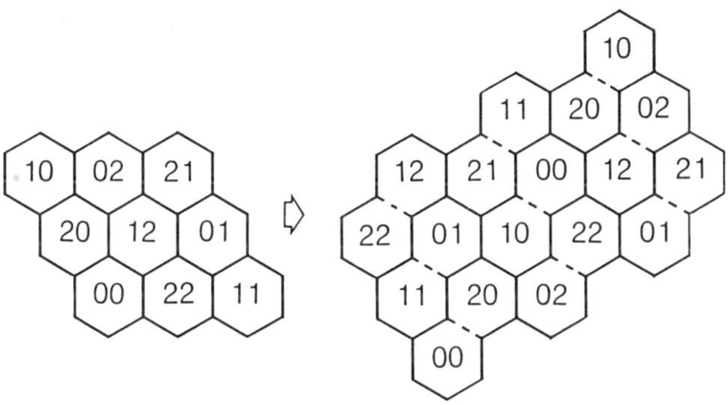

Figure 6. The rotated division colouring f'^+_3.

This concept is illustrated in Fig. 6. The following property of f'^+ is essential. Since a division step of the Korn–Spalding pattern yields a 60° clockwise rotation of the sequence $a^+ b^+ c^+ a^- b^- c^-$ in each cell, the same rotation of **C** around ⟨0, 0⟩ is an isomorphism between **C** and the array of daughter cells in a strong sense: it preserves all colours, edge labels, and edge orientations. Thus, we obtain

Lemma 5. The rotated division colouring f'^{+} does not depend on the number of generation steps. []

Now we are ready to prove the main result of this section.

Theorem 6. A t-periodic colouring $(t \geq 2)$ which, apart from rotation, is isomorphic to t^*, can be generated by a region-labelled system with one edge label, t^2 region labels, and $3t^2$ production rules.

Proof. Consider the infinite hexagonal array **C** with its t-periodic t^*-colouring. Label and orient all edges of **C** in accordance with the first cell of Fig. 1, i.e., two edges have the same label and orientation if and only if they are geometrically parallel. In each cell, the six edge labels follow each other in the order $a^+ b^+ c^+ a^- b^- c^-$ in a clockwise direction, and this property holds in every generation step. Moreover, each edge e belongs to two cells, one of which contains e in positive orientation, the other in negative orientation.

Label each cell C by its colour $f_t(C)$. We assume that the colouring scheme of the daughter cells is obtained by a rotation of 60° from the previous colouring. In order to find production rules for edges labelled a, let A and A′ be the labels of the two regions containing such an edge. (The pair A, A′ may be distinct for distinct edges labelled a.) Suppose that the daughters of the cells labelled A and A′ have colours A_1, A_2 and A_1', A_2', respectively, in the rotated t^*-colouring. By Lemmas 4 and 5, the A_i and A_i' $(i = 1, 2)$ are determined by A and A′. Thus, the first rule of the Korn–Spalding system can be written in the form

$$(A, a, A') \rightarrow c^- \downarrow(A_1', a^+, A_2')\, b^- \uparrow(A_1, a^-, A_2)\, c^-.$$

For edges labelled b or c, the production rules are

$$(B, b, B') \rightarrow (D, a^+, D') \quad \text{and} \quad (C, c, C') \rightarrow (E, b^+, E'),$$

where D, D′, E, E′ are determined by B, B′, C, C′ (according to Lemmas 4 and 5).

By the definition of f'^{+}, and by the fact that rotation maps each colour class onto some colour class, the arrangement of cell labels is isomorphic to the t^*-colouring in every generation step. Since f'^{+} does not depend on the number of steps, we do not need more than t^2 production rules for the splitting edges in the region-labelled system. Each of the edge labels a, b, c can appear in just t^2 distinct combinations with pairs of cell labels, because the label of the cell to the left of the edge determines the label of the cell to the right. Hence, each of a, b, and c appears on the left-hand side of t^2 production rules in the corresponding edge- and region-labelled system. Thus, altogether $3t^2$ production rules are needed. Finally, we observe that for $t \geq 2$, distinct pairs of cell labels always occur together with the same edge label (either a or b or c). This means that a single label can be substituted for a, b, and c. []

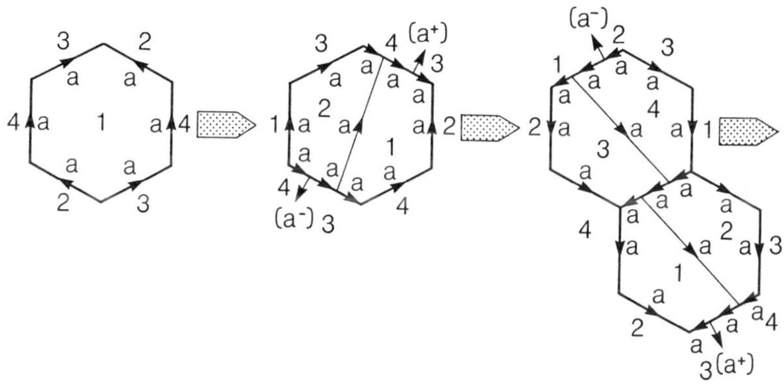

Figure 7. Proper colouring with 4 colours, as generated by the system $M(2^*)$.

Remark. If, for a certain application, the t^*-colouring itself should be generated, rather than some of its rotations, then this can also be done by a map D0L-system. In this case the division colouring f' should be applied, instead of f'^+. One can observe that after six division steps the same type of division colouring and the same edge orientation is obtained. This results in a map D0L-system with $18t^2$ edge labels (or with $6t^2$ region labels). For $t = 3$ it means 162 edge labels, while by the application of f'^+ this number is reduced to $3t^2 = 27$.

We illustrate the above ideas by the particular cases $t = 2, 3$. Let $M(t^*)$ denote the region-labelled system generating the rotated t^*-colouring, as constructed in the proof of Theorem 6.

Example 6.1. For $t = 2$, the system $M(2^*)$ has $2^2 = 4$ cell labels (1, 2, 3, 4), a single edge label (a), and 12 production rules as follows:

$$(2, a, 1) \rightarrow a^- \downarrow(2, a^+, 1)\, a^- \uparrow(3, a^-, 4)\, a^-$$
$$(3, a, 4) \rightarrow a^- \downarrow(2, a^+, 1)\, a^- \uparrow(3, a^-, 4)\, a^-$$
$$(4, a, 3) \rightarrow a^- \downarrow(4, a^+, 3)\, a^- \uparrow(1, a^-, 2)\, a^-$$
$$(1, a, 2) \rightarrow a^- \downarrow(4, a^+, 3)\, a^- \uparrow(1, a^-, 2)\, a^-$$

$$(2, a, 3) \rightarrow (3, a^+, 4) \qquad (1, a, 4) \rightarrow (1, a^+, 2)$$
$$(3, a, 2) \rightarrow (3, a^+, 4) \qquad (4, a, 1) \rightarrow (1, a^+, 2)$$
$$(1, a, 3) \rightarrow (1, a^+, 4) \qquad (3, a, 1) \rightarrow (3, a^+, 2)$$
$$(4, a, 2) \rightarrow (1, a^+, 4) \qquad (2, a, 4) \rightarrow (3, a^+, 2)$$

The first three clones of the resulting coloured map sequence are shown in Fig. 7. We have also indicated the colours of neighbour cells of those clones when they are embedded into the 2^*-coloured **C**.

Example 6.2. For $t = 3$, we have either an edge-labelled system of 27 edge labels, or a cell-labelled system with 9 cell labels. Here we show the latter. We use an edge label a and cell labels ij, $0 \le i, j \le 2$.

Let the cells of **C** be labelled according to the pattern f_3 as in Fig. 4(b). Then the rotated division colouring $f'^+ = f'^+{}_3$ is $f'^+(00) = (00, 20)$, $f'^+(01) = (12, 02)$, $f'^+(02) = (21, 11)$, $f'^+(10) = (22, 12)$, $f'^+(11) = (01, 21)$, $f'^+(12) = (10, 00)$, $f'^+(20) = (11, 01)$, $f'^+(21) = (20, 10)$, $f'^+(22) = (02, 22)$, as it can be seen from Fig. 7. Hence, the cell-labelled system $M(3^*)$ has the following production rules:

$$(20, a, 00) \to a^- \downarrow(20, a^+, 00)\, a^- \uparrow(11, a^-, 01)\, a^-$$
$$(10, a, 20) \to a^- \downarrow(01, a^+, 11)\, a^- \uparrow(22, a^-, 12)\, a^-$$
$$(00, a, 10) \to a^- \downarrow(12, a^+, 22)\, a^- \uparrow(00, a^-, 20)\, a^-$$
$$(12, a, 22) \to a^- \downarrow(22, a^+, 02)\, a^- \uparrow(10, a^-, 00)\, a^-$$
$$(02, a, 12) \to a^- \downarrow(00, a^+, 10)\, a^- \uparrow(21, a^-, 11)\, a^-$$
$$(22, a, 02) \to a^- \downarrow(11, a^+, 21)\, a^- \uparrow(02, a^-, 22)\, a^-$$
$$(01, a, 11) \to a^- \downarrow(21, a^+, 01)\, a^- \uparrow(12, a^-, 02)\, a^-$$
$$(21, a, 01) \to a^- \downarrow(02, a^+, 12)\, a^- \uparrow(20, a^-, 10)\, a^-$$
$$(11, a, 21) \to a^- \downarrow(10, a^+, 20)\, a^- \uparrow(01, a^-, 21)\, a^-$$

$$(01, a, 22) \to (12, a^+, 22) \qquad (01, a, 12) \to (12, a^+, 00)$$
$$(10, a, 01) \to (22, a^+, 02) \qquad (20, a, 01) \to (11, a^+, 02)$$
$$(22, a, 10) \to (02, a^+, 12) \qquad (02, a, 10) \to (21, a^+, 12)$$
$$(11, a, 02) \to (01, a^+, 11) \qquad (10, a, 21) \to (22, a^+, 10)$$
$$(20, a, 11) \to (11, a^+, 21) \qquad (11, a, 22) \to (01, a^+, 22)$$
$$(02, a, 20) \to (21, a^+, 01) \qquad (00, a, 11) \to (00, a^+, 21)$$
$$(00, a, 21) \to (00, a^+, 10) \qquad (12, a, 20) \to (10, a^+, 01)$$
$$(12, a, 00) \to (10, a^+, 20) \qquad (21, a, 02) \to (20, a^+, 11)$$
$$(21, a, 12) \to (20, a^+, 00) \qquad (22, a, 00) \to (02, a^+, 20)$$

This system generates a sequence of coloured maps whose colouring is isomorphic to $f'^+{}_3$ in every step.

Remark. Though the system $M(3^*)$ is presented with oriented edges (a^+ and a^-), the orientation signs are not needed. They are needed, however, for the system $M(2^*)$ with four cell labels.

The previous two constructions and, in general, Theorem 6 demonstrate the first non-trivial examples supporting the conjecture of Cheung (mentioned by Culik and Wood [2]) concerning graph generating systems (PG0L-systems). The conjecture states that node labelling (that corresponds to region labelling in the case of maps) is sufficient to generate every PG0L-language, whenever the degrees of vertices are bounded. This is the first time that the conjecture is supported by an infinite family of non-trivial examples. Now, in analogy to Cheung's conjecture for PG0L-systems, we can pose the following open problem.

Problem 7. Suppose that a map sequence M is generated by an edge- and region-labelled D0L-system. Can M be generated by such a system with only region labelling, i.e., with a single edge label, provided that every cell has a bounded number of edges?

The colouring pattern generated by $M(3^*)$ contains various pairs of colour classes whose regions do not share an edge, so that this colouring provides patterns of proper colourings with less than 9 colours as well. Some of these possibilities will be discussed below. Now the following related question arises.

Problem 8. Find the number $n(t, k)$ of t-periodic proper colourings of **C** with k colours, $t \geq 2$, $t^2 \geq k \geq 3$. (Two colourings are considered to be identical if they can be obtained from each other by renumbering — but not identifying — colour classes.)

A similar question can be raised for arbitrary colourings. A further variant occurs when we do not distinguish between two colourings if they can be obtained from each other by some automorphism of **C** and renumbering of colour classes.

The values of the $n(t, k)$ are related to some frequently studied combinatorial functions on partitions, such as e.g. Stirling numbers, but there is no immediate correspondence between them, for several reasons. First, the restriction that we do not consider non-proper colourings means that some partitions of the t^2 colours are excluded. On the other hand, even if all types of colourings (not necessarily proper ones) are allowed, not all partitions of the t^2 colours will yield non-isomorphic colourings, since periodicity results in isomorphism in some cases. Finally, not all partitions of the *number* t^2 can be verified as proper colourings. For example, in the case $t = 3$, one cannot define a proper colouring with 4 colours by partitioning the 9 colours into sets of respective cardinalities 3, 2, 2, 2; similarly, the (2, 2, 2, 2, 1)-partition does not correspond to any proper 5-colouring. (Each of the other 10 partitions of the 9 colours into classes of cardinalities ≤ 3 can be verified as a 3-periodic proper colouring.)

4. Applications

An interesting biological problem is how the different cell types arise properly placed and in proper ratios in the receptor layer of the retina of the vertebrate eye. The type and arrangement of receptor cells in the retina of chicken has been described and analysed by Morris [12]. She found the receptor cell layer to be a more or less regular hexagonal array of cells with the arrangement of the five main types being quite uniform. For example, in samples of peripheral retina, the percentage of rods, double cones, and single cones of types 1, 2, 3 are about 33, 30, 23.5, 9, 4.5 per cent, respectively. Roughly speaking, we can consider a distribution such that 1/3 of the cells are rods, 1/3 are double cones, and 1/3 are single cones within which type-1 cells are nearly twice as frequent as types 2 and 3 together. Furthermore, there is a highly regular distribution of rods forming a triangular lattice, the double cones are arranged between them in their own triangular lattice, and a third lattice is formed by single cones (see [12, Fig. 6]). This arrangement of cells can be generated by the 3^*-colouring f_3, described in Example 6.2, as follows. The colouring pattern f_3 gives a proper 9-colouring of regular hexagonal arrays by colours ij, $0 \le i, j \le 2$. The main cell type is determined by the first component i:

$i = 0 \quad \rightarrow \quad$ rod

$i = 1 \quad \rightarrow \quad$ single cone

$i = 2 \quad \rightarrow \quad$ double cone

According to our remark on $f_{3,1}$ at the end of Section 3.2, each of these three classes of cells forms a triangular lattice (since $f_{3,1}$ is a proper colouring). Among single cones, type 1 can be distinguished by the second component j:

$i = 1, \ j = 0$ or $1 \quad \rightarrow \quad$ single cone type 1

$i = 1, \ j = 2 \quad \rightarrow \quad$ single cone type 2 or 3

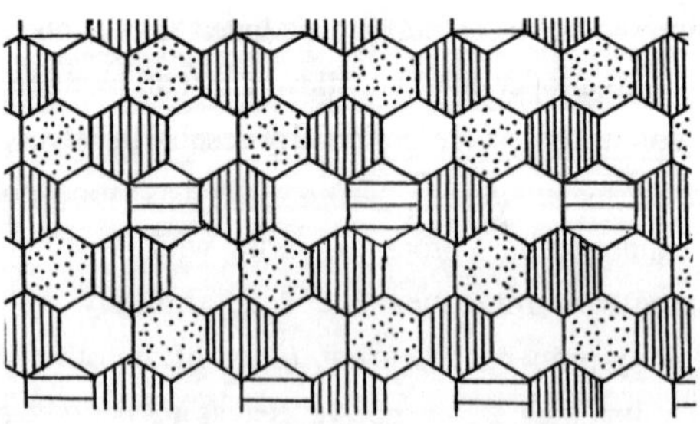

Figure 8. The 3-periodic colouring pattern derived from f_3 by identifying colours 00, 01, 02 with rods (vertical shading), colours 20, 21, 22 with double cones (white), colours 10, 11 with single cones of type 1 (dotted), and colour 12 with single cones of types 2 and 3 (horizontal shading).

Since any refinement of a proper colouring is again a proper colouring, each of the four cell types is uniformly distributed with ratios $3 : 3 : 2 : 1$. The colouring pattern obtained is shown in Fig. 8. (While there are several ways to partition 9 colours into 4 cell types with proportions $1/3 : 1/3 : 2/9 : 1/9$, not all of them would yield a proper colouring. One might ask why we did not use the simpler 2-periodic 2^*-colouring — Example 6.1 — for generating the four cell types. The reason is that in this case the proportion of the distinct cell types would be the same — i.e., $1 : 1 : 1 : 1$ — which does not agree with observed ratios.

When the map generating system $M(3^*)$ is used, cell differentiation is governed by not more than 9 cellular states. Empirical support for this mechanism would have to come from the experimental identification of different components which determine those states of cells.

In a further refinement of this model, we could distinguish between types 2 and 3 of single cones, either by taking a system $M(t^*)$ with some larger t or by introducing a stochastic model where each cell type occurs with a given probability. In the main types (rods and double cones), however, such a stochastic model cannot be applied because rods — as well as double cones — form a regular triangular lattice that can never be the case in a random system.

The mechanism we have proposed here for the differentiation and arrangement of receptor cell types has the important advantage that it is a size-independent pattern generating system. This follows from the fact that after an arbitrary number of division steps, the colour classes of f_3 are uniformly distributed. Thus, starting from a single cell, we can allow cell divisions to produce an arbitrarily large number of cells, at which point cell division stops and differentiation occurs, yielding the required arrangement of cell types.

Beside the development of the retina, there are several other cell layers where various cell types are produced in a fixed ratio and in a regular (non-random) arrangement. On a surface, all such arrangements are approximately hexagonal. It is possible to extend our method to produce colouring patterns in hexagonal arrays for an arbitrary number of cell types and for arbitrary (rational) proportions of them. The generating system $M(t^*)$ can distinguish between t^2 colours or their various partitions. The advantage of this construction is that the control factors operating on the edges remain constant, while the number of differentiation factors involving genes or cytoplasmic components increases. In this way the information necessary for differentiation of various cell types is contained within each cell, in agreement with the prevalent view of differentiation and gene expression. The factors responsible for the directions of cell divisions are, on the other hand, attached to the edges (the edges have polarity). This type of models corresponds to developmental mechanisms controlled by cell lineages. The other main type is that of interactive models in which cell-to-cell interactions play a role. It depends on experimental evidence which mechanism is responsible for a particular kind of development.

References

1. Culik II, K., and Lindenmayer, A.: Parallel graph generating and graph recurrence systems for multicellular development. Intern. J. General Systems 3, 53–66 (1976)
2. Culik II, K., and Wood, D.: A mathematical investigation of propagating graph 0L-systems. Information and Control 43, 50–82 (1979)
3. de Does, M., and A. Lindenmayer, A.: Algorithms for the generation and drawing of maps representing cell clones. In: Graph Grammars and Their Applications to Computer Science (H. Ehrig, M. Nagl and G. Rozenberg, eds.). Lecture Notes in Computer Science, Vol. 153, pp. 39–57. Springer 1983
4. Korn, R.W., and Spalding, R.M.: The geometry of plant epidermal cells. New Phytologist 72, 1357–1365 (1973)
5. Lewis, F.T.: The effect of cell division on the shape and size of hexagonal cells. Anatomical Record 33, 331–355 (1926)
6. Lindenmayer, A.: An introduction to parallel map generating systems. In: Graph Grammars and Their Applications to Computer Science (H. Ehrig, M. Nagl, A. Rosenfeld and G. Rozenberg, eds.). Lecture Notes in Computer Science, Vol. 291, pp. 27–40. Springer 1987
7. Lindenmayer, A., and Culik II, K.: Growing cellular systems: Generation of graphs by parallel rewriting. Intern. J. General Systems 5, 45–55 (1979)
8. Lindenmayer, A., Malitz, J., and Tuza, Zs.: Self-generating hexagonal cell division patterns. Geometriae Dedicata 34, 19–34 (1990)
9. Lindenmayer, A., and Rozenberg, G.: Parallel generation of maps: Developmental systems for cell layers. In: Graph Grammars and Their Applications to Computer Science (V. Claus, H. Ehrig and G. Rozenberg, eds.). Lecture Notes in Computer Science, Vol. 73, pp. 301–316. Springer 1979
10. Lück, J., Lindenmayer, A., and Lück, H.B.: Models for cell tetrads and clones in meristematic cell layers. Botanical Gazette 149, 127–141 (1988)
11. Lück, J., and Lück, H.B.: Proposition d'une typologie de l'organisation cellulaire des tissus vegetaux. In: Actes du 1er Seminaire de l'Ecole de Biologie Theorique du C.N.R.S. (H. Le Guyader and Th. Moulin, eds.). pp. 335–371. Paris: Ecole Normale Sup. Techn. Avan. 1981
12. Morris, V.B.: Symmetry in the receptor mosaic demonstrated in the chick from the frequencies, spacing and arrangement of the types of retinal receptor. J. Comp. Neurol. 140, 359–387 (1970)
13. Nakamura, A., Lindenmayer, A., and Aizawa, K.: Some systems with map generation. In: The Book of L (G. Rozenberg and A. Salomaa, eds.), pp. 323–332. Berlin: Springer 1986
14. Sierro, P.L.J., Rozenberg, G., and Lindenmayer, A.: Cell division patterns: Syntactical description and implementation. Conputer Graphics and Image Processing 18, 329–346 (1982)

A model for cellular development in morphogenetic fields

Martin J. M. de Boer, F. David Fracchia and Przemyslaw Prusinkiewicz

Department of Computer Science, University of Regina, Regina, Saskatchewan, Canada S4S 0A2

Abstract: This paper introduces *cell systems* as a new framework for the simulation of pattern and shape formation in two-dimensional cellular structures. Cell systems describe development in terms of lineage rules, with the orientations of division walls specified relative to a directional morphogenetic field. The shapes of cells result from mechanical cell interactions due to osmotic pressure and wall tension. Cell systems offer a highly intuitive method for describing the development of cellular structures, and capture their natural variability in a straightforward way. The concept is illustrated using abstract patterns and a developmental model of a fern gametophyte.

Keywords: developmental models in biology, computer simulation, cell system, map L-system, cell division pattern, morphogenetic field, fern gametophyte

1 Introduction

An important issue in developmental biology is the exploration of *morphogenesis*, or mechanisms responsible for the development of complex forms found in living organisms. Cellular structures in plants are determined by the underlying *cell division patterns*, that is, the spatial and temporal distributions of cell divisions. Their analysis reveals the recursive nature of many cell lineages [15].

Map L-systems were proposed by Lindenmayer and Rozenberg [14] as a formal description of developmental sequences of cell patterns. They operate on graphs with cycles, called *maps* [19], with regions representing cells, and edges representing cell walls. The topology of a generated structure is determined by production rules operating on the walls. The cells may grow, pushing their neighbors until an equilibrium state is reached [5, 8, 18]. This allows for the investigation of the relationship between the cell division pattern and the overall shape of the structure [4]. Unfortunately, since observations usually capture the behavior of cells rather than walls, they must be translated to wall productions using a laborious, nonintuitive procedure.

In order to describe the development of fern gametophytes in a more straightforward way, de Boer and de Does introduced the notion of *cell division systems* [4]. They employ two types of rewriting rules,

$$A \to \frac{B}{C} \qquad \text{and} \qquad A \to B \mid C,$$

to represent the division of cell A into daughter cells B and C. The horizontal and vertical bars indicate the approximate orientation of the division wall, which can be parallel (*periclinal*) or perpendicular (*anticlinal*) to the closest section of the apical front of the thallus. Cell division systems specify patterns at a sufficient level of detail for growth analysis, but abstract from the precise positions of division walls needed to simulate development.

This paper introduces *cell systems* as a new framework for the simulation of pattern and shape formation in two-dimensional structures. They can be viewed as an extension and a formalization of cell division systems, allowing for a precise specification of the orientation of division walls. This is accomplished by referring to a *vector field*, regarded as a *morphogenetic field* that provides directional information to the cells. Vector fields may be specified independently of the structure or be controlled locally by the structure itself. Examples include light direction, gravity, and concentration gradient of a morphogen. As in map L-systems, the geometry of the generated structures is determined by the mechanical interaction of cells due to cell pressure and wall tension.

This paper focuses on the general concept of context-free cell systems, while leaving the detailed characterization and mathematical analysis for future study. Section 2 provides the definition of cell systems, introduces a notation for rule specification, and presents simple examples that explain their operation. As some cell systems are equivalent to known map L-systems, the natural question of the relationship between both formalisms is addressed in Section 3. Section 4 presents the construction of a cell system for the development of the fern gametophyte *Microsorium linguaeforme*. Finally, Section 5 evaluates the inherent properties of cell systems from a biological perspective, and proposes future extensions.

2 Context-free cell systems

This section formalizes the notion of cell systems as rewriting systems operating on cellular structures.

Definition 2.1. A *cellular structure* is a finite set of cells, each characterized by its *shape* and *state*. The cells are bounded, simple, non-overlapping polygons that tile a surface, thus forming a *polygon mesh* [7, page 473].

The discussion is limited to two-dimensional *cellular layers*, in which cells are characterized by their outlines, as seen under a microscope. The shapes of cells result from the action of mechanical forces: internal cell pressure and tension of cell walls. In multicellular structures, neighboring cells exercise pressure on both sides of the shared walls, thus pushing each other until a global equilibrium is reached. Details of this model are given below.

Definition 2.2. The *pressure-tension model* of physical cell structure consists of the following postulates:

- the modeled structure is represented as a single layer of cells,
- the layer is represented as a two-dimensional network of masses, corresponding to cell vertices, connected by springs that correspond to cell walls,

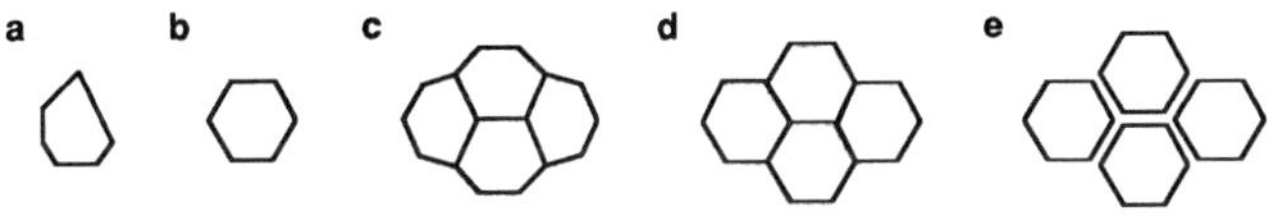

Figure 1: The operation of the pressure-tension model. Description in text

- the springs are always straight and obey Hooke's law, $F = -k(x - x_0)$,
- each cell exerts pressure on its bounding walls directly proportional to the wall length and inversely proportional to the cell area,
- the pressure on a wall is divided evenly between its vertices,
- the motion of the masses is damped, and
- no other forces (for example, due to friction or gravity) are considered.

The shape of a structure defined using the pressure-tension model can be found by integrating the equations of motion of the masses [8, 18].

Example 2.1. Figure 1 illustrates the operation of the pressure-tension model. The shape (b) results from the initial shape (a), provided that all rest lengths x_0 and all spring constants k are equal. Identical springs are also assumed in structure (c), while (d) has internal walls two times stronger than the external walls, $k_i = 2k_e$. The latter case can be conceptualized as a juxtaposition of separate cells (e), with pairs of adjacent walls fused into single walls. This interpretation explains the regular shape of cells in (d).

Cell division is the engine of the development of cellular structures. The positions of the division walls are specified using the assumption that the development takes place in a vector field.

Definition 2.3. A vector field $\mathcal{Z}$ over a surface X is an assignment of a vector $\vec{Z}$ to each point of X.

Sample vector fields are shown in Figure 2. In the case of a uniform field (a), the reference vectors $\vec{Z}$ point in the same direction for all points in the field. Case (b) is similar, except that the vectors $\vec{Z}$ change direction between the developmental steps. In case (c), the reference vectors indicate the direction towards the closest point on the boundary of the structure, thus the field lines are mutually orthogonal to the family of curves equidistant from the border. In case (d), the reference vectors are specified during the process of cell division as normal to the division wall for each of the daughter cells. This vector field is frequently used throughout this paper and is referred to as the *polarity vector field.*

A multitude of other fields can be found easily, hence their classification is needed. The fields that do not change over time are referred to as *stationary* (a). The complementary class of *non-stationary* fields contains those that do not depend on the embedded structure (b), those that depend on global properties of the structure (c) and those that depend on its local properties (d). Any non-stationary field can be considered formally as a function

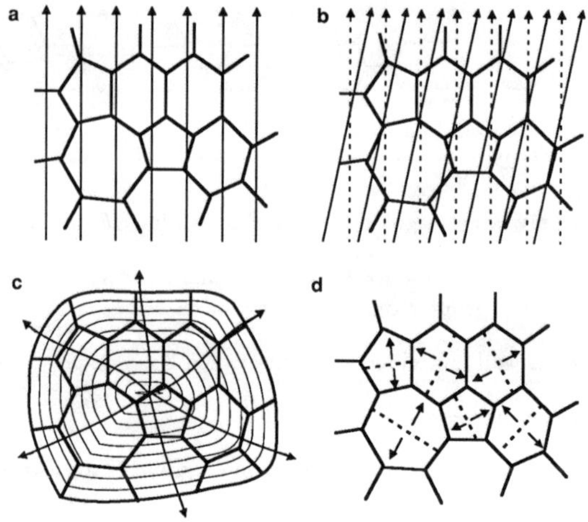

Figure 2: Sample vector fields embedding a cellular structure

of time. However, such a definition may obscure possible explicit dependencies between the model and the field (c and d).

Definition 2.4. A *context-free cell system* $\mathcal{H}$ over a vector field $\mathcal{Z}$ consists of:

- a finite *alphabet* of cell states (types) Σ,
- a *starting cellular structure* ω with cell labels from Σ, and
- a finite *set of cell productions* (developmental rules)

$$P \in (\Sigma \times \Sigma) \cup (\Sigma \times \Sigma \times \Re \times (0,1) \times \Sigma)\,,$$

where $\Re$ denotes the set of real numbers. It follows that productions may have two forms. A production $(A, B) \in P$, written $A \longrightarrow B$, changes the cell label from A to B. A production $(A, B, \alpha, r, C) \in P$, written $A \longrightarrow B \uparrow (\alpha, r)\, C$, splits cell A into two cells labeled B and C. The orientation of the division wall is specified by the angle α between the direction of the reference vector $\vec{Z}$, measured at the center of gravity of A immediately before the division, and the normal vector to the wall. Cell B lies on the side of the wall opposite to the direction of the normal, while C lies on the side pointed to by the normal. The areas of cells B and C are fractions r and $1 - r$ of the area of the mother cell A (Figure 3). In the frequently occurring case where the areas of cells B and C are the same ($r = 0.5$), the notation $A \longrightarrow B \uparrow (\alpha, r)\, C$ is shortened to $A \longrightarrow B \uparrow (\alpha)\, C$.

Definition 2.5. A cell system $\mathcal{H}$ operates in a sequence of *derivation steps*, noted $\mu_i \Longrightarrow \mu_{i+1}$, during which the *predecessor structure* μ_i yields the *successor structure* μ_{i+1} for the subsequent values of the index i. Each derivation step consists of three phases:

- A structure μ'_{i+1} is obtained from μ_i by applying productions from P *simultaneously* to all cells in μ_i. The vector field at the time immediately preceding production application is noted $\mathcal{Z}_i$.

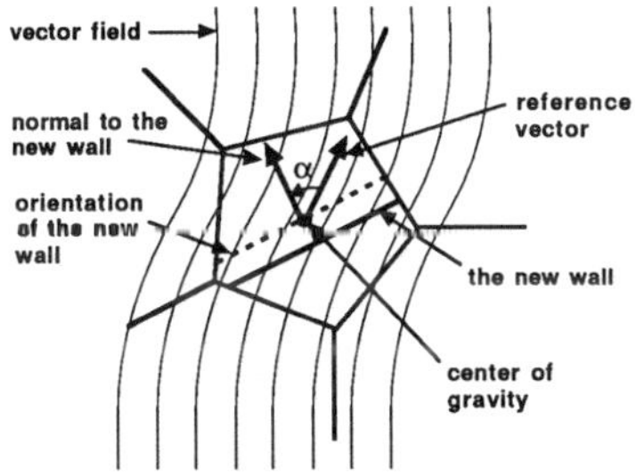

Figure 3: Specification of the division wall with respect to the vector field

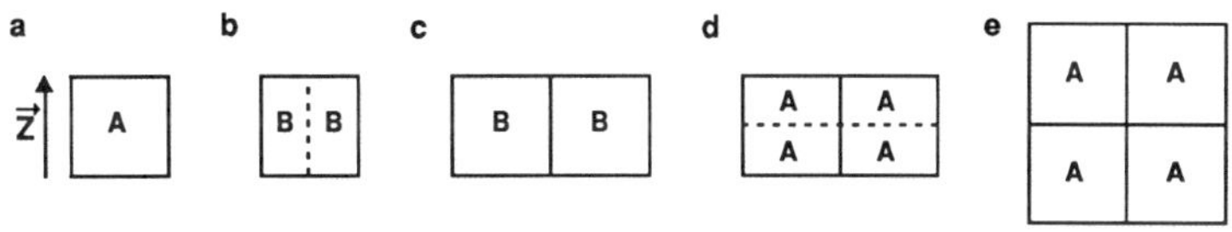

Figure 4: The development of a simple two-dimensional pattern. The original shape (a) undergoes a succession of divisions (b, d), followed by the application of the pressure-tension model (c, e).

- A structure μ_{i+1} is obtained from μ'_{i+1} by applying the pressure-tension model to determine the steady-state geometry of μ_{i+1}. The rest length and the spring constant of the walls, as well as the coefficient relating cell pressure to area, are global parameters of the model.

- The vector field is recomputed (in the non-stationary case), yielding the new field $\mathcal{Z}_{i+1}$, to be used in the subsequent derivation step.

The sequence of structures starting with $\mu_0 = \omega$ is called the *developmental sequence* generated by $\mathcal{H}$.

The above definitions are interpreted as follows. The state of a cell comprehends all features, other than shape, that characterize it at a given moment. At fixed intervals, cells divide or change state without a division. The fate of each cell is determined by the applicable production. If a cell undergoes a division, the production specifies the position of the division wall in addition to the states of the daughter cells. The orientation of the division wall is given by the angle α between the reference vector $\vec{Z}$ and the wall normal. By default, the division wall splits the mother cell into two cells of equal area. Unequal cell division is specified in terms of the relative areas of the daughter cells. After production application, the structure undergoes a period of growth. The resulting geometry is determined by the mechanical interaction between cells. At the end of a derivation step the vector field is updated, providing new reference directions for the subsequent cell divisions. The updating algorithm depends on the modeled structure and is not included in the definition of cell systems.

Example 2.2. Figure 4 shows a simple structure that develops in a uniform stationary field $\mathcal{Z}$, with the vector $\vec{Z}$ pointing up at all points in the field. The alphabet Σ of the cell system consists of two symbols, A and B. The initial structure ω is a square cell labeled A. The development is governed by two productions that divide cells alternately by vertical and horizontal walls:

$$p_1: \mathtt{A} \rightarrow \mathtt{B} \uparrow (\mathtt{90})\, \mathtt{B}$$
$$p_2: \mathtt{B} \rightarrow \mathtt{A} \uparrow (\mathtt{0})\, \mathtt{A}$$

The pressure-tension model operates under the following assumptions:

- the relation between cell area and pressure on the walls is the same for all walls, $pressure = C/area$, where C is a constant,
- the rest length of all walls is equal to 0, and
- the spring constants k_i and k_e for the internal and external walls satisfy the equation $k_i = 2k_e$.

Example 2.3. Figure 5 shows the development of a pattern first studied by Korn and Spalding as an idealized model for cell divisions in the epidermis of onion [10]. According to a subsequent formalization [5, 14], this pattern can be constructed as follows. The initial structure is a single hexagonal cell. During the first derivation step the division wall splits two opposite walls at 1/3 and 2/3 of their length (a). Simple trigonometry shows that the normal to the division wall is at an angle $\alpha \approx -80.89°$ with respect to the reference vector $\vec{Z}$. In subsequent steps, the cells divide in a similar manner, except that the reference direction is defined locally as being perpendicular to the previous division wall (thus, the polarity vector field is used). It is convenient to assume that, from the topological point of view, cell divisions take place on the surface of a torus, thus all walls are shared by two cells (b). Consequently, at each stage of development, the structure consists entirely of hexagons, which are made regular by the application of the pressure-tension model (c).

According to the above description, the Korn-Spalding pattern can be captured by a cell system with a single production,

$$p: \mathtt{A} \rightarrow \mathtt{A} \uparrow (-\mathtt{80.89})\, \mathtt{A}\,.$$

Initially, the field $\mathcal{Z}$ consists of diagonal vectors $\vec{Z}$ (Figure 5a). Subsequently, the reference vectors are specified for pairs of sister cells as normal to the division wall (polarity vector field – Figure 2d).

For the purpose of comparison, Figure 6 shows two patterns obtained by the same cell system operating in the plane rather than on a torus. The border cells usually have less than six walls. This affects divisions of the internal cells, and breaks the regularity of the original pattern.

Example 2.4. Figure 7 shows a slightly more complex pattern composed of cells with five, six, and seven walls [2]. The corresponding cell system is given below.

$$\omega: \mathsf{5}$$
$$p_1: \mathsf{5} \rightarrow \mathsf{6} \uparrow (-\mathsf{95.5}, \mathsf{0.67})\, \mathsf{5}$$
$$p_2: \mathsf{6} \rightarrow \mathsf{7} \uparrow (-\mathsf{109.0}, \mathsf{0.78})\, \mathsf{5}$$
$$p_3: \mathsf{7} \rightarrow \mathsf{7} \uparrow (-\mathsf{93.5}, \mathsf{0.67})\, \mathsf{6}$$

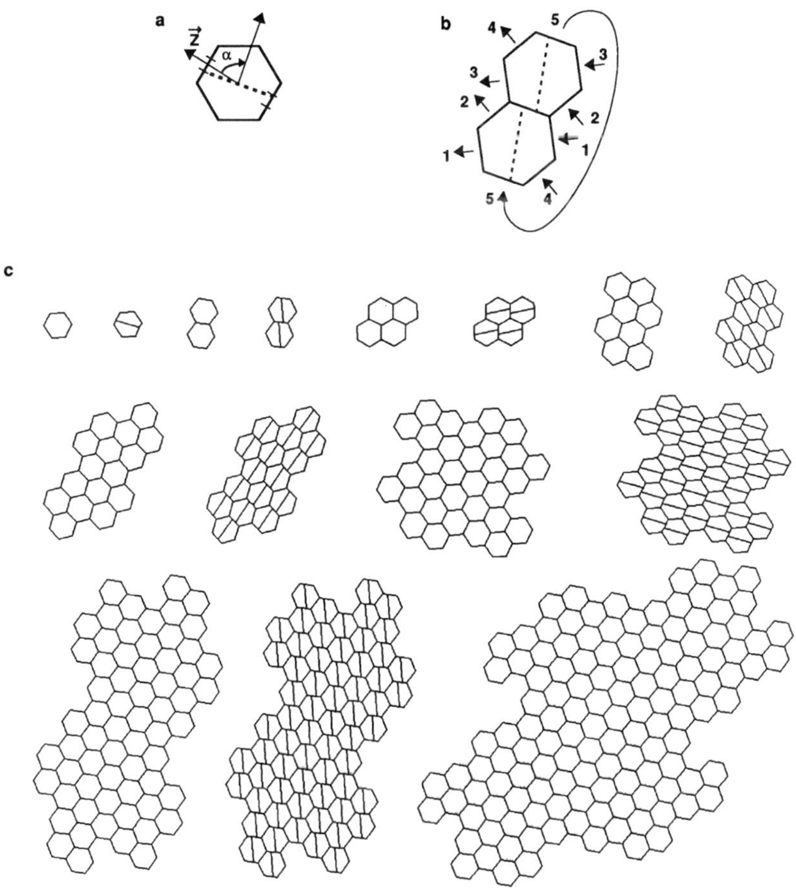

Figure 5: Construction of the *Korn-Spalding* pattern. (a) Geometry of wall insertion. (b) Neighborhood relations between walls on the surface of a torus. Arrows indicate pairs of walls 'fused' together. (c) The developmental sequence.

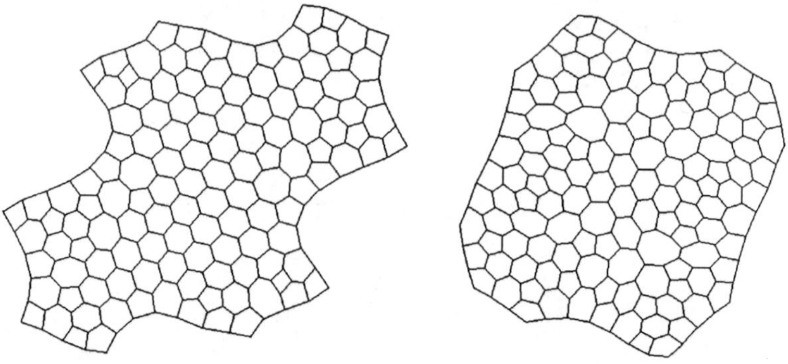

Figure 6: Derivatives of the Korn-Spalding pattern generated in the plane. (a) The spring constants for the interior and exterior walls are not equal, $k_i = 2k_e$. (b) The spring constants for all walls are equal.

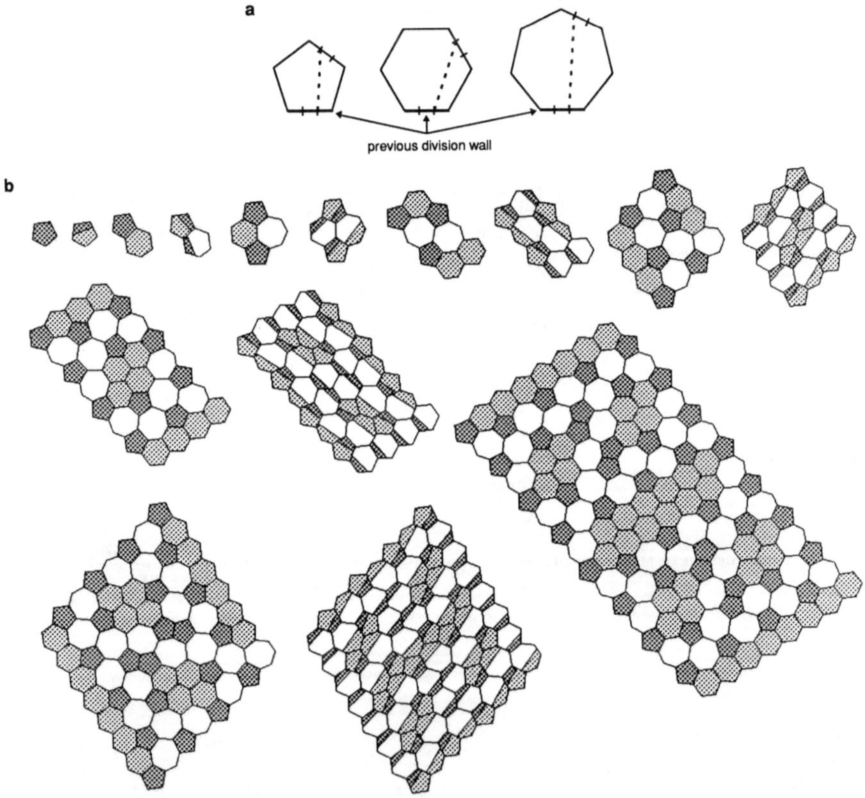

Figure 7: A pattern consisting of cells with five, six, and seven walls. (a) Postulated positions of division walls. (b) The developmental sequence.

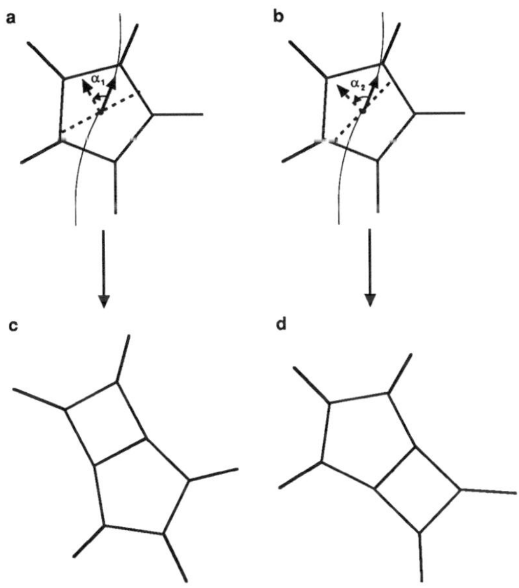

Figure 8: The interlacing of topology and geometry in cell system operation. The angle α determines structure topology after the division wall has been inserted (a,b). The new topology determines geometry at equilibrium (c,d).

The orientation and positions of the division walls were estimated using Figure 7a. As in the case of the Korn-Spalding pattern, the reference direction is the polarity vector field, and the development takes place on a torus.

The structures presented in the above examples can be also generated using map L-systems. It is therefore interesting to compare these two formalisms.

Both map L-systems and cell systems are parallel rewriting systems. Cell systems can be viewed as an extension of string L-systems [12] to polygon meshes, while map L-systems extend string L-systems to maps. The fundamental difference is that a map is a purely topological notion, while a mesh combines topological and geometric aspects (Figure 8). Positions of the inserted walls are specified in geometric terms, yet they determine the topology of the mesh after each step of rule application. The new topology determines, in turn, the geometry of the cellular structure, reached at equilibrium.

The definition of the vector field is a key component of cell system specification and has a paramount impact on its properties. The next section shows that, for the polarity vector field, the generative powers of cell systems and map L-systems are closely related.

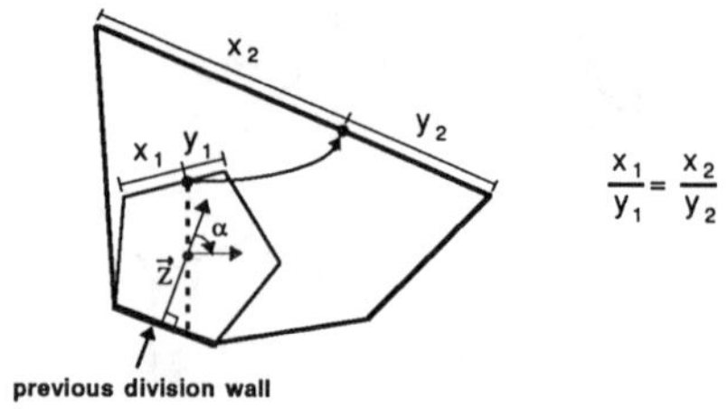

Figure 9: Cell division with regularization.

3 A comparison of cell systems to map L-systems

Strictly speaking, map L-systems and cell systems cannot be compared, since they operate on two different types of objects: maps, which are defined in topological terms, and cellular structures, defined in geometric terms. Nevertheless, it is possible to abstract from the geometry of cells, and view each cellular structure as a representation of a map. Then the classes of maps generated by map L-systems and cell systems can be compared. The following discussion employs binary propagating context-free map L-systems with markers (mBPMOL-systems) [17] as the representative type of map L-systems.

A production in a map L-system can subdivide an existing wall by the insertion of vertices, without attaching a new wall to them. Consequently, a map L-system can introduce vertices of order two into the generated map. In contrast, a cell system creates vertices as attachment points for new division walls, and can only generate vertices of order three or higher. In view of this discrepancy, maps that differ only by the number and the placement of order-two vertices are considered equivalent.

In order to abstract from cell geometry, it is assumed that, prior to division, each cell is transformed into a regular polygon irrespective of other cells (Figure 9). In this process, the previous division wall preserves its position with respect to the vector field. Once a new division wall is inserted, the cell is transformed back to its original shape. This inverse transformation preserves proportions at which two previously existing walls are split by the attachment points of the new division wall. As a result, all cells with the same label are divided in the same way, regardless of shape, as long as they have the same orientation with respect to the vector field $\mathcal{Z}$.

Theorem. Consider a map L-system $\mathcal{G}$ that generates a sequence of maps $\omega_{\mathcal{G}}, M_1, M_2, \ldots$, and assume that all regions have a number of walls bounded by a constant N. It is then possible to construct a context-free cell system $\mathcal{H}$ that generates a sequence of equivalent cellular structures $\omega_{\mathcal{H}}, S_1, S_2, \ldots$.

Proof. The construction of cell system $\mathcal{H}$ is straightforward:

1. Consider the map sequence $\omega_{\mathcal{G}}, M_1, M_2, \ldots$. Find the set of distinct circular words w_i that describe possible sequences of wall labels characterizing regions of these maps. Since the number of productions in the map L-system $\mathcal{G}$ is finite, and the number of walls per cell is bounded by a constant, all circular words will appear in the finite sequence of maps $\omega_{\mathcal{G}}, M_1, \ldots, M_n$ for some n.

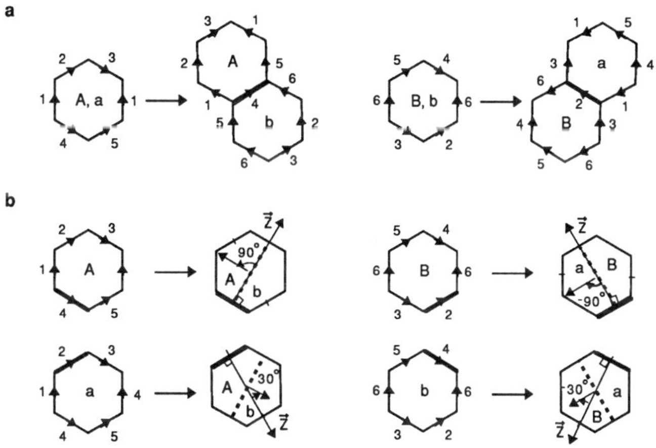

Figure 10: Construction of a cell system from the map L-system for the *Lewis* pattern. (a) Finding the circular words and mapping them to cell labels. (b) Determining the angle between the reference vector and division wall normal for each cell.

2. For each circular word w_i, find the set of marked words $\underline{w}_{ij}$, where the marker indicates the position of the previous division wall. Clearly, the same circular word may result from the partition of various regions, which is why it may admit different markings. If a word occurs only in the initial map $\omega_{\mathcal{G}}$, choose its marking arbitrarily. Since all possible circular words appear in maps $\omega_{\mathcal{G}}, M_1, \ldots, M_n$, the complete set of their successors, and therefore marked words, can be identified in maps $\omega_{\mathcal{G}}, M_1, \ldots, M_{n+1}$.

3. Assign a unique cell label to each marked circular word, and determine the cell production associated with each word using maps $\omega_{\mathcal{G}}, M_1, \ldots, M_{n+1}$. Use the polarity vector field $\mathcal{Z}$ and assume that the cells are regularized prior to division.

4. Transform the initial map $\omega_{\mathcal{G}}$ to the initial cellular structure $\omega_{\mathcal{H}}$ by matching a previously identified marked circular word to each region of $\omega_{\mathcal{G}}$, and assigning the corresponding label to it. The initial structure $\omega_{\mathcal{H}}$ and the cell productions found in Step 3 constitute the cell system $\mathcal{H}$ equivalent to the map L-system $\mathcal{G}$. □

Example 3.1. The following map L-system, specified using notation from [18, pages 146–148], generates the sequence of patterns proposed by Lewis [11] and described formally by Lindenmayer and Rozenberg [14]:

$$
\begin{array}{rlcl}
\omega: & \overrightarrow{1}\overrightarrow{2}\overrightarrow{3}\overleftarrow{1}\overleftarrow{5}\overleftarrow{4} & & \\
p_1: & \overrightarrow{1} & \rightarrow & \overrightarrow{2} \\
p_2: & \overrightarrow{2} & \rightarrow & \overrightarrow{3}\,[+\overrightarrow{2}]\,\overleftarrow{1} \\
p_3: & \overrightarrow{3} & \rightarrow & \overleftarrow{5}\,[-\overleftarrow{4}]\,\overleftarrow{6} \\
p_4: & \overrightarrow{4} & \rightarrow & \overleftarrow{1}\,[+\overrightarrow{4}]\,\overleftarrow{5} \\
p_5: & \overrightarrow{5} & \rightarrow & \overleftarrow{6}\,[-\overleftarrow{2}]\,\overrightarrow{3} \\
p_6: & \overrightarrow{6} & \rightarrow & \overrightarrow{4}
\end{array}
$$

Two circular words are generated by this map L-system (Figure 10a): $\overline{\overrightarrow{1}\overrightarrow{2}\overrightarrow{3}\overleftarrow{1}\overleftarrow{5}\overleftarrow{4}}$ and $\overrightarrow{6}\overrightarrow{5}\overrightarrow{4}\overleftarrow{6}\overleftarrow{2}\overleftarrow{3}$.
The production rules for the circular words with underscores indicating the inserted division wall are:

$$\begin{array}{lcl}
\overrightarrow{1}\overrightarrow{2}\overrightarrow{3}\overleftarrow{1}\overleftarrow{5}\overleftarrow{4} & \to & \overrightarrow{1}\overrightarrow{2}\overrightarrow{3}\overleftarrow{1}\overleftarrow{5}\overleftarrow{\underline{4}} \quad \overrightarrow{6}\overrightarrow{5}\overrightarrow{\underline{4}}\overleftarrow{6}\overleftarrow{2}\overleftarrow{3} \\
\overrightarrow{6}\overrightarrow{5}\overrightarrow{4}\overleftarrow{6}\overleftarrow{2}\overleftarrow{3} & \to & \overrightarrow{1}\overrightarrow{\underline{2}}\overrightarrow{3}\overleftarrow{1}\overleftarrow{5}\overleftarrow{4} \quad \overrightarrow{6}\overrightarrow{5}\overrightarrow{4}\overleftarrow{6}\overleftarrow{\underline{2}}\overleftarrow{3}
\end{array}$$

The marked circular words can be mapped to cell labels as follows:

$$\begin{array}{lcl}
\overrightarrow{1}\overrightarrow{2}\overrightarrow{3}\overleftarrow{1}\overleftarrow{5}\overleftarrow{\underline{4}} & \mapsto & \mathtt{A} \\
\overrightarrow{1}\overrightarrow{\underline{2}}\overrightarrow{3}\overleftarrow{1}\overleftarrow{5}\overleftarrow{4} & \mapsto & \mathtt{a} \\
\overrightarrow{6}\overrightarrow{5}\overrightarrow{\underline{4}}\overleftarrow{6}\overleftarrow{2}\overleftarrow{3} & \mapsto & \mathtt{b} \\
\overrightarrow{6}\overrightarrow{5}\overrightarrow{4}\overleftarrow{6}\overleftarrow{\underline{2}}\overleftarrow{3} & \mapsto & \mathtt{B}
\end{array}$$

The cell system equivalent to the given map L-system has initial structure A and four productions with predecessors A, a, B, and b. The positions of the division walls are found assuming that the structure develops on a torus, thus all cells are regular hexagons (Figure 10b). As stipulated by the proof of the theorem, these positions are expressed with respect to the polarity field vectors.

$$\begin{array}{rclcl}
\omega & : & \mathtt{A} & & \\
p_1 & : & \mathtt{A} & \to & \mathtt{b} \uparrow (90)\, \mathtt{A} \\
p_2 & : & \mathtt{a} & \to & \mathtt{A} \uparrow (30)\, \mathtt{b} \\
p_3 & : & \mathtt{B} & \to & \mathtt{B} \uparrow (-90)\, \mathtt{a} \\
p_4 & : & \mathtt{b} & \to & \mathtt{a} \uparrow (-30)\, \mathtt{B}
\end{array}$$

The sequence of cellular structures generated by this cell system is shown in Figure 11.

The inverse problem, regarding the existence of an mBPMOL-system equivalent to a given cell system, remains open. As shown in [3], if the cell system generates a sequence of structures with a bounded number of cell walls, it is possible to find an equivalent double-wall map L-system [15, 16]. It can be converted to a map L-system with markers only if the number of configurations of labels on both sides of the walls is finite (the double-wall system has *parity* [3, 15]). At this point it is not clear whether bounded cell systems equivalent to double-wall systems without parity exist.

4 A model of a fern gametophyte

This section presents a cell system that simulates the development of a fern gametophyte *Microsorium linguaeforme*. Its developmental sequence had been previously studied and simulated using map L-systems [2, 4, 18]. The modeling process consisted of the following steps:

1. microscopic observations of the development of sample gametophytes,
2. the expression of observations using a cell division system,
3. translation of the cell division system to a map L-system, and

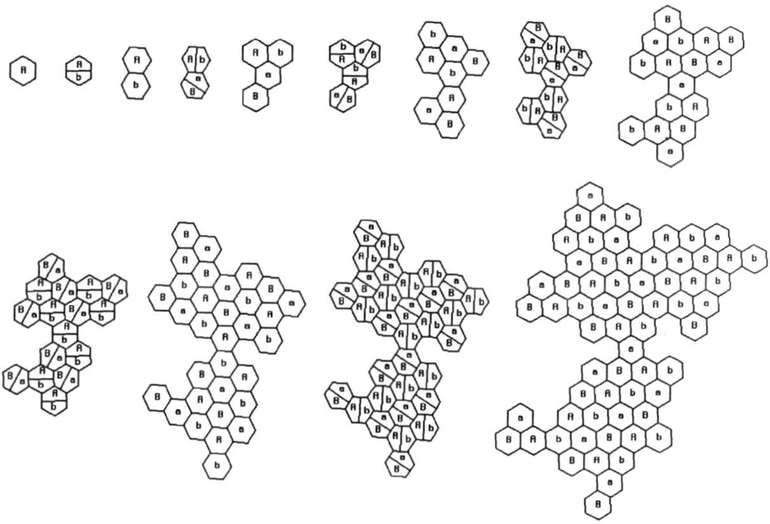

Figure 11: The *Lewis* pattern

4. simulation of the development using this map L-system.

The translation of a cell division system to a map L-system is a laborious process which may introduce rigid assumptions regarding offsets in cell tetrads [15], and yields a nonintuitive description of the modeled structure. We show that a cell system provides a model which captures the observed phenomena in a straightforward way.

A gametophyte of *Microsorium linguaeforme* forms a plant body called a *thallus* that is not differentiated into stem, leaf and root. The development of a thallus can be described conveniently in terms of the activity of the *apical cell* and the development of *segments.* The apical cell is the originator of the gametophyte structure. It divides repetitively, giving rise each time to a new apical cell and a primary (initial) segment cell. The division walls are oriented alternately to the left and to the right, yielding two columns of segments separated by a zig-zag dividing line (Figure 12): The recursive nature of the apical activity is captured by the following cell division system.

$$A_L \to Q_1 \mid A_R \qquad\qquad A_R \to A_L \mid Q_1$$

The primary segment cells Q_1 subdivide further, yielding multicellular segments. Microscopic observations reveal that most segments follow the same developmental sequence, shown diagrammatically in Figure 13. The cell Q_1 is first divided by a periclinal wall into two cells, Q_2 and Q_3. Subsequently, the cell Q_3 is divided by another periclinal wall into two terminal cells T that do not undergo further divisions. At the same time, the cell Q_2 lying on the thallus border is divided by an anticlinal wall into two cells of type Q_1. Each of these cells divides in the same way as the primary cell. Consequently, the recursive

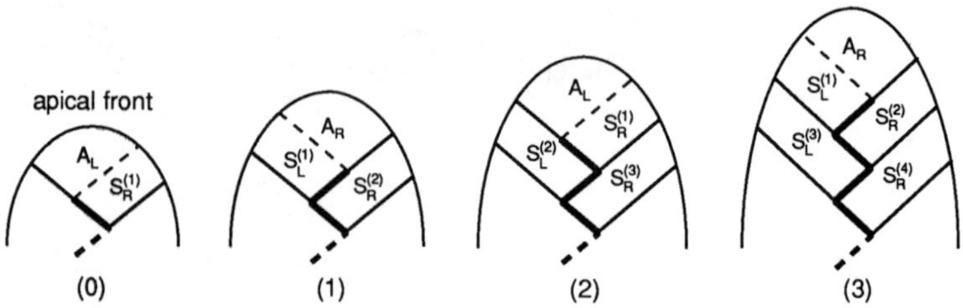

Figure 12: Apical production of segments. The labels A_R and A_L denote apical cells producing right segment S_R and left segment S_L, respectively. Dashed lines indicate the newly created division wall. The superscripts represent segment age. The internal structure of segments is not shown.

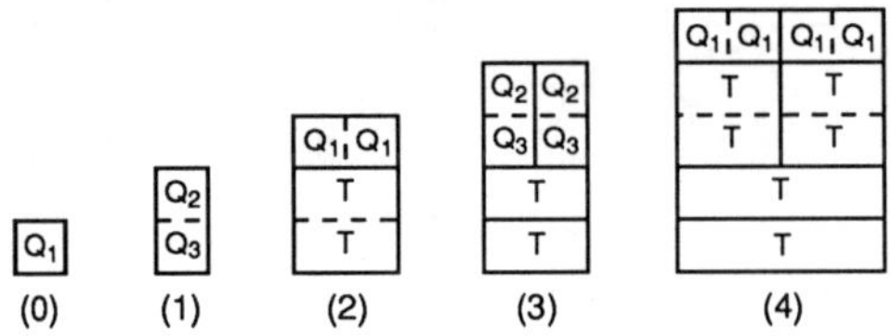

Figure 13: Developmental sequence of a *Microsorium* segment

nature of segment development can be captured by the following cell division system:

$$Q_1 \rightarrow \frac{Q_2}{Q_3} \qquad Q_2 \rightarrow Q_1 \mid Q_1 \qquad Q_3 \rightarrow \frac{T}{T}$$

The two oldest segments, situated at the thallus base, form a modified pattern with less extensive cell divisions. The developmental sequence of the right basal segment is shown in Figure 14. The corresponding cell division system is given below:

$$B_1 \rightarrow \frac{B_2}{T} \qquad B_2 \rightarrow B_1 \mid T$$

The left basal segment follows a symmetric pattern:

$$C_1 \rightarrow \frac{C_2}{T} \qquad C_2 \rightarrow T \mid C_1$$

All of the above productions form a cell division system that captures the recursive pattern of cell divisions in the gametophyte, but lacks precise geometric information needed for simulation. In order to formulate a corresponding cell system, we define a vector field closely related to the organism description in terms of periclinal and anticlinal walls. For any pair of sister cells, the reference vector $\vec{Z}$ is normal to the division wall. In contrast to the polarity vector field of Figure 2d, the vector $\vec{Z}$ has the same orientation in both sister cells. The resulting cell system is given below.

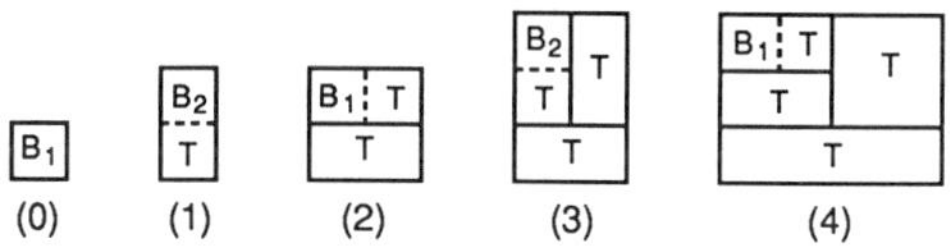

Figure 14: Developmental sequence of the right basal segment

$$
\begin{array}{llllllll}
p_1: & \mathtt{A_L} & \rightarrow & \mathtt{A_R} \uparrow (-45.0, 0.3)\, \mathtt{Q_1} & p_6: & \mathtt{B_1} & \rightarrow & \mathtt{B_2} \uparrow (-90.0)\, \mathtt{T} \\
p_2: & \mathtt{A_R} & \rightarrow & \mathtt{Q_1} \uparrow (45.0, 0.7)\, \mathtt{A_L} & p_7: & \mathtt{B_2} & \rightarrow & \mathtt{B_1} \uparrow (90.0)\, \mathtt{T} \\
p_3: & \mathtt{Q_1} & \rightarrow & \mathtt{Q_2} \uparrow (-98.0)\, \mathtt{Q_3} & p_8: & \mathtt{C_0} & \rightarrow & \mathtt{C_1} \\
p_4: & \mathtt{Q_2} & \rightarrow & \mathtt{Q_1} \uparrow (90.0)\, \mathtt{Q_1} & p_9: & \mathtt{C_1} & \rightarrow & \mathtt{C_2} \uparrow (-90.0)\, \mathtt{T} \\
p_5: & \mathtt{Q_3} & \rightarrow & \mathtt{T} \uparrow (0.0)\, \mathtt{T} & p_{10}: & \mathtt{C_2} & \rightarrow & \mathtt{T} \uparrow (90.0)\, \mathtt{C_1}
\end{array}
$$

The cell system closely follows the cell division system with the exception of production p_8, which delays the development of the left basal segment by one step with respect to the right basal segment. The values of angles and relative areas are obtained by comparing simulations to the microscopic observations, and adjusting parameters accordingly. The developmental sequence generated using this cell system is shown in Figure 15. Different gray levels indicate the apical cell, the alternating segments close to the apex, and the pair of basal segments. The model shows good correspondence with the shape of the real organism and with the model expressed using a map L-system [18, page 161]. Although all segments originating from the apical cell develop according to the same cell production rules, they exhibit slightly different division patterns (*segment-to-segment variability*), due to different cell shapes.

The angles at which the division walls are inserted into the mother cells have a pronounced effect on the generated structure. For example, Figure 16 illustrates the result of setting the angles in productions p_1 and p_3 to $-90°$, and p_2 to $90°$. The variability of structures generated using cell systems with different parameters in productions coincides with observations of real gametophyte structures. The ease of capturing *specimen-to-specimen variation* is an asset of cell systems. In contrast, a small change in the developmental sequence simulated using a map L-system may require substantial modifications of many productions.

5 Discussion

This paper presents a new method for the modeling of cell division and expansion in single-layered cellular structures. Cell divisions are described by productions capturing cell lineage. The orientations of the division walls are specified relative to a vector field (morphogenetic field), as opposed to a topological description of the attachment sites. Consequently, cell systems provide a more intuitive description of cell division patterns than map L-systems. The geometry of cells, determined by a model that takes into account internal cell pressure and wall tension, influences the topology of the generated structures. This accounts for a variability in cell division patterns, consistent with that observed in nature.

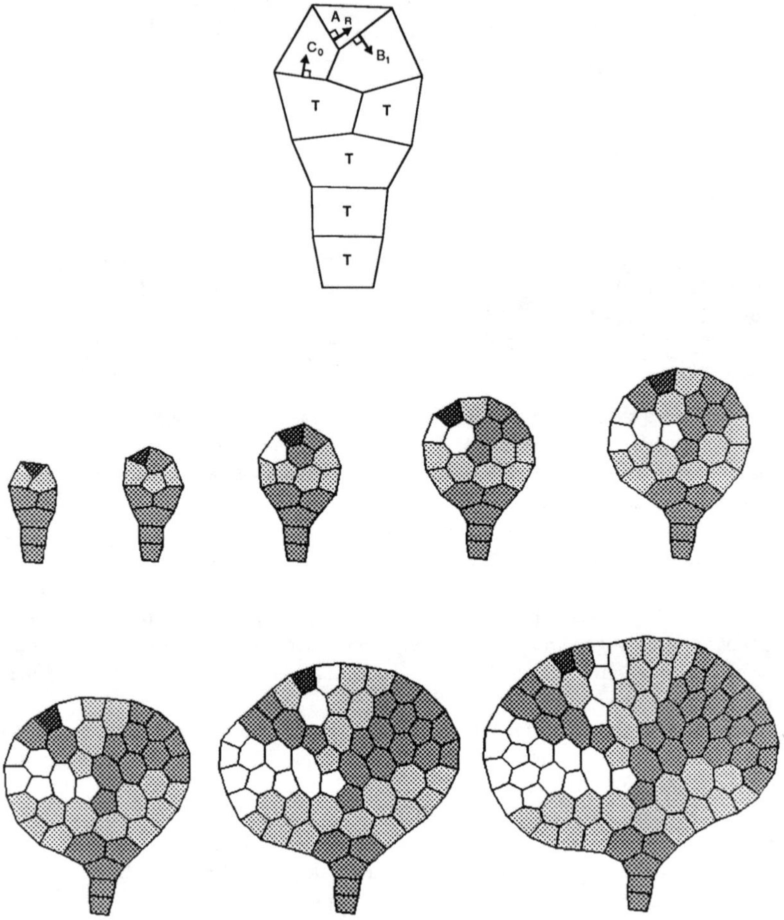

Figure 15: Simulated development of *Microsorium linguaeforme*. (a) The initial structure. Arrows indicate the initial reference vectors. (b) The developmental sequence.

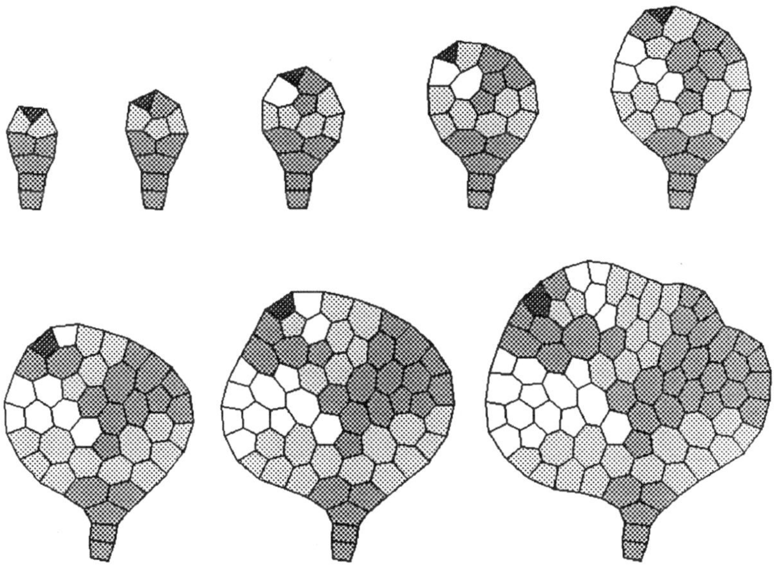

Figure 16: A variant of *Microsorium linguaeforme* thallus

Cell systems operate under the assumption that positions of division walls are controlled from inside the cells. In contrast, map L-systems employ a control mechanism that resides in cell walls. Which mechanism is more justified from a biological point of view is an open problem. Cytological observations of developing tissues reveal that microtubule structures called *phragmosomes*, which radiate from the cell nucleus, coincide with the position and orientation of the division walls [1, 6]. This supports the assumption that the position of the division wall is controlled from inside the cell. Moreover, the polarity vector field corresponds to the notion of *cytoplasmic inheritance* [9], in which the orientation of the phragmosome of a mother cell determines that of the daughter cells. On the other hand, observations do not exclude the possibility that cell walls specify autonomously the attachment sites for the new division walls [9, 13].

The influence of cell shape on microtubule organization, and hence on the position of the division wall, is also supported by observations [6]. This influence may account for topological irregularities exhibited by many cell patterns found in nature. In addition to the variability due to different positions of cells in a structure (Figure 15), cell systems can also capture topological variations resulting from different timings of cell divisions (Figure 17).

Many problems related to cell systems are open for further research. For example, globally defined parameters incorporated in the pressure-tension model impose identical physical properties on all cells. Cell systems with the properties specified locally for individual cells could lead to more faithful models of some structures.

Furthermore, the exchange of information between cells may play as important a role in the development of cellular structures as the lineage mechanism considered in this

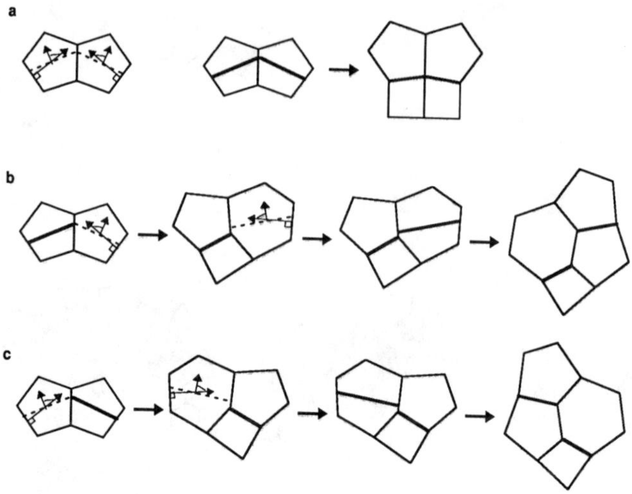

Figure 17: The relationship between the order of cell divisions and the topology of the resulting structure. (a) Left and right cells divide simultaneously. (b) Left cell divides first. (c) Right cell divides first.

paper. A convenient framework for the investigation of patterns that result from the interactions between cells in developing cell layers may be offered by cell systems with context-sensitive productions.

Acknowledgements

We are indebted to Prof. Aristid Lindenmayer for having inspired our research into mechanisms governing cellular development. The reported research has been supported in part by an operating grant and equipment grants from the Natural Sciences and Engineering Research Council of Canada, and by a research grant from Apple Computer, Inc. Facilities of the Department of Computer Science, University of Regina, were also essential. All support is gratefully acknowledged.

References

[1] P. W. Barlow. From cell wall networks to algorithms: The simulation of cell division patterns in plants. *Protoplasma*, 1991. In press.

[2] M. J. M. de Boer. *Analysis and computer generation of division patterns in cell layers using developmental algorithms*. PhD thesis, University of Utrecht, the Netherlands, 1989.

[3] M. J. M. de Boer. Construction of map L-systems for developmental sequences of cell layers. In H. Ehrig, H. J. Kreowski, and G. Rozenberg, editors, *Graph-grammars and their application to computer science; Fourth International Workshop*, Lecture Notes in Computer Science. Springer-Verlag, Berlin, 1991. In press.

[4] M. J. M. de Boer and M. de Does. The relationship between cell division pattern and global shape of young fern gametophytes I. A model study. *Botanical Gazette*, 151:423–434, 1990.

[5] M. de Does and A. Lindenmayer. Algorithms for the generation and drawing of maps representing cell clones. In H. Ehrig, M. Nagl, and G. Rozenberg, editors, *Graph-grammars and their application to computer science*, Lecture Notes in Computer Science 153, pages 39–57. Springer-Verlag, 1983.

[6] D. J. Flanders, D. J. Rawlins, P. J. Shaw, and C. W. Lloyd. Nucleus-associated microtubules help determine the division plane of plant epidermal cells: Avoidance of four-way junctions and the role of cell geometry. Submitted to *J. Cell Biol.*, 1989.

[7] J. D. Foley, A. Van Dam, S. K. Feiner, and J. F. Hughes. *Computer graphics: Principles and practice.* Addison-Wesley, Reading, Massachusetts, 1990.

[8] F. D. Fracchia, P. Prusinkiewicz, and M. J. M. de Boer. Animation of the development of multicellular structures. In N. Magnenat-Thalmann and D. Thalmann, editors, *Computer Animation '90*, pages 3–18, Tokyo, 1990. Springer-Verlag.

[9] B. E. S. Gunning. The cytokinetic apparatus: Its development and spatial regulation. In C. W. Lloyd, editor, *The cytoskeleton in plant growth and development*, pages 229–292. Academic Press, London, 1982.

[10] R. W. Korn and R. M. Spalding. The geometry of plant epidermal cells. *New Phytologist*, 72:1357–1365, 1973.

[11] F. T. Lewis. The geometry of growth and cell division in epithelial mosaics. *Amer. J. Bot.*, 30:766–776, 1943.

[12] A. Lindenmayer. Mathematical models for cellular interaction in development, Parts I and II. *Journal of Theoretical Biology*, 18:280–315, 1968.

[13] A. Lindenmayer. Models for plant tissue development with cell division orientation regulated by preprophase bands of microtubules. *Differentiation*, 26:1–10, 1984.

[14] A. Lindenmayer and G. Rozenberg. Parallel generation of maps: Developmental systems for cell layers. In V. Claus, H. Ehrig, and G. Rozenberg, editors, *Graph grammars and their application to computer science; First International Workshop*, Lecture Notes in Computer Science 73, pages 301–316. Springer-Verlag, Berlin, 1979.

[15] J. Lück, A. Lindenmayer, and H. B. Lück. Models for cell tetrads and clones in meristematic cell layers. *Botanical Gazette*, 149:1127–141, 1988.

[16] J. Lück and H. B. Lück. Sur la structure de l'organisation tissulaire et son incidence sur la morphogenèse. In Hervé Le Guardier, editor, *Actes du deuxième séminaire de l'Ecole de Biologie Théorique du CNRS*, pages 385–397. Publications de l'Université de Rouen, Abbaye de Solignac, 1982.

[17] A. Nakamura, A. Lindenmayer, and K. Aizawa. Some systems for map generation. In G. Rozenberg and A. Salomaa, editors, *The Book of L*, pages 323–332. Springer-Verlag, Berlin, 1986.

[18] P. Prusinkiewicz and A. Lindenmayer. *The Algorithmic Beauty of Plants.* Springer-Verlag, New York, 1990. With J. Hanan, F. D. Fracchia, D. R. Fowler, M. J. M. de Boer and L. Mercer.

[19] W. T. Tutte. *Graph theory.* Addison-Wesley, Reading, Massachusetts, 1982.

SPLICING SCHEMES AND DNA

Tom Head

Department of Mathematical Sciences, State University of New York at Binghamton, Binghamton, New York 13901, U.S.A.

Abstract: Splicing schemes are formal devices which act on sets of strings in a manner that is suggested by the action of sets of restriction enzymes and a ligase on double stranded DNA molecules. Previous work on splicing systems is reviewed and problems are suggested that concern the generative capacities of splicing schemes acting on both linear and circular strings.

Keywords: Splicing systems, formal language theory, automata, regular languages, DNA, genetic algorithms

1. Introduction

Aristid Lindenmayer enlarged the scope of formal language theory and the class of topics to which this theory applies when he initiated the use of parallel rewriting of strings to model biological development in [18]. Major developments followed in both formal language theory and the life sciences [15,22,21]. Is it possible that formal language theory can also be enriched by developing models for representing the behaviors of the informational macromolecules that have been studied in molecular biology? With the hope for an affirmative answer, we introduced a new class of generative devices in [14]. These devices, called splicing systems, were defined to allow certain DNA recombinant behaviors to be directly represented as operations on pairs of strings over a finite alphabet. In the case of double stranded DNA, the natural alphabet D consists of the four hydrogen-bonded deoxyribonucleotide pairs. Thus D = {[A/T], [C/G], [G/C], and [T/A]}. Virtually the whole of [14] was devoted to explaining how the splicing model is related to specified recombinant behaviors of linear DNA molecules. In the present article an expansion of this formalism is proposed that includes the modeling of recombinant behaviors involving both circular and linear DNA molecules. Details of the modeling aspect can be derived from a reading of [14].

Three additional papers have appeared that concern splicing systems. K. Culik II & T. Harju proved an important regularity result in [4,5] through the use of their previously developed domino concept [3]. K. Denninghoff & R. Gatterdam enriched the splicing concept through the use of sets with multiplicity and proved in [6] that every recursively enumerable language can be generated by a splicing system when multiplicities are treated appropriately. Valuable examples of splicing operations and languages appear in [10].

In this article the definition of the splicing concept is expanded in order to provide a system that is more natural from a mathematical point of view. The multiplicity concept introduced in [6] is used and the splicing operation is extended to include circular strings as well as linear strings. The results in the literature discussed above are reviewed and new problems are proposed.

As background references on DNA recombinant behavior [25,17,13] are recommended. For formal language theory see [16,23,8]. The present work may suggest formal ways to specify restrictions on the cross-over operations used in the design of genetic algorithms [12].

2. Linear & circular strings, involutions, and multiplicities

Let A be a finite set. Then A^* is the set of all strings of symbols in A having finite length. This includes the empty string that has length zero and is denoted by 1. By a circular string of symbols in A we mean, intuitively, a string that has its left end butted against its right end to form a circle of symbols. The circular string consisting of four symbol occurrences with the symbols in order being a, b, c, d will be written: ^abcd. The symbol ^ will be reserved to indicate circularity. In writing a circular string in this linearized form, the choice of the left-most symbol is arbitrary. Thus ^abcd, ^bcda, ^cdab, & ^dabc are all notations for the same circular string. The set of all circular strings of symbols in A will be denoted by A^. Formally speaking, A^ is the set equivalence classes in A^* with respect to the binary relation ~ defined by: x~y if x is a cyclic permutation of y.

The four symbol alphabet D for modeling double stranded DNA has a crucial involution that will be denoted with a prime: [A/T]'=[T/A], [C/G]'=[G/C], [G/C]'=[C/G], [T/A]'=[A/T]. Because we need an involution to model DNA, we assume in splicing theory that the alphabet A is equipped with an involution that is denoted by the prime '. We extend this involution to an anti-automorphism of A^* by defining $(a_1...a_n)' = a_n'...a_1'$. In an abstract case where no specific involution is required, the identity involution a'=a may be assumed. In this case the action of ' on A^* is merely to reverse strings.

Following [8] and [6], sets with multiplicity will be used. Multiplicities will be counted in a semiring R augmented with unary operations of incrementation and decrementation. Only two examples will be considered. Let B be the two element Boolean semiring for which 0+0=0; 0+1=1+0=1; 1+1=1; 0*0=0*1=1*0=0; 1*1=1; incrementing means adding 1; and decrementing both 0 and 1 leaves them unchanged. Since decrementing 1 gives 1, some may prefer to replace the 1 in B by the symbol ∞. Let N be the usual semiring of non-negative integers with the symbol ∞ for infinity adjoined. Addition and multiplication are extended to N in the usual way; incrementing means adding 1; and decrementing means subtracting 1 except that decrementing 0 leaves 0 unchanged.

An R-subset I of a set S is a function from S into R. For each x in S, I(x) is said to be the multiplicity of the occurrence of x in I. The support of I is the set {x in S : I(x)≠0}. When x is in the support of I we say, informally, that x is in I. When the support of I is finite we say that I is finite.

3. Splicing Schemes and Their Actions on A^*

Definition. A splicing scheme is a triple S=(A,T,P), where A is a finite set, T is a finite subset of $A^*XA^*XA^*$, and P is a binary relation on T which satisfies the following condition: If p,x,q, u,y,v are in A^*, (p,x,q) & (u,y,v) are in T, and (p,x,q) P (u,y,v) then x=y. The set A is called the alphabet, T the triples, and P the pairing relation of the scheme.

Informally expressed, the splicing scheme (A,T,P) acts on each pair of strings of the form hpxqk and wuxvz, where each of the letters is a string in A^* and (p,x,q) P (u,x,v), to produce the pair hpxvz and wuxqk. The strings p,q,u, and v serve to allow context restrictions to be applied. The formal statement of the action of (A,T,P) on A^* must be given in terms of subsets of A^* and must take multiplicity into account. This is done in the next paragraph which is based on [6].

Let S=(A,T,P) be a splicing scheme and R a semiring with incrementation and decrementation. Then S defines the following binary relation => in the family of R-subsets of A^*: For R-subsets I and I', I => I' holds if and only if I' can be obtained from I by the following seven steps: Set I'=I. Choose a string hpxqk in I'. Decrement I'(hpxqk). Choose a string wuxvz in I' for which (p,x,q) P (u,x,v). Decrement I'(wuxvz). Increment I'(hpxvz). Increment I'(wuxqk).

Splicing as understood in [14, 10, 4, 5] is subsumed under the above definition as the special case in which R=B. In this fundamental case the two decrementation steps may be deleted and the description of the relation => may be given in the following simplified form: I=>I' holds if and only if there are strings hpxqk and wuxvz in I for which (p,x,q) P (u,x,v) and I' is the union of I with {hpxvz, wuxqk}.

The action of (A,T,P) on I, as defined above, may differ from the intent of [6] when the splicing of two copies of the same string is considered. If R=N and I(s)=1 for a string s then two copies of s cannot be spliced since there do not exit two copies of s in the N-set I.

4. The Action of Splicing Schemes on A^

The behavior of pairs of circular DNA molecules under certain recombination processes ([17,Fig.32.5], [1,Fig.e]) suggests the following: Let S=(A,T,P) be a splicing system and let ^hpxq & ^wuxv be circular strings in A^ for which (p,x,q) P (u,x,v). Then S acts on these two circular strings to produce the single longer circular string ^hpxvwuxq. In the next paragraph we will give our last fully formal definition of a splicing action. It is assumed later that formal definitions can be produced from the informal statements given.

Let S=(A,T,P) be a splicing scheme and R a semiring with incrementation and decrementation. Then S defines the following binary relation => in the family of R-subsets of A^: For R-subsets I and I', I => I' holds if and only if I' can be obtained from I by the following six steps: Set I'=I. Choose an ^hpxq in I'. Decrement I'(^hpxq). Choose a ^wuxv in I' for which (p,x,q) P (u,x,v). Decrement I'(^wuxv). Increment I'(^hpxvwuxq). For the fundamental case R=B, the following is an adequate summary: When (p,x,q) P (u,x,v), {^hpxq,^wuxv} => {^hpxvwuxq}.

The behavior of a single circular DNA molecule having a so called "direct repeat" ([17,Fig.32.6], [1,Fig.1B]) suggests: S acts on each circular string ^hpxqwuxv, for which (p,x,q) P (u,x,v), to produce the pair of circular strings ^hpxv and ^wuxq.

The behavior of a single circular DNA molecule having a so called "inverted repeat" ([1,Fig.1D]) suggests: S acts on each circular string ^hpxqwv'x'u', for which (p,x,q) P (u,x,v), to produce the single circular string ^hpxvw'q'x'u'.

5. The Action of Splicing Schemes on the Union of A^* and A^

The behavior of linear and circular DNA molecules under certain recombination processes ([17,Figs.32.18,32.19], [1,Fig.3]) suggests: Let S=(A,T,P) be a splicing system. Let hpxqk be a linear string in A^* and ^wuxv a circular string in A^ for which (p,x,q) P (u,x,v). Then S acts on this pair to produce the single linear string hpxvwuxqk.

The behavior of a single linear DNA molecule having a "direct repeat" ([17,Figs.32.18,33.9], [1,Fig.3]) suggests: S acts on each linear string hpxqkuxvz for which (p,x,q) P (u,x,v) to produce the pair consisting of hpxvz and ^kuxq. Note that the linear string hpxvz can also be produced by the action of S on two copies, h(pxq)kuxvz and hpxqk(uxv)z, of the given string where the parentheses indicate the triple where the crossover is required.

For the fundamental case R=B, we may summarize these two mixed actions: When (p,x,q) P (u,x,y), {hpxqk,^wuxv} => {hpxvwuxqk} and {hpxqkuxvz} => {hpxvz,^kuxq}. In formalizing the definitions of actions I=>I', when both circular and linear strings are involved, the appropriate domain of both I and I' is the union A^* ∪ A^.

6. Unary Actions of Splicing Schemes

The behavior of a single DNA molecule having an "inverted repeat" ([17,Fig.33.10]) suggests: S acts on each string having the form hpxqkv'x'u'z (resp., ^hpxqkv'x'u'z), with (p,x,q) P (u,x,v), to produce the single string hpxvk'q'x'u'z (resp., ^hpxvk'q'x'u'z). Note that the net effect of this action is to invert the middle of the original string, i.e., in the context hpx___x'u'z', qkv' may be replaced by (qkv')'=vk'q'. When q=v, or more generally, when (p,x,v) P (u,x,q), the string qkv' can be rotated arbitrarily many times when it lies in a context hpx___x'u'z. This is not only a fascinating formal fact, but its DNA interpretation can have the effect of switching a gene off and on. Note that this string hpxvk'q'x'u'z is also obtainable in two steps using the binary action of S: From two copies of s=h(pxq)kv'x'u'z, with the second copy denoted in inverted form s=s'=z'(uxv)k'q'x'p'h', we obtain the desired string hpxvk'q'x'p'h' and also the string z'uxqkv'x'u'z.

7. Splicing Systems and the Languages They Generate

Definition. Let $\underline{R}$ be a semiring with incrementation and decrementation. An $\underline{R}$-splicing system is a quintuple G=(A,T,P,I,J), where (A,T,P) is a splicing scheme, I is an $\underline{R}$-subset of A^*, and J is an $\underline{R}$-subset of A^. The language generated by G is the set (note: not $\underline{R}$-subset) L(G)={s in $A^* \cup$ A^: s is in an $\underline{R}$-subset K of $A^* \cup$ A^ for which $I \cup J =>^* K$}, where $=>^*$ is the transitive closure of =>.

In all previous work on splicing systems the $\underline{R}$-set I has been assumed to be finite and J has not appeared at all. When we wish to totally disallow the occurrence or production of circular strings, as has been the case in all previous literature, we will speak of splicing systems G=(A,T,P,I). In this case L(G) will be a subset of A^*. When we wish to allow the production of circular strings by systems for which the initial set of circular strings is empty, we will speak of splicing systems G=(A,T,P,I,∅). Thus L(A,T,P,I,∅) may properly contain L(A,T,P,I).

8. Summary of Existing Results in Splicing Theory

The achievement of [6] was to introduce the consideration of multiplicity and to prove:

Theorem 1 (Denninghoff & Gatterdam). For every recursively enumerable language L in A^* there is an $\underline{N}$-splicing system G=(A,T,P,I), with I finite, for which L=L(G).

This result was obtained by constructing a system G for which I contains a string s with I(s)=1 and no action is possible without involving the string s. Moreover, G was constructed so that this property is maintained throughout any sequence of actions. Thus if $I=>^*I'$ then I' contains a string s' with I(s')=1 and no action on I' is possible without involving s'. The effect of the maintenance of this property is to impose a sequentiality on the actions of G. Since sequentiality can be imposed in $\underline{R}$-systems, these authors called splicing systems that use multiplicity <u>sequential splicing systems</u>.

Theorem 1 has as a consequence the fact that the membership problem for <u>N</u>-splicing languages is undecidable. From the outset of the study of splicing, a fundamental problem has been to characterize the generating power of splicing systems. This author's intuition, which is based on the case <u>R</u>=<u>B</u>, has always been that this power is extremely low. The result of [6] requires that this intuition be restricted to <u>B</u>-splicing.

In [4,5] Culik and Harju used their previous work on dominoes over a free monoid [3] to make a direct study of linked strings that have the power to represent the so called "restriction fragments" when DNA is acted on by restriction enzymes. (See [14] and either [25] or [17].) From their results on semigroups of such paired strings, they derived the following fundamental result on splicing languages:

<u>Theorem 2 (Culik II & Harju)</u>. For each regular language I and each <u>B</u>-splicing system G=(A,T,P,I), the language L=L(G) is regular.

This Theorem supports the intuition that the generating power of splicing schemes is low when <u>R</u>=<u>B</u>. When I is finite it might be necessary that L=L(A,T,P,I) lie in a highly restricted subclass of the regular languages. Such subclasses have been studied systematically in [19] and [20]. It may not be easy to specify exactly which regular languages are <u>B</u>-splicing languages since in [10] Gatterdam has observed that the non-counting (=aperiodic) language $b(aa)^*$ has the form L(A,T,P,I) with I finite, but the non-counting language $(aa)^*$ does not. In [9] Gatterdam provides algorithms for dealing with <u>B</u>-splicing systems (A,T,P,I) with I finite and he has directed the production of related computer programs.

Although the main accomplishment of [14] was to establish the splicing model as it pertains to linear strings, the class of strictly locally testable languages ([19,20]) was characterized in several ways via splicing systems. A splicing system (A,T,P,I,J) in which every triple in T has the form (1,x,1) is called a null context splicing system. It was shown in [14] that a language L is generated by a null context splicing system (A,T,P,I) with I finite if and only if L is strictly locally testable. This characterization follows directly from A.DeLuca & A.Restivo's characterization [7] of strictly locally testable languages which used Schutzenberger's concept of a constant in a semigroup [24].

9. Problems for Further Study

A prime goal of the present article is to attract interest to circular strings and their interaction with linear strings. The first conjecture requires a preliminary definition:

Definition. A subset S of A^ is regular if there exists a regular language L in A^* for which $\hat{}a_1...a_n$ is in S if and only if at least one cyclic permutation of $a_1...a_n$ lies in L.

It appears that the regularity result of Culik & Harju should hold not only for linear strings, but for the combination of circular and linear strings as well. Thus we conjecture: For every B-splicing system G=(A,T,P,I,J) with I and J regular, both $A^* \cap L(G)$ and $A\hat{} \cap L(G)$ are regular.

In the generation of linear strings from linear strings, it may not be necessary to consider intermediate circular strings. This intuition is formulated precisely as the conjecture: For every scheme (A,T,P) and every subset I of A^*, $L(A,T,P,I,\emptyset) \cap A^* = L(A,T,P,I)$.

The most effective way of investigating splicing actions may be to express these actions in terms of familiar construction procedures that have been studied previously in formal language theory. The class of regular languages is closed under virtually all standard language combining constructions. This may be the source of an alternate proof of the Culik & Harju regularity result. This suggests the problem of generalizing this regularity result by finding classes of languages in A^* such that L(A,T,P,I) belongs to the class whenever I belongs to the class. More generally, for which classes W & X of linear strings and Y & Z of circular strings do we have $L(A,T,P,I,J) \cap A^*$ in X and $L(A,T,P,I,J) \cap A^\wedge$ in Z whenever I is in W and J is in Y? Introductions to the literature on classes of languages that are closed under collections of construction processes appear in [23], [2], and [11].

There remains from [14] the <u>problem</u> of characterizing the languages L(A,T,P,I) where I is finite. The corresponding new version of this problem is to characterize the languages L(A,T,P,I,J) where both I & J are finite. The purely circular languages L(A,T,P,∅,J) have been neglected here. Surely these must be regular whenever J is regular? Even so, the <u>problem</u> remains to characterize those for which J is finite. Finally, the present study of splicing systems may be regarded as only the first step in a larger program of the development and study of formal representations of the behaviors of the informational macromolecules and the enzymes that act on them [14,Sec.6].

10. Special Conditions for Splicing Schemes Modeling DNA Behavior

As explained in [14], in modeling DNA the strings $x_1 \ldots x_n$ and $x_n' \ldots x_1'$, with the x_i in D, must be regarded as notations for the <u>same</u> molecule because molecules lie in three dimensional space and not on a line. Likewise, $^\wedge x_1 \ldots x_n$ and $^\wedge x_n' \ldots x_1'$ must be regarded as notations for the <u>same</u> molecule. On the other hand, following the explanation in [14] in terms of 3' and 5' bounds, $^\wedge x_1 \ldots x_n$ and $^\wedge x_n \ldots x_1$ need not represent the same molecule for the same reason that $x_1 \ldots x_n$ and $x_n \ldots x_1$ needn't. As a consequence, in modeling DNA behavior with <u>R</u>-splicing schemes, the basic set of linear strings is not A^* itself but the set of equivalence classes formed in A^* when each pair of strings x and x' are equated. When <u>R</u>=<u>N</u> this fact must be kept clearly in mind, especially in understanding the effect of incrementation and decrementation. When <u>R</u>=<u>B</u> no problem arises over incrementation and decrementation due to the triviality of these operations. As was pointed out in Sec.2 for splicing schemes in general, A^ consists of equivalence classes of strings in A^* that are circular permutations of each other. In modeling DNA behavior with R-splicing schemes, these equivalence classes must undergo one further collapse, namely, for each ^x, the class containing ^x and the class containing ^x' must be united to form a single class. As in the linear case, this collapse must be kept clearly in mind when multiplicity relative to N is used, but causes little problem when B is used.

We note that in the biological literature the involution ' on D is not always made explicit. The result is that symbols denoting subsegments of DNA molecules in such literature are sometimes not easily read with complete accuracy. One must keep in mind that x and x' can not in general be substituted for each other within a context. If a is a symbol for which a≠a', as is the case for all four symbols in D, then axa'≠ax'a' unless x is a palindrome, i.e., unless x=ww' for some string w or x=wbw' for a symbol b for which b=b'.

In addition to the altered understanding of A^* and A^, that is appropriate for dealing with a splicing scheme S=(A,T,P) modeling DNA behavior, T and P may be assumed to satisfy additional conditions. For each triple (p,x,q) in T, (q',x',p') may be assumed to be in T. This is a consequence of the identity of strings hpxqk and k'q'x'p'h' and is explained in [14,p.744]. The definitions of the actions of a scheme make it possible to add the assumption that P is symmetric without effecting the generating power of schemes. For modeling the behaviors discussed in [14], P must be assumed to be reflexive.

It would be reasonable to restrict ones study of splicing schemes to those that have the extra properties appropriate for modeling DNA behaviors. An interesting problem is to determine whether there are significant structural features that hold for splicing languages based on such restricted schemes that do not hold for splicing languages in general.

The principal focus of [14] was on the generative capacity of the iteration of the process of cutting DNA with a finite set of restriction enzymes followed by restoration through ligation. Each cut produces either a 5' overhang, a 3' overhang, or no overhangs = blunt ends. Ligation is possible exactly between (Chargaff) matching overhangs of the same type. This allowed the triples T to be subdivided into two subsets B and C where B consisted of the triples associated with enzymes that produce either 5' overhangs or blunt ends and C consisted of the triples associated with enzymes producing 3' overhangs. In these terms the binary relation P in T may be specified by: (p,x,q) P (u,y,v) if and only if x=y and either (p,x,q) & (u,x,v) both lie in B or they both lie in C. The use of the sets B and C in [14] is natural in enzymatic terms, but not in mathematical terms. For a formal mathematical theory of splicing the present formulation in terms (A,T,P) may be expected to yield results of greater generality having proofs that are more naturally presented.

References

[1] Berg,D.E.: Genomic rearrangements in prokaryotes. In: Gene Rearrangement, B.D.Hames & D.M.Glover, Eds. Oxford: Oxford Univ. Press 1990

[2] Berstel,J.: Transductions and Context-Free Languages. Stuttgart: Teubner 1979

[3] Culik II,K. and T.Harju: Dominoes over a free monoid. Theoretical Computer Science, 18, 279-300(1982)

[4] Culik II,K. and T.Harju: The regularity of splicing systems and DNA. In: 16th International Colloquium on Automata Languages and Programming. Lecture Notes in Computer Science, Vol. 372,pp.222-233. Berlin-Heidelberg-New York: Springer 1989

[5] Culik II,K. and T.Harju: Splicing semigroups of dominoes and DNA. Discrete Applied Mathematics, (to appear).

[6] Denninghoff,K.L. and R.W.Gatterdam: On the undecidability of splicing systems. International Journal of Computer Mathematics, 27, 133-145(1989)

[7] De Luca,A. and A.Restivo: A characterization of strictly locally testable languages and its application to subsemigroups of a free semigroup. Information and Control, 44, 300-319(1980)

[8] Eilenberg,S.: Automata, Languages, and Machines. Vol.A. New York: Academic Press 1974

[9] Gatterdam,R.W.: Algorithms for splicing systems. Univ. of Alaska, Math.Dept. Technical Report

[10] Gatterdam,R.W.: Splicing systems and regularity. International Journal of Computer Mathematics, 31, 63-67(1989)

[11] Ginsburg,S.: Algebraic and Automata-Theoretic Properties of Formal Languages. New York: North Holland/Elsevier 1975

[12] Goldberg,D.E.: Genetic Algorithms. Reading, Mass.: Addison-Wesley 1989

[13] Hames,B.D. and D.M.Glover,Eds.: Gene Rearrangement. Oxford: Oxford Univ. Press 1990

[14] Head,T.: Formal language theory and DNA: an analysis of the generative capacity of specific recombinant behaviors. Bulletin of Mathematical Biology, 49, 737-759(1987)

[15] Herman, G.T. and G. Rozenberg: Developmental Systems and Languages. New York: North Holland/American Elsevier 1975

[16] Hopcroft,J.E. and J.D.Ullman: Introduction to Automata Theory, Languages, and Computation. Reading, Mass: Addison-Wesley 1979

[17] Lewin,B.: Genes IV. New York: Oxford Univ. Press 1990

[18] Lindenmayer,A.: Mathematical models for cellular interactions in development I,II: Journal of Theoretical Biology, 18, 280-315(1968)

[19] McNaughton,R. and S.Papert: Counter-Free Automata. Cambridge, Mass.: MIT Press 1971

[20] Pin,J.E.: Varieties of Formal Languages. New York: Plenum 1986

[21] Prusinkiewicz,P. and A.Lindenmayer: The Algorithmic Beauty of Plants. New York: Springer-Verlag 1990

[22] Rozenberg,G. and A.Salomaa: The Mathematical Theory of L-Systems. New York: Academic Press 1980

[23] Salomaa,A.: Formal Languages. New York: Academic Press 1973

[24] Schutzenberger,M.P.: Sur certaines operations de fermeture dans les langages rationnels. Symposia Mathematica, 15, 245-253 (1975)

[25] Watson,J.D., J.Tooze, and D.T.Kurtz: Recombinant DNA: A Short Course. New York: Freeman 1983

CELLWORKS : AN APPLICATION TO PLANT MORPHOGENESIS

Jacqueline Lück and Hermann B. Lück

Laboratoire de Botanique analytique et Structuralisme végétal
Faculté des Sciences et Techniques de St-Jérôme, 13397
Marseille cedex 13, France

Abstract: Cellwork-systems with double face labeling are used to investigate the division possibilities of autoreproductive tetrahedral cells. In plant morphogenesis, such cells, in apical position, are at the origin of the cell-rows which compose the shoots. The difference between 1, 2, or 3-rowed elementary theoretical shoots, with compact or tunica-like cell assemblage, can be traced back to specific orientations of the apical cells in their filiation.

Keywords: Cellwork-systems, morphogenesis, plant development

1. Introduction

Parallel rewriting dw-cellwork L-systems (dw for double-wall labeling) have been proposed [3] as topological models simulating the development of the cellular pattern of plant meristems. Such systems can be used to elucidate the apical organization of plants. Especially, in simple apices like those of ferns and mosses, they describe the behaviour of the derivatives of the helicoidal segmentation of initial apical tetrahedral cells. The aim is to investigate if deterministic or quasi-deterministic rules of cell-wall siting are operative and if such rules are sufficient to distinguish the shoot organization of different ferns and mosses. For this purpose, the present study shows which pattern can be expected from consecutive divisions of an initial tetrahedral cell. In the 2-dimensional case, several map systems have proved that such apical cells are capable not only to generate moss- and fern-like structures, but even phyllotactic patterns of higher plants (cf. bibliographie in [3]).

The double labeling of the polyhedral cell-faces are adopted in analogy to real plant cell walls which are composed of the walls of two contiguous cells (Fig.1). Instead of an unique wall label, we attribute a label to each of the sides of a

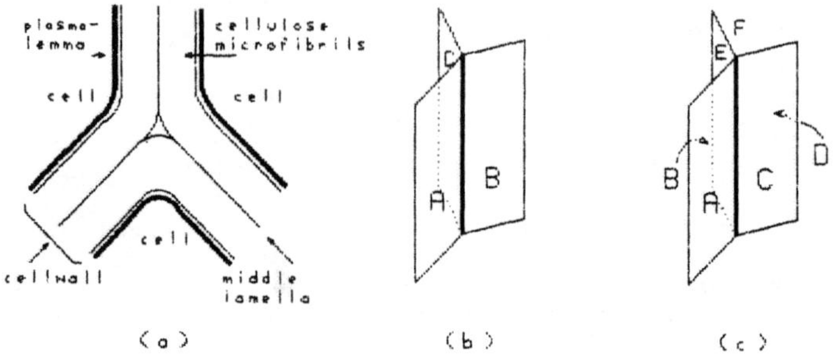

Fig.1: The double-labeling of plant cell walls; (a): optical 2D-cut through a plant tissue, 3 cells meet at a corner; (b) and (c): in 3D-cellworks 3 cells meet at an edge; (b): single-wall labeling: one label on each wall; (c): double-wall labeling: one label for each side of a wall.

wall. Other 3D-rewriting-systems based on edge labels have been proposed by [1,2].

2. Representation of a tetrahedral cell

A tetrahedron represents the wall boundary of a 4-faced cell with face labels viewed from within the polyhedron (Fig.2a): the label A always represents the previous division wall (otherwise division wall labels have to be underlined, cf. [3]) and its neighbour faces are labeled clockwise B, C, and D. A cell wall is shared either with a neighbour cell or is a free surface exposed to the environment of the shoot.

A polyhedral cell face, viewed from within the cell, is described by its label and, in parentheses, by the clockwise circular sequence of neighbour wall labels, called wall environment: e.g. A(BCD) represents the division wall. If the label of a neighbour wall is not specified, it is replaced by the label e. The polyhedral wall boundary of a cell, when cut off at the previous division wall, becomes representable by a wall-map which gives the topological relationship of the walls as viewed from within the cell (Fig.2b). A wall-map, which is a planar view of the opened cell boundary, is designated by a cell label x_y with

x: the label of the cell type characterized by the map, and

y: subscript of x, which indicates the number of cell-faces.

The wall-map of the tetrahedron is named accordingly by the corresponding cell label a_4. The division wall A is figured on

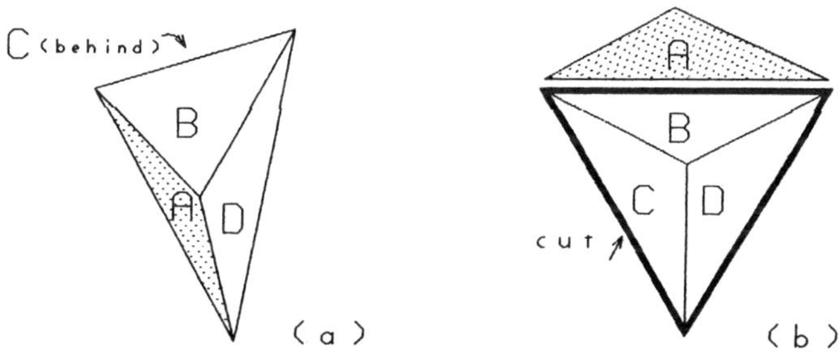

Fig.2: Tetrahedral cell: (a) 3D-view with inner face labels; (b) wall-map: opened planar view of the walls of a cell having 4 faces.

the byside; the boundary of the map is the cut, a sequence of edges by which the division-wall A is attached to other walls.

A wall-map is also defined by the set of its wall labels. The following list of walls describes the wall-map of the cell a_4:

$$\text{wall-map} \quad a_4 = \left\{ \begin{array}{l} A^{a4} \ (BCD) \\ B^{a4} \ (ADC) \\ C^{a4} \ (ABD) \\ D^{a4} \ (ACB) \end{array} \right\} .$$

The exponent of each wall symbol indicates the cell type to which it belongs. Using cell labels, exponents are not absolutely necessary. They are omitted if the attribution of

3. Cell division

The cell division is considered as a partition of a polyhedron by introducing a partition or division wall. This wall splits each polyhedral face, on which it is anchored, into two parts. Accordingly, the growth of the cellwork is performed by addition of new walls and bipartition of existing walls. We suppose that the new division wall is inserted at a determined position with respect to the wall formed during the previous cell division and submitted to constraints of the polyhedral shape.

In our systems, a cell division is defined as a bipartition of a wall-map (Fig.3). For this purpose, a sequence of new edges dividing some of the walls, is introduced by means of wall production rules. These edges will be the anchor for the new wall. Furthermore, these new edges induce by contact, on the complementary neighbour faces, also wall partitionings. As a consequence, the number of walls in neighbour cells is increased. Map bipartitions for the polyhedral division, and map transformations for induced wall divisions are both specified by sets of wall production rules. Such a set can also be considered as a map production rule.

3.1. MAP BIPARTITIONS

During a bipartitioning of a wall map (Fig.3) which defines a cell division,

- a wall (i.e. a region of the map) is split at most once,
- the cut of the map is realized by a sequence of edges which, in 3D, is a cycle (a closed line).

After splitting, the labeling of the new walls is arbitrary. However, we use the following rule:

- the label A is reserved for the division wall;
- the walls on each side along the cut are labeled first and counter-clockwise,
- the remaining walls are labeled afterwards.

In the presently studied example, the wall-map a_4 is split as indicated in Fig.3b (it represents the division pattern d_1 discussed in chapter 6). It results in a daughter map a_4 to which belong three regions lying on one side of the cut to which will be added the new division wall A^{a4}. The second daughter map, b_5, results from four regions situated on the opposite side of the cut, plus a division wall A^{b5} (Fig.3b,c).

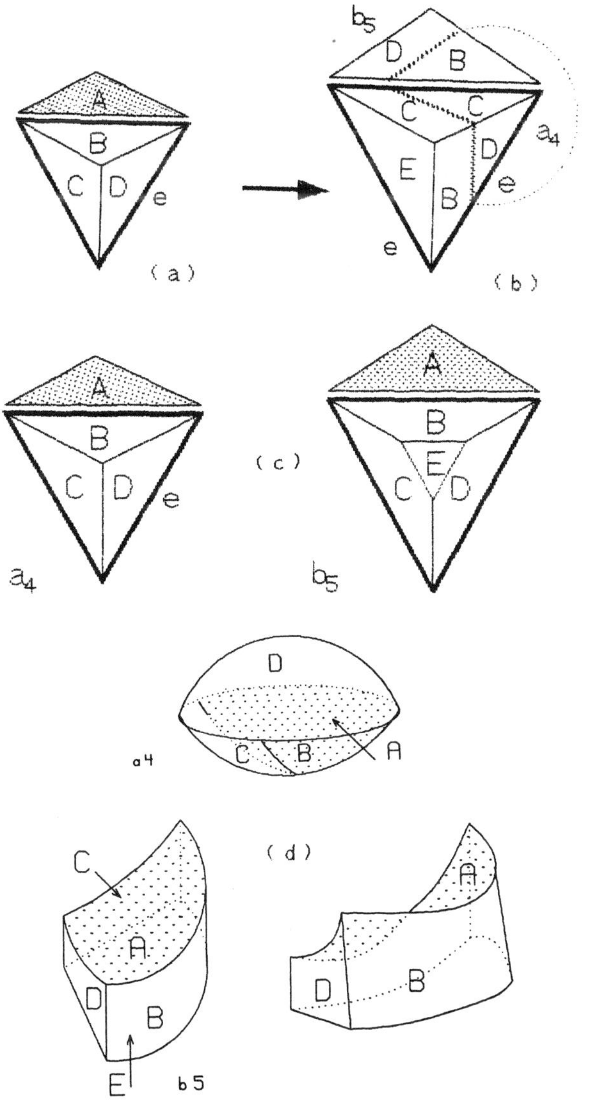

Fig.3: Wall-map representations; (a): of the tetrahedral cell a_4; (b): map-splitting, stippled lines for anchor edges of the division wall; (c): a_4 and b_5, daughter cells resulting from the splitting; (d): 3D-representation of the cells; free geometrical interpretation of the topological figures; a_4 with one free face D, the other 3 faces embedded in the tissue; b_5 viewed under 2 different perspectives; stippled surface: division wall by which sister cells are glued together.

A set of wall productions describes this map splitting:

$$P_{a4} = \left\{ \begin{array}{lll} A(BCD) \rightarrow & B(ADC)\ {}^{a4}/_{b5} & D(ACEB) \\ B(ADC) \rightarrow & C(ABD)\ {}^{a4}/_{b5} & C(ABED) \\ C(ABD) \rightarrow & & E(DCB) \\ D(ACB) \rightarrow & D(ACB)\ {}^{a4}/_{b5} & B(ADEC) \end{array} \right\} .$$

In a wall production, the two parts which result from the splitting are separated by a slash or marker. The names of the new daughter cell types to which belong the two walls are indicated on both sides of the marker (here a_4 on the left of the marker, and b_5 on its right). The following wall-maps represent the daughter cells (their map representation in Fig.3c) indicated by their list of walls:

$$a_4 = \left\{ \begin{array}{l} A(BCD) \\ B(ADC) \\ C(ABD) \\ D(ACB) \end{array} \right\} \text{ and } b_5 = \left\{ \begin{array}{l} A(BCD) \\ B(ADEC) \\ C(ABED) \\ D(ACEB) \\ E(DCB) \end{array} \right\} .$$

The polygonal degree of a wall is indicated by the length of its wall environment (in parentheses). So, the daughter cell a_4 has four 3-sided walls; it is a tetrahedral cell, identical to its mother cell. In other words, the tetrahedral cell is autoreproductive like apical cells in plant meristems. The cell b_5 has two 3-sided walls and three 4-sided ones. The division-wall A has 3 sides because its label occurs three times in the environment of the other walls.

The more realistic 3D-outline (Fig.3d), which is a free geometrical interpretation of the topological generated figures, respects the fact that the division of the tetrahedron is parallel to the face C. In order to draw b_5, the production D() → D()/B() gives the orientation with respect to the cell a_4. It has to be noted that if walls are viewed from the outside of the cells, their inner wall neighbourhood is read counter-clockwise.

3.2. INDUCED WALL DIVISIONS

The division of the tetrahedral cell can be indicated by the following cell production showing that it reproduces itself with the set P_{a4} of wall productions:

$$a_4 \xrightarrow[P_{a4}]{} a_4 b_5 \quad .$$

During its further divisions, there are contact induced wall splittings in its derivatives. It follows that the cell b_5 will be transformed into a cell b_6, with one wall more.

By touching a neighbour division wall, the face A() of b_5 is split into faces $A^{b6}()$ and $\bar{A}^{b6}()$ specified by the first production in P_{b5}. The marker has no indications as both walls resulting from the splitting remain in the same cell. The complementarity is underlined by dashes (A and $\bar{A}$, for example: in consecutive induced splittings of a same wall, we use double, triple, etc. dashes). The wall environments have to be corrected in the other productions according to the increase in wall numbers (As the wall C only figures in the environment of $\bar{A}$, it cannot appear in the environment of C). The set of wall transformations becomes:

$$P_{b5} = \left\{ \begin{array}{lll} A^{b5}(DBC) & \to & A^{b6}(\bar{A}^{b6}DB) \ / \ \bar{A}^{b6}(A^{b6}BCD) \\ B(ADEC) & \to & B(\bar{A}ADEC) \\ C(ABED) & \to & C(\bar{A}BED) \\ D(ACEB) & \to & D(A\bar{A}CEB) \\ E(DCB) & \to & E(DCB) \end{array} \right\} \quad .$$

It results in the cell type b_6 with its wall-map:

$$\text{wall-map } b_6 = \left\{ \begin{array}{l} A(\bar{A}BC) \\ \bar{A}(ABCD) \\ B(\bar{A}ADEC) \\ C(\bar{A}BED) \\ D(A\bar{A}CEB) \\ E(DCB) \end{array} \right\} \quad .$$

The cell boundary has two 3-sided faces, two 4-sided faces, and two 5-sided ones (3D-outline of b in Fig.5). The cell transformation is noted: $b_5 \xrightarrow[P_{b5}]{} b_6$.

A second induced wall division transforms a cell b_6 into a cell b_7 with the set P_{b6} of wall productions:

$$b_6 \xrightarrow[P_{b6}]{} b_7 \quad .$$

It is the wall A^{b6} which is divided by induction into A^{b7} and $\bar{\bar{A}}^{b7}$, as formulated by the following set of productions (the division of A() and its repercussion on the wall environments of all other walls has been considered):

$$P_{b6} = \left\{ \begin{array}{lcl} A^{b6}(\bar{A}^{b6}DB) & \rightarrow & A^{b7}(\bar{\bar{A}}^{b7}B\bar{A}^{b7}) \;/\; \bar{\bar{A}}^{b7}(A^{b7}\bar{A}^{b7}DB) \\ \bar{A}^{b6}(A^{b6}BCD) & \rightarrow & \bar{A}^{b7}(\bar{\bar{A}}ABCD) \\ B(\bar{A}ADEC) & \rightarrow & B(\bar{A}A\bar{\bar{A}}DEC) \\ C(\bar{A}BED) & \rightarrow & C(\bar{A}^{b7}BED) \\ D(A\bar{A}CEB) & \rightarrow & D(\bar{\bar{A}}\bar{A}CEB) \\ E(DCB) & \rightarrow & E(DCB) \end{array} \right\} .$$

(Wall A does not appear in the environment of $D(\bar{\bar{A}}\bar{A}CEB)$ because $A^{b7}(\bar{\bar{A}}^{b7}B\bar{A}^{b7})$ does not contain D in its wall environment). It results in the cell type b_7 (3D-representation in Fig.5) with the list of its walls:

$$\text{wall-map } b_7 = \left\{ \begin{array}{l} A(\bar{\bar{A}}B\bar{A}) \\ \bar{A}(\bar{\bar{A}}ABCD) \\ \bar{\bar{A}}(A\bar{A}DB) \\ B(\bar{A}A\bar{\bar{A}}DEC) \\ C(\bar{A}BED) \\ D(\bar{\bar{A}}\bar{A}CEB) \\ E(DCB) \end{array} \right\} .$$

The cell boundary has two 3-sided faces, two 4-sided faces, two 5-sided faces, and one 6-sided face. Afterwards, there are no further induced wall divisions in the derivatives. The cell b_5 is the youngest merophyte and b_7 the eldest merophyte. From the state b_6 on, the merophyte can undergo further cell divisions, which are not considered here.

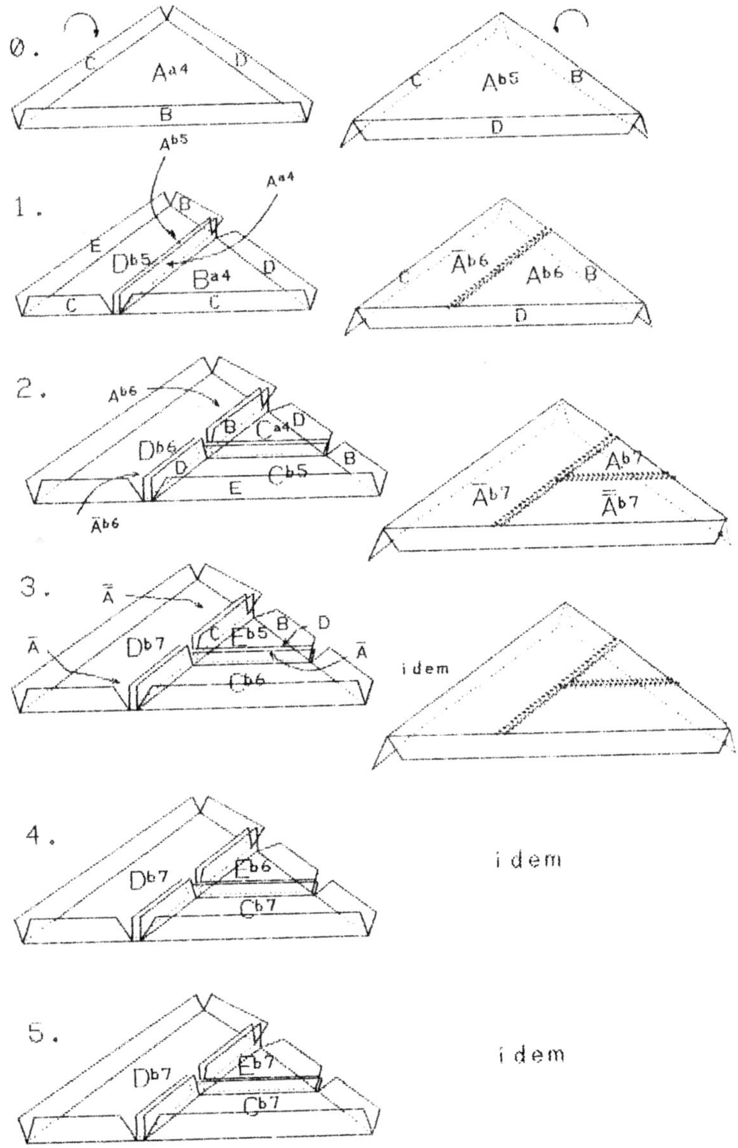

Fig.4: Five map derivation steps (0. ... 5.) of a division wall; initial map 0.: division wall; its two sides dissociated, A^{a4} belonging to the cell above (cell environments read clockwise as looked from inside the cell), A^{b5} belonging to the cell underneath (wall-environments read counter-clockwise as viewed from the ouside of the cell). Further explanations in text.

4. Wall matching

The division-wall derivation is a map sequence representing the development of a division wall. The cellwork system generates a map as the initial state is the map representation of a division wall with its two face labels (Fig.4). Each map of this sequence is an assemblage of walls (regions) belonging to different cells which divide during the development of the tissue. Therefore, the neighbour regions of a wall do not represent its wall-environment. This is represented by walls going upwards or downwards with respect to the map (Fig.4).

For investigating the fate of each of the two complementary sides of the division wall, they are dissociated in their map representation: one of the faces is considered as a map with clockwise labeling, the other as a map with counter-clockwise labeling.

The 3D-conformation of cells have been easily deduced from the wall-maps, but the accurate assemblage of the different cell types needs wall matching rules which are implicite to the division wall derivation. Accordingly, we can construct for the given example the elementary shoot (Fig.5). It is composed of a tetrahedral apical cell in contact with three derivatives respectively 5, 6, and 7-sided (Fig.5a). Beneath this apical complex, there are three cell-rows with helicoidal packing of their constituants (Fig.5b).

The maps of the division-wall derivation contain all edges used for the construction of the tissue. They can be used either to define faces used in a single wall cellwork system with face labels, or to define edges in an edge-label cellwork system.

5. Cellwork-system

The dw-cellwork system is defined by a wall-map alphabet, subsets of map division and map transformation rules, and an axiom which consists at least of a cell whose boundary is described by a wall-map. In addition, face alignement rules are needed to specify the contact between neighbour-cells. They are defined by wall matching during the division wall derivation.

The cellwork derivation is obtained by rewriting in parallel all wall-maps contained in a cellwork. New walls are inserted according to the circular sequence of edges specified by the

$$a_4 \rightarrow a_4\ b_5$$
$$b_5 \rightarrow b_6$$
$$b_6 \rightarrow b_7$$
$$b_7 \rightarrow b_7$$

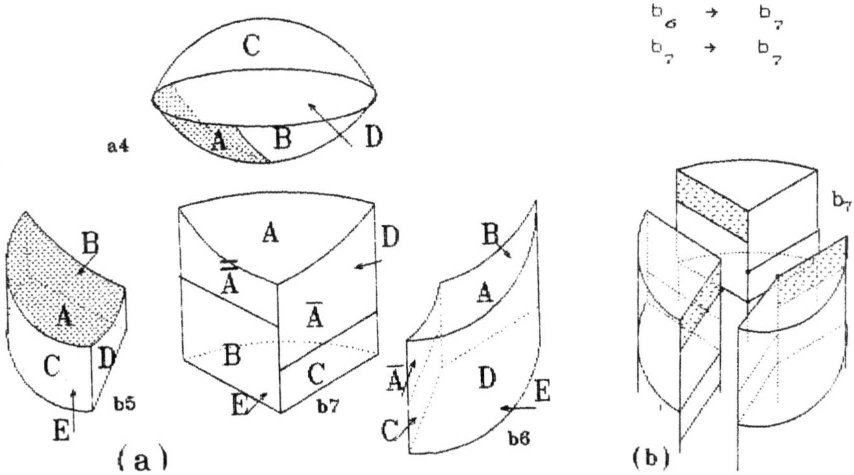

Fig.5: Theoretical elementary 'plant shoot' resulting from the activity of a tetrahedral apical cell: (a): realistic representation; apical cell a_4 with 3-sided helicoidal segmentation; b_5 and b_6: intermediary cell-states of derivatives; b_7: final cell type of derivatives; (b): the shoot is build up by 3 vertical cell-rows.

cut of the splitted maps. The exact topological relationship of cells is obtained by deriving in parallel the two complementary faces of a division wall. This furnishes the derived alphabet over double-walls. The cellwork system generates the topological cell organization of a 3-dimensionnal cellular tissue.

For a same division device of the tetrahedral apical cell, different wall matchings are possible. We consider the division rules as given genetically and related to the cortical structure of a cell. According to the production rules, this structure is reestablished in autoreproductive cells notwithstanding a global reorientation. Nevertheless, we suppose that the wall matching is random in the framework of the possibilities, introducing some variability to the cellular assemblage.

In the given example, there is an alphabet of four wall-maps (wp), $\Sigma = \{wp\ a_4, wp\ b_5, wp\ b_6, wp\ b_7\}$, and a set of wall-map divisions and transformations $P = \{P_4, P_5, P_6, P_7\}$. P_{a7} states

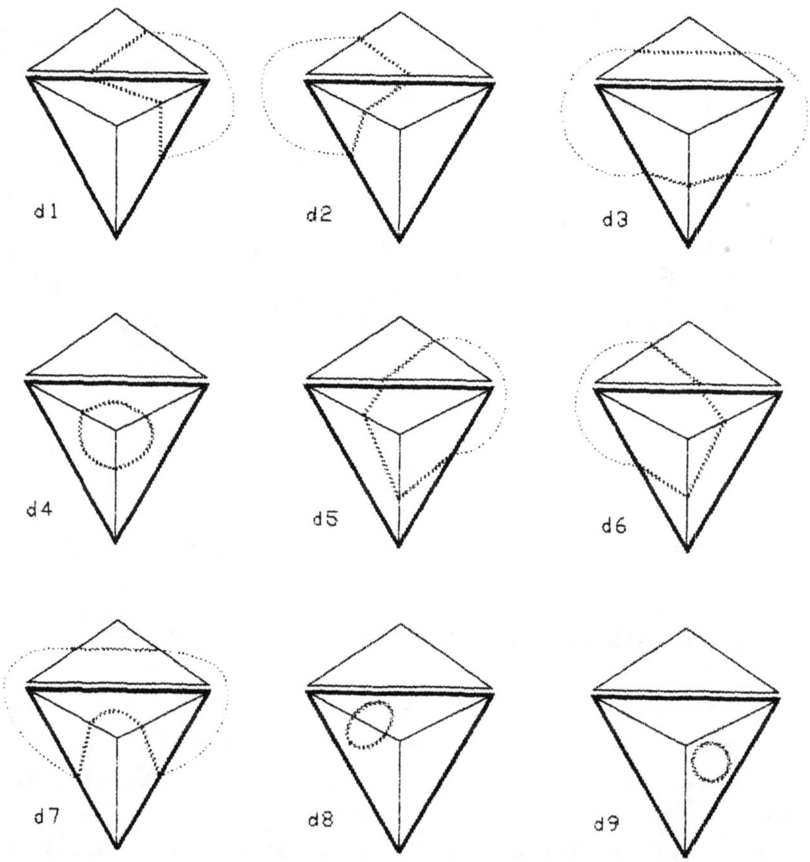

Fig.6: Different possibilities of map splitting (stippled line: cut) through the wall-map of a tetrahedral cell (under constraints indicated in the text).

the transformation of $b_7 \rightarrow b_7$. As in our example the derivatives of the tetrahedral cell do not divide further on, the wall matching rules give only one possibility of cell assemblage, indicated by Figs.4 and 5.

6. Division possibilities of the tetrahedral cell

Theoretically, there are as many division patterns of the tetrahedral cell than possible partitionings of its wall-map.

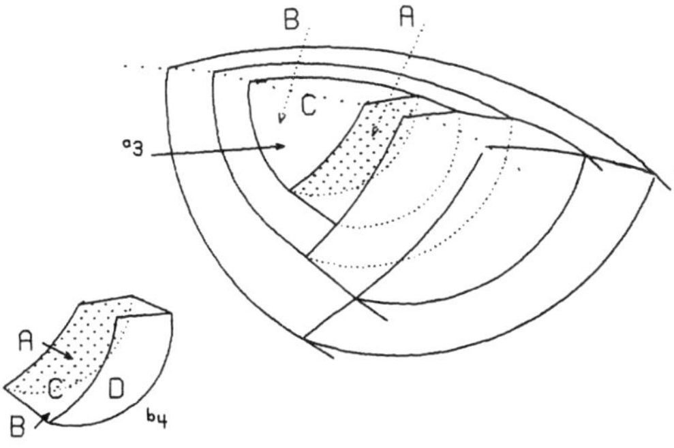

(a)

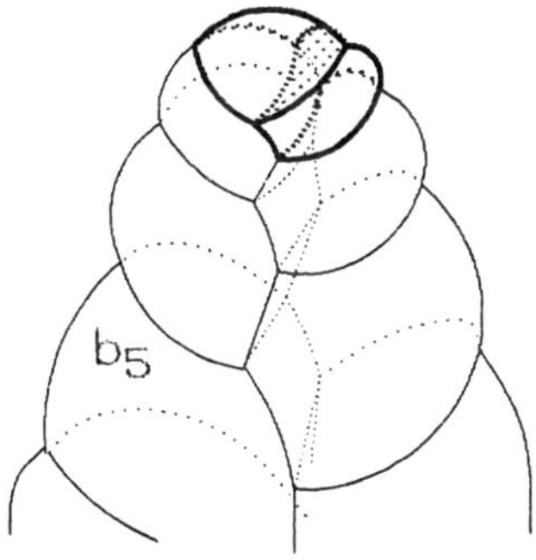

(b)

Fig.7: Theoretical elementary 'plant shoot' resulting from the activity of a 3-sided apical cell, a_3, with 2-sided alternate segmentation; realistic 3D-representation; b_4: intermediary cell-state of derivatives of a_3; b_5: final cell type; the shoot build up by 2 vertical cell-rows.

Most of them are shown in Fig.6. Nevertheless, the distribution of wall labels on the splitted regions of a wall-map also controls the form of the generated cellular body.

6.1 WALL-MAP PARTITIONS

In the example of the helicoidal 3-sided segmentation of the tetrahedral apical cell (partition d_1 in Fig.6), the division goes through three walls including the previous division wall A

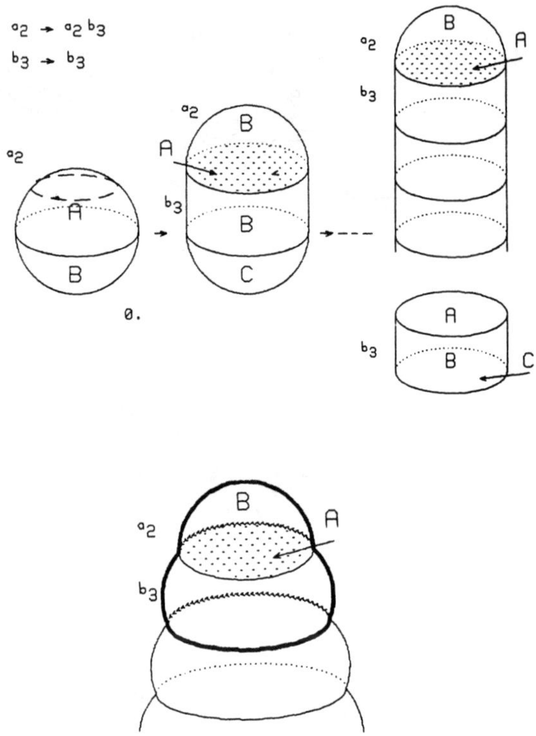

Fig.8: Theoretic a l elementary 'plant shoot' reduced to a filament with a 2-sided apical cell and 1 cell-row.

and the division wall is parallel to the fourth remaining wall. If only two of the four walls are split (case d_2, Fig.6), it follows a 3-sided apical cell which generates segments, alternately in two opposite directions, in parallel to the two embedded walls (Fig.7a). The cell productions are

$$a_3 \longrightarrow a_3 b_4 \quad ,$$

$$b_4 \longrightarrow b_5 \quad , \qquad \text{and} \qquad b_5 \longrightarrow b_5 \quad .$$

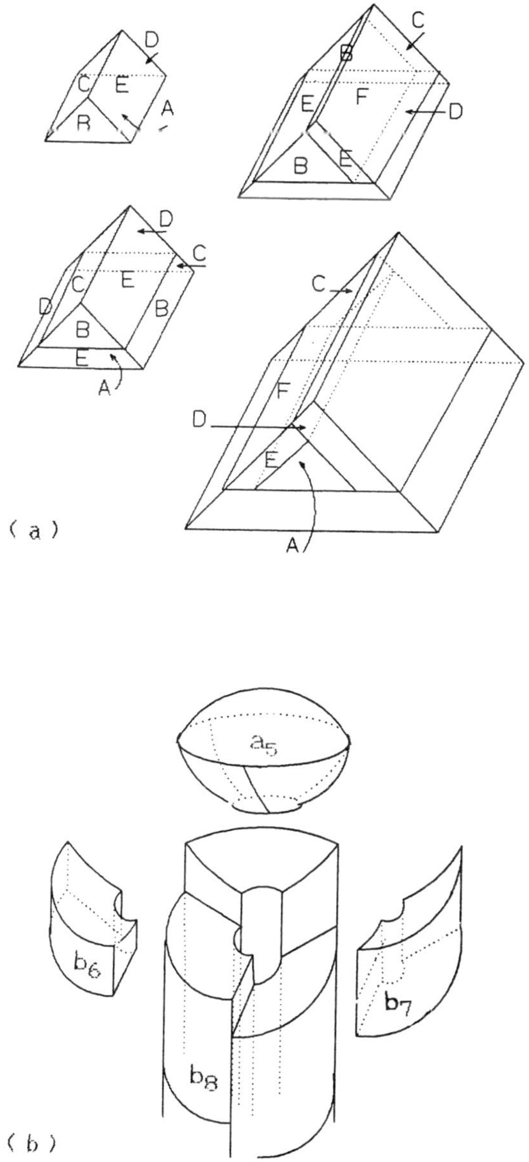

Fig.9: Elementary 'plantshoot'with a prismatic 5-sided apical cell; (a): helicoidal 3-sided segmentation; (b): realistic 3D-representation, b_5, b_6, b_7, and b_8 cell types of derivatives; 'tunica'-like cell assemblage with 3 cell-rows.

In a geometrical realistic representation (Fig.7b), the merophytes are disposed in two vertical rows.

If the partitioning of the tetrahedron only affects one wall (case d_{10}, Fig.6), the corresponding cellwork system generates a cellular filament through the division activity of an apical 2-sided cell (Fig.8). The pattern only needs two types of cells as indicated by the cell productions:

$$a_2 \longrightarrow a_2 b_3 \text{ , and } b_3 \longrightarrow b_3 \text{ .}$$

The case in which the cut of a wall-map goes through all the four walls of the tetrahedron (d_5, Fig.6) leads to a 5-sided apical cell; its shape and division behaviour are indicated in Fig.9. It derives segments in the helicoidal 3-sided manner shown in Fig.5. However, the 3 vertical rows of cells (called merophytes if they undergo further divisions) do not form a compact cell body. They form a hole like a finger of a glove (Fig.9). In plant shoots, the apical organization shows frequently several of such layers called 'tunica'. The corresponding cell productions are:

$$a_5 \longrightarrow a_5 b_6 \text{ , } b_6 \longrightarrow b_7 \text{ , } b_6 \longrightarrow b_7 \text{ , and } b_6 \longrightarrow b_7 \text{ .}$$

6.2 DISTRIBUTION OF WALL LABELS

The distribution of wall labels over the walls created during a wall-map partition plays an important role in the reorientation of the daughter cells with respect to the mother cell. This is especially important for the autoreproductive apical cells. For the case d_1 of splitting (Fig.6), concerning the example of Fig.5, there are 6 possibilities to label the daughter cell a_4 which relays the apical mother cell (Fig.10). In four of these cases, the corresponding cellwork-systems generate a 3-sided helicoidal segmentation (like in Fig.5), in two of them the helix is dextre, and in the other two, it is senestre. The other two remaining cases lead to the 2-sided alternate segmentation pattern (like in Fig.7). Nevertheless, this last pattern is said 'amplified' as realized with a 4-sided apical cell instead of the 3-sided one; and two walls remain always associated as a unity. From the morphological point of view, the cellular body has the same 'Bauplan' as that realized with the minimal system based on a 3-sided apical cell.

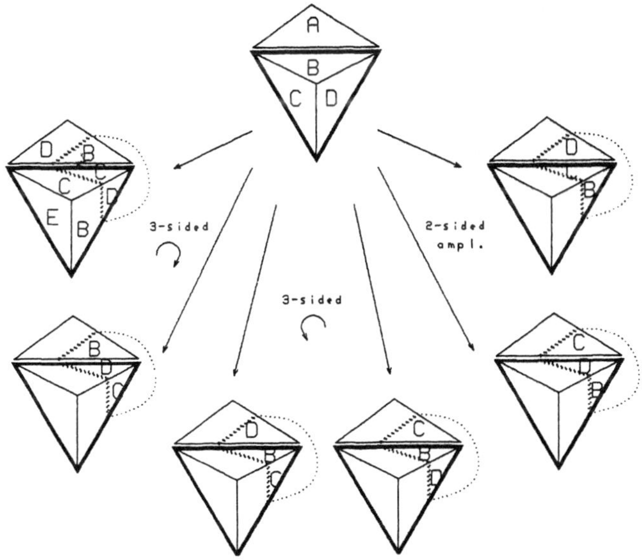

Fig.10: Different possibilities of labeling for a wall-map bipartition.

7. CONCLUSION

Morphogenesis is closely related to the disposition of cells in their filiations and points to the existence of deterministic laws for the positional laydown of new walls. We suppose that this possible basic deterministic behaviour in the cell organization is genetically programmed but (1) soon overwritten by physiological influences such as dorsiventrality, hormone gradiants, etc., (2) also masqued by the fact that on the shared wall of two contigous dividing cells, there is some freedom in the relative positioning of the two anchor edges. In other words, the relative position of cell families (segments or merophytes) determines the overall morphology and depends on the division behaviour of the apical cell. Nevertheless, there is some variability over the cell assemblage which is reinforced by variable lengths of cell cycles. This hypothesis is a consequence deduced from the behaviour of cellwork-systems and can be tested by simulation through them.

REFERENCES

1. Fracchia, F.D. & Prusinkiewicz, P.: Physically-based graphical interpretation of marker cellwork L-systems. In: Graph grammars and their application to computer science (H. Ehrig, H.-J. Kreowski, and G. Rozenberg, eds). Lecture Notes in Computer Science. Berlin: Springer 1991(to appear)
2. Lindenmayer, A.: Models for plant tissue development with cell division orientation regulated by preprophase bands of microtubules. Differentiation 26: 1-10 (1984)
3. Lück,J. & Lück, H.B.: Double-wall cellwork systems for plant meristems. In: Graph grammars and their application to computer science (H. Ehrig, H.-J. Kreowski, and G. Rozenberg, eds). Lecture Notes in Computer Science, pp. 564-581. Berlin: Springer 1991 (to appear)

6. Crossroads and Branching Path:

Variations and Generalizations of L Systems, Related Models

Parallel Communicating Systems of L Systems

Gheorghe Păun

Institute of Mathematics, Str. Academiei 14, 70109 Bucuresti, Romania

Abstract: The generative capacity of various types of (centralized) parallel communicating systems of L systems is investigated, comparing it with the power of usual systems.

Keywords: L Systems, parallel communicating grammar systems.

1 Introduction

The parallelism is a very modern topic in computer science, important from the practical point of view and appealing from the theoretical one. In formal language theory, the parallelism is studied for many years, mainly in the frame of L-systems, originated by Aristid Lindenmayer in [11]. There are also Chomsky context-free grammars with parallel derivations (the so-called Russian and Indian grammars; see references in [4]). In these generative devices, the parallelism is localized at the level of rewriting the current sentential form: all its symbols, in L-systems, and all occurrences of a rule left hand nonterminal, in Russian/Indian grammars, are rewritten at each derivation step.

Another type of parallelism occurs in the frame of *grammar systems*. There we do not have *one* generative device for producing a language, as in the classical formal language theory, but a *system* of generative devices, working together in some well defined way for producing a language. When the components of the system work sequentially (at each moment only one is enabled and rewrites the common sentential form) we obtain the cooperating/distributed grammar systems, first considered in [12], and then investigated in a series of papers, [1], [2], [3], [5], [6], [7], [9], under various computer science, artificial intelligence, psychology, etc. motivations. However, the components of a grammar system can also work in parallel (each having its own sentential form) and communicate in some way each other (sending the currently generated string to other component); thus the parallel communicating grammar systems are obtained. They were introduced in [16] and investigated in [13], [14], [15], [17].

Here we bring together the two types of parallelism and consider parallel communicating grammar systems whose components are L-systems (of various types). This is mainly motivated by theoretical reasons, but seems to have also some biological "applications" (some kinds of symbiosis, parasitism, simultaneous interrelated growing of plants and animals can probably be modelled by such systems).

We examine here only the generative capacity of these mechanisms. As it is expected (similar results where obtained for Chomsky grammars, [16], [17]), the power of parallel systems is larger than that of corresponding type of components. This is true for OL, DOL, EOL, EDOL, TOL, DTOL, EDTOL, ETOL systems. For instance, a parallel communicating system with three OL components or with two DTOL components can generate languages which are not ETOL, the largest family of interactionless L languages (but the systems with deterministic EOL components cannot generate languages not in EDTOL, a result which says that, in the deterministic case, the parallel work of EOL systems can be simulated by tables, also deterministic).

2 Parallel communicating systems

The reader is assumed familiar with basic notions of formal language theory (from [19] for instance) and of L system theory (from [10], [18]). In particular, we use without defining again the notations OL, DOL, EOL, EDOL, TOL, DTOL, ETOL, EDTOL. When specifying an EOL system $G = (V, T, w, P)$ or an ETOL system $G = (V, T, w, \{P_1, \dots, P_r\})$, V is the vocabulary, $T \subseteq V$ is the terminal vocabulary, w is the axiom, P and P_i, $1 \leq i \leq r$, are the sets of rules (tables) whose completion elements $a \to a$ are not explicitely given.

Some further general notations: V^* is the free monoid generated by V, λ is the null element of V^*, $|x|$ is the length of $x \in V^*$, $|x|_U$ is the length of the string obtained by erasing from $x \in V^*$ all symbols not in U, alph(x) is the set of symbols appearing in the string x.

We give now the definition of parallel communicating systems (PCS, for short) with EOL components. For different types of components, the modifications are obvious.

A PCS (of degree n, $n \geq 1$) of EOL systems is a construct

$$\gamma = (G_1, G_2, \dots, G_n),$$

where $G_i = (V_i, T_i, w_i, P_i)$, $1 \leq i \leq n$, are EOL systems such that V_1 contains the special symbols $Q_2, Q_3, \dots, Q_n$ (*query symbols*; denote by K their set) used in derivations as follows.

For $(x_1, \dots, x_n), (y_1, \dots, y_n)$, $x_i, y_i \in V_i^*$, $1 \leq i \leq n$, we write $(x_1, \dots, x_n) \Rightarrow (y_1, \dots, y_n)$ if one of the next two cases holds:

(i) $|x_1|_K = 0$ and $x_i \Longrightarrow y_i$ in the EOL system G_i, $1 \leq i \leq n$;

(ii) $x_1 = z_1 Q_{i_1} z_2 Q_{i_2} \dots z_t Q_{i_t} z_{t+1}$, $t \geq 1$, $z_i \in (V_1 - K)^*$, $1 \leq i \leq t+1$, and $y_1 = z_1 x_{i_1} z_2 x_{i_2} \dots z_t x_{i_t} z_{t+1}$, $y_{i_j} = w_{i_j}$, $1 \leq j \leq t$, $y_s = x_s$ for $s \neq 1$, $s \neq i_j$, $1 \leq j \leq t$.

In words, an n-tuple $(x_1, \dots, x_n)$ directly yields $(y_1, \dots, y_n)$ if either no query symbol appears in x_1 and then we have a componentwise derivation, $x_i \Longrightarrow y_i$ in G_i, $1 \leq i \leq n$, or x_1 contains query symbols and then a ***communication step*** is performed, as these symbols impose. Specifically, each symbol Q_{i_j} is replaced by the corresponding component x_{i_j}, whereas the component G_{i_j} resumes working from its axiom. The communication has priority over rewritting. (This implies the query symbols have a special position, as they are not rewritten, but replaced by the strings generated in that moment by the system components with the same subscripts.) The language generated by γ is:

$$L(\gamma) = \{x \in (T_1 - K)^* | \, (w_1, \dots, w_n) \Longrightarrow^* (x, \alpha_2, \dots, \alpha_n), \alpha_i \in V_i^*, 2 \leq i \leq n\}.$$

In other terms, a derivation consists of repeated rewriting and communication steps, starting from the n-tuple of axioms; we retain in $L(\gamma)$ the string generated in this way on the first component, G_1 (which is considered the *master* of the system), without containing query symbols.

Note that only G_1 may ask for the strings generated by other components; we say that γ is a *centralized* PCS. In [15], [17] also non-centralized PCS (with context-free components) are considered, but here we discuss only the above defined particular case. (Similarly, we deal here only with *returning to axiom* PCS, although this restriction can be removed too: after communicating, the components continue to rewrite the current strings, [17].)

We shall denote by $\mathrm{PC}_n X$ the family of languages generated by PCS of degree at most n, $n \geq 1$, with components of type X; by X we denote also the family of languages generated by L systems of type X. We consider here the following cases: X$\in$ {OL, DOL, EOL, EDOL, TOL, DTOL, ETOL, EDTOL}.

For systems without auxiliary symbols (the non-extended case), we admit $Q_2, \dots, Q_n$ be in V_1, although they are not allowed to appear in the generated strings ($Q_2, \dots, Q_n$ are neither terminals, nor auxiliary symbols in the usual sense, as they cannot be rewritten).

Here is an EOL PCS of degree 3, $\gamma = (G_1, G_2, G_3)$:

$$\begin{aligned} G_1 &= (\{a, b, c, Q_2, Q_3\}, \{a, b\}, d, \{d \rightarrow d, d \rightarrow Q_2, c \rightarrow aQ_3\}) \\ G_2 &= (\{c\}, c, \{c \rightarrow cc\}) \\ G_3 &= (\{b\}, b, \{b \rightarrow bb\}) \end{aligned}$$

A terminal derivation in γ is as follows:

$$\begin{aligned}(d,c,b) \Longrightarrow^* \; & (d, c^{2^k}, b^{2^k}) \Longrightarrow (Q_2, c^{2^{k+1}}, b^{2^{k+1}}) \Longrightarrow \\ & (c^{2^{k+1}}, c, b^{2^{k+1}}) \Longrightarrow ((aQ_3)^{2^{k+1}}, cc, b^{2^{k+2}}) \Longrightarrow \\ & ((ab^{2^{k+2}})^{2^{k+1}}, cc, b),\; k \geq 0.\end{aligned}$$

Therefore

$$L(\gamma) = \{(ab^{2^{n+1}})^{2^n} \mid n \geq 1\}.$$

which is not an ETOL language (see [8] and Theorems 1, 2 below).

3 The generative capacity

Clearly,

$$X = \mathrm{PC}_1X \subseteq \mathrm{PC}_2X \subseteq \ldots$$

for all $X \in \{$ OL, DOL, EOL, EDOL, TOL, DTOL, ETOL, EDTOL$\}$. The problem whether these hierarchies are infinite (whether $\mathrm{PC}_nX \subseteq \mathrm{PC}_{n+1}X$ is proper) remains *open*. We prove here only that, in most cases, $X \subset \mathrm{PC}_2X$ (two components are strictly more powerful than one, that is the parallelism is useful indeed).

First, look for families PC_nX containing languages not in ETOL, the largest family of interactionless L languages.

Theorem 1 *$PC_3OL-\ ETOL \neq \emptyset$.*

Proof. Consider the system $\gamma = (G_1, G_2, G_3)$, with

$$\begin{aligned}G_1 = \; & (\{a,b,c,d,e,Q_2,Q_3\}, d, \{d \to d, d \to Q_2, c \to Q_3\}) \\ G_2 = \; & (\{c,e\}, ec, \{e \to ec\}) \\ G_3 = \; & (\{a,b\}, a, \{a \to ab\})\end{aligned}$$

The derivations in γ are of the following form:

$$\begin{aligned}(d,ec,a) \Longrightarrow^* \; & (d, ec^{k+1}, ab^k) \Longrightarrow (Q_2, ec^{k+2}, ab^{k+1}) \Longrightarrow \\ & (ec^{k+2}, ec, ab^{k+1}) \Longrightarrow (eQ_3^{k+2}, ec^2, ab^{k+2}) \Longrightarrow \\ & (e(ab^{k+2})^{k+2}, ec^2, a),\; k \geq 0.\end{aligned}$$

Therefore

$$L(\gamma) = \{d\} \cup \{ec^n \mid n \geq 2\} \cup \{e(ab^n)^n \mid n \geq 2\}.$$

For the regular language $R = \{e\}\{a,b\}^*$ we obtain

$$L(\gamma) \cap R = \{e(ab^n)^n |\, n \geq 2\}$$

which is not an ETOL language (see Theorem 1 and Corollary 2 in [8]: the symbol a is rare in $L(\gamma) \cap R$, but it is also frequent). As ETOL is a full AFL, it follows that $L(\gamma) \notin$ ETOL. □

Corollary 1

$$(i)\quad OL \subset PC_3OL, \qquad (ii)\quad TOL \subset PC_3TOL,$$
$$(iii)\quad EOL \subset PC_3EOL, \qquad (iv)\quad ETOL \subset PC_3ETOL.$$

Stronger forms of some of these inclusions are true, as consequences of the next results.

Theorem 2 $PC_2DTOL - ETOL \neq \emptyset$.

Proof. Consider the system $\gamma = (G_1, G_2)$ with

$$G_1 = (\{a,b,c,d,e,Q_2\}, ec, \{\{e \to ec\}, \{e \to d, c \to Q_2\}\})$$
$$G_2 = (\{a,b\}, a, \{\{a \to ab\}\})$$

The derivations in γ are of the following form:

$$(ec, a) \Longrightarrow^* (ec^{k+1}, ab^k) \Longrightarrow (dQ_2^{k+1}, ab^{k+1}) \Longrightarrow$$
$$(d(ab^{k+1})^{k+1}, a), k \geq 0.$$

Therefore

$$L(\gamma) = \{ec^n |\, n \geq 1\} \cup \{d(ab^n)^n |\, n \geq 1\}.$$

Again, this is not an ETOL language. □

Corollary 2

$$(i)\quad DTOL \subset PC_2DTOL, \qquad (ii)\quad TOL \subset PC_2TOL,$$
$$(iii)\quad EDTOL \subset PC_2EDTOL, \qquad (iv)\quad ETOL \subset PC_2ETOL.$$

Theorem 3 PC_2OL-$EOL \neq \emptyset$.

Proof. Consider the system $\gamma = (G_1, G_2)$, with

$$G_1 = (\{a,b,c,Q_2\}, c, \{c \to c, c \to Q_2, a \to ab\}),$$
$$G_2 = (\{a\}, a, \{a \to aa\}).$$

The derivations in γ are of the following forms:

$$\begin{aligned}(c,a) \Longrightarrow^* \ & (c,a^{2^k}) \Longrightarrow (Q_2, a^{2^{k+1}}) \Longrightarrow \\ & (a^{2^{k+1}}, a) \Longrightarrow^* ((ab^m)^{2^{k+1}}, a^{2^r}),\ k,r \geq 0.\end{aligned}$$

Therefore

$$L(\gamma) = \{c\} \cup \{a^{2^n} \mid n \geq 1\} \cup \{(ab^m)^{2^n} \mid m \geq 0, n \geq 1\}.$$

The EOL family is closed under intersection by regular sets. Take $R = \{ab^*\}^*$. We obtain

$$L(\gamma) \cap R = \{(ab^m)^{2^n} \mid m \geq 0, n \geq 1\}$$

which is not an EOL language: $\{a\}$ is numerically dispersed in $L(\gamma) \cap R$, but is not clustered in $L(\gamma) \cap R$ (see Theorem 3 and Corollary 5 in [8]). It follows that $L(\gamma)$ is not an EOL language. □

Corollary 3 *(i)OL ⊂ PC₂OL, (ii)EOL ⊂ PC₂EOL.*

Remark 1. The above language $L(\gamma)$ is in TOL, as it can be generated by the TOL system

$$(\{a,b,c\}, c, \{\{c \to a^2\}, \{a \to a^2, b \to \lambda\}, \{a \to ab\}\}).$$

It is an *open problem* whether PC_2OL-TOL is nonempty or not.

Theorem 4 $PC_2DOL - OL \neq \emptyset$.

Proof. Consider the system $\gamma = (G_1, G_2)$, with

$$\begin{aligned} G_1 = &\ (\{a,b,c,Q_2\}, a, \{a \to cb, b \to cQ_2\}) \\ G_2 = &\ (\{b\}, b, \{b \to bb\}) \end{aligned}$$

Let us examine some derivation steps in γ:

$$\begin{aligned}(a,b) \Longrightarrow \ & (cb, bb) \Longrightarrow (ccQ_2, bbbb) \Longrightarrow (ccbbbb, b) \Longrightarrow \\ & (cccQ_2cQ_2cQ_2cQ_2, bb) \Longrightarrow (cccbbcbbcbbcbb, b) \Longrightarrow \\ & (ccccQ_2cQ_2ccQ_2cQ_2ccQ_2cQ_2ccQ_2cQ_2, bb) \Longrightarrow \\ & (ccccbbcbbccbbcbbccbbcbbccbbcbb, b) \Longrightarrow \ldots\end{aligned}$$

After generating a, cb and $ccbbbb$, the subsequent steps are obtained by using the homomorphism $h : \{b,c\}^* \to \{b,c\}^*$, defined by $h(c) = c$, $h(b) = cbb$ (indeed, in G_1 one uses $c \to c$, $b \to cQ_2$, whereas in G_2 one uses $b \to bb$; in total we obtain $b \to cbb$). Thus,

$$L(\gamma) = \{a, cb\} \cup \{x_i \mid i \geq 0\}$$

for

$$x_0 = ccbbbb, x_i = h(x_{i-1}), i \geq 1.$$

The following properties of strings x_i, $i \geq 0$, will be useful:

1. If $x_i = ub^r v$, then $r = 1$ or $r = 2$ for all $i \geq 1$.

Proof. Obvious form the definition of h.

2. $|x_i|_b = |x_i|_c + 2$, $i \geq 0$.

Proof. The relation is trivial for x_0. Assume that it is true for all x_i, $i \leq n$, and take the string x_{n+1}. As $x_{n+1} = h(x_n)$, we obtain $|x_{n+1}|_b = 2|x_n|_b$, $|x_{n+1}|_c = |x_n|_c + |x_n|_b$. But $|x_n|_b = |x_n|_c + 2$, hence $|x_{n+1}|_b = 2|x_n|_b = |x_n|_b + |x_n|_c + 2 = |x_{n+1}|_c + 2$, which completes the induction argument.

3. $x_i = c^{i+2}bby_i$, $i \geq 0$.

Proof. Obvious (the leftmost occurrence of b introduces by $h(b) = cbb$ a further occurrence of c in the prefix of x_i compared with x_{i-1}).

Suppose now that $L(\gamma)$ ∈OL, $L(\gamma) = L(G)$ for an OL system $G = (\{a, b, c\}, w, P)$. Let us examine the possible form of rules in P.

If $c \to xby$ is in P, then we distinguish two cases:

(i) There is a rule $b \to z$, $|z| \geq 1$, in P. Then, rewriting a string x_i with arbitrarily large i, by these rules, we obtain a string of the form $xbyu$, with $|xbyu| \geq |x_i|$, but with a prefix c^s with $s \leq |x|$, which contradicts the above property 3 (the prefix c^t of x_i has $t = i + 2$).

(ii) There is a rule $b \to \lambda$ in P. Rewriting cb, we obtain xby, rewriting $ccbbbb$, we obtain $xbyxby$. However, there is no string z in $L(\gamma)$ such that zz is in $L(\gamma)$ too.

Consequently, the c rules must be of the form $c \to c^r$, $r \geq 0$. At least a rule $b \to x$ is in P with $|x|_b > 0$.

Apply these rules both to cb and $ccbbbb$. We obtain

$$cb \Longrightarrow c^r z, ccbbb \Longrightarrow c^{2r} z^4.$$

Assume $|z|_b = k$, $|z|_c = j$. Therefore,

$$|c^r z|_c = r + j, \ |c^r z|_b = k$$
$$|c^{2r} z^4|_c = 2r + 4j, |c^{2r} z^4|_b = 4k.$$

As $c^r z$ and $c^{2r} z^4$ must be string of the form x_i, $i \geq 1$, we have for them property 2, that is

$$r + j + 2 = k,$$
$$2r + 4j + 2 = 4k.$$

Substituting k from the former equation to the latter one, we obtain

$$2r + 4j + 2 = 4r + 4j + 8$$

which implies $r = -3$, a contradiction with $r \geq 0$.

In conclusion, no OL system can generate $L(\gamma)$. □

Corollary 4 *(i)$DOL \subset PC_2DOL$, (ii)$OL \subset PC_2OL$.*

The above results show that the necessary conditions known in L systems theory cannot reject languages with respect to PC_nX families. However, these families are not *too large*. The next result proves that, in some sense, PC_nTOL has the same limitations as TOL.

Theorem 5 *The language $L = \{a^2, a^3\}$ is not in PC_nTOL, $n \geq 1$.*

Proof. Suppose $L = L(\gamma)$ for a PCS $\gamma = (G_1, \ldots, G_n)$ with TOL components, $G_i = (V_i, w_i, \{P_{1,i}, \ldots, P_{k_i,i}\})$, $1 \leq i \leq n$, $V_1 = \{a\}$. Clearly, $w_1 \in \{a^2, a^3\}$. If $a^2 \Longrightarrow a^3$ or $a^3 \Longrightarrow a^2$ can be obtained in G_1, then $L(\gamma) \neq L$: in the former case we need a rule $a \rightarrow \lambda$ and a rule $a \rightarrow a^3$ (and then λ and a^6 can be obtained), or $a \rightarrow a$ and $a \rightarrow a^2$ (and then a^4 can be obtained); in the later case we need a rule $a \rightarrow \lambda$ (and λ can be produced). In conclusion, at least a communication step is necessary. A rule $a \rightarrow x$, $x \in \{a, Q_2, \ldots, Q_n\}^*$ can be used both in $aa \Longrightarrow xx$ and in $aaa \Longrightarrow xxx$. Therefore, the strings which are communicated to G_1 must be equal to a: if a string a^t, $t \neq 1$, is communicated, then either λ is obtained (for $t = 0$) or a^{2t}, a^{3t}, respectively, are obtained, and these strings are not in L for $t > 1$. In this way, from w_1 we can only obtain w_1, in spite of the fact that L contains two strings. In conclusion, $L \neq L(\gamma)$, hence L is not in PC_nTOL. □

Corollary 5 *(i)$EOL - PC_nTOL \neq \emptyset, n \geq 1$, (ii)$EDTOL - PC_nTOL \neq \emptyset, n \geq 1$.*

The next result is perhaps the most significant of this paper, as it says that in the deterministic (extended) case, the parallel working in a system can be simulated by tables (deterministic, too).

Theorem 6 *$PC_nEDOL \subseteq EDTOL$, $n \geq 1$.*

Proof. Consider a PCS $\gamma = (G_1, \ldots, G_n)$ with EDOL components $G_i = (V_i, T_i, w_i, P_i)$, $1 \leq i \leq n$. When $L(\gamma)$ is finite, the assertion in theorem is trivial. Assume that $L(\gamma)$ is infinite.

A graph Γ can be associated to γ, $\Gamma = (N, E)$, with

$$\begin{aligned} N = \; & 2^{V_1} \times 2^{V_2} \times \ldots \times 2^{V_n} \\ E = \; & \{((M_1, \ldots, M_n), (M_1', \ldots, M_n')) \mid \text{there are } (x_1, \ldots, x_n), \\ & (y_1, \ldots, y_n), x_i, y_i \in V_i^*, M_i = \text{alph}(x_i), M_i' = \text{alph}(y_i), \\ & 1 \leq i \leq n, \text{ and } (x_1, \ldots, x_n) \Longrightarrow (y_1, \ldots, y_n) \text{ in } \gamma\} \end{aligned}$$

As all components of γ are deterministic EOL systems, when $((M_1, \ldots, M_n), (M_1', \ldots, M_n')) \in E$, $((M_1, \ldots, M_n), (M_1'', \ldots, M_n'')) \in E$, then $M_i' = M_i''$, $1 \leq i \leq n$, that is no branching there is in Γ. Therefore, the actual derivations in γ correspond to paths in Γ starting form $(\text{alph}(w_1), \ldots, \text{alph}(w_n))$, following a sequence of different vertices in N, $((M_{1,2}, \ldots, M_{n,2}), (M_{1,3}, \ldots, M_{n,3}), \ldots, (M_{1,k}, \ldots, M_{n,k}))$ and then (because N is finite and $L(\gamma)$ is infinite) entering a cycle $((M_{1,k}, \ldots, M_{n,k}), (M_{1,k+1}, \ldots, M_{n,k+1}), \ldots, (M_{1,k+r}, \ldots, M_{n,k+r}), (M_{1,k}, \ldots, M_{n,k}))$. Each derivation in γ corresponds to an initial part of this unique path in Γ.

Call *age* of a string $x \in L(G_i)$, $2 \leq i \leq n$, the number of derivation steps in G_i necessary for obtaining it, that is

$$\text{age}(x, G_i) = p \text{ iff } w_i \Longrightarrow z_1 \Longrightarrow z_2 \Longrightarrow \ldots z_p = x.$$

In a PCS, the age of a string is the number of derivation steps from the very beginning of the derivation in γ or from the previous communication from the corresponding component G_i.

Clearly, the age of all communicated strings in a derivation in γ is bounded by $k+r$, the above constants associated to the path in Γ describing the work of γ. Note also that $k + r \leq \text{card}(2^{V_1} \times \ldots \times 2^{V_n})$.

These remarks can be used in order to simulate a PCS of EDOL systems by an ETOL in the following way. Add to a string generated in G_1 $n-1$ counters which memorize the age of strings generated in $G_2, \ldots, G_n$ (at each step, they increase by one). When one counter becomes larger than $k+r$, it is erased and no further communication is possible from the corresponding component. If a query symbol is introduced in G_1, then a string in the corresponding component G_i is generated from this symbol, by a derivation with as many steps as the associated counter says (now, at each step, the associated counter decreases by one, the other remains unchanged). After completing this derivation, the corresponding counter is set to zero and a new string age is memorized. The process continues. At some moment, the auxiliary symbols are removed and a string in $L(\gamma)$ is obtained.

Formally, we construct the ETOL system

$$G = (V, T, w', \mathcal{P})$$

as follows. For a vocabulary U, a set $M \subseteq U$ and a set P of rules over U, denote

$$\text{alph}(P, M) = \{a \in U \mid a \in \text{alph}(z), b \Longrightarrow z \text{ in } P, b \in M\}.$$

Denote also

$$W = V_1 \cup \{(a, i) \mid a \in V_i, 2 \leq i \leq n\}$$

and consider the homomorphism $h_i : V_i^* \to \{(a,i) | \ a \in V_i\}^*$ defined by $h_i(a) = (a,i)$, $a \in V_i$, $2 \le i \le n$.

Then

$$\begin{aligned} V = \ & W \cup \{(M),[M] | \ M \subseteq W\} \cup \{Z\} \cup \\ & \{(i,t_i),[i,t_i] | \ 2 \le i \le n, 0 \le t_i \le k+r+1\} \end{aligned}$$

((i,t_i) and $[i,t_i]$ are the above mentioned counters, Z is a trap symbol)

$$\begin{aligned} T = \ & T_1 \\ w' = \ & w(2,0)(3,0)\dots(n,0)(\text{alph}(w)) \end{aligned}$$

(each generated string will be of the form $x(2,t_2)\dots(n,t_n)(\text{alph}(x))$, possibly with (i,t_i) replaced by $[i,t_i]$ and $(\text{alph}(x))$ replaced by $[\text{alph}(x)]$, x being a string generated in γ and (i,t_i), $[i,t_i]$ indicating the age of strings generated in G_i at the current moment, $2 \le i \le n$).

The set $\mathcal{P}$ contains the following tables.

$$\begin{aligned} 1) \quad & P_1 \cup \{(i,t_1) \to (i,t_i+1) | \ 2 \le i \le n, 0 \le t_i \le k+r\} \cup \\ & \{[i,t_i] \to Z | \ 2 \le i \le n, 0 \le t_i \le k+r+1\} \cup \\ & \{(M) \to (\text{alph}(P_1,M))\} \cup \\ & \{(M') \to Z | \ M' \subseteq W, M \neq M'\} \cup \\ & \{[M'] \to Z | \ M' \subseteq W\}, \end{aligned}$$

for $M \subseteq \{a \in V_1 | \ a \to x \in P_1 \text{ and } |x|_K = 0\}$. (In the presence of a symbol (M), a derivation in P_1, without introducing query symbols is simulated. The age-components of symbols (i,t_i) are increased by one and the symbols occurring in the current string are memorized by $(\text{alph}(P_1,M))$.)

$$\begin{aligned} 2) \quad & P_1 \cup \{(i,t_i) \to (i,t_i+1) | \ 2 \le i \le n, \text{ no rule } a \to x, \\ & |x|_{Q_i} > 0 \text{ is in } P_1 \text{ with } a \in M, 0 \le t_i \le k+r\} \cup \\ & \{(i,t_i) \to [i,t_i+1] | \ 2 \le i \le n, \text{ there is } a \to x \in P_1, \\ & a \in M, |x|_{Q_i} > 0, 0 \le t_i \le k+r\} \cup \\ & \{(M) \to [\text{alph}(P_1,M)]\} \cup \\ & \{(M') \to Z | \ M' \subseteq W, M \neq M'\} \cup \\ & \{[M'] \to Z | \ M' \subseteq W\}, \end{aligned}$$

for $M \subseteq V_1$ such that there is $a \in M$, $a \to x \in P_1$, $|x|_K > 0$. (When a query symbol Q_i is introduced, the symbol memorizing the set of current symbols will be bracketed and the same for the corresponding age-symbols.)

$$3)\quad \{Q_i \rightarrow h_i(w_i) \mid 2 \le i \le n\} \cup$$
$$\{(M') \rightarrow Z \mid M' \subseteq W\} \cup$$
$$\{[M] \rightarrow [\mathrm{alph}(\{Q_i \rightarrow h_i(w_i) \mid 2 \le i \le n\}, M)]\} \cup$$
$$\{[M'] \rightarrow Z \mid M' \subseteq W, M \neq M'\},$$

for $M \subseteq W$, $M \cap K \neq \emptyset$. (The modified axioms $h_i(w_i)$ replace the symbols Q_i, the age symbols are not modified, but the wittness symbol $[M]$ memorizes this change. The symbols (a, i) in $h_i(w_i)$ are introduced in order to not use rules of P_j in rewriting components x_i with $i \neq j$, or symbols in G_1.)

$$4)\quad \{h_i(a) \rightarrow h_i(x) \mid a \rightarrow x \in P_i, 2 \le i \le n\} \cup$$
$$\{[i, t_i] \rightarrow [i, t_i - 1] \mid 2 \le i \le n, 1 \le t_i \le k + r\} \cup$$
$$\{[M] \rightarrow [\mathrm{alph}(\{h_i(a) \rightarrow h_i(x) \mid a \rightarrow x \in P_i, 2 \le i \le n\}, M]\} \cup$$
$$\{[M'] \rightarrow Z \mid M' \subseteq W, M \neq M'\} \cup$$
$$\{(M') \rightarrow Z \mid M' \subseteq W\},$$

for $M \subseteq W - K$ containing symbols (a, i), $a \in V_i$, $2 \le i \le n$. (One simulates the derivations in components G_i for which query symbols Q_i were introduced. The corresponding age symbols are decreased, the other symbols remain unchanged.)

$$5)\quad \{(a, i) \rightarrow a, [i, 0] \rightarrow (i, 0) \mid a \in V_i, (a, i) \in M\} \cup$$
$$\{[i, t_i] \rightarrow Z \mid 1 \le t_i \le k + r + 1\} \cup$$
$$\{[M] \rightarrow [\mathrm{alph}(\{(a, i) \rightarrow a \mid a \in V_i\}, M)]\} \cup$$
$$\{[M'] \rightarrow Z \mid M' \subseteq W, M \neq M'\} \cup$$
$$\{(M') \rightarrow Z \mid M' \subseteq W\},$$

for $2 \le i \le n$, $M \subseteq W - K$ containing symbols (a, i), $a \in V_i$. (When some communications are terminated, the corresponding age symbols $[i, 0]$ are replaced by $(i, 0)$. The change is registered also in the wittness symbol $[M]$, where (a, i) are replaced by a, $a \in V_i$.)

$$6)\quad \{[M] \rightarrow (M)\}, \text{ for } M \subseteq V_1$$

(When no symbol (a, i) is present, that is no symbol $[i, t_i]$ is present, this means the communication is accomplished. A new derivation in P_1 can be performed.)

$$7)\quad \{(i, k + r + 1) \rightarrow \lambda \mid 2 \le i \le n\}$$

(When the age of a string associated to a component G_i, $2 \leq i \leq n$, is greater than $k+r$, this component will never be asked to communicate.)

$$8) \quad \{(M) \to \lambda\} \cup \{(i,t_i) \to \lambda \mid 2 \leq i \leq n, 0 \leq t_i \leq k+r\},$$

for $M \subseteq T_1$. (The derivation is terminated, no symbol in $W - T_1$ is present.)

Clearly, G is a deterministic ETOL and, from the above explanations, we obtain $L(\gamma) = L(G)$, hence $L(\gamma) \in$ EDTOL. □

4 Final remarks

Denote by REG, CF, CS, RE the families of regular, context-free, context sensitive and recursively enumerable languages, respectively. From Theorem 5 we find that REG-PC_nTOL$\neq \emptyset$, whereas Theorem 6 implies CF-PC_nEDOL$\neq \emptyset$ (it is known that CF-EDTOL$\neq \emptyset$, [18]).

Clearly, PC_nCF$\subseteq$ PC_nEOL (each CF component becomes an EOL component by adding rules $a \to a$ for each symbol, terminal or nonterminal), and PC_nETOL$\subseteq$RE (follows from Turing-Church thesis, or can be proved directly). The inclusion is probably proper. If the λ-rules could be eliminated from PCS with ETOL components (this can be done for usual ETOL systems [10]), then we would obtain PC_nETOL$\subseteq$CS.

Denote also by PR the class of λ-free programmed grammars with appearance checking. It is known that ETOL$\subseteq$PR. Is the corresponding inclusion PC_nETOL$\subseteq$$PC_n$PR true? Anyway, it seems to have PC_2EOL-PR$\neq \emptyset$. For instance, the language

$$L = \{xcx \mid x \in D_{a,b}\} - \{c\}$$

$D_{a,b}$ being the Dyck language over $\{a,b\}$, is probably not in PR (see open problem 1.2.3, page 42 in [4]), but it can be generated by the PCS $\gamma = (G_1, G_2)$ with EOL components

$$\begin{aligned} G_1 &= (\{a,b,c,A,Q_2\}, \{a,b,c\}, A, \{A \to A, A \to Q_2cQ_2\}) \\ G_2 &= (\{a,b,S\}, \{a,b\}, S, \{S \to SS, S \to ab, S \to aSb\}) \end{aligned}$$

Of course, many other questions remain to be investigated. In fact, most problems studied for usual L systems can be formulated for PCS of L systems too (the role of erasing and of coding, closure properties – is PC_nETOL a full AFL similar to ETOL? – and so on); we do not refer here to them.

References

[1] Csuhaj-Varju, E., Dassow, J.: On cooperating/distributed grammar systems, *Elektr. Inform. Kybern.*, to appear.

[2] Csuhaj-Varju, E., Dassow, J., Păun, Gh.: Dynamically controlled cooperating/distributed grammar systems, *Inform. Sci.*, to appear.

[3] Csuhaj-Varju, E., Kelemen, J.: Cooperating grammar systems: a syntactical framework for blackboard model of problem solving, *Conf. A.I.Control Syst. Robotics*, 1989.

[4] Dassow, J., Păun, Gh.: *Regulated rewriting in formal language theory*, Akademie-Verlag, Berlin, 1989, Springer-Verlag, Berlin, 1990.

[5] Dassow, J., Păun, Gh.: On some variants of cooperating/distributed grammar systems, *Stud. Cerc. Matem.*, 42, 2(1990), 153-165.

[6] Dassow, J., Păun, Gh.: Cooperating/distributed grammar systems with registers, *Found. Control. Eng.*, 15, 1(1990), 19-38.

[7] Dassow, J., Păun, Gh.: On the succincteness of describing of context-free languages by cooperating/distributed grammar systems, *Computers and A.I.*, submitted.

[8] Ehrenfeucht, A., Rozenberg, G.: Three useful results concerning L languages without interactions, in vol. *L Systems, Lect. Notes in Computer Sci., 15* (ed. G.Rozenberg, A.Salomaa), Springer-Verlag, Berlin, Heidelberg, 1974, 72-77.

[9] Gheorghe, M., Păun, Gh.: Further remarks on cooperating/distributed grammar systems, *Bull. Math. Soc. Sci. Math. Roumanie*, 34(82), 3(1990), 231-245.

[10] Herman, G., Rozenberg, G.: *Developmental systems and languages*, North-Holl. Publ. Comp., Amsterdam, 1975.

[11] Lindenmayer, A.: Mathematical models for cellular interactions in development, *J. Theoret. Biol.*, Part I-18(1968), 280-299, Part II-18(1968), 300-315.

[12] Meersman, R., Rozenberg, G.: Cooperating grammar systems, *Lect. Notes Computer Sci.*, 64, Springer-Verlag, Berlin, 1978, 364-373.

[13] Păun, Gh.: Parallel communicating grammar systems: the context-free case, *Found. Control Eng.*, 14, 1(1989), 39-50.

[14] Păun, Gh.: On the power of synchronization in parallel communicating grammar systems, *Stud. Cerc. Matem.*, 41, 3(1989), 191-197.

[15] Păun, Gh.: On the syntactic complexity of parallel communicating grammar systems, *Stud. Cerc. Matem.*, to appear.

[16] Păun, Gh., Santean, L.: Parallel communicating grammar systems: the regular case, *An. Univ. Buc. Ser. Mat.-Inform,*, 37, 2(1989), 55-63.

[17] Păun, Gh., Santean L.:, Further remarks on parallel communicating grammar systems, *Intern. J. Computer Math.*, 34(1990), 187-203.

[18] Rozenberg, G., Salomaa, A.: *The mathematical theory of L systems*, Academic Press, New York, 1980.

[19] Salomaa, A.: *Formal languages*, Academic Press, New York, London, 1973.

"FORGETFUL" L SYSTEMS

K.Krithivasan[+], M.V. Nagendra Prasad[*1], Meena Mahajan[+]

[+]Department of Computer Science and Engineering,
Indian Institute of Technology, Madras 600 036, India.

[*]Intelligent Engineering Systems Laboratory, Dept. of Civil Engg.,
Massachusetts Institute of Technology, Cambridge MA 02139 U.S.A.

Abstract : This paper presents "forgetful" L systems as a new variant of the classical L systems. In this variant, apart from the state of a cell, its ancestry is also allowed to influence its future development. A cell in a particular state will, depending on its ancestry, remember either all rules or only special marked rules. Such systems are compared with the nonforgetful L systems. Various decidability issues are considered and some closure properties are presented. These systems attempt to provide a more realistic model for modelling certain developmental processes.

Keywords : L systems, generative power of rewrite systems, decidability, closure properties.

1. Introduction

L systems have proved to be invaluble tools in studying various aspects of developmental theory. They have been studied widely since the time Lindenmayer proposed them [5]. Besides being used as models for biological development, they have also been used in computer generated imagery for biological patterns [7] and folk art patterns [6]. An extensive study of the formal language theoretic aspects of these systems has also been carried out. An excellent summary of most of the important results can be found in [1, 8].

OL systems were the first systems to be introduced and studied. Gradually additional features like extension and tables were introduced to model more sophisticated systems. While most of the studies were done

---

[1]This work was done when the second author was at the Department of Computer Science and Engineering, Indian Institute of Technology, Madras 600 036, India.

in terms of controlling the way productions are applied or strings in the language are chosen, little has been done with respect to the effect of ancestry of a cell on its development. It is quite natural that in some situations two cells of one kind under identical conditions may develop in a different manner when their ancestries are different. In this paper we try to build "forgetfulness" into L systems. A cell in a particular state will, depending on its ancestry, remember either all rules or only some "special" rules. While the definition of the forgetful L systems given below may not directly reflect any real life systems, it is an attempt towards studying more complex L systems whose developmental strategies also depend on ancestry, besides other things.

Similar investigations have been done by Vollmar [10] with respect to Turing machines and by Hollerer and Vollmar [3] with respect to cellular automata. Furthermore, Salomaa [9] and Krithivasan et al. [4] also investigated formal languages with a variable (time-variant) structure.

In section 2, we give the preliminary definitions and results from L systems theory. In section 3, forgetful L systems are introduced and examples are given. In section 4, we give some results which highlight the comparative generative powers of nonforgetful and forgetful systems. Decidability issues are then addressed in section 5, and section 6 gives a glimpse of some of the closure properties.

2. Definitions and Notation

In the following discussion it is assumed that every step in a derivation takes one time unit and that the derivation starts at time $t = 0$. Time and number of steps in a derivation are interchangeably used.

Definition 2.1 : A forgetful 0L system (henceforth abbreviated as Fr–0L) is a 4–tuple $< \Sigma, (\overline{P}, P), \omega, \phi>$ where Σ is the alphabet and ω is the axiom or the starting string. The production tuple and the function ϕ are described below. Such a system generates an Fr–0L language.

With respect to the modifications of L systems we define a mapping $\tau : \Sigma \times \Sigma \times \mathbb{N} \longrightarrow \mathbb{N} \cup \{\infty\}$, where $\tau(a,b,t)$ indicates, for letter 'a', at time t, the last occurrence of 'b' before time t in its ancestry. The letter 'd' is a proper ancestor of a letter 'a' in a derivation D iff $d \overset{+}{\Rightarrow} \alpha_1 a \alpha_2$ is a sub-derivation of derivation D. Hence, in terms of the derivation tree, 'd' is a proper ancestor of 'a' iff it lies on the path from the root to the father of 'a'. The ancestry of a letter 'a' is the set of its proper ancestors. If letter 'b' has not occurred between time 0 and t - 1, then $\tau(a,b,t) = \infty$. The letter 'a' is in state 1 or 0 as described below :

If 'a' at time t has occurred within a period $\phi(t)$ where ϕ is a function $\phi : \mathbb{N} \longrightarrow \mathbb{N}^+$, 'a' goes into state 1. If S is the state of letter 'a' then

$$S = 1 \text{ if } t - \tau(a,a,t) \le \phi(t)$$
$$= 0 \text{ otherwise.}$$

If S = 0 then the letter can apply only certain productions i.e. the productions in $\overline{P} \subseteq P$ (for convenience these productions will also be called "the marked productions").

If S = 1 then it can apply productions from P (i.e. develop along any production - marked or unmarked).

Informally, this means that a letter 'a' at a particular stage in a derivation will be in state S = 1 (or nonforgetful state) if it has occurred at least once in the "recent past" (i.e. at least once in the past $\phi(t)$ steps) or if it has never occurred in its ancestry. The letter 'a' will be in state S = 0 (or forgetful state) if it has occurred at least once in its ancestry but not within the past $\phi(t)$ steps. The case when $\tau(a,a,t) = \infty$ takes care of the case when the letter 'a' occurs for the first time. Since $\tau(a,a,t) = \infty$, $t - \tau(a,a,t) = -\infty < \phi(t)$, implying that S = 1. In the state S = 0 the letter forgets certain productions - the unmarked ones. The set $\overline{P}$ contains at least one rule per letter (because at no stage should the letter, whether in state 0 or 1, be left with no alternative to develop further).

The 0L system < Σ, P, ω> is called the underlying system. It is the system obtained by removing the forgetfulness property of the Fr-0L system.

Remarks :

1. In this paper we shall speak of only one class of functions i.e. $\phi(t) = k$, where k is a constant. Other types of functions are possible [3], but we will not consider them here.
2. In general, the forgetful version of L systems will not produce the same results as the non-forgetful version. If $\phi(t) \ge t$, or if $\overline{P} \subseteq P$, forgetting will not occur; therefore L systems may be regarded as a special case of Fr-L systems.

Example : Consider the Fr–OL system $G = <\Sigma, (\overline{P}, P), \omega, \phi>$ where $\Sigma = \{a,b,c,d\}$, $\omega = d$, $\phi(t) = 2$, and $\overline{P}$, P are as given below.

1. $d \rightarrow a^2$
*2. $d \rightarrow a$
*3. $a \rightarrow b$
*4. $b \rightarrow c$
*5. $c \rightarrow d$

Here the rules marked with * are those that belong to $\overline{P}$.
$L(G) = \{a,b,c,d,a^2,b^2,c^2,d^2\}$. A typical derivation is shown in Figure 1. But the underlying OL-system gives the language $\{a^n,b^n,c^n,d^n \mid n \geq 1\}$.

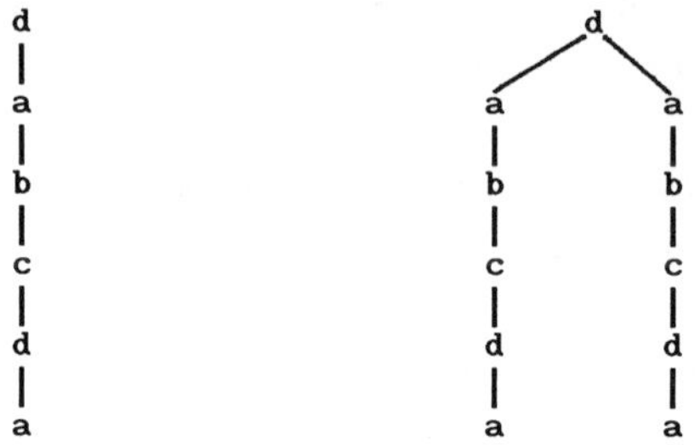

Figure 1

Definition 2.2 : A forgetful EOL system (henceforth abbreviated as Fr–EOL) is a 5–tuple $< \Sigma, (\overline{P},P), \omega, \Delta, \phi>$ where Σ is the alphabet, ω is the axiom or the starting string and Δ is the extension set. It performs the same function as in ordinary L systems. The production tuple and the function ϕ are as described in definition 2.1.

Definition 2.3 : A forgetful TOL system (henceforth abbreviated as Fr–TOL) is a 4–tuple $< \Sigma, (\overline{\mathbb{P}}, \mathbb{P}), \omega, \phi>$ where Σ is the alphabet and ω is the axiom or the starting string. The function ϕ is as described above. $\overline{\mathbb{P}}$ and $\mathbb{P}$ are sets of tables such that there is a one-to-one correspondence between the tables of the two sets. Let us call a corresponding pair $(\overline{P}_i, P_i)$. Then $\overline{P}_i \subseteq P_i$, and at any step, all letters of a string have to use rules from some P_i or $\overline{P}_i$ according to whether the letter is in state 1 or 0 respectively.

Example : Consider the Fr–TOL $G = < \Sigma, (\overline{\mathbb{P}}, \mathbb{P}), \omega, \phi>$ where $\Sigma = \{a,b\}$, $\omega = a$, $\phi(t) = 1$, and the productions are as follows:

$\overline{P}_1$	$a \rightarrow b$	P_1	$a \rightarrow b$
			$a \rightarrow b^2$
	$b \rightarrow a$		$b \rightarrow a$
			$b \rightarrow a^2$

$\overline{P}_2$	$a \longrightarrow b$	P_2	$a \longrightarrow b$
			$a \longrightarrow b^3$
	$b \longrightarrow a$		$b \longrightarrow a$
			$b \longrightarrow a^3$

$L(G) = \{ a, a^2, a^3, a^4, a^5, a^6, a^7, a^9, b, b^2, b^3, b^4, b^5, b^6, b^7, b^9 \}$

Definition 2.4 : In an Fr–OL (Fr–EOL) system $G = < \Sigma, (\overline{P}, P), \omega, \phi>$ the maximum number of rules in $\overline{P}$ for any letter is called the degree of the system. Thus it represents the maximum number of "marked" rules for any letter in Σ.

Remark: Nonforgetful L systems are a special case of forgetful L systems where every rule is marked.

3. Generative Powers of Forgetful L Systems

It will be interesting to compare the powers of L systems and Fr–L systems with certain restrictions on $\phi(t)$ and on the degree of the system. Obviously, if $\phi(t) > t$, or even if $\phi(t) = t$, then the Fr–L system produces the same language as the underlying L-system, because if the "look back" power grows at a rate faster than the counter itself then no forgetting can occur. Similarly, if there is no restriction on the degree of the system, then by marking all the rules in the forgetful system we can produce a system which works in the same way as the non-forgetful one. However it will be interesting to study the power of the forgetful L systems when there is a restriction on their degree.

We need to introduce some concepts before we can study the comparative generative powers.

Definition 3.1 : A rule graph of a OL (EOL) system $< \Sigma, P, \omega>$ is a directed graph $< \Sigma, \Gamma >$, where $\Gamma \subseteq \Sigma \times \Sigma$. An edge (a_1, a_2) belongs to Γ iff there exists a production $a_1 \longrightarrow \alpha$ in P such that $a_2 \in alph(\alpha)$. [$alph(\alpha)$ is the smallest $\overline{\Sigma} \subseteq \Sigma$ such that $\alpha \in \overline{\Sigma}^*$]. Weight of an edge (a_1, a_2), i.e. $w(a_1, a_2)$, is the number of productions of the form $a_1 \longrightarrow \alpha \in P$ such that $a_2 \in alph(\alpha)$. In order to take the effect of λ rules into consideration, we shall adopt a colouring scheme for the edges. If there is a rule of the form $a_1 \longrightarrow \lambda$ in P, then, for any a_2, the edge (a_1, a_2), if it exists, is coloured red. If there is no rule of the form $a_1 \longrightarrow \lambda$ in P, then, for any a_2, the edge (a_1, a_2), if it exists, is coloured blue. Thus for any OL (EOL) system we can obtain its rule graph from its production set P.

For a TOL system the rule graph is defined in a slightly different manner. If $a_i \longrightarrow \alpha \in P_k$ and $a_j \in alph(\alpha)$, then edge (a_i, a_j) is introduced with label k, and weight $w(a_i, a_j, k)$ is the number of productions $a_i \longrightarrow \alpha \in P_k$ such that $a_j \in alph(\alpha)$. Thus if $a_r \longrightarrow \alpha \in P_k$ and $a_r \longrightarrow \beta \in P_m$ and $a_t \in alph(\alpha) \cap alph(\beta)$, then there are two distinct edges from a_r to a_t, one labeled k and the other labeled m, with their corresponding weights. Here if $a_1 \longrightarrow \lambda \in P_k$ then any edge (a_1, a_2) labeled k will be coloured red. Otherwise an edge will be a blue edge.

Definition 3.2: Consider any two cycles in the rule graph G, with at least one vertex in common. If either of the two cycles has a vertex v satisfying the following :

(i) v is not a common vertex, and

(ii) v has outdegree $od(v) > m$, where

$$od(v) = \begin{cases} \sum_{edge\ vu} w(v,u) & \text{if edges vu are coloured blue} \\ \left[\sum_{edge\ vu} w(v,u)\right] + 1 & \text{if edges vu are coloured red} \end{cases}$$

then this vertex is called an m–forgetful vertex.

Theorem 3.1 : If the rule graph of a OL system $< \Sigma, P, \omega>$ has no m-forgetful vertex, then an arbitrary derivation in the OL system can be "simulated" by an Fr–OL system with degree $\le m$, for a given m.

Proof : If the rule graph of a OL system has no m–forgetful vertex, then there are three possibilities:

1) The rule graph of the OL system has no cycles.

In this case, since no letter needs to be "recalled", no forgetting can take place. Hence any forgetful OL system with the same set of productions as the given OL-system can be used for the simulation.

2) The rule graph of the OL system has cycles but they are disjoint.

In this case consider a forgetful OL system with $\phi(t) = k$, where k is the maximum length of any cycle in the rule graph. The set of productions is the same as in the non-forgetful system. Any rules may be marked. Any letter that occurs at a particular stage of derivation again after its first occurrence would have anyway occurred within the previous k steps, since it belongs to only one cycle, and so cannot go into forgetful state.

3) There are pairs of connected cycles in the rule graph of the OL system but these cycles do not have any m–forgetful vertex.

In this case, if a vertex u has outdegree $od(u) > m$ then it must satisfy one of the following:

(i) u does not lie on a cycle.

(ii) u lies on a single cycle which is disjoint from all other cycles.

(iii) u is a common vertex of two or more cycles which do not intersect any cycle not containing u.

This is because if a vertex u has outdegree od(u) > m and belongs to cycle $\mathcal{C}$, then it must belong to all cycles which intersect $\mathcal{C}$, otherwise it will be m–forgetful.

In case (i), u never needs to be recalled. So any of the rules for u may be marked, and no forgetting will occur. In case (ii), provided $\phi(t)$ is greater than or at least equal to the length of the cycle, any of the rules for u may be marked, and no forgetting will occur. In case (iii) also, traversal of any of the these cycles will always pass through u, and hence, by setting $\phi(t)$ to be the maximum length of any of the cycles containing u, u would have always occurred in the past $\phi(t)$ steps after its first occurrence. So no rule of the form $u \longrightarrow \alpha$ can be forgotten. Thus all three cases can be handled by setting $\phi(t)$ to be the maximum length of any cycle in the rule graph.

For any vertex v of outdegree at most m, there are at most m rules of the form $v \longrightarrow \alpha$. So they can all be marked, and no forgetting will occur. ■

Definition 3.3: Consider any two cycles in the rule graph G, with at least one vertex in common. If either of the two cycles has an edge (v,x) satisfying the following :

(i) v is not a common vertex, and

(ii) the edge (v,x) is blue and has $w(v,x) > m$, or

the edge (v,x) is red and has $w(v,x) \geq m$,

then this edge is called an m–forgetful edge.

Theorem 3.2 : If the rule graph of a 0L system has m-forgetful edges, then an arbitrary derivation in the system $< \Sigma,\ P,\ \omega>$ cannot be "simulated" by a Fr–0L system with degree $\leq$ m, for a given m, .

Proof : If the rule graph has m-forgetful edges, then consider the following traversal in the rule graph:

On reaching any vertex u shared by two or more cycles, one of which has an m–forgetful edge vx, traverse that cycle which has edge vx first. On reaching u again, traverse the other cycle sufficiently often so that if the first cycle is traversed again now, v will be reached in a forgetful state. Clearly, there are more than m rules for developing v. In any forgetful 0L system that is trying to simulate a derivation following this sequence, there will be at least one act of forgetfulness, because, for v, the degree permits at most m rules to be marked. ■

Corollary 3.1 : The above theorems applies for EOL and TOL systems also.

Theorem 3.1 gives a condition $\mathcal{P}$ under which a OL system can be simulated by a forgetful one. Theorem 3.2 gives a condition Q, which is stronger than $\neg\mathcal{P}$, under which such a simulation is not possible. What happens when neither $\mathcal{P}$ nor Q are satisfied is yet to be studied.

Now we will see that introduction of context can "improve" the situation.

Theorem 3.3 : Every derivation in an EIL system can be simulated by an Fr–EIL system with $\phi(t) = k$.

Proof : At each stage of a derivation in the EIL system, the "equivalent stage" in the Fr–EIL system should be in a nonforgetful state. (For deterministic systems Fr–EIL = EIL because by definition at least one rule per letter has to be marked). We will describe a method by which the letter which is going to produce the next stage of development would have repeated itself in "the recent past", so that on looking back we can always see the letter as having occurred within $\phi(t)$ time.

Let the given <k,l> system be $< \Sigma, g, P, \omega, \Delta>$. We will construct an Fr–<k,l> EIL system $<\Sigma', g, (\bar{P}', P'), \omega', \Delta', \phi>$ where $\omega' = \omega$, $\Delta' = \Delta$, $\Sigma' = \Sigma \cup \bar{\Sigma} \cup \bar{\bar{\Sigma}} \cup \{F,\$\}$, where $\bar{\Sigma} = \{ \bar{a} \mid a \in \Sigma \}$ and $\bar{\bar{\Sigma}} = \{ \bar{\bar{a}} \mid a \in \Sigma \}$, $\{F,\$\} \cap (\Sigma \cup \bar{\Sigma} \cup \bar{\bar{\Sigma}}) = \Phi$, $\phi(t) = 1$, and

(i) if $< \alpha_1, a, \alpha_2 > \longrightarrow \alpha \in P$ and $\alpha \neq \lambda$, then $< \bar{\alpha}_1, \bar{a}, \bar{\alpha}_2 > \longrightarrow \bar{\bar{\alpha}} \in P'$.

Here if $\alpha = a_1 a_2 \ldots a_m$ then $\bar{\alpha} = \bar{a}_1 \bar{a}_2 \ldots \bar{a}_m$ and $\bar{\bar{\alpha}} = \bar{\bar{a}}_1 \bar{\bar{a}}_2 \ldots \bar{\bar{a}}_m$.

(ii) if $< \alpha_1, a, \alpha_2 > \longrightarrow \lambda \in P$, then $< \bar{\alpha}_1, \bar{a}, \bar{\alpha}_2 > \longrightarrow \$ \in P'$.

For $\gamma_1 \in \Sigma^k$, $\gamma_2 \in \Sigma^l$, rules of type (iii), (iv), (v) are included in $\bar{P}'$.

(iii) For all $a \in \Sigma$, $< \gamma_1, a, \gamma_2> \longrightarrow \bar{a}$.

(iv) For all $a \in \Sigma$, $< \bar{\gamma}_1, \bar{a}, \bar{\gamma}_2> \longrightarrow \bar{a}$.

(v) For all $a \in \Sigma$, $< \bar{\bar{\gamma}}_1, \bar{\bar{a}}, \bar{\bar{\gamma}}_2> \longrightarrow a$.

Further, for $\gamma_1 \in \Sigma'^k$, $\gamma_2 \in \Sigma'^l$, the following rules are included in $\bar{P}'$.

(vi) $< \gamma_1, \bar{a}, \gamma_2> \longrightarrow F$ if $alph(\gamma_1) \cap (\Sigma \cup \bar{\bar{\Sigma}} \cup \{\$\}) \neq \Phi$
or $alph(\gamma_2) \cap (\Sigma \cup \bar{\bar{\Sigma}} \cup \{\$\}) \neq \Phi$

(vii) $< \gamma_1, \bar{\bar{a}}, \gamma_2> \longrightarrow F$ if $alph(\gamma_1) \cap (\Sigma \cup \bar{\Sigma}) \neq \Phi$
or $alph(\gamma_2) \cap (\Sigma \cup \bar{\Sigma}) \neq \Phi$

(viii) $< \gamma_1, \$, \gamma_2> \longrightarrow \lambda$

(ix) $< \gamma_1, F, \gamma_2> \longrightarrow F$

Note : In an interacting system there is at least one marked rule for every context for every letter.

We will give an informal proof here. The Fr–EIL system works as follows. For every string $\alpha \in \Sigma^*$ such that $\omega \overset{*}{\Rightarrow} \alpha$, any letter $a \in alph(\alpha)$ goes to $\bar{a}$ using $<\alpha_1, a, \alpha_2> \longrightarrow \bar{a}$ as this is the only rule available to it. At $\bar{a}$ the production $<\bar{\alpha}_1, \bar{a}, \bar{\alpha}_2> \longrightarrow \bar{a}$ is used, as, at the previous step, $\bar{a}$ would not have occurred, and hence only the marked rule can be used (because $\phi(t) = 1$). The string in the previous step has only letters belonging to Σ. After this step of deriving $\bar{a}$ from $\bar{a}$, on looking back one step an $\bar{a}$ is found, and hence any rule of the form $<\bar{\alpha}_1, \bar{a}, \bar{\alpha}_2> \longrightarrow \bar{\bar{\alpha}}$ or $<\bar{\alpha}_1, \bar{a}, \bar{\alpha}_2> \longrightarrow \bar{a}$ (which is the marked rule) can be used. If some letters use rules of the form (i) or (ii) whereas some of them use rules of the form (iv) then an F is produced somewhere which will make the subsequent derivations useless. The \$ symbol is introduced so that it "tells" other symbols which kind of rule has been used. In the subsequent step it disappears as it should and hence it obviously does not change the course of the derivation. So four steps are needed in this system for every step in the Fr-EIL system. ■

Even when OL and EOL systems are being simulated by forgetful systems, the simulating system still needs extension symbols and a right or left context of at least one symbol so that at every stage it is possible to check that the derivation is proceeding in a synchronised manner i.e. each of the steps in the derivation contains strings belonging to only Σ or only $\bar{\Sigma}$ or only $\bar{\bar{\Sigma}} \cup \{\$\}$. At any time if in a string there are letters from more than one of the sets Σ or $\bar{\Sigma}$ or $\bar{\bar{\Sigma}} \cup \{\$\}$, then the string becomes useless by deriving an F somewhere.

Similarly the proof can be extended to ETIL, TIL and TOL systems by using an Fr-ETIL system.

Theorem 3.4 : For every Fr–EIL system G with $\phi(t) = k$, there exists a Turing Machine T such that L(T) = L(G).

Proof : Let $G = <\Sigma, g, (\bar{P}, P), \omega, \Delta, \phi>$ be an Fr–EIL system with $\phi(t) = k$. Let $|\Sigma| = n$. Then a non-deterministic Turing machine which accepts L(G) can be constructed as follows.

Consider a multi-tape Turing machine with k+1 tapes. The last tape has three tracks and the first tape has n+1 tracks. On the last tape we use the uppermost of the three tracks to store the given string x. The middle track is used to store the word y, which is the latest word in the derivation. The lower track is for temporary storage. We initialize the middle track to ω and the last track to λ. The kth tape is initialized to $r\omega r$ and the rest of the tapes are blank. $r \notin \Sigma$.

At any step of the derivation, let $b_1b_2\ldots b_m$ be the string on the middle track of the last tape. We check the k tapes above it for the ancestry of b_i (this checking will be explained shortly) and then accordingly apply a rule $(\gamma_1, b_i, \gamma_2) \longrightarrow \alpha_i$ to b_i and then write the string on the last track in the following way :

$$r\,\alpha_1\,r\,\alpha_2\,\ldots\,r\,\alpha_m\,r$$

The first k tapes contain the ancestry information for the previous k time steps. In the first tape, the last of the n+1 tracks contains the string derived k time steps back. The upper n tracks of the first tape contain the occurrence of a letter in the ancestry beyond the k previous steps. This is necessary because the state of a letter at a particular stage depends, by definition, on whether it has ever occurred before or not. So the n tracks contain, for each symbol on the $n+1^{th}$ track of the first tape, binary information about whether a letter has occurred in its ancestry or not. So on deriving a new string we need to now update the information on these tapes. Thus we will overwrite the string on the last track of the first tape. But before we do this, we must extract the ancestry information in it for the symbols in it. For every letter that is being displaced, a '1' is written on the corresponding track. Then the string on tape 1 is overwritten by the one on tape 2, the string on tape 2 is overwritten by the one on tape 3 and so on. Finally the string on tape k is overwritten by the one on the last track of tape k+1. Then the string is cleaned of r symbols in the last track and copied to the middle track. The ancestry information stored in the first k tapes can be scanned as follows :

On scanning a string from left to right if we arrive at a symbol 'a' after passing m 'r' symbols then it means that the immediate ancestor of symbol 'a' is the mth symbol in the string above, counting from left to right and without counting the symbol 'r'. Counting of the 'r' symbols can be done by counters maintained on each tape at one end of the string.

For example consider tape 1 and tape 2 in the following status :
$\Sigma = \{a,b\}$, n = 2.

```
                 1  0  0     1  1  0        Track for letter a
Tape 1           1  1  0     1  1  1        Track for letter b
              r  a  b  a  r  a  b  a  r
Tape 2  r  a  a  r  r  a  a  r  a  a  r  b  b  r  a  a  r
```

The immediate ancestor for the third 'a' in the last string above is the second 'a' in the upper string. This is because we have to cross three 'r's in the last string to reach the third 'a', and hence the third letter in the upper string, without counting the 'r's, is its ancestor. Thus for the above strings the tree segment in the derivation tree will be as shown in Figure 2. On overwriting the upper string by the lower string we have the following contents in the first tape.

```
  1 1     1 1   1 1   1 1   1 1
  1 1     0 0   1 1   1 1   1 1
r a a r r a a r a a r b b r a a r
```

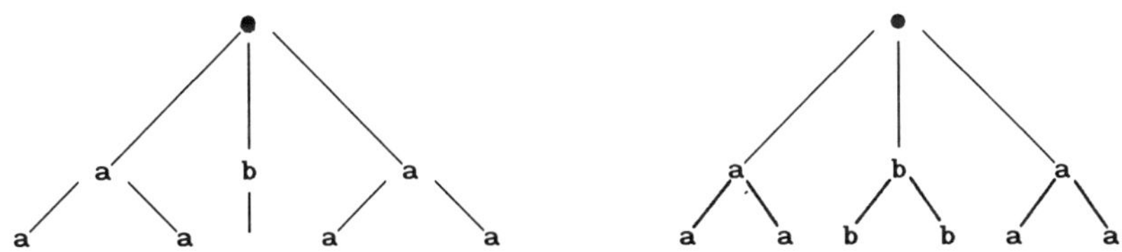

Figure 2

The machine T nondeterministically guesses a derivation for x. It then simulates this derivation, retaining ancestry information as described above. If the strings on the first and second tracks of the $k+1^{th}$ tape match, then T halts, accepting the input. ■

It is a well known theorem [1] that EIL systems generate exactly the class of recursively enumerable languages, i.e. $\mathcal{F}$(EIL) = $\mathcal{F}(\mathcal{RE})$. This, along with Theorems 3.3 and 3.4, gives us the following result.

Theorem 3.5 : $\mathcal{F}$(EIL) = $\mathcal{F}$(Fr–EIL) with $\phi(t) = k$.

Now we will consider an interesting theorem on the relationship between forgetful and nonforgetful interactionless L systems.

Theorem 3.6: There exists an algorithm which for an arbitrary Fr–0L system G with $\phi(t) = k$ produces a 0L system $\hat{G}$ (nonforgetful one) and a homomorphism h such that $L(G) = h(L(\hat{G}))$.

Before we go into the proof of the above theorem we will first define some structures which will be used later.

Definition 3.4 : Let Σ be a finite alphabet, $\Sigma = \{a_1, a_2, \ldots, a_n\}$.

(i) A Σ–vector is a vector $\mathbf{v} = \langle v_1, \ldots, v_n\rangle$ such that for each i, $1 \le i \le n$, $v_i = 0$ or 1. If $w \in \Sigma^*$, then the w–vector (denoted as B(w)) is defined as a Σ–vector $\mathbf{v} = \langle v_1, \ldots, v_n\rangle$ such that for each i, $1 \le i \le n$, $v_i = 1$ iff $w = \bar{w} a_i \bar{\bar{w}}$ for some $\bar{w}$ and $\bar{\bar{w}}$ in Σ^*.

(ii) If $\mathbf{v} = \langle v_1, \ldots, v_n\rangle$ and $\mathbf{v'} = \langle v'_1, \ldots, v'_n\rangle$ are two Σ–vectors, then we say that **v** is contained in **v'** iff for all i, $1 \le i \le n$, if $v_i = 1$ then $v'_i = 1$. We denote this by $\mathbf{v} \le \mathbf{v'}$. **0** denotes the zero Σ–vector; it is $\mathbf{0} = \langle 0, \ldots, 0\rangle$.

(iii)Given two Σ–vectors **v** and **v'**, their sum (denoted as **v** + **v'**) is defined as $\mathbf{v} + \mathbf{v'} = \langle z_1, \ldots, z_n\rangle$, where, for $1 \le i \le n$, $z_i = 1$ iff $v_i = 1$ or $v'_i = 1$.

Proof of theorem 3.6 : The alphabet set of our new OL system G is a set of 3-tuples. Every 3-tuple has the following structure. The first component is a letter from Σ (Σ is the alphabet set of the Fr–OL system). The second component is a queue of length k and the third component is a bit vector of length $|\Sigma|$. If the first component of the three tuple is a letter $a \in \Sigma$, then the second component stores its ancestry k steps back and the third component represents the occurrences of a letter $b \in \Sigma$ in the ancestry of 'a' beyond the k steps. This is needed, by definition of Fr–OL systems, for finding the present state of 'a'.

Let $G = \langle \Sigma, (\bar{P}, P), \omega, \phi\rangle$ be the Fr–OL system with $\phi(t) = k$. Let $0 \notin \Sigma$. Define $\hat{G} = \langle \Sigma, \hat{P}, \hat{\omega}\rangle$, where

$\hat{\Sigma} = \{[u, \mathbf{v}, \mathbf{w}] \mid u \in \Sigma,\ \mathbf{v} = \langle a_1, a_2, \ldots, a_k\rangle \text{ for } a_i \in \Sigma,\ \mathbf{w} \text{ is a } \Sigma\text{–vector}\}$.

For a three tuple $[a, \mathbf{v}, \mathbf{w}]$, if **w** has a '1' in the place corresponding to 'a' and the queue **v** has no 'a' in it then for every rule of the form $a \rightarrow b_1 b_2 \ldots b_m \in \bar{P}$ the following rule belongs to $\hat{P}$.

$$[a, \mathbf{v}, \mathbf{w}] \rightarrow [b_1, \mathbf{v}_1, \mathbf{w}_1][b_2, \mathbf{v}_2, \mathbf{w}_2] \ldots [b_m, \mathbf{v}_m, \mathbf{w}_m]$$

where if $\mathbf{v} = \langle a_1, a_2, \ldots, a_k\rangle$ then $\mathbf{v}_i = \langle a_2, a_3, \ldots, a_k, a\rangle$ and $\mathbf{w}_i = \mathbf{w} + B(a_1)$.

If $a \rightarrow \lambda \in \bar{P}$ then $[a, \mathbf{v}, \mathbf{w}] \rightarrow \lambda \in \hat{P}$.

For a three tuple $[a, \mathbf{v}, \mathbf{w}]$, if the queue **v** has an 'a' in it, or if **w** has a '0' in the place corresponding to 'a', then for every rule of the form $a \rightarrow b_1 b_2 \ldots b_m \in P$ the following rule belongs to $\hat{P}$.

$$[a, \mathbf{v}, \mathbf{w}] \rightarrow [b_1, \mathbf{v}_1, \mathbf{w}_1][b_2, \mathbf{v}_2, \mathbf{w}_2] \ldots [b_m, \mathbf{v}_m, \mathbf{w}_m]$$

where if $\mathbf{v} = \langle a_1, a_2, \ldots, a_k\rangle$ then $\mathbf{v}_i = \langle a_2, a_3, \ldots, a_k, a\rangle$ and $\mathbf{w}_i = \mathbf{w} + B(a_1)$.

If $a \rightarrow \lambda \in P$ then $[a, \mathbf{v}, \mathbf{w}] \rightarrow \lambda \in \hat{P}$.

If $\omega = a_1 a_2 \ldots a_k$ then $\hat{\omega} = [a_1, \mathbf{v}_1, \mathbf{w}_1][a_2, \mathbf{v}_2, \mathbf{w}_2] \ldots [a_k, \mathbf{v}_k, \mathbf{w}_k]$ where $\mathbf{v}_i = \langle 0, 0, \ldots, 0 \rangle$ and $\mathbf{w}_i = \langle 0, 0, \ldots, 0 \rangle$.

Thus we can capture the forgetfulness information in the components of the 3-tuples instead of the function $\phi(t)$.

Let h be a homomorphism $h : \hat{\Sigma} \longrightarrow \Sigma$ such that $h([a, \mathbf{v}, \mathbf{w}]) = a$ for each $a \in \Sigma$. It is clear that $h(L(\hat{G})) = L(G)$. ∎

Similarly we can show the following.

Theorem 3.7 : There exists an algorithm which for an arbitrary Fr–EOL (Fr–TOL) system G with $\phi(t) = k$ produces an EOL (TOL) system $\hat{G}$ (non-forgetful one) and a homomorphism h such that $L(G) = h(L(\hat{G}))$.

4. Decidability

Decidability results form an important part of any formal language theoretic study. Hence we will tackle some of these issues now.

Definition 4.1 : The membership problem for Fr–OL systems is said to be decidable if there is there an algorithm which, given an arbitrary Fr–OL system G and an arbitrary word x, decides whether or not x is in L(G).

Clearly it is desirable that the problem be decidable, for otherwise the use of such systems in practical applications would be limited. It is decidable for the classes $\mathcal{F}$(Fr–OL), $\mathcal{F}$(Fr–EOL) and $\mathcal{F}$(Fr–TOL) with certain restrictions on $\phi(t)$, as will be shown later.

Definition 4.2 : The equivalence problem for Fr–OL systems is said to be decidable if there is an algorithm which, given two arbitrary Fr–OL systems G and H, decides whether or not $L(G) = L(H)$.

These definitions can be directly extrapolated to other kinds of L systems also.

Theorem 4.1 : The membership problem for Fr–OL systems is decidable.

Proof : By theorem 3.6, we know that every Fr–OL system G with $\phi(t) = k$ can be "simulated" by a OL system $\hat{G}$ and a homomorphism h such that $h(L(\hat{G})) = L(G)$. In the proof of theorem 3.6 we see that the homomorphism h extracts the first component from every 3–tuple of the string of 3–tuples obtained in $L(\hat{G})$. Hence in the OL system $\hat{G}$ we have to find out if there exists a string of 3–tuples whose first components, when extracted and concatenated in the same order, form the given string x. There are only finitely many such strings, and they must be checked for membership in

$L(\hat{G})$. The membership problem for OL systems is known to be decidable [1]. Hence it is easy to see that membership problem for Fr–OL systems with $\phi(t) = k$, where k is a constant, is decidable. ■

Theorem 4.2 : The membership problem for Fr–EOL systems is decidable.

Proof : Let $G = < \Sigma, (\overline{P},P), \omega, \Delta, \phi>$ be the given Fr–EOL system. Consider the Fr–OL system $\hat{G} = < \Sigma, (\overline{P},P), \omega, \phi>$. A string x is in L(G) iff x is in $L(\hat{G}) \cap \Delta^*$. It is trivial to decide whether or not x is in Δ^*. Then it follows from the previous theorem that the membership problem for Fr-EOL systems is also decidable. ■

Theorem 4.3 : The membership problem for Fr–TOL systems with $\phi(t) = k$ is decidable.

Proof : This follows from theorem 3.7 and an argument similar to that used in proving theorem 4.1. ■

Next we consider the equivalence problems. These are undecidable for non-forgetful L systems [1]. Since such systems are a special case of Fr–L systems with $\overline{P} = P$, we immediately have :

Theorem 4.4 : There is no algorithm which, given an arbitrary pair of Fr–OL (Fr–EOL, Fr–TOL) systems G and H, decides whether or not L(G) = L(H).

Note : Even if we restrict Fr–L systems to have $\overline{P} \neq P$, the equivalence problem remains undecidable. This can be shown by following the proof of undecidability of the equivalence problem for OL systems and then applying Theorem 3.1 in a straightforward fashion.

5. Closure Properties

In this section we will deal with some closure results for forgetful L systems. No attempt is being made to be exhaustive and we will only have a glimpse at some of the results.

Theorem 5.1 : $\mathcal{F}$(Fr-OL) is not closed under union and concatenation.

Proof : (i) Consider the languages $\{ a^2 \}$ and $\{ a^3 \}$. They are generated by the Fr–OLs $G_1 = < \{a\}, (\{a \rightarrow a\}, \{a \rightarrow a\}), a^2, \phi_1>$ and

$G_2 = < \{a\}, (\{a \rightarrow a\}, \{a \rightarrow a\}), a^3, \phi_2>$ respectively.

(ϕ_1 and ϕ_2 can be any functions.) But their union $L = \{ a^2, a^3 \}$ cannot be generated by any Fr–OL system with any $\phi(t)$ and any degree, as can be seen below.

Let an Fr–OL system $G = < \Sigma, (\overline{P}, P), \omega, \phi >$ generate L. $\omega \in L$, so ω is either a^2 or a^3.

If $a^3 = \omega$, then there must be a rule of the form $a \longrightarrow \lambda$, and on using this rule at the first step we get λ, which does not belong to L. (At the first step, by definition, any rule can be used).

If $a^2 = \omega$, then $a^2 \overset{+}{\Rightarrow} a^3$. So rules of the form $a \longrightarrow a^3$ and $a \longrightarrow \lambda$ or $a \longrightarrow a^2$ and $a \longrightarrow a$ must exist, which, on being used at the first step, can also generate strings not in L.

Thus L cannot be generated by any Fr–0L system.

(ii) Consider the languages $\{ a^2 \}$ and $\{ \lambda, a \}$. Both are Fr–0L languages but their concatenation gives $\{ a^2, a^3 \}$ which is not an Fr–0L language. ■

Theorem 5.2 : $\mathcal{F}$(Fr–T0L) is not closed under union and concatenation.

Proof : The proof for this theorem is identical to the previous proof and makes use of the fact that $\{ a^2, a^4 \}$ is not an Fr–T0L language for any $\phi(t)$ and any degree. ■

We next present a theorem on the "collapse of degree hierarchy". It will be later used in proving some closure results for Fr–E0L systems.

Theorem 5.3 : For every Fr–E0L system G with degree k + 1, k > 1, there exists an Fr–E0L system G' with degree k such that L(G) = L(G').

Proof : Let $G = < \Sigma, (\bar{P}, P), \omega, \Delta, \phi >$ be an Fr–E0L system with arbitrary $\phi(t)$ and degree k + 1.

We construct an Fr–E0L system $G' = < \Sigma', (\bar{P}', P'), \omega', \Delta', \phi' >$ with degree k and $\omega' = \omega$, $\Delta' = \Delta$, $\Sigma' = \Sigma \cup \{ \bar{a}, \bar{\bar{a}} \mid a \in \Sigma \}$. If $\alpha = a_1 a_2 \ldots a_r$, then $\bar{\alpha} = \bar{a}_1 \bar{a}_2 \ldots \bar{a}_r$ and $\bar{\bar{\alpha}} = \bar{\bar{a}}_1 \bar{\bar{a}}_2 \ldots \bar{\bar{a}}_r$. If $\alpha = \lambda$, then $\bar{\alpha}, \bar{\bar{\alpha}} = \lambda$. For any $a \in \Sigma$, if there are k + 1 productions for 'a' in $\bar{P}$ (i.e. there are k+1 marked productions for 'a'), then select any two productions $a \longrightarrow \alpha$ and $a \longrightarrow \beta$. For any production $a \longrightarrow \gamma$ among the rest of the productions of 'a' in $\bar{P}$, include $a \longrightarrow \bar{\gamma}$ in $\bar{P}'$. Also include in $\bar{P}'$ the production $a \longrightarrow \bar{\bar{a}}$. Include $\bar{\bar{a}} \longrightarrow \alpha$ and $\bar{\bar{a}} \longrightarrow \beta$ in $\bar{P}'$. For all those letters 'b' with at most k marked productions, we include, for $b \longrightarrow \alpha$ in $\bar{P}$, a rule $b \longrightarrow \bar{\alpha}$ in $\bar{P}'$. $\bar{P}'$ also contains a rule $\bar{c} \longrightarrow c$ for each $c \in \Sigma$. By definition the productions in $\bar{P}'$ are also in P'. Further, for each of those productions $a \longrightarrow \alpha$ in P but not in $\bar{P}$, a production of the form $a \longrightarrow \bar{\alpha}$ is included in P'. Let $\phi'(t) = 2\phi(t)$. This system has degree k and produces the same language as G. Informally this is how the system works. In order to reduce the degree of the system, we perform, in two steps in the new system, what is done in a single derivation step in G. Instead of having k+1 alternate development strategies for a letter in state S = 1, we allow it to develop along any of k-1 strategies and if it has to develop along the other two it has to first go to a letter $\bar{\bar{a}}$ and from there develop along any of the two srategies. Meanwhile those that developed along the k-1 initial strategies make the transition from $\bar{a}$ to a. Since every step in a derivation in G now takes two steps in G', $\phi'(t)$ becomes $2\phi(t)$. ■

Corollary 5.3 : For every Fr–EOL system G with degree k + m, k > 1, m > 0, there exists an Fr–EOL system G' with degree k such that L(G) = L(G').

Definition 5.1 : An Fr–EOL system $G = < \Sigma, (\bar{P}, P), \omega, \Delta, \phi >$ is said to be synchronised iff for every symbol $a \in \Delta$ and string $\alpha \in \Sigma^*$, if $a \overset{+}{\Rightarrow} \alpha$, then α is not in Δ^*. This means that if a string contains an element of Δ, then all other strings derivable from it contain some element not in Δ.

Theorem 5.4: There is an algorithm which for any Fr–EOL system G with degree k, k > 1, produces a synchronised Fr–EOL system $\hat{G}$ with the same degree, such that $L(G) = L(\hat{G})$.

Proof : Let $G = < \Sigma, (\bar{P}, P), \omega, \Delta, \phi >$ be an Fr–EOL system with degree k. For any symbol $a \in \Delta$, let $\bar{a}$ be a symbol not in Σ, such that if $a \neq b$, then $\bar{a} \neq \bar{b}$. Let $\bar{\Delta} = \{ \bar{a} \mid a \in \Delta \}$, let S and F be symbols not in $\Sigma \cup \bar{\Delta}$, and let $\hat{\Sigma} = \Sigma \cup \bar{\Delta} \cup \{S, F\}$.

With any α in Σ^* we associate a string $\bar{\alpha}$ in $\hat{\Sigma}^*$ as follows. If $\alpha = \lambda$, then $\bar{\alpha} = \lambda$. If $\alpha = a_1a_2...a_n$, then $\bar{\alpha} = b_1b_2...b_n$, where $b_i = \bar{a}_i$ if a_i is in Δ, and $b_i = a_i$ if a_i is not in Δ.

Let $\hat{G}$ be the Fr–EOL system $< \hat{\Sigma}, (\bar{\hat{P}}, \hat{P}), S, \hat{\Delta}, \hat{\phi} >$, where

$$\begin{aligned}\bar{\hat{P}} = &\{ a \rightarrow \bar{\alpha} \mid a \in \Sigma - \Delta,\ a \rightarrow \alpha \in \bar{P} \} \cup \\ &\{ \bar{a} \rightarrow \bar{\alpha} \mid a \in \Delta,\ a \rightarrow \alpha \in \bar{P} \} \cup \\ &\{ S \rightarrow \bar{\omega} \} \cup \\ &\{ \bar{a} \rightarrow a \mid a \in \Delta \} \cup \\ &\{ a \rightarrow F \mid a \in \Delta \cup \{F\} \}\end{aligned}$$

$$\begin{aligned}\hat{P} = \bar{\hat{P}} \cup &\{ a \rightarrow \bar{\alpha} \mid a \in \Sigma - \Delta,\ a \rightarrow \alpha \in P \} \cup \\ &\{ \bar{a} \rightarrow \bar{\alpha} \mid a \in \Delta,\ a \rightarrow \alpha \in P \}\end{aligned}$$

$\hat{\Delta} = \Delta$ and $\hat{\phi}(t) = \phi(t)$.

Clearly $\hat{G}$ is synchronised and is of degree k+1. It can also be proved that $L(\hat{G}) = L(G)$. By Theorem 5.3 we can reduce the degree to k, and from the proof it is obvious that the synchronisation property will not be lost. Thus the theorem is proved. ■

Theorem 5.5 : If L_1 and L_2 are Fr–EOL languages with degree m > 1, then $L_1 \cup L_2$, L_1L_2, L_1^+ and L_1^* are also Fr–EOL languages with the same degree m.

Proof : If L_1 and L_2 are Fr–EOL languages then there exist Fr–EOL systems $G_1 = < \Sigma_1, (\bar{P}_1, P_1), \omega_1, \Delta_1, \phi_1>$ and $G_2 = < \Sigma_2, (\bar{P}_2, P_2), \omega_2, \Delta_2, \phi_2>$ with $\phi_1(t) = k_1$ and $\phi_2(t) = k_2$ such that $L(G_1) = L_1$ and $L(G_2) = L_2$. Without loss of generality we can assume that the given systems are synchronised, and that $(\Sigma_1 - \Delta_1) \cap \Sigma_2 = \Phi$ and $(\Sigma_2 - \Delta_2) \cap \Sigma_1 = \Phi$. Let S be a symbol not in $\Sigma_1 \cup \Sigma_2$.

(i) $L(G) = L_1 \cup L_2$, where $G = < \Sigma, (\overline{P}, P), S, \Delta_1 \cup \Delta_2, \phi_3 >$.

$\Sigma = \Sigma_1 \cup \Sigma_2 \cup \Sigma_r \cup \Sigma_s \cup \{S\}$, $\phi_3(t) = \mathrm{LCM}(k_1, k_2)$ where LCM denotes Least Common Multiple, and $\overline{P}$, P are as specified below.

$$\text{Let } k'_1 = \frac{\mathrm{LCM}(k_1,k_2)}{k_1} \text{ and } k'_2 = \frac{\mathrm{LCM}(k_1,k_2)}{k_2} .$$

Let $\Sigma_r = \{ a^r_i \mid i = 1 \text{ to } k'_1 - 1,\ a \in \Sigma_1\}$ and

$\Sigma_s = \{ b^s_i \mid i = 1 \text{ to } k'_2 - 1,\ b \in \Sigma_2\}$

All the a^r_i and b^s_i are distinct, and $(\Sigma_r \cup \Sigma_s) \cap (\Sigma_1 \cup \Sigma_2) = \Phi$.

If $\alpha = abcd... \in \Sigma_1^+$, then $\alpha^r_i = a^r_i b^r_i c^r_i d^r_i ...$, and if $\beta = abcd... \in \Sigma_2^+$, then $\beta^s_i = a^s_i b^s_i c^s_i d^s_i ...$.

If $a \longrightarrow \alpha \in \overline{P}_1$ then $a \longrightarrow \alpha^r_1 \in \overline{P}$

If $a \longrightarrow \beta \in \overline{P}_2$ then $a \longrightarrow \beta^s_1 \in \overline{P}$

If $a \longrightarrow \alpha \in P_1$ then $a \longrightarrow \alpha^r_1 \in P$

If $a \longrightarrow \beta \in P_2$ then $a \longrightarrow \beta^s_1 \in P$

The following productions belong to both $\overline{P}$ and P.

$a^r_i \longrightarrow a^r_{i+1}$ for all $a \in \Sigma_1$, $i = 1, 2, \ldots, k'_1 - 2$

$a^r_i \longrightarrow a$ for all $a \in \Sigma_1$, $i = k'_1 - 1$

$b^s_i \longrightarrow b^s_{i+1}$ for all $b \in \Sigma_2$, $i = 1, 2, \ldots, k'_2 - 2$

$b^s_i \longrightarrow b$ for all $b \in \Sigma_2$, $i = k'_2 - 1$

Also include $S \longrightarrow (\omega_1)^r_1$ and $S \longrightarrow (\omega_2)^s_1$ in $\overline{P}$ and P.

The intuition behind the proof is that $\phi_3(t)$ has to be a function "agreeable" to both the systems. So we first perform a derivation that produces a step "equivalent" to a derivation in either G_1 or G_2, and then fill in dummy steps until we reach the step which occurs at the "consensus time" i.e. $\mathrm{LCM}(k_1, k_2)$.

(ii) The above proof also holds for concatenation with minor modifications.

(iii) and (iv) The proofs for these are almost identical to those for nonforgetful systems and are hence omitted here. ■

6. Conclusion

This paper presents "forgetful" L systems as a new variant of the classical L systems. Various decidability issues, comparison of these systems with respect to the nonforgetful L systems, and some closure properties are presented. This model is an attempt towards more realistic L systems which make use of ancestry information during their developmental process.

7. References

[1] G. T. Herman and G. Rozenberg, "Developmental Systems and Languages" North Holland Publishers, Amsterdam, 1975.

[2] J. E. Hopcroft and J. D. Ullman, "Introduction to Automata Theory, Languages and Computation", Addison Wesley 1979.

[3] W. O. Hollerer and R. Vollmar, On "forgetful" cellular automata, Journal of Computer and System Sciences **11** (1975), 237-251.

[4] Krithivasan, K. and Das, A., Terminal weighted grammars and picture description, Computer Vision, Graphics and Image Processing **30**, 1985, pp 534–549.

[5] A. Lindenmayer, Mathematical models for cellular interaction in development: I and II, Journal of Theoretical Biology **18** (1968), 280-315.

[6] Prusinkiewicz, P., Krithivasan K. and Vijayanarayana M.G., Applications of L systems to algorithmic generation of South Indian folk art patterns and Karnatik music, in "A Perspective in Theoretical Computer Science - Commemorative Volume for Gift Siromoney", ed. R. Narasimhan, World Scientific Publishers, Singapore, 1989, pp 229–247.

[7] Prusinkiewicz, P., Applications of L systems to Computer Imagery, Graph Grammars and their applications to Computer Science, LNCS 291, 1986, pp 534 - 549.

[8] G. Rozenberg and A. Salomaa, "The Mathematical Theory of L systems", Academic Press, New York, 1980.

[9] Salomaa, A., "Formal Languages", Academic Press Incorporated, 1973.

[10] R. Vollmar, Uber Turingmaschinen mit variablem Speicher, Elektron Informationsverarbeit Kybernetik 9 (1973), 3-13.

SKEW-MORPHISMS AND SYSTEMS

M. Ito[1] *and G. Thierrin*[2]

[1]*Faculty of Science, Kyoto Sangyo University, Kyoto 603, Japan*
[2]*Department of Mathematics, The University of Western Ontario, London, Ontario, Canada*

Abstract: *Skew-morphisms are generalizations of endomorphisms. Sequential functions, automaton mappings and length-preserving word functions are special cases of skew-morphisms. Properties of skew-morphisms of free monoids are given, as well as a general method to construct them. Images of skew-morphisms and inverse skew-morphisms are considered. Replacing morphisms by skew-morphisms, generalizations of DOL systems, called SKDOL systems are introduced naturally. Some relations between generalized sequential functions and skew-morphisms are studied. In particular, a characterization of skew-morphisms that are generalized sequential functions is given and it is shown that in general skew-morphisms do not preserve regularity of languages.*

Keywords: *language, morphism, system, mapping, function, sequential, automaton.*

O. Introduction

A mapping ϕ of a semigroup M into M such that $\phi(xy) = \phi(x)z$ for some $z \in M$ is called a *skew − morphism*. Obviously, skew-morphisms are generalizations of endomorphisms of semigroups.

If the semigroup M is the free monoid X^* generated by the set X, then skew-morphisms occur naturally in several situations. For example, generalized sequential functions (see [2]), automaton mappings (see [3]) and length-preserving word functions (see [6]) are all special cases of skew-morphisms. Every infinite sequence of mappings of X into X^* induces a skew-morphism of X^*. Also skew-morphisms can be associated with every element of left semaphore codes and hence with left ideals of X^*.

In section 1, we give some examples and elementary properties of skew-morphisms and in section 2, a general method for constructing skew-morphisms in X^* is described.

In section 3, images of skew-morphisms and inverse skew-morphisms are considered. In particular, we show that every language containing the empty word 1 is the image of X^* by some skew-morphism, which indicates the difficulty of constructing an algebraic theory of skew-morphisms.

DOL systems (see [4]) are defined as constructs of the form (X, ϕ, w) where X is an alphabet, ϕ is a morphism of X^* and w is a nonempty word. The language associated with this system is the set $\{\phi^n(w) | n \geq 0\}$. It is then natural to consider generalizations of DOL systems called SKDOL systems where the morphism ϕ is replaced by a skew-morphism. The purpose of section 4 is to characterize languages generated by several classes of SKDOL systems.

Every generalized sequential function is a skew-morphism, but the converse is not true. In sections 5 and 6, we show the relations between generalized sequential functions and skew-morphisms and in particular we characterize the skew-morphisms that are generalized sequential functions.

It is known that a generalized sequential function preserves the regularity of languages (see [2] for example). In the last section of this paper, we show that this is no more the case for skew-morphisms. We show also the existence of length-preserving skew-morphisms preserving the regularity of languages without being sequential functions. The characterization of skew-morphisms preserving regularity is an open problem.

I. Definitions and elementary properties

In this section, we give the definition of skew-morphism and then provide some examples and elementary properties of skew-morphisms.

A mapping ϕ of a semigroup M into M such that $\phi(xy) = \phi(x)z$ for some $z \in M$ is called a *skew-morphism.*

Examples.

1. Every morphism is a skew-morphism.

2. Let X^* be the free monoid with identity 1 generated by X, let Σ be the set of all mappings ϕ of X into X^* and let Σ^ω be the set of all infinite sequences of elements of Σ. If $x = x_1x_2...x_n$ with $x_i \in X$ and if $\Phi \in \Sigma^\omega$ with $\Phi = \phi_1\phi_2...\phi_i...\phi_n...$, then $\Phi(1) = 1$ and

$$\Phi(x_1x_2...x_i...x_n) = \phi_1(x_1)\phi_2(x_2)...\phi_i(x_i)...\phi_n(x_n), \; x_i \in X$$

Clearly $\Phi(xy) = \Phi(x)z$ for some $z \in X^*$ and hence Φ is a skew-morphism.

Remark that not all skew-morphisms of X^* are of the form Φ with $\Phi \in \Sigma^\omega$. For example, let $X = \{a, b\}$ and $\phi(a) = a, \phi(b) = b, \phi(ax) = ax, \phi(bx) = bx'$ where $a' = b, b' = a$ and $x' = x'_1...x'_n$ if $x = x_1...x_n$.

In the following section, we will determine the structure of skew-morphisms in X^*.

3. Recall that a nonempty set $C \subseteq X^+$ is called a *code* (see [1]) over X if for

$$x_1, x_2, ..., x_n, y_1, y_2, ..., y_m \in C \text{ the equality } x_1x_2...x_n = y_1y_2...y_m$$

implies that $n = m$ and $x_i = y_i$, $i = 1, 2, ..., n$ (for details, see [1]). One of the typical codes is a prefix (suffix) code. A nonempty set $C \subseteq X^+$ is called a *prefix (suffix) code* over X if $CX^+ \cap C = \emptyset$ ($X^+C \cap C = \emptyset$).

A left semaphore code (see [5]) Q over X is a suffix code such that for every $x \in X^*$ and every $u \in Q$, there exists $v \in Q$ such that $ux = yv$ for some $y \in X^*$ (see [4]).

It is easy to see that the mapping ϕ_u defined on X^* by $\phi_u(x) = y$ where $ux = yv$ is a skew-morphism.

It is immediate that in a group, every mapping is a skew-morphism and consequently this notion is of no interest. The next proposition shows that the algebraic structures having this property are rather restricted as far as semigroups are concerned.

Proposition 1.1. Let M be a semigroup. Then every mapping of M is a skew-morphism if and only if M is a right simple or a left zero semigroup.

Proof. ($\Leftarrow$) Let $x, y \in M$. If M is right simple, then $\phi(x)M = M$. Hence $\phi(xy) = \phi(x)z$ for some $z \in M$. Now let M be left zero. Then $\phi(xy) = \phi(x)z$ for any $z \in M$. Therefore ϕ is a skew-morphism.

($\Rightarrow$) Assume that M is neither right simple nor left zero. Since M is not right simple, there exist $a, b \in M$ such that $b \notin aM$. Moreover, since M is not left zero, there exist $c, d \in M$ such that $cd \neq c$. Consider a mapping ϕ satisfying the conditions that $\phi(cd) = b$ and $\phi(c) = a$. Then $\phi(cd) \neq \phi(c)z$ for any $z \in M$, a contradiction. Therefore M is right simple or left zero. □

Proposition 1.2. The set of all skew-morphisms of a semigroup M is a monoid with the composition of mappings.

Proof. Immediate. □

Proposition 1.3. Let ϕ be a skew-morphism of the semigroup M and let a be a fixed element of M. The mapping ϕ_a of M into M defined by $\phi_a(x) = a\phi(x)$ is a skew-morphism.

Proof. Immediate. □

Proposition 1.4. If P is a left semaphore code over X, then the set $\phi_P = \{\phi_u | u \in P\}$ is a free subsemigroup of X^* under the composition of mappings and it is isomorphic to the free semigroup generated by P.

Proof. Immediate. □

II. The structure of skew-morphisms in X^*

In this section, we give a general method to construct the skew-morphisms of a free monoid.

Let $L \subseteq X^*$. Then the set $pr(L) = \{x \in X^* | \exists y \in X^*,\ xy \in L\}$ is called the *pre-closure* of L. If $pr(L) = L$, then L is called a *pre-closed set.*

Let ϕ be a mapping of a pre-closed set L into X^*. Then ϕ is called a *sskew-morphism (semi-skew-morphism)* of L into X^* if ϕ satisfies the following condition:

$$\phi(xy) = \phi(x)z \text{ where } x, y, z \in X^* \text{ and } xy \in L.$$

Construction of skew-morphisms in X^*: Let ϕ be a skew-morphism of X^* into X^* and $\hat{\phi}_n = \phi|_{X^{[n]}}, n \geq 0$ where $X^{[n]} = \bigcup_{i=0}^{n} X^i$ and $\phi|_{X^{[n]}}$ means the restriction of ϕ to $X^{[n]}$. Then $\hat{\phi}_n$ is a sskew-morphism of $X^{[n]}$ into X^*. Now consider the relation between $\hat{\phi}_n$ and $\hat{\phi}_{n+1}$. First $\hat{\phi}_{n+1}|_{X^{[n]}} = \hat{\phi}_n$. Let $u \in X^{[n+1]} \setminus X^{[n]}$. Then $u = u'a$ where $u' \in X^n$ and $a \in X$. Since $\hat{\phi}_{n+1}$ is the restriction of ϕ to $X^{[n+1]}$, $\hat{\phi}_{n+1}(u) = \phi(u) = \phi(u'a) = \phi(u')w_{u',a} = \hat{\phi}_n(u')w_{u',a}$ where $w_{u',a}$ is an element of X^*. This observation suggests a contruction method of skew-morphisms in X^*.

Step 1. Let ϕ be a sskew-morphism of $X^{[n]}$ into X^*. We define a mapping $\bar{\phi}$ of $X^{[n+1]}$ into X^* as follows:

$$\bar{\phi}(u) = \phi(u) \text{ if } u \in X^{[n]}$$

and

$$\bar{\phi}(u) = \phi(u')w_{u',a} \text{ if } u \in X^{[n+1]} \setminus X^{[n]}.$$

where $u = u'a, u' \in X^n, a \in X, w_{u',a}$ an arbitrary element of X^*.

Then $\bar{\phi}$ is well defined as a mapping of $X^{[n+1]}$ into X^*. Moreover, it is easy to see that $\bar{\phi}$ is a sskew-morphism of $X^{[n+1]}$ into X^*.

Step 2. We start by constructing a sskew-morphism ϕ_0 of $X^{[0]}$ into X^*. We can choose any element of X^* as $\phi_0(1)$. Using the procedure in Step 1, we can get a series of sskew-morphisms $\phi_0, \phi_1, \phi_2, \ldots$ such that

$$\phi_0 = \phi_1|_{X^{[0]}}, \phi_1 = \phi_2|_{X^{[1]}}, \phi_2 = \phi_3|_{X^{[2]}}, \ldots$$

Now define a mapping of X^* into X^* as follows: $\phi(u) = \phi_i(u)$ if $u \in X^{[i]}$. Then ϕ is well defined as a mapping of X^* into X^*.

Proposition 2.1. ϕ is a skew-morphism of X^* into X^*. Moreover $\hat{\phi}_n = \phi|_{X^{[n]}} = \phi_n$.

Proof. That $\hat{\phi}_n = \phi_n$ is obvious by definition. Let $u, uv \in X^*$. Then, since $X^* = \bigcup_{i=0}^{\infty} X^{[i]}$ and $X^{[i]} \subseteq X^{[i+1]}$, there exists $i \geq 1$ such that $u, uv \in X^{[i]}$. Note that ϕ_i is a sskew-morphism. Hence $\phi(uv) = \phi_i(uv) = \phi_i(u)z = \phi(u)z$ for some $z \in X^*$. This shows that ϕ is a skew-morphism of X^* into X^*. □

Therefore the above method can provide a skew-morphism. Moreover, by the observation at the beginning, any skew-morphism of X^* into X^* can be constructed by the above method.

III. Images of skew-morphisms and inverse skew-morphisms

In this section, images of skew-morphisms and inverse skew-morphisms are considered. As a result, we can show that a skew-morphism does not always preserve the Chomsky hierarchy.

Let $u, v \in X^*$. By $u \leq v$ we mean that $v = ux$ for some $x \in X^*$. If $u \neq v$, then inequality is strict and is denoted by $<$. It is easy to see that $\leq$ is a partial order and that every skew-morphism is a $\leq$-preserving mapping.

Proposition 3.1. Let $L \subseteq X^*$ with $1 \in L$. Then there exists a skew-morphism ϕ of X^* into X^* such that $\phi(X^*) = L$.

Proof. For any $x \in X^*$, we define $\phi(x)$ as follows:

$$\phi(x) = u \ (u \in L, u \leq x \text{ and } \ (u < v \leq x \Rightarrow v \notin L)).$$

Clearly $\phi(X^*) = L$. Now we show that ϕ is a skew-morphism. Let $x, y \in X^*$. Then there exists a chain

$$v_1 \leq x < v_2 < v_3 < ... < v_r \leq xy$$

such that $v_i \in L$, $i = 1, 2, ..., r$, $(v_i < w < v_{i+1} \Rightarrow \ w \notin L)$, $i = 1, 2, ..., r$ and $(v_r < w < xy \ \Rightarrow \ w \notin L)$. By the definition of ϕ, $\phi(x) = v_1$ and $\phi(xy) = v_r$. Since $v_1 \leq v_r$, $\phi(xy) = \phi(x)z$ for some $z \in X^*$. This completes the proof of the proposition. □

Corollary 1. A skew-morphism does not always preserve the Chomsky hierarchy. □

Remark 3.1. Proposition 3.1 indicates the difficulty of constructing the algebraic theory of skew-morphisms. (Note that a skew-morphism does not always preserve even a monoid.)

Remark 3.2. In the case $1 \notin L$, the assertion of Proposition 3.1 does not hold (see Proposition 3.3). In this case, we have the following:

Proposition 3.2. Let $1 \notin L \subseteq X^*$ and let $|m(L)| < +\infty$ where $m(L) = \{x \in L | x$ is a minimal element of L related to the order $\leq\}$ and $|m(L)|$ is the cardinality of $m(L)$. Then there exist a finite set $K \subseteq X^*$ and a skew-morphism ϕ such that $\phi(X^* \setminus K) = L$.

Proof. Let $K = \{y \in X^* | \exists x \in m(L), y < x\}$ and let $T = \{y \in X^* | y \notin K$ and $\forall x \in m(L), x \not\leq y\}$. Since $|m(L)| < +\infty, K$ is a finite set. Moreover let $z \in L$. We define ϕ as follows:

$\phi(u) = z$ if $u \in T$,

$\phi(u) = 1$ if $u \in K$,

$\phi(u) = v$ ($v \in L, v \leq u$ and ($v < w < u \Rightarrow w \notin L$)) if $v \in X^* \setminus (T \cup K)$.

Then it is easy to verify that ϕ is a skew-morphism and $\phi(X^* \setminus K) = L$. □

Remark 3.3. In the above proposition, the condition $|m(L)| < +\infty$ is necessary (see the following proposition).

Proposition 3.3. Let $L \subseteq X^+$ be an infinite prefix code. Then $\phi(X^* \setminus K) \neq L$ for any skew-morphism ϕ and any finite set $K \subseteq X^*$.

Proof. Suppose that $\phi(X^* \setminus K) = L$ for some skew-morphism ϕ and finite set K. Since K is finite, there exists a positive integer $n \geq 1$ such that $X^n \subseteq X^* \setminus K$. Let $u \in X^* \setminus K$. Then there exists $w \in X^n$ such that $w \leq u$ or $u \leq w$. Note that $\phi(w) \leq \phi(u)$ or $\phi(u) \leq \phi(w)$ and $\phi(w), \phi(u) \in L$. Since L is a prefix code, $\phi(w) = \phi(u)$. Consequently, $\phi(X^* \setminus K) \subseteq \{\phi(w) | w \in X^n\}$. This contradicts the infiniteness of L. □

Let ϕ be a skew-morphism of X^*. If $\phi(x) < \phi(xy)$ for every $x \in X^*, y \in X^+$, then ϕ is called a *stskew-morphism* (strong skew-morphism).

By Proposition 3.2, for general skew-morphisms ϕ, ϕ^{-1} does not necessarily preserve the property of prefix code. However, the following is obvious.

Proposition 3.4. Let ϕ be a stskew-morphism and let $L \subseteq X^*$ be a prefix code. Then $\phi^{-1}(L)$ is a prefix code. □

IV. Languages generated by SKDOL systems

In this section, we define the SKDOL scheme (system, language) and deal with the problem of determining which language are SKDOL languages.

A *SKDOL scheme (system)* S is a pair (X, ϕ) (a triple (X, ϕ, w)) ϕ with a skew-morphism. A *SKDOL language* generated by S is the set $\{\phi^n(w) | n = 0, 1, 2, ...\}$. A DOL scheme (system, language) is the special case where ϕ is a morphism.

Lemma 4.1. Let $L \subseteq X^*$ be a pre-closed set and let ϕ be a sskew-morphism of L into X^*. Then there exists a skew-morphism $\bar{\phi}$ of X^* into X^* such that $\bar{\phi}|_L = \phi$.

Proof. Let $u \in X^*$. By $\alpha_L(u)$ we denote the element $u' \in L$ such that $u' \leq u$ and $u' < v \leq u$ implies $v \notin L$. Now we define $\bar{\phi}(u)$ as follows:

$$\bar{\phi}(u) = \phi(\alpha_L(u)) \text{ for any } u \in X^*$$

Obviously $\bar{\phi}$ is well defined as a mapping of X^* into X^* and $\bar{\phi}(u) = \phi(u)$ for any $u \in L$. Now let $u, uv \in X^*$. Since $u = \alpha_L(u)x$ for some $x \in X^*$, $uv = \alpha_L(u)xv$. Thus $\alpha_L(u) \leq \alpha_L(uv)$ and $\phi(\alpha_L(u)) \leq \phi(\alpha_L(uv))$, i.e. $\bar{\phi}(u) \leq \bar{\phi}(uv)$.

This completes the proof of the lemma. □

Let $S = (X, \phi, w)$ be a SKDOL system. Then S is called a *SSKDOL system* (strict SKDOL system) if S satisfies the following condition:

$\phi^i(w) < \phi^j(w)$ if $\phi^i(w)$ and $\phi^j(w)$ are comparable with respect to $\leq$ and $i < j$.

An example of a SSKDOL system is a system $S = (X, \phi, w)$ such that

$$|w| < |\phi(w)| < |\phi^2(w)| < ... < |\phi^i(w)| < |\phi^{i+1}(w)| < ...$$

By the construction method of skew-morphisms, such an example can easily be provided.

Proposition 4.2. Let $L \subseteq X^*$. Then L is generated by a SSKDOL system if and only if L satisfies one of the following conditions:

(1) L is an infinite prefix code,

(2) $L = L_0 \cup L_1 \cup L_2 \cup ... \cup L_n$ where L_0 is a finite prefix code ($L_0 = \emptyset$, possible), $|L_1| = |L_2| = ... = |L_n| = +\infty$, each L_i, $i = 1, 2, ..., n$ is an infinite ascending chain (with respect to $\leq$) and u and v are incomparable (with respect to $\leq$) for any $(u, v) \in L_i \times L_j$, $i \neq j$, $i, j = 0, 1, 2, ..., n$.

Proof. ($\Rightarrow$) Let $S = (X, \phi, w)$ be a SSKDOL system. First we show that it is impossible to have the following inequalities:

$$\phi^i(w) \leq \phi^{i+r}(w), \quad \phi^i(w) \leq \phi^{i+t}(w)$$

where $\phi^{i+r}(w)$ and $\phi^{i+t}(w)$ are incomparable.

Suppose that the above inequalities hold. Let $u = \phi^i(w)$. Then $\phi^{i+r}(w) = \phi^r(u) = ux$ and $\phi^{i+t}(w) = \phi^t(u) = uy$. Then $\phi^{rt}(u) = \phi^r\phi^r...\phi^r(u) = uxx'$ for some $x' \in X^*$. On the other hand, $\phi^{rt}(u) = \phi^t\phi^t...\phi^t(u) = uyy'$ for some $y' \in X^*$. Therefore $ux \leq uy$ or $uy \leq ux$, a contradiction.

Let $L \subseteq X^*$ be the language generated by S. Then we have the following two cases:

(1) L contains $\{u, ux\}, x \neq 1$,

(2) L is a prefix code.

Case (1). Let $u = \phi^i(w)$ and $ux = \phi^{i+p}(w), x \neq 1$. We can assume that i is the smallest possible such integer and that $uX^* \cap \phi^{i+s}(w) = \emptyset$ for any $1 < s < p$. Let $L_0 = \{w, \phi(w), ..., \phi^{i-1}(w)\}$ ($L_0 = \emptyset$ if $w = u$). Obviously L_0 is a prefix code. Let $v = \phi^{i+k}(w)$ where $1 \leq k \leq p$. Then for any $n \geq 1$, we have $\phi^{i+np+k}(w) = vx'$ for some $x' \in X^+$. By the remark given at the beginning, there exist two distinct infinite ascending chains contained in L to which u and v are belonging respectively. Let $L_u(\ni u)$ and $L_v(\ni v)$ be those chains and let $u' \in L_u$ and $v' \in L_v$. If $u' \leq v'$ or $v' \leq u'$, then obviously $u \leq v$ or $v \leq u$. This yields a contradiction. Hence u' and v' are incomparable. Note that the number of distinct infinite ascending chains contained in L is less then p. Therefore L can be splitted into $L_0, L_1, L_2, ..., L_n$, i.e. $L = L_0 \cup L_1 \cup L_2 \cup ... \cup L_n$ where L_0 is a finite prefix code and each L_i, $i = 1, 2, ..., n$ is an infinite ascending chain and u, v are incomparable if $(u, v) \in L_i \times L_j$, $i \neq j$, $i, j = 0, 1, 2, ..., n$.

Case (2). In this case, since S is a SSKDOL system, L becomes an infinite prefix code.

($\Leftarrow$) Let $L \subseteq X^*$ be an infinite prefix code, $L = \{u_1, u_2, u_3, ...\}$. First we define $\phi(u_i) = u_{i+1}$ for $i = 1, 2, ...$. Moreover, $\phi(u) = 1$ for $u < u_i$, $i = 1, 2, ...$. Since L is a prefix code, ϕ can be well defined as a mapping of $pr(L)$ into X^*. Note that $pr(L)$ is a pre-closed set. By Lemma 4.1, ϕ can be extended to a skew-morphism $\bar{\phi}$ of X^* into X^*. Then $(X, \bar{\phi}, u_1)$ is a SSKDOL system and generates L.

Now assume that $L = L_0 \cup L_1 \cup L_2 \cup ... \cup L_n$ where L_0 is a finite prefix code ($L_0 = \emptyset$, possible), each L_i, $i = 1, 2, ..., n$, is an infinite ascending chain and u, v are incomparable for any $(u, v) \in L_i \times L_j$, $i \neq j$, $i, j = 0, 1, 2, ..., n$. Let $L_0 = \{u_1, u_2, ..., u_r\}$ and $L_i = \{u_1^{(i)} < u_2^{(i)} < u_3^{(i)} < ...\}, i = 1, 2, ..., n$. First we define

$$\phi(u_i) = u_{i+1},\ i = 1, 2, ..., r-1,\ \phi(u_r) = u_1^{(1)},$$
$$\phi(u_1^{(1)}) = u_1^{(2)},\ \phi(u_1^{(2)}) = u_1^{(3)}, ...,\ \phi(u_1^{(n-1)}) = u_1^{(n)},$$
$$\phi(u_1^{(n)}) = u_2^{(1)},\ \phi(u_2^{(1)}) = u_2^{(2)}, ...,\ \phi(u_2^{(n-1)}) = u_2^{(n)},$$
$$\phi(u_2^{(n)}) = u_3^{(1)}, ...$$

Moreover, $\phi(u) = 1$ for $u < u_i$, $i = 1, 2, ..., r$, $\phi(u) = 1$ for $u < u_1^{(k)}$, $k = 1, 2, ..., n$, $\phi(u) = u_i^{(k+1)}$ for $u_i^{(k)} < u < u_{i+1}^{(k)}$, $i = 1, 2, ..., k = 1, 2, ..., n-1$ and $\phi(u) = u_{i+1}^{(1)}$ for $u_i^{(n)} < u < u_{i+1}^{(n)}$, $i = 1, 2, ...$.

It can be easily checked that ϕ is well defined as a mapping of $pr(L)$ into X^* from the fact that u, v are incomparable for any $(u, v) \in L_i \times L_j$, $i \neq j$, $i, j = 0, 1, 2, ..., n$. It remains to show that ϕ is a sskew-morphism of $pr(L)$ into X^*. Let $u, uv \in pr(L)$. If $u < u_i$, $i = 1, 2, ..., r$ or $u < u_1^{(k)}$, $k = 1, 2, ..., n$, then $1 = \phi(u) \leq \phi(uv)$. In other cases, $u, uv \in pr(L)$ means one of the following two cases:

(1) $u_i^{(k)} \le u \le uv < u_{i+1}^{(k)}$,

(2) $u_i^{(k)} \le u < u_{i+1}^{(k)} \le u_{i+j}^{(k)} \le uv < u_{i+j+1}^{(k)}$, $j \ge 1$

Case (1). $\phi(u) = u_i^{(k+1)} = \phi(uv)$ if $k = 1, 2, ..., n-1$.

$\phi(u) = u_{i+1}^{(1)} = \phi(uv)$ if $k = n$

Case (2). $\phi(u) = u_i^{(k+1)} < u_{i+j}^{(k+1)} = \phi(uv)$ if $k = 1, 2, ..., n-1$,

$\phi(u) = u_{i+1}^{(1)} < u_{i+j+1}^{(1)} = \phi(uv)$ if $k = n$.

In every case, $\phi(u) \le \phi(uv)$. This means that ϕ is a sskew-morphism of $pr(L)$ into X^*. By Lemma 4.1, there exists a skew-morphism $\bar{\phi}$ of X^* into X^* such that $\bar{\phi}|_{pr(L)} = \phi$. Obviously $(X, \bar{\phi}, u_1)$ ($(X, \bar{\phi}, u_1^{(1)})$ if $L_0 = \emptyset$) becomes a SSKDOL system and generates L. □

Let $S = (X, \phi, w)$ be a SKDOL system. Then S is called a *QSSKDOL system* (quasi-SSKDOL system) if S satisfies the following condition:

$$\phi^i(w) \le \phi^j(w)$$

if $\phi^i(w)$ and $\phi^j(w)$ are comparable with respect to $\le$ and $i < j$.

An example of QSSKDOL system is a system $S = (X, \phi, w)$ such that $|w| \le |\phi(w)| \le |\phi^2(w)| \le ... \le |\phi^i(w)| \le |\phi^{i+1}(w)| \le ...$.

Lemma 4.3. Let $S = (X, \phi, w)$ be a QSSKDOL system and let $\phi^i(w)$ and $\phi^{i+p}(w)$ be comparable, i.e. $\phi^i(w) \le \phi^{i+p}(w)$. Moreover let p be the smallest integer with this property. Then any $\alpha, \beta \in \{\phi^{i+1}(w), \phi^{i+2}(w), ..., \phi^{i+p-1}(w)\}, \alpha \ne \beta$, are incomparable.

Proof. Let $u = \phi^i(w)$ and $ux = \phi^{i+p}(w)$. Suppose there exist

$$\alpha, \beta \in \{\phi^{i+1}(w), ..., \phi^{i+p-1}(w)\}$$

such that $\alpha < \beta$. Let $\alpha = \phi^{i+s}(w), s \ge 1$ and $\beta = \alpha y = \phi^{i+t}(w), s < t < p,\ y \ne 1$. Then $\alpha = \phi^s(u)$ and $\alpha y = \phi^t(u)$. Note that $\phi^{p-t}(\alpha y) = \phi^{i+p}(w) = ux$. On the other hand, $\phi^{p-t}(\alpha y) = \phi^{p-t}(\alpha)y' = \phi^{p-t+i+s}(w)y'$ for some $y' \in X^*$. Consequently, u and $\phi^{p-t}(\alpha)$ are comparable.

Since $i < p-t+s+i < i+p$, $u < \phi^{p-t}(\alpha) < ux$. This contradicts the minimality of p. Hence the lemma holds. □

Proposition 4.4. Let $L \subseteq X^*$. Then L is generated by a QSSKDOL system $S = (X, \phi, w)$, but not by a SSKDOL system, if and only if L satisfies one of the following conditions:

(1) L is a finite prefix code,

(2) $L = L_0 \cup L_1 \cup L_2 \cup ... \cup L_n$ where L_0 is a finite prefix code ($L_0 = \emptyset$, possible), $|L_1| = |L_2| = ... = |L_n| < +\infty$, each L_i, $i = 1, 2, ..., n$ is an ascending chain with respect to $\leq$ and u, v are incomparable for any $(u, v) \in L_i \times L_j$, $i \neq j$, $i, j = 0, 1, 2, ..., n$.

Proof. ($\Rightarrow$) The proof can be carried out almost in the same way as in Proposition 4.2 except that L is finite.

We show first that L is finite. Since $S = (X, \phi, w)$ is a QSSKDOL system and not a SSKDOL system, there exist $i \geq 0$, $p \geq 1$ such that $\phi^i(w) = \phi^{i+p}(w)$. Therefore $L = \{w, \phi(w), \phi^2(w), ..., \phi^{i+p-1}(w)\}$, i.e. L is finite. As before, we have the following two cases:

(1) L contains $\{u, ux\}, x \neq 1$,

(2) L is a prefix code.

Since L is finite, it is sufficient to consider the Case (1). The notations are the same as before. By Lemma 4.3, $|L_u| = |L_v| < +\infty$. And $|L_0| < +\infty$. The remaining part of the proof is the same as before too.

($\Leftarrow$) Let $L \subseteq X^+$ be a finite prefix code and let $L = \{u_1, u_2, ..., u_r\}$. We define $\phi(u_1) = u_2, \phi(u_2) = u_3, \phi(u_3) = u_4, ..., \phi(u_{r-1}) = u_r$ and $\phi(u_r) = u_1$. Moreover, we define $\phi(u) = 1$ for $u < u_i, i = 1, 2, ..., r$. Then ϕ can be well defined as a mapping of $pr(L)$ into X^*. Furthermore, it can easily be seen that ϕ is a sskew-morphism of $pr(L)$ into X^*. Therefore, by Proposition 4.1, ϕ can be extended to a skew-morphism $\bar{\phi}$. Then $(X, \bar{\phi}, u_1)$ is obviously a QSSKDOL system, but not a SSKDOL system, generating L.

Now assume that $L = L_0 \cup L_1 \cup L_2 \cup ... \cup L_n$. Let $L_0 = \{u_1, u_2, ..., u_r\}$ and $L_i = \{u_i^{(i)} < u_2^{(i)} < ... < u_t^{(i)}\}, i = 1, 2, ..., n$.

We define ϕ in the following way:

$$u_1 \xrightarrow{\phi} u_2 \xrightarrow{\phi} ... \xrightarrow{\phi} u_{r-1} \xrightarrow{\phi} u_r \xrightarrow{\phi} u_1^{(1)} \xrightarrow{\phi} u_1^{(2)} \xrightarrow{\phi} ... \xrightarrow{\phi} u_1^{(n-1)} \xrightarrow{\phi} u_1^{(n)}$$
$$\xrightarrow{\phi} u_2^{(1)} \xrightarrow{\phi} u_2^{(2)} \xrightarrow{\phi} ... \xrightarrow{\phi} u_2^{(n-1)} \xrightarrow{\phi} u_2^{(n)} \xrightarrow{\phi} u_3^{(1)} \xrightarrow{\phi} ...$$
$$\xrightarrow{\phi} u_{t-1}^{(n)} \xrightarrow{\phi} u_t^{(1)} \xrightarrow{\phi} u_t^{(2)} \xrightarrow{\phi} ... \xrightarrow{\phi} u_t^{(n-1)} \xrightarrow{\phi} u_t^{(n)} \text{ and } u_t^{(n)} \xrightarrow{\phi} u_t^{(1)}$$

Moreover $\phi(u) = 1$ for $u < u_i$, $i = 1, 2, ..., r$, $\phi(u) = 1$ for $u < u_1^{(k)}$, $k = 1, 2, ..., n$, $\phi(u) = u_i^{(k+1)}$ for $u_i^{(k)} < u < u_{i+1}^{(k)}$, $i = 1, 2, ..., t-1$, $k = 1, 2, ..., n-1$ and $\phi(u) = u_{i+1}^{(1)}$ for $u_i^{(n)} < u < u_{i+1}^{(n)}$, $i = 1, 2, ..., t-1$. It is not difficult to see that ϕ is a sskew-morphism of $pr(L)$ into X^*. Therefore by Proposition 4.1 there exists a skew-morphism $\bar{\phi}$ of X^* into X^* such that $\bar{\phi}|_{pr(L)} = \phi$. It is also easy to see that $(X, \bar{\phi}, u_1)$ becomes a QSSKDOL system, but not a SSKDOL system, and generates L. □

Summing up the results, we have:

Proposition 4.5. Let $L \subseteq X^*$. Then L is generated by a QSSKDOL system if and only if L satisfies one of the following conditions:

(1) L is a prefix code,

(2) $L = L_0 \cup L_1 \cup L_2 \cup ... \cup L_n$ where L_0 is a finite prefix code ($L_0 = \emptyset$ possible), $|L_1| = |L_2| = ... = |L_n|$, each $L_i, i = 1, 2, ..., n$ is an ascending chain and u and v are incomparable for any $(u, v) \in L_i \times L_j$, $i \neq j$, $i, j = 0, 1, 2, ..., n$. □

V. Relations between generalized sequential functions and skew-morphisms

For generalized sequential machines and generalized sequential functions, see for example S. Eilenberg (Vol. A). In what follows, we assume that the set of input symbols and the set of output symbols are the same, i.e. X.

Let ϕ be a generalized sequential function on X^*. Then:

(1) ϕ is a skew-morphism,

(2) $\exists k \geq 1, \forall x \in X^*, |\phi(x)| \leq k|x|$.

However a function on X^* satisfying the conditions (1) and (2) is not necessarily a generalized sequential function.

Example. Let $X = \{a, b\}$ and $\Phi = \phi\psi^2\phi^3\psi^4...\phi^{2i-1}\psi^{2i}...$ where $\phi(a) = \phi(b) = a$ and $\psi(a) = \psi(b) = b$. Obviously Φ satisfies (1) and (2). Now suppose that Φ is a generalized sequential function. Then there exists a generalized sequential machine $\mathcal{M} = (Q, q_0, \lambda)$ (by Eilenberg's notation) such that $q_0 a^i = q_0 a^{i+p}$. Let $\phi(a^i x) = \phi(a^i)z$ and $\phi(a^{i+p}x) = \phi(a^{i+r})z'$ where $x \in X^*$. Since $z = (q_0 a^i, x)\lambda$ and $z' = (q_0 a^{i+p}, x)\lambda$, $z = z'$. On the other hand, by the structure of Φ, $z \neq z'$ when $|x|$ is sufficiently large, a contradiction. Therefore Φ is not a sequential function.

We consider now a skew-morphism of the form $\Phi = \phi_1\phi_2....$ where ϕ_i, $i = 1, 2, ...$ is a mapping of X into X^*.

Proposition 5.1. Φ is a generalized sequential function if and only if $\Phi = \phi_1\phi_2... \phi_k(\phi_{k+1}\phi_{k+2}... \phi_{k+p})^\omega$.

Proof. ($\Rightarrow$) Let $\Phi = \phi_1\phi_2\phi_3...$. Since Φ is a generalized sequential function, there exists a generalized sequential machine $\mathcal{M} = (Q, q_0, \lambda)$. Since $|Q| < +\infty$, there exist $k, p \geq 1$ such that $q_0 a^k = q_0 a^{k+p}$ where $a \in X$. Let $x \in X^*$. Put $\Phi(a^k x) = \Phi(a^k)z$ and $\Phi(a^{k+p}x) = \Phi(a^{k+p})z'$. Then $z = z'$. This means that $\Phi_k(x) = \Phi_{k+p}(x)$ for any $x \in X^*$ where $\Phi_t = \phi_{t+1}\phi_{t+2}...$ for $t \geq 1$. Therefore

$$\Phi = \phi_1\phi_2...\phi_k(\phi_{k+1}\phi_{k+2}...\phi_{k+p})^\omega$$

($\Leftarrow$) Let $\Phi = \phi_1\phi_2...\phi_k(\phi_{k+1}\phi_{k+2}...\phi_{k+p})^\omega$. We define a generalized sequential machine $\mathcal{M} = (Q, q_0, \lambda)$ as follows:

(1) $Q = \{0, 1, 2, ..., k + p\}$, $q_0 = 0$,

(2) $ia = i + 1$ ($i = 0, 1, 2, ..., k + p - 1$, $a \in X$),

$(k+p)a = k \ (a \in X)$,

$(i,a)\lambda = \phi_{i+1}(a) \ (i = 0,1,2,...,k+p-1, \ a \in X)$.

Then it is easy to see that $\Phi(a_1a_2...a_r) = \phi_1(a_1)\phi_2(a_2)...\phi_r(a_r) = (0, a_1a_2...a_r)\lambda$, i.e. Φ is a generalized sequential function. □

VI. Characterization of generalized sequential functions

Not all generalized sequential functions are of the form $\Phi \in \Sigma^\omega$. As a counterexample, consider the following sequential machine $\mathcal{M} = (\{q_0, q_1, q_2\}, q_0, \lambda)$ where:

$$X = \{a,b\}, q_0a = q_1, \ q_0b = q_2, q_1x = q_1, q_2x = q_2, \ x \in X$$
$$(q_0, x)\lambda = x, \ (q_1, x)\lambda = a, (q_2, x)\lambda = b, \ x \in X.$$

Let ϕ be the generalized sequential function induced by $\mathcal{M}$. Then $\phi(a^n) = a^n$ and $\phi(ba^{n-1}) = b^n$. Suppose that $\phi \in \Sigma^\omega$, i.e. $\phi = \phi_1\phi_2...$. Then $\phi_1(a) = \phi(a) = a$ and $\phi_1(a)\phi_2(a) = \phi(a^2) = a^2$. Therefore $\phi_2(a) = a$. On the other hand, since $\phi(ba) = \phi_1(b)\phi_2(a) = b^2, \phi_2(a) \neq a$. This is a contradiction. Thus $\phi \notin \Sigma^\omega$.

Let ϕ be a sskew-morphism. Then by $D(\phi)$ we denote the domain of ϕ. Let $u, v \in X^*$. Then $u^{-1}v = w$ means $v = uw$.

Let I be a nonempty index set and let ϕ_i, $i \in I$, be a sskew-morphism. Then it is easy to see that $\bigcup_{i\in I} D(\phi_i)$ is a pre-closed set. If $\phi_i(u) = \phi_j(u)$ for $u \in D(\phi_i) \cap D(\phi_j)$, $i,j \in I$, then the mapping $\bigoplus_{i\in I} \phi_i$ is defined as follows:

$(\bigoplus_{i\in I} \phi_i)(u) = \phi_i(u)$ if $u \in D(\phi_i)$

Lemma 6.1. $\bigoplus_{i\in I} \phi_i$ is a sskew-morphism of $\bigcup_{i\in I} D(\phi_i)$ into X^*.

Proof. Let $u, uv \in \bigcup_{i\in I} D(\phi_i)$. Then there exists $i \in I$ such that $uv \in D(\phi_i)$. Since $D(\phi_i)$ is a pre-closed set, $u \in D(\phi_i)$. Hence $(\bigoplus_{i\in I} \phi_i)(u) = \phi_i(u) \leq \phi_i(uv) = (\bigoplus_{i\in I} \phi_i)(uv)$. □

Given ϕ_i, $i \in I$, then $\bigoplus_{i\in I} \phi_i$ is called the *amalgamation* of the sskew-morphisms ϕ_i. If I is a finite set, $I = \{i_1, i_2, ..., i_r\}$, then we write $\phi_{i_1} \oplus \phi_{i_2} \oplus ... \oplus \phi_{i_r}$.

Let $\{A_i | i = 1,2,...,r\}$ be a partition of X^*. Then it is called an *admissible partition* if for any $i \geq 1$ and $a \in X$ there exists $j \geq 1$ such that $A_ia \subseteq A_j$.

Now we give a characterization of generalized sequential functions.

Proposition 6.2. Let ϕ be a skew-morphism. Then ϕ is a generalized sequential function if and only if

$$\phi = \phi_0 \oplus \phi_1 \oplus ... \oplus \phi_r$$

where

$D(\phi_i) = pr(A_i)$, $i = 0, 1, ..., r$, $\phi_i(1) = 1$, $i = 0, 1, ..., r$ and $\{A_i | i = 0, 1, ..., r\}$ is an admissible partition of X^* and $\phi_j(u)^{-1}\phi_j(ua) = \phi_j(v)^{-1}\phi_j(va)$ for any $u, v \in A_i$ and $a \in X$ such that $A_i a \subseteq A_j$.

Proof. ($\Rightarrow$) Assume that ϕ is induced by a generalized sequential machine $\mathcal{M} = (Q, q_0, \lambda)$. Hence $\phi(u) = (q_0, u)\lambda$. Let $Q = \{q_0, q_1, q_2, ..., q_r\}$ and let $A_i = \{u \in X^* | q_0 u = q_i\}$, $i = 0, 1, 2, ..., r$. Obviously $\{A_i | i = 0, 1, 2, ..., r\}$ is an admissible partition of X^*. Moreover let $\phi_i = \phi|_{pr(A_i)}$, $i = 1, 2, ..., r$. Since $\phi(1) = 1, \phi_i(1) = 1$ for any $i = 0, 1, 2, ..., r$. Assume that $u, v \in A_i$ and $A_i a \subseteq A_j$ for $a \in X$. Then $\phi_j(ua) = (q_0, ua)\lambda = (q_0, u)\lambda(q_0 u, a)\lambda = \phi_j(u)(q_0 u, a)\lambda$. In the same way, $\phi_j(va) = \phi_j(v)(q_0 v, a)\lambda$. Note that $q_0 u = q_0 v = q_j$. Hence $\phi_j(u)^{-1}\phi_j(ua) = \phi_j(v)^{-1}\phi_j(va) = (q_j, a)\lambda$.

($\Leftarrow$) We can assume that $1 \in A_0$ without loss of generality. We construct a generalized sequential machine $\mathcal{M} = (\{A_0, A_1, ..., A_r\}, A_0, \lambda)$ as follows:

$$A_i a = A_j \text{ and } (A_i, a)\lambda = \phi_j(w)^{-1}\phi_j(wa)$$

if $A_i a \subseteq A_j, a \in X, w \in A_i$.

Now compute $(A_0, u)\lambda$ for $u \in X^*$. Let $u = a_1 a_2 ... a_t$, $a_i \in X$, $i = 1, 2, ..., t$. Then we have:

$$\begin{aligned} A_0 a_1 &= A_{i_1},\ (A_0, a_1)\lambda = \phi_{i_1}(a_1), \\ A_{i_1} a_2 &= A_{i_2},\ (A_i, a_2)\lambda = \phi_{i_2}(a_1)^{-1}\phi_{i_2}(a_1 a_2), \\ A_{i_2} a_3 &= A_{i_3},\ (A_{i_2}, a_3)\lambda = \phi_{i_3}(a_1 a_2)^{-1}\phi_{i_3}(a_1 a_2 a_3), \\ &\ldots \\ A_{i_{t-1}} a_t &= A_{i_t},\ (A_{i_{t-1}}, a_t)\lambda = \phi_{i_t}(a_1 a_2 ... a_{t-1})^{-1}\phi_{i_t}(a_1 a_2 ... a_t). \end{aligned}$$

Therefore

$$\begin{aligned} (A_0, u)\lambda &= \phi_{i_1}(a_1)(\phi_{i_2}(a_1)^{-1}\phi_{i_2}(a_1 a_2))(\phi_{i_3}(a_1 a_2)^{-1}\phi_{i_3}(a_1 a_2 a_3))... \\ &\quad \phi_{i_{t-1}}(a_1 a_2 ... a_{t-1})(\phi_{i_t}(a_1 a_2 ... a_{t-1}^{-1})\phi_{i_t}(a_1 a_2 ... a_t)) \\ &= \phi(a_1)(\phi(a_1)^{-1}\phi(a_1 a_2))(\phi(a_1 a_2)^{-1}\phi(a_1 a_2 a_3))... \\ &\quad \phi(a_1 a_2 ... a_{t-1})(\phi(a_1 a_2 ... a_{t-1})^{-1}\phi(a_1 a_2 ... a_t) \\ &= \phi(a_1 a_2 ... a_t) = \phi(u). \end{aligned}$$

Consequently ϕ is a generalized sequential function. □

Remark. Let ϕ be a generalized sequential function. Then ϕ_i, $i = 1, 2, ..., r$ in Proposition 6.2 becomes a partial generalized sequential function. (For the definition of a partial generalized sequential function, see S. Eilenberg, Vol. A).

VII. Skew-morphisms preserving the regularity of languages

Let $\Phi \in \Sigma^\omega$ and let $\Phi = \phi_1\phi_2...$. Then $\Phi \in \mathcal{C}(\Sigma^\omega)$ means that there exists $n \geq 1$

such that $\{\phi_n(a), \phi_{n+1}(a), ...\}$ is a code for every $a \in X$. Note that $\mathcal{C}(\Sigma^\omega)$ is a huge class. For example, every $\Phi \in \Sigma^\omega$ such that $|\Phi(u)| = |u|$ for any $u \in X^*$ is in $\mathcal{C}(\Sigma^\omega)$.

Proposition 7.1. Let $\Phi \in \mathcal{C}(\Sigma^\omega)$. Then Φ preserves the regularity of languages if and only if Φ is a generalized sequential function.

Proof. ($\Leftarrow$) It is proved in S. Eilenberg (Vol.A) that a generalized sequential function preserves the regularity of languages.

($\Rightarrow$) Let $\Phi = \phi_1\phi_2...$ and let $X = \{a_1, a_2, ..., a_r\}$. For each s, $1 \leq s \leq r$, since $\Phi(a_s{}^*)$ is regular, there exists a deterministic automaton $\mathcal{A}_s = (Q_s, i_s, T_s)$ (following S. Eilenberg's notation) which accepts $\Phi(a_s{}^*)$. By the finiteness of Q_s, $1 \leq s \leq r$, there exist $t, p \geq 1$ such that

$$i_s\phi_1(a_s)\phi_2(a_s)...\phi_t(a_s) = i_s\phi_1(a_s)\phi_2(a_s)...\phi_t(a_s)\phi_{t+1}(a_s)...\phi_{t+p}(a_s)$$

for any s, $1 \leq s \leq r$. Note that we can choose $p \leq |Q_1| \times |Q_2| \times ... \times |Q_r|$ and t such that $\{\phi_{t+1}(a_s), \phi_{t+2}(a_s), ...\}$ is a code. Since $\phi_1(a_s)\phi_2(a_s)...\phi_t(a_s) \in \Phi(a_s{}^*)$ and $i_s\phi_1(a_s)...\phi_t(a_s) = i_s\phi_1(a_s)...\phi_t(a_s)(\phi_{t+1}(a_s)...\phi_{t+p}(a_s))^{n+1}$ for any $n \geq 1$, $\phi_1(a_s)...\phi_t(a_s)(\phi_{t+1}(a_s)...\phi_{t+p}(a_s))^{n+1} \in \Phi(a_s{}^*)$.
Hence there exists $m > p$ such that

$$\phi_1(a_s)...\phi_t(a_s)(\phi_{t+1}(a_s)...\phi_{t+p}(a_s))^{n+1} = \phi_1(a_s)...\phi_t(a_s)\phi_{t+1}(a_s)...\phi_{t+m}(a_s).$$

From the facts that

$$(\phi_{t+1}(a_s)...\phi_{t+p}(a_s))^n = \phi_{t+p+1}(a_s)...\phi_{t+m}(a_s)$$

and $\{\phi_{t+1}(a_s), ..., \phi_{t+p}(a_s)\}$ is a code, we have $m = np$ and

$$\phi_{t+d}(a_s) = \phi_{t+d+(m-1)p}(a_s)$$

for any d, $1 \leq d \leq p$ and $n \geq 1$. This means that

$$\Phi = \phi_1\phi_2...\phi_t(\phi_{t+1}\phi_{t+2}...\phi_{t+p})^\omega.$$

By Proposition 5.1, Φ is a generalized sequential function. □

A generalized sequential function is called a *sequential function* if $|\phi(u)| = |u|$ for any $u \in X^*$.

Corollary 1. Let $\Phi \in \Sigma^\omega$ and let Φ be length-preserving. Then Φ preserves the regularity of languages if and only if Φ is a sequential function. □

Corollary 2. Let $\Phi \in \mathcal{C}(\Sigma^\omega)$ and assume that Φ preserves the regularity of languages. Then the related code $\{\phi_n(a), \phi_{n+1}(a), ...\}$ is finite for any $a \in X$. □

The structure of skew-morphisms which preserve regularity and are not in $\mathcal{C}(\Sigma^\omega)$ seems to be complicated. One might expect a periodicity for a skew-morphism preserving regularity as in the case of skew-morphisms in $\mathcal{C}(\Sigma^\omega)$. However we can show that this is not the case.

Example. Let $p_1, p_2, \ldots$ be an infinite sequence of positive integers such that $p_{i+1} - p_i > i$, $i = 1, 2, 3, \ldots$ and in which there is no periodicity. Let $a \in X$. We define $\Phi = \phi_1\phi_2 \ldots \in \Sigma^\omega$ as follows:

$\phi_k(x) = a$ if $x \in X$ and $k = p_i$ for $i \geq 1$,

$\phi_k(x) = 1$ otherwise.

Now we show that Φ preserves the regularity of languages. Let $L \subseteq X^*$ be a regular language. If L is finite, then $\Phi(L)$ is finite and hence regular. Assume that L is infinite. By the pumping lemma for regular languages, there exist $u, w \in X^*$ and $v \in X^+$ such that $uv^+w \subseteq L$. By the definition of Φ, $a^{|uvw|}a^* \subseteq \Phi(uv^+w) \subseteq \Phi(L)$. Consequently, $\Phi(L) = a^{|uvw|}a^* \cup F$ where $F \subseteq X^*$ is a finite set, and thus $\Phi(L)$ is regular. Therefore, in any case, Φ preserves the regularity of languages. However Φ does not have any periodicity.

A function ϕ on X^* is called *rational* if the relation ϕ^{-1} on X^* preserves regularity. A rational skew-morphism preserves regularity (see S. Eilenberg, Vol. A). Furthermore, it is shown in S. Eilenberg (Vol A., p.317) that a rational skew-morphism is a generalized sequential function if and only if there exists $k \geq 1$ such that $|\phi(u)| \leq k|u|$ for any $u \in X^*$. Therefore, Φ in the preceding example is a skew-morphism preserving regularity without being rational. Hence the set of all rational skew-morphisms is properly contained in the set of all skew-morphisms preserving regularity. Consequently, this example suggests that the structure of the set of skew-morphisms preserving regularity is more complicated than that of the set of all rational skew-morphisms.

We will now show that even a length-preserving skew-morphism preserving regularity is not in general a sequential function.

Let ϕ be a length-preserving skew-morphism of a^* into a^*. Then it is easy to verify that the following conditions are equivalent:

(1) ϕ is a morphism,

(2) ϕ is a sequential function,

(3) ϕ is rational,

(4) ϕ preserves regularity.

Therefore, there is nothing to do for the case $|X| = 1$. However, the situation is completely different for the case $|X| \geq 2$.

Proposition 7.2. Let $|X| \geq 2$. Then there exists a length-preserving skew-morphism preserving the regularity without being a sequential function.

Proof. Let $a, b \in X, a \neq b$. We define ϕ as follows:

$$\phi(u) = a^{|u|} \text{ if } u = a^i u', u' \notin aX^* \text{ and } |u'| \leq i,$$
$$\phi(u) = a^{2i} b^{|u'|-i} \text{ if } u = a^i u', u' \notin aX^* \text{ and } |u'| > i,$$

and

$$\phi(u) = u \text{ if } u \notin aX^*.$$

It is easy to see that ϕ is a length-preserving skew-morphism.
First, we show that ϕ is not a sequential function. Suppose that ϕ is induced by a sequential machine $\mathcal{M} = (Q, q_0, \lambda)$. Let $i \geq |Q|$. Then $q_0 a^i b^j = q_0 a^i b^{j+r}$ where $r \geq 1$ and $1 \leq j < j + r \leq |Q|$. For any $p \geq 1$, $q_0 a^i b^j = q_0 a^i b^{j+pr}$. Let $j + pr \geq i$. Then

$$\phi(a^i b^{j+pr+1}) = (q_0, a^i b^{j+pr+1})\lambda = (q_0, a^i b^{j+pr})\lambda(q_0 a^i b^{j+pr}, b)\lambda =$$

$$(q_0, a^i b^{j+pr})\lambda(q_0 a^i b^j, b)\lambda = \phi(a^i b^{j+pr})\overline{\phi(a^i b^{j+1})}$$

where $\overline{\phi(x)}$ is the last letter of $\phi(x)$ for $x \in X^+$. Since $\phi(a^i b^{j+1}) = a^{i+j+1}$, $\overline{\phi(a^i b^{j+1})} = a$, i.e. $\phi(a^i b^{j+pr+1}) \in X^* a$, a contradiction.

We prove now that ϕ preserves the regularity of languages. Let $K \subseteq X^*$ be a regular language. If K is finite, then $\phi(K)$ is finite and thus regular. Assume that K is infinite. Let $K = (K \cap aX^*) \cup (K \backslash aX^*)$. Since $\phi(K \backslash aX^*) = K \backslash aX^*$ and $K \backslash aX^*$ is regular, to prove that $\phi(K)$ is regular, it is enough to show that $\phi(K \cap aX^*)$ is regular. Therefore, we can assume that $K \subseteq aX^*$ without loss of generality.

Now we consider the mapping ρ of K into $a^+ b^*$ defined as:

$$\rho(a^i x) = a^i b^{|x|} \text{ if } a^i x \in K \text{ and } x \notin aX^*.$$

Let $\mathcal{M}(\{q_0, q_1\}, q_0, \lambda)$ be the following sequential machine:

$$q_0 a = q_0,\ q_0 x = q_1 \text{ if } x \in X, x \neq a, q_1 y = q_1 \text{ if } y \in X,$$

and

$$(q_0, a)\lambda = a, (q_0, x)\lambda = b \text{ if } x \in X, x \neq a, (q_1, y)\lambda = b \text{ if } y \in X.$$

By θ we denote the sequential function induced by $\mathcal{M}$. Then $\rho(K) = \theta(K)$. Hence $\rho(K)$ is regular because a sequential function preserves the regularity. Moreover, $\phi(K) = \phi(\rho(K))$. Therefore, we can assume that $K \subseteq a^+ b^*$ without loss of generality. Let $\mathcal{A}$ be an automaton with the state set Q and the initial state q_0 which accepts K. Let $K = K_1 \cup K_2$ where $K_1 = \{u \in K \mid |u|_a \geq |u|_b\}$ ($|u|_x$ means the number of occurences of x in u) and $K_2 = \{u \in K || u|_a < |u|_b\}$. First, we show that $\phi(K_1)$ is regular. Note that there exist $0 \leq p < |Q|, 1 \leq r < |Q|$ such that $q_0 a^p = q_0 a^{p+r}$. Let $u = a^i b^j \in K_1$. If $i \geq p$, then $a^i (a^r)^* b^j \subseteq K_1$. This means that $a^{i+j}(a^r)^* \subseteq \phi(K_1)$ if $a^{i+j} \in \phi(K_1)$. Let $G = \{a^i \in \phi(K_1) | i \geq 2|Q| \text{ and } (a^j \in \phi(K),\ i \equiv j \pmod r \Rightarrow i \leq j)\}$. Then G is a finite set and $\phi(K_1) = F \cup (\bigcup_{g \in G} g(a^r)^*)$ where F is a finite set. Therefore $\phi(K_1)$ is regular.

We consider now K_2.

Case (1). $u = a^i b^j \in K_2$ and $i > |Q|^2$. Then there exists $0 \leq p < |Q|^2, 1 \leq r < |Q|^2$ such that $q_0 a^p = q_0 a^{p+r}$ and $q_0 a^i b^p = q_0 a^i b^{p+r}$. Therefore

$$a^i b^j \in \bigcup_{n=0}^{\infty} a^p (a^r)^n a^q b^p (b^r)^n b^q (b^r)^* b^s \subseteq K_2$$

where $0 \leq p, q, r < |Q|^2, p + q \equiv i \pmod r$, $s \leq r$ and $j - i \equiv s \pmod r$. Hence $\phi(u) = a^{2i} b^{j-i} \in a^{2p}(a^{2r})^* a^{2q} (b^r)^* b^s \subseteq \phi(K_2)$.

By $G(p, q, r, s)$ we denote the set $a^{2p}(a^{2r})^* a^{2q}(b^r)^* b^s$.

Case (2). $u = a^i b^j \in K_2$ and $i \leq |Q|^2$. As in the Case (1), we can show the following: $a^i b^j \in a^i b^i b^p (b^r)^* b^q \subseteq K_2$ where $0 \leq p, q < |Q|^2, 1 \leq r < |Q|^2$ and $p + q \equiv j - i \pmod r$.

Hence $\phi(u) = a^{2i} b^{j-i} \in a^{2i} b^p (b^r)^* b^q \subseteq \phi(K_2)$.

By $H(i, p, q, r)$ we denote the set $a^{2i} b^p (b^r)^* b^q$.

From Case (1) and Case (2), $\phi(K_2)$ can be represented as follows:

$$\phi(K_2) = F \cup (\bigcup_{\alpha \in A} G(p_\alpha, q_\alpha, r_\alpha, s_\alpha)) \cup (\bigcup_{\beta \in B} H(i_\beta, p_\beta, q_\beta, r_\beta))$$

where F is finite and A, B are finite index sets.

Since F, $G(p_\alpha, q_\alpha, r_\alpha, s_\alpha)$, $\alpha \in A, H(i_\beta, p_\beta, q_\beta, r_\beta)$, $\beta \in B$, are regular, ϕ preserves the regularity of languages. □

The characterization of skew-morphisms preserving regularity is an open problem.

REFERENCES

[1] J. Berstel and D. Perrin, *Theory of Codes*, Academic Press, New York (1985).

[2] S. Eilenberg, *Automata, Languages and Machines.* Vol. A, Academic Press (1974).

[3] F. Gécseg and I. Péak, *Algebraic Theory of Automata*, Academiai Kiado, Budapest (1972).

[4] G.T. Herman and G. Rozenberg, *Developmental Systems and Languages*, North-Holland Publ., Amsterdam (1975).

[5] Y.Q. Guo, G. Thierrin and S.H. Zhang, *Semaphore codes and ideals*, J. Information and Optimization Sci. 9 (1988), 73-83.

[6] P.H. Starke, *Abstract Automata*, Akademie-Verlag (1977).

This research has been supported by Natural Science and Engineering Council of Canada, Grant No. A7877.

An Algebraic Approach to Incomparable Families of Formal Languages*

Peter R.J. Asveld

Department of Computer Science, Twente University of Technology, P.O. Box 217, 7500 AE Enschede, The Netherlands (e-mail: infprja@cs.utwente.nl)

Abstract: We extend some results of Ginsburg and Spanier [21] on incomparable full AFL's (full Abstract Families of Languages) to similar structures like full hyper-AFL's, full hyper(1)-AFL's and prequasoids. A general approach based on universal algebra and lattice theory enables us to establish Ginsburg and Spanier-like theorems for several types of language families. It turns out that a considerable part of the proofs is algebraic or lattice-theoretic in nature. On the other hand formal language theory provides language families that may serve as (counter)examples in algebra.

Keywords: lattice, universal algebra, family of languages, prequasoid, quasoid, full trio, full AFL (full Abstract Family of Languages), full super-AFL, full hyper-AFL.

1. Introduction

The advent of L system theory gave rise to new types of language families such as full hyper-AFL's [35,36], full hyper(1)-AFL's [35,2,13], quasoids [35] and prequasoids [1,2]. An important tool for investigating the structure of, e.g., a full hyper-AFL K is the description of the class of all full hyper-AFL's that are included in K. Or to phrase it algebraically, considering K as an algebra, we are interested in the class of all subalgebras of K.

A classic result in universal algebra says that the class of subalgebras $Sa(A)$ of a given algebra A constitutes a so-called algebraic or compactly generated lattice with set-theoretical inclusion as partial order [10,11,22,23,29]; cf. Theorem 2.8 below. So problems concerning

(i) the existence of a maximal or the largest proper subalgebra of A that includes a given proper subalgebra B, and

(ii) the existence of a maximal or the largest subalgebra of A which is incomparable with a given subalgebra B in $Sa(A)$,

possess a lattice-theoretic character. In this paper we study problems like (i) and (ii) at the level of algebraic lattices and, particularly, for the algebraic lattice $Sa(A)$. Then we apply our results to formal language theory, i.e., we take A equal to the family of all languages. In this

* This research has been supported by the Netherlands Organization for Scientific Research (NWO) under grant No. 62-58.

way it is sufficient to prove certain results only once at the abstract level of lattice theory and of universal algebra, instead of modifying the arguments for each slightly different signature (set of operations) with which we want to provide A. It also becomes clear which parts of our proofs belong in essence to lattice theory, to universal algebra or to formal language theory.

This paper originates from an attempt to generalize results on incomparable full AFL's (full Abstract Family of Languages) due to Ginsburg and Spanier [21] to other different types of related language families. Instead of adapting the proofs from [21] for each new type of language family, we followed the uniform and algebraic approach sketched above. At appropriate places in the sequel we will indicate which of our propositions may be considered as generalizations in the areas of lattice theory or universal algebra, based on theorems obtained by Ginsburg and Spanier [21]. Apart from their own algebraic interest the main results of this paper also show how little of the full AFL-properties one actually needs (cf. Definition 4.2) in order to establish those Ginsburg and Spanier-like theorems.

The present introductory section is followed by five other sections. In Section 2 we survey and slightly extend the "folklore" of the theory, i.e., we deal with basic definitions and results on algebraic lattices, the class of subalgebras $Sa(A)$ of a given algebra A, and their connection. In that section we also establish some straightforward results on lattice-theoretic formulations of the problems (i) and (ii) (Theorems 2.1-2.6) and their interpretation in the algebraic lattice $Sa(A)$; cf. Theorems 2.9 and 2.10.

In Section 3 we define A1-lattices [A2-lattices], i.e., [strongly] algebraic lattices which satisfy two additional axioms (Definition 3.1). These lattices turn out to be uncountable and each of their elements – apart from the zero and the unit – possesses an incomparable element (Theorem 3.3). Our main results on A1- and A2-lattices (Theorems 3.5 and 3.7) characterize those elements that have a largest incomparable element: (1) x has a largest incomparable element if and only if x is strongly compact and unequal to the zero, and (2) x' is the largest element incomparable with x if and only if x is the smallest element incomparable with x'.

In Section 4 we introduce two special kinds of algebras, called A1- and A2-algebras, and we show that for each such algebra A, the lattice of subalgebras $Sa(A)$ is an A1- and an A2-lattice, respectively. Section 5 is devoted to the application of these algebraic results to formal language theory, i.e., to full hyper-AFL's, full super-AFL's [26], prequasoids, etc. Finally, Section 6 consists of some concluding remarks.

In this paper we use the phrases "algebra" and "algebraic structure" synonymously. Other equivalent names are "algebraic system" [30] and "universal algebra" [11,23,29,34]; the latter also refers to the theory of algebraic structures.

2. Algebraic Lattices and Universal Algebra

In this section we recall some basic definitions in order to fix notation and terminology. Then we quote a few standard results concerning the algebraic lattice of subalgebras $Sa(A)$ of a given algebra A. We also establish some elementary facts on incomparable elements in algebraic lattices.

The reader is assumed to be familiar with set theory, e.g., with the contents of [28]. Recall that an element s of a nonempty partially ordered set $(S, \leq)$ is *maximal*, if for each t in S, $s \leq t$ implies $s = t$. An element s of $(S, \leq)$ is called the *largest* element of S if $t \leq s$ for all t

in S. Clearly, a partially ordered set may have several maximal elements but at most one largest element. Sometimes we need the Axiom of Choice in the sequel; from its various equivalent formulations we use

Zorn's Lemma. *If each chain in a nonempty partially ordered set has an upper bound, then this set has a maximal element.* □

With respect to lattice theory and universal algebra this paper is almost self-contained. For more material on lattice theory and universal algebra the interested reader is referred to [9,10,12,15,24] and to [10,11,22,23,29,30,34], respectively.

A binary operation $\cdot$ on a set S is called *associative* [*commutative, idempotent*] if for all $a,b,c \in S$ we have $(a \cdot b) \cdot c = a \cdot (b \cdot c)$ [$a \cdot b = b \cdot a$, $a \cdot a = a$, respectively].

A *lattice* L is a nonempty set provided with two binary operations $\vee$ – called *join* or *lub*, i.e., least upper bound – and $\wedge$ – called *meet* or *glb*, i.e., greatest lower bound – which are both associative, commutative, idempotent and which are coupled by the following two "absorption laws": $a \wedge (a \vee b) = a$ and $a \vee (a \wedge b) = a$ for all $a,b \in L$. A lattice is partially ordered by $\leq$, where for all $a,b \in L$, $a \leq b$ holds whenever $a \wedge b = a$ or, equivalently, whenever $a \vee b = b$. As usual we write $a < b$ if $a \leq b$ and $a \neq b$. If a lattice L has a smallest [largest] element with respect to $\leq$, then this unique element, denoted by 0 [1], is called the *zero* [*unit*] of L. Clearly, we have for each a in L, $a \wedge 0 = 0$, $a \wedge 1 = a$, $a \vee 0 = a$ and $a \vee 1 = 1$.

Let S be an arbitrary subset of a lattice L. An element a of L is called an *upper bound* [a *lower bound*] of S, when $s \leq a$ [$s \geq a$, respectively] for all s in S. As usual, an element a of L is the *join* or *lub* [*meet* or *glb*] of S, denoted by $a = \bigvee S$ [$a = \bigwedge S$], if a is an upper bound [a lower bound] of S and $a \leq b$ [$a \geq b$] for each upper bound [lower bound] b of S. Clearly, $\bigvee\{a,b\} = a \vee b$ and $\bigwedge\{a,b\} = a \wedge b$ for all $a,b \in L$. A lattice L is *complete* when for each subset S of L, $\bigvee S$ and $\bigwedge S$ exist. Each complete lattice L possesses a zero and a unit defined by $0 = \bigwedge L$ and $1 = \bigvee L$, respectively.

An element a of a lattice L is called [*strongly*] *compact* if, whenever $a \leq \bigvee S$ for some nonempty subset S of L, there exists a finite subset F of S [an element b of S] with $a \leq \bigvee F$ [$a \leq b$]. An element a of L is called *coprime* if, whenever $a \leq \bigvee F$ for some nonempty finite subset F of L, there exists an element b in F with $a \leq b$. Obviously, a in L is strongly compact if and only if a is both compact and coprime. Note that the zero 0 in a lattice is strongly compact. A lattice L is said to be [*strongly*] *algebraic* or [*strongly*] *compactly generated* if L is complete and each element in L is the join of some set of [strongly] compact elements of L.

Two elements a and b of a partially ordered set are called [*in*]*comparable*, which we will denote by $a : b$ [$a \not: b$], if either $a \leq b$ or $b \leq a$ [neither $a \leq b$ nor $b \leq a$, respectively]. Note that $a : b$ if and only if $b : a$, and that $a \not: b$ if and only if $b \not: a$.

The existence of maximal [largest] incomparable elements is rather simple to establish for those [strongly] compact elements, which already possess an incomparable element, since we have (cf. [21] Theorem 3.1):

Theorem 2.1. *Let c and d be incomparable elements in a complete lattice L.*

(1) *If c is compact, then there exists a maximal c' in L, incomparable with c and $c' \geq d$.*

(2) *If c is strongly compact, then there exists in L a largest element incomparable with c.*

Proof: (1) We apply Zorn's Lemma to the nonempty partially ordered set $P = \{a \mid a \in L,\ a \not: c,\ d \leq a\}$. Let A be a chain in P. We show that $\bigvee A$ belongs to P.

Clearly, $d \le \bigvee A$. Suppose $c : \bigvee A$. If $\bigvee A \le c$, then $d \le c$ contradicting $c \not\lessgtr d$. Therefore we have $c \le \bigvee A$. Since c is compact, there exists a finite subchain F of A with $c \le \bigvee F$. Let f be the largest element of F. Then $c \le f$; but this is impossible as f is in A. Consequently $c \not\lessgtr \bigvee A$, and hence $\bigvee A$ is in P. By Zorn's Lemma there exists a maximal element c' incomparable with c and $c' \ge d$.

(2) Let $B = \{b \mid b \in L,\ b \not\lessgtr c\}$. Obviously, B is partially ordered and nonempty because $d \in B$. Suppose $c : \bigvee B$. If $\bigvee B \le c$, then $d \le c$, contradicting $c \not\lessgtr d$. Therefore we have $c \le \bigvee B$. Since c is strongly compact there exists an element b_0 in B with $c \le b_0$. But this contradicts $b_0 \not\lessgtr c$. Hence $c \not\lessgtr \bigvee B$ and, consequently, $\bigvee B \in B$. □

An element e in a lattice L *splits* [27] if there exist $a, b \in L$ such that $a \not\lessgtr b$ and $e = a \vee b$. Clearly, a coprime element does not split. However, in general the converse does not hold; cf. [21] p. 97 and Proposition 2.5 below.

We now turn to maximal or largest elements properly smaller than a given element. These elements also play an important part in the study of largest incomparable elements; cf. Lemma 3.4 and Theorem 3.5 below.

Theorem 2.2. *Let b and c be elements in a complete lattice L such that* $b < c$.

(1) *If c is compact, then there exists a maximal element* c_b *in L such that* $b \le c_b < c$.

(2) *If c is compact and does not split, then there exists a largest* c_l *in L such that* $c_l < c$.

Proof: (1) We apply Zorn's Lemma to the nonempty and partially ordered set $P = \{x \mid x \in L,\ b \le x < c\}$. Let A be a chain in P. Certainly we have $b \le \bigvee A \le c$. If $c = \bigvee A$, then the compactness of c implies that there is a finite subchain F of A such that $c \le \bigvee F$. Let f be the largest element of F, then $c \le f$, contrary to the fact that $f \in P$. Thus $\bigvee A < c$ and $\bigvee A \in P$. From Zorn's Lemma we conclude that P contains a maximal element c_b with $b \le c_b < c$.

(2) Let $B = \{x \mid x \in L,\ x < c\}$. Then B is a nonempty partially ordered set and $\bigvee B \le c$. If $c = \bigvee B$, then the compactness of c implies the existence of a finite subset F of B with $c \le \bigvee F \le \bigvee B = c$, i.e., $c = \bigvee F$. Suppose that F has a largest element f. Then $c = f$ and thus $c \in B$, contradicting the fact that $c \notin B$. Therefore F has a finite number of maximal elements $f_1, \dots, f_n$ $(n \ge 2)$ such that $c = f_1 \vee \dots \vee f_n$. But this implies that c splits. Hence $\bigvee B < c$. From the definition of B it follows that $\bigvee B = c_l$. □

Corollary 2.3. *For each strongly compact element c in a complete lattice L with* $c \ne 0$, *there exists a largest element* c_l *in L such that* $c_l < c$. □

For algebraic lattices we have the following converse of Theorem 2.2(2).

Theorem 2.4. *Let a be an element of an algebraic lattice L. If there exists a largest element* a_l *in L such that* $a_l < a$, *then a is compact and does not split.*

Proof: Consider $Q = \{x \in L \mid x < a\}$ and $C = \{c \in L \mid c \le a,\ c \text{ is compact}\}$. Clearly, $a_l = \bigvee Q$, and since L is algebraic we have $a = \bigvee C$ and $a_l = \bigvee (C \cap Q)$. As $a_l < a$, we have $C \cap Q \subset C$. Therefore there exists an element in $C - Q$ which has to be a. Hence a is compact.

Suppose a splits, i.e., $a = d \vee e$ and $d \not\lessgtr e$. Then $d < a$, $e < a$ and $d \vee e \le a_l < a = d \vee e$. Hence a does not split. □

For a special class of algebraic lattices a converse of Theorem 2.1(2) will be established in Section 3; cf. Theorem 3.5.

A lattice L is called *distributive* if $(a \wedge b) \vee (a \wedge c) = a \wedge (b \vee c)$ or, equivalently, $a \vee (b \wedge c) = (a \vee b) \wedge (a \vee c)$ for all $a, b, c \in L$. A complete lattice L is called *completely distributive* if, for each doubly indexed family a_{ij} ($i \in I$, $j \in J$) in L we have

$$\bigwedge\{\bigvee\{a_{ij} \mid j \in J\} \mid i \in I\} = \bigvee\{\bigwedge\{a_{i\phi(i)} \mid i \in I\} \mid \phi \in J^I\}$$

where J^I denotes the set of all functions from I to J.

In distributive lattices the notions of coprime and non-splitting element coincide. The proof is straightforward ([10] pp. 11-12, [12] pp. 19-21) and it is therefore omitted.

Proposition 2.5. *The following statements are equivalent for each lattice L.*

(i) *L is distributive.*

(ii) *An element in L is coprime if (and only if) it does not split.* □

For algebraic lattices we have the following result which characterizes complete distributivity; cf. [8,15].

Theorem 2.6. *An algebraic lattice is completely distributive if and only if it is strongly algebraic.*

Proof: The "if" part has been established by Raney [31] for arbitrary complete lattices; cf. [7] and [9] p. 245. In order to prove the converse implication, it suffices to show that each compact element c in a completely distributive algebraic lattice L is the join of strongly compact elements.

If c equals 0, then c is strongly compact. So we assume $c > 0$. Let C be the set of all maximal elements m in L such that $0 \le m < c$. By Theorem 2.2(1) C is nonempty. If C is a singleton, then c possesses a largest element c_l such that $c_l < c$. From Theorem 2.4 and Proposition 2.5 it follows that c is strongly compact.

Assume now that C is not a singleton and write $C = \{m_i \mid i \in I\}$. For each i in I define $M_i = \{x \in L \mid x < c,\ x \not\le m_i\}$. Since C contains at least two elements, each M_i is nonempty.

Suppose $\bigwedge M_i \le m_i$. Then $m_i = m_i \vee \bigwedge M_i = \bigwedge\{m_i \vee x \mid x < c,\ x \not\le m_i\}$. From the maximality of m_i, $m_i \le m_i \vee x \le c$ and $x \not\le m_i$ ($x \in M_i$), we obtain $m_i \vee x = c$, and hence $m_i = \bigwedge c = c$. However, this contradicts $m_i < c$. Therefore $\bigwedge M_i \not\le m_i$, and hence $m_i \wedge \bigwedge M_i < \bigwedge M_i$.

Let y be an element with $y < \bigwedge M_i$. Then $y \notin M_i$ and so $y \le m_i$. Hence $y \le m_i \wedge \bigwedge M_i$. But this means that $m_i \wedge \bigwedge M_i$ is the largest element of $\{y \in L \mid y < \bigwedge M_i\}$. Theorem 2.4 and Proposition 2.5 imply that $\bigwedge M_i$ is strongly compact.

Since $\bigwedge M_i \not\le m_i$ for each i in I, we have $\bigvee\{\bigwedge M_i \mid i \in I\} \not\le m_j$ for each j in I. Hence $c = \bigvee\{\bigwedge M_i \mid i \in I\}$, i.e., c is the join of strongly compact elements. □

Let L be a lattice. A function $\Gamma : L \to L$ is a *closure operator* on L if, for all a and b in L,

- $a \le \Gamma(a)$ (Γ is *extensive*),
- $a \le b$ implies $\Gamma(a) \le \Gamma(b)$ (Γ is *monotone*),
- $\Gamma(\Gamma(a)) = \Gamma(a)$ (Γ is *idempotent*).

A [*completely*] *additive closure operator* Γ on L is a closure operator which satisfies $\Gamma(\bigvee S) = \bigvee\{\Gamma(s) \mid s \in S\}$ for each finite subset [arbitrary subset] S of L.

If L is a [strongly] algebraic lattice, then a closure operator Γ on L is called [*strongly*] *algebraic* if for each [strongly] compact element c in L with $c \le \Gamma(b)$ for some b in L, there exists a [strongly] compact element b' with $b' \le b$ and $c \le \Gamma(b')$.

Theorem 2.7. (1) *If Γ is a closure operator on a complete lattice L, then $\Gamma(L) = \{\Gamma(a) \mid a \in L\}$ is a complete lattice in which the meet coincides with the meet in L, whereas the join*

ϑ *in* $\Gamma(L)$ *satisfies* $\vartheta S = \Gamma(\bigvee S)$ *for each subset* S *of* $\Gamma(L)$, *where* $\bigvee$ *is the join in* L.

(2) *If* Γ *is a [completely] additive closure operator on a [completely] distributive complete lattice* L, *then* $\Gamma(L)$ *is [completely] distributive.*

(3) *If* Γ *is a [strongly] algebraic closure operator on a [strongly] algebraic lattice* L, *then* $\Gamma(L)$ *is a [strongly] algebraic lattice.*

Proof: (1) is well known; cf. [10] p. 112, or [9] Theorem II.4.12.

(2) Let a_{ij} ($i \in I$, $j \in J$) be in L, where I and J are finite [arbitrary] index sets. By (1), the [complete] distributivity, and the fact that Γ is [completely] additive, we have

$$\bigwedge\{\vartheta\{a_{ij} \mid j \in J\} \mid i \in I\} = \bigwedge\{\bigvee\Gamma(\{a_{ij} \mid j \in J\}) \mid i \in I\} = \bigwedge\{\bigvee\{\Gamma(a_{ij}) \mid j \in J\} \mid i \in I\} =$$
$$\bigvee\{\bigwedge\{\Gamma(a_{i\phi(i)}) \mid i \in I\} \mid \phi \in J^I\} = \bigvee\{\Gamma(\bigwedge\{a_{i\phi(i)} \mid i \in I\}) \mid \phi \in J^I\} =$$
$$\Gamma(\bigvee\{\bigwedge\{a_{i\phi(i)} \mid i \in I\} \mid \phi \in J^I\}) = \vartheta\{\bigwedge\{a_{i\phi(i)} \mid i \in I\} \mid \phi \in J^I\},$$

i.e., $\Gamma(L)$ is [completely] distributive.

(3) The algebraic case is standard – cf. [10] VII, [29], [34] §4 Lemma 2 – whereas the proof of the strongly algebraic variant can be obtained as a simple modification. □

We now show how these lattice-theoretic concepts and results can be used in universal algebra. An *algebra* (A, Ω) is a pair consisting of a set A – called the *carrier set* – and a *set of operations* $\Omega = \{f_\xi \mid 0 \leq \xi < \alpha\}$ (α is an arbitrary ordinal number), such that A is closed under each f_ξ from Ω. Let $\sigma(f_\xi)$ denote the arity or rank of the operation f_ξ. We call an algebra (A, Ω) [*pseudo*]*unary* if $\sigma(f_\xi) = 1$ [$\sigma(f_\xi) \leq 1$] for each f_ξ in Ω, i.e., (A, Ω) only possesses unary operations [and distinguished elements]. In contravention of standard texts – cf. [9,10,11,23,29,30,34] – we allow the carrier set A to be empty provided that $\sigma(f_\xi) \neq 0$ for each f_ξ in Ω; cf. [22].

An algebra (B, Ω') is a *subalgebra* of the algebra (A, Ω) if $B \subseteq A$, Ω' consists of the restrictions g_ξ of f_ξ from Ω with respect to B – i.e., $g_\xi = f_\xi \mid B$ for each ξ ($0 \leq \xi < \alpha$) – and B is closed under each g_ξ from Ω'.

Whenever Ω – and consequently each Ω' – is understood, we will identify an algebra with its carrier set. Thus we use the symbol A to denote both the algebra (A, Ω) and its carrier set A; from the context it should always be clear which one is meant.

Let A be an algebra. We denote the class of all subalgebras of A by $Sa(A)$. For B_1 and B_2 in $Sa(A)$ we write $B_1 \leq B_2$ whenever B_1 is a subalgebra of B_2 and so $Sa(A)$ is obviously provided with a partial order.

With an algebra A we associate a closure operator Γ which maps each subset X of the carrier set A into the carrier set of the smallest subalgebra B of A such that $X \subseteq \Gamma(X) = B$. We also say that B is the subalgebra generated by X. The algebra B is said to be *finitely generated* [*cyclic*] if there exists a finite set [singleton] X such that $\Gamma(X) = B$.

Let A be an arbitrary but fixed algebra and consider the power set of the carrier set of A. It is well known that this power set (just as each other power set) is a completely distributive complete lattice with set-theoretic inclusion as partial order, $0 = \varnothing$, and $1 = A$. It is easy to verify that the [strongly] compact elements of this lattice are exactly the finite sets [singletons together with $\varnothing$] and that this lattice is strongly algebraic.

In view of Theorem 2.7 the major part of the following basic result is not surprising.

Theorem 2.8. (1) *For each algebra A, the set* $Sa(A)$ *of all subalgebras of A is an algebraic lattice with for each S with* $S \subseteq Sa(A)$, $\bigwedge S = \bigcap S$, $\bigvee S = \Gamma(\bigcup S)$, $1 = A$, *and* $0 = \Gamma(\varnothing)$ *where* Γ *is the closure operator associated with A. Moreover*

(i) Γ *is an algebraic closure operator satisfying* $\Gamma(X) = \bigcup\{\Gamma(X_f) | X_f \subseteq X, X_f \text{ is finite}\}$,

(ii) *an element B in* $Sa(A)$ *is compact if and only if B is finitely generated.*

(2) *If A is a pseudo-unary algebra, then* $\bigvee S = \bigcup S$ *for each S with* $S \subseteq Sa(A)$, *and*

(iii) Γ *is a completely additive strongly algebraic closure operator,*

(iv) $Sa(A)$ *is a strongly algebraic lattice.*

Proof: The main part is (up to a reformulation) equal to standard results proved in [10] VIII.5-4, [11] II.5, [23] 6 & 9, and [29] 3.6 & 3.8. The only new proposition is (iv) which directly follows from the complete distributivity of $Sa(A)$ and Theorem 2.6. □

A statement similar to Theorem 2.8(1)(ii) for strongly compact elements and cyclic subalgebras respectively, does not hold in general; cf. Theorems 5.4.2 and 5.5.4 below.

By Theorem 2.8 terminology and results on (algebraic) lattices become applicable to $Sa(A)$. So phrases like incomparable (sub)algebras, non-splitting algebra, coprime algebra, and so on, are meaningful. From Theorems 2.1 and 2.8 we obtain

Theorem 2.9. *Let C and D be incomparable subalgebras of an algebra A.*

(1) *If C is finitely generated, then there exists a maximal subalgebra of A which includes D and which is incomparable with C.*

(2) *If C is finitely generated and coprime, then there exists a largest subalgebra of A which is incomparable with C.* □

An element a of the carrier set A of an algebra is said to be a *non-generator* of the algebra, if for each subset X of A, the condition that $X \cup \{a\}$ generates A, implies that X generates A. Let $N(A)$ denote the set of all non-generators of the algebra A.

Theorem 2.10. *Let C be an algebra.*

(1) *If C is finitely generated, then each proper subalgebra of C is included in a maximal proper subalgebra of C.*

(2) $N(C)$ *is the intersection of all maximal proper subalgebras of C. Consequently, if C has a proper subalgebra, then* $N(C)$ *is a proper subalgebra of C.*

(3) *Let C have a proper subalgebra. Then the following propositions are equivalent.*

(i) *C is finitely generated and does not split.*

(ii) $N(C)$ *is the largest proper subalgebra of C.*

Proof: (1) directly follows from Theorems 2.2(1) and 2.8 (cf. [29] pp. 80-83), whereas (2) has been established in [29] pp. 80-83. Finally, (3) is obtained from (2), Theorems 2.2(2), 2.4 and 2.8. □

3. A1- and A2-lattices

In order to characterize largest incomparable elements in algebraic lattices we introduce in this section two particular types of algebraic lattices (Definition 3.1; cf. [21] Lemma 1.1) for which we establish our main results (Theorems 3.5 and 3.7). But first we consider the following obvious generalization of the notion of compactness; cf. [22].

An element a in a lattice L is called *weakly compact* if, whenever $a \leq \bigvee S$ for some nonempty subset S of L, there exists a countable subset S_0 of S with $a \leq \bigvee S_0$. Clearly, each compact element is weakly compact.

Definition 3.1. An *A1-lattice* is an algebraic lattice L satisfying the following properties.

(1) For each weakly compact element a in L, $x \leq a$ implies that x is weakly compact.

(2) For each pair of weakly compact elements a and b in L, there exists a weakly compact element c in L such that $a < c$ and $c \wedge b = a \wedge b$.

A strongly algebraic A1-lattice is called an *A2-lattice*. □

Property 3.1(2) says that for each pair of weakly compact elements a and b, a can be properly enlarged to c without changing its meet with b. Thus in an A1-lattice the unit is not weakly compact. In Section 4 we show that in the algebraic lattice $Sa(A)$ of subalgebras of (A, Ω) property 3.1(2) is a consequence of the fact that Ω consists of a countable number of operations.

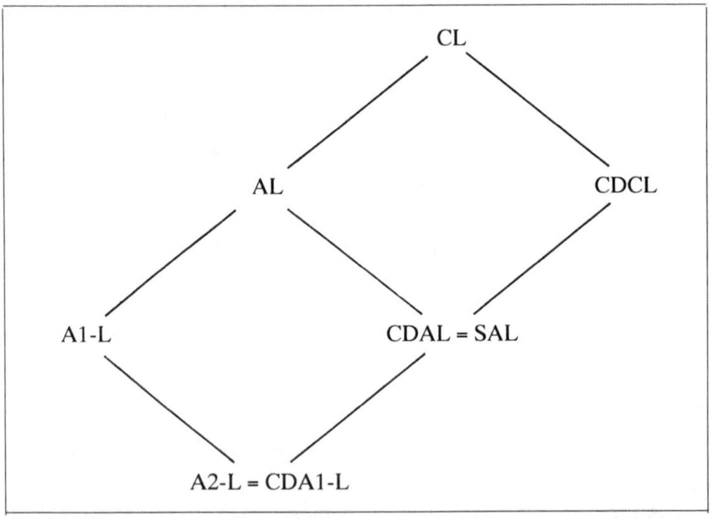

CL = complete lattices; AL = algebraic lattices; CDCL = completely distributive complete lattices; A1-L = A1-lattices; CDAL = completely distributive algebraic lattices; SAL = strongly algebraic lattices; A2-L = A2-lattices; CDA1-L = completely distributive A1-lattices.

Figure 1.

The main types of lattices discussed in Sections 2 and 3 are surveyed in Figure 1. A line in this figure means inclusion; in Section 6 we show that these inclusions are proper.

Lemma 3.2. (cf. [21] Lemma 1.2). *Let x be an element of an A1-lattice L with $x \neq 1$. If b is weakly compact in L with $x \not\leq b$, then there exists a weakly compact element y in L such that $b < y$ and $x \not\leq y$.*

Proof: Since 1 is not weakly compact we have $b \neq 1$. Distinguish the following two cases.

Case 1: x is not weakly compact. Since L is algebraic we have for each e in L, $e = \bigvee\{c \mid c$ is compact; $c \leq e\}$. Hence, $b \neq 1$ and $x \neq 1$ imply the existence of compact elements d and z, such that $d \not\leq b$ and $z \not\leq x$. Define y by $y = b \vee d \vee z$. Being the join of weakly compact elements, y is also weakly compact. Furthermore, we have $b < y$; otherwise $d \leq b$ holds which contradicts $d \not\leq b$. Since $z \leq y$ and $z \not\leq x$, we have $y \not\leq x$. On the other hand 3.1(1) yields $x \not\leq y$. Hence $x \parallel y$.

Case 2: x is weakly compact. By 3.1(2) there exists a weakly compact element y with $b < y$ and $y \wedge x = b \wedge x$. Suppose $y \leq x$. Then $y = y \wedge x = b \wedge x$. Since $b \leq y \leq x$ we have $b \wedge x = b$ and, consequently, $y = b$, which contradicts $b < y$. Suppose $x \leq y$. Then $x = y \wedge x = b \wedge x$ or, equivalently, $x \leq b$, which contradicts $x \not\leq b$. Hence $x \parallel y$. □

Theorem 3.3. (cf. [21] Theorem 1.1).

(1) *Each A1-lattice is uncountable.*

(2) *Each element x in an A1-lattice which satisfies $0 < x < 1$ possesses a weakly compact element incomparable with x.*

Proof: (1) Let L be an A1-lattice with unit 1. Since 1 is not weakly compact, and $1 = \bigvee L$, L is uncountable. (2) follows from Lemma 3.2 with $b = 0$. □

We are now almost able to establish the first characterization result (Theorem 3.5) for which we need the following lemma and the notion of relative pseudo-complement.

Lemma 3.4. (cf. [21] Proposition 2.1 and Corollary). *Let x be an element of an A1-lattice L.*

(1) *If x has a largest incomparable element x', then $x \wedge x'$ is the largest element in L strictly smaller than x.*

(2) *If x has a largest incomparable element, then x is compact and does not split.*

Proof: (1) From $x \parallel x'$ it follows that $x \wedge x' \neq x$ and hence $x \wedge x' < x$. Let z be in L such that $z < x$. In order to prove that $z \leq x \wedge x'$ it suffices to show that $c \leq x \wedge x'$ for each compact element c with $c \leq z$, because L is algebraic, i.e., each element in L is the join of compact elements. Obviously, c is weakly compact, $x \neq 1$ and $x \not\leq c$. Thus by Lemma 3.2 there exists an element y in L with $c < y$ such that $x \parallel y$. This implies $c < y \leq x'$, since x' is the largest element incomparable with x. Together with $c \leq z \leq x$, this yields $c \leq x \wedge x'$.

(2) directly follows from (1) and Theorem 2.4. □

Let a and b be elements in a lattice L. If there exists a largest element x in L such that $a \wedge x \leq b$, then this element is denoted by $a \blacklozenge b$ and it is called the *relative pseudo-complement* of a with respect to b.

Theorem 3.5. (cf. [21] Theorem 2.1). *Let x be an element of an A1-lattice L. Then the following statements are equivalent.*

(1) *x is strongly compact and $x > 0$.*

(2) *There exists a largest element x' in L such that $x \parallel x'$.*

(3) *There exists a largest element x_l in L such that $x_l < x$, whereas $x \blacklozenge x_l$ exists and $x \blacklozenge x_l$ is (the largest element) incomparable with x.*

Proof: (1) $\Rightarrow$ (2). By Theorems 2.1(2), 3.3(2) and the fact that 1 is not strongly compact.

(2) $\Rightarrow$ (3). According to Lemma 3.4(1) we have for the largest element x_l satisfying $x_l < x$, $x \wedge x' = x_l$. Hence $x \blacklozenge x_l$ exists and $x' \leq x \blacklozenge x_l$.

Let z satisfy $x \wedge z \leq x_l$. Suppose $x \leq z$. Then $x \wedge z = x < x_l$, which is impossible. Therefore, either $z \leq x_l < x$ or $z \not\leq x$, from which in both cases we obtain $z \leq x'$. Hence we have $x \blacklozenge x_l = \bigvee\{z \mid x \wedge z \leq x_l\} \leq x'$.

(3) $\Rightarrow$ (1). Since there is an element incomparable with x, we have $0 < x < 1$. Now by Theorem 2.4, x is compact. So it remains bo prove that x is coprime.

Suppose x is not coprime, i.e., there exists a finite set F such that $x \leq \bigvee F$ and $x \not\leq f$ for each f in F. Then $x \wedge f \leq x_l$ and so $f \leq x \blacklozenge x_l$ for each f in F. Consequently, $x \leq \bigvee F \leq x \blacklozenge x_l$, which contradicts $(x \blacklozenge x_l) \not\leq x$. Hence x is coprime. □

Besides this result we give another characterization of largest incomparable elements in an A1-lattice (Theorem 3.7) for which we need the following lemma.

Lemma 3.6. (cf. [21] Lemma 2.2). *If an element x' of an A1-lattice has a smallest incomparable element x, then*

(1) *x is compact, and*

(2) *x' is not weakly compact.*

Proof: (1) Since L is algebraic, there exists a compact element c such that $c \leq x$ and $c \not\leq x'$. But then $c \not\leq x'$; otherwise we have $x' \leq c \leq x$ which contradicts $x \not\leq x'$. Hence $x \leq c$, because x is the smallest element incomparable with x'. But this means $x = c$, i.e., x is compact.

(2) Suppose x' is weakly compact. Then $x \vee x'$ and – by 3.1(1) – $x \wedge x'$ are also weakly compact, and $x \vee x' < 1$. Applying 3.1(2) to $x \wedge x'$ and $x \vee x'$ yields the existence of a weakly compact element w, such that $x \wedge x' < w$ and $w \wedge (x \vee x') = (x \wedge x') \wedge (x \vee x') = x \wedge x'$.

Consider w and x'. First, $w \leq x'$ is impossible, since this implies $w \leq x \vee x'$ and hence $w = w \wedge (x \vee x') = x \wedge x'$, which contradicts the fact that $x \wedge x' < w$. On the other hand the supposition $x' \leq w$ implies $x' \leq w \wedge (x \vee x') = x \wedge x'$ or, equivalently, $x' \leq x$, which contradicts $x \not\leq x'$. Hence $w \not\leq x'$.

Since x is the smallest element incomparable with x', we have $x \leq w$. But this implies $x \leq w \wedge (x \vee x') = x \wedge x'$, or $x \leq x'$, which contradicts $x \not\leq x'$.

Hence x' is not weakly compact. □

Theorem 3.7. (cf. [21] Theorem 2.2). *Let x and x' be elements in an A1-lattice L. Then the following propositions are equivalent.*

(1) *x' is the largest element in L incomparable with x.*

(2) *x is the smallest element in L in comparable with x'.*

Proof: (1) $\Rightarrow$ (2). Let y be incomparable with x'. We show that $x \leq y$.

Suppose $x \not\leq y$. Then $y \leq x'$ because x' is the largest element incomparable with x. This contradicts $y \not\leq x'$. Suppose $y < x$. From Lemma 3.4(1) we obtain $y \leq x_l = x \wedge x' \leq x'$, which contradicts $y \not\leq x'$. Hence we have $x \leq y$.

(2) $\Rightarrow$ (1). Define $E(x) = \{c \mid c \text{ is compact; } x \not\leq c\}$. Since L is algebraic we have for each element y incomparable with x that $y \leq \bigvee E(x)$. In particular, $x' \leq \bigvee E(x)$.

We show that $\bigvee E(x) \leq x'$. Let c be an element from $E(x)$. Then $x \not\leq c$. Since x is the smallest element incomparable with x', we have $c : x'$. By Lemma 3.6, x' is not weakly compact and according to 3.1(1) this implies $x' \not\leq c$. Hence $c \leq x'$, and consequently $\bigvee E(x) \leq x'$. By $x \not\leq x'$ and the definition of $E(x)$, it follows that x' is the largest element incomparable with x. □

4. A1- and A2-algebras

In this section we introduce two particular types of algebras called A1- and A2-algebras; cf. Definition 4.2. The main result of this section (Theorem 4.3) says that for each A1-algebra [A2 algebra] A the lattice of subalgebras $Sa(A)$ is an A1-lattice [A2-lattice, respectively]. However, first we establish a preliminary result and we recall some auxiliary concepts.

Proposition 4.1. *Let A be an algebra with a countable set of operations Ω.*

(1) *A subalgebra B of A is weakly compact in $Sa(A)$ if and only if B has a countable carrier set.*

(2) *$Sa(A)$ satisfies property* 3.1(1).

Proof: (1) Since $Sa(A)$ is algebraic we have $B = \bigvee\{C_\lambda | \lambda \in \Lambda\}$, where each C_λ is the carrier set of a finitely generated algebra; cf. Theorem 2.8. If B is weakly compact, then there exists a countable subset Λ_ω of Λ such that $B = \bigvee\{C_\lambda | \lambda \in \Lambda_\omega\}$. By Theorem 2.8(1) we have $B = \Gamma(\bigcup\{C_\lambda | \lambda \in \Lambda_\omega\})$ where Γ is the closure operator associated with A. As Ω is countable, each C_λ and, consequently, $\bigvee\{C_\lambda | \lambda \in \Lambda_\omega\}$ is countable. According to [23] p. 47 or [29] p. 78 the algebra B has a countable carrier set.

Conversely, suppose B has a countable carrier set. Let $B \subseteq \bigvee\{D_\lambda | \lambda \in \Lambda\}$ where each D_λ is a subalgebra of A. Since Ω is countable, for each element b in B there exists a countable subset Λ_b of Λ such that $b \in \Gamma(\bigcup\{D_\lambda | \lambda \in \Lambda_b\})$. Define $\Lambda_\omega = \bigcup\{\Lambda_b | b \in B\}$ which is countable. Moreover, $B \subseteq \Gamma(\bigcup\{D_\lambda | \lambda \in \Lambda_\omega\}) = \bigvee\{D_\lambda | \lambda \in \Lambda_\omega\}$, i.e., B is weakly compact.

(2) follows from (1) and the fact that each countable algebra can only possess countable subalgebras. □

Let f be an n-ary operation on a collection of sets. Then f is called *compact* if

$$f(S_1, \ldots, S_n) = \bigcup\{f(F_1, \ldots, F_n) | F_i \subseteq S_i;\ \text{each } F_i \text{ is finite } (1 \le i \le n)\},$$

and *monotone* if $S_i \subseteq T_i$ for each i $(1 \le i \le n)$ implies $f(S_1, \ldots, S_n) \subseteq f(T_1, \ldots, T_n)$.

Definition 4.2. Let O be a collection of sets – called the *objects* of O – that contains a distinguished countably infinite set S_0, called the *central object* of O, with the property that

(0) each (countably) infinite subset of the central object S_0 also belongs to O.

An *A1-algebra over O* is an algebra (A, Ω) satisfying the following conditions.

(1) $A \subseteq O$, i.e., the carrier set A is a subcollection of O.

(2) Ω is a countable set of operations which are all monotone as well as compact.

(3) Ω contains a 0-ary operation f_0 such that f_0 equals the central object of O, i.e., $f_0 = S_0$.

An *A2-algebra over O* is a pseudo-unary A1-algebra over O. □

Remark. In the sequel we always assume that the collection O is given. Therefore, we usually omit all references to O and, in particular, we use the phrase "A1-algebra" instead of "A1-algebra over O". □

Theorem 4.3. *If (A, Ω) is an A1-algebra [A2-algebra] over O of which the carrier set A satisfies $A = O$, then $Sa(A)$ is an A1-lattice [A2-lattice].*

Proof: According to Theorem 2.8, Proposition 4.1, Definitions 3.1 and 4.2 it only remains to show that for each A1-algebra A, the lattice $Sa(A)$ satisfies 3.1(2). The argument is a straightforward generalization of a proof technique due to Ginsburg and Spanier ([21] Lemma 1.1; cf. [17] pp. 208-211).

By Theorem 2.8 and Proposition 4.1 it suffices to prove that for each pair of countable proper subalgebras B_1 and B_2 of A, there exists an object S not in the carrier set of B_2 such that $\Gamma(B_2 \cup \{S\}) \cap B_1 = B_2 \cap B_1$, where Γ is the closure operator associated with A.

Let Ψ be the set of all operations from Ω augmented with the identity operation and all unary constant operations $c_Q(X) = Q$ for each object Q in B_2. Since both B_2 and Ω are countable, so is Ψ.

Define Φ_0 to be the set of all operations ϕ defined by $\phi(X) = \psi(X,...,X)$, where ψ is an m-ary operation from Ψ. For each n ($n \geq 0$), define Φ_{n+1} to be the set of all operations $\phi(X) = \psi(\phi_1(X),...,\phi_m(X))$, where ψ is an m-ary operation from Ψ and $\phi_1,...,\phi_m \in \Phi_n$. Finally, let $\Phi = \bigcup\{\Phi_n | n \geq 0\}$. As the composition of monotone [compact] operations is monotone [compact], each operation in Φ is both monotone and compact. Moreover, we have that $\Gamma(B \cup \{a\}) = \{\phi(a) | \phi \in \Phi\}$ for each a in the carrier set A.

Let $\{S_i\}_{i \geq 1}$, $\{\phi_i\}_{i \geq 1}$ and $\{Q_i\}_{i \geq 1}$ be enumerations of the objects in $B_1 - B_2$, the elements of Φ, and the objects in B_2, respectively. We define an object S that satisfies

(i) S is a countably infinite subset of the central object S_0,

(ii) $S \neq Q_k$ for each k ($k \geq 1$), and

(iii) for each pair (i,j) from $\mathbb{N} \times \mathbb{N}$, $\phi_i(S) \neq S_j$, where $\mathbb{N}$ is the set of nonnegative integers.

From (ii) it follows that $S \notin B_2$, and by (iii) we obtain

$$\Gamma(B_2 \cup \{S\}) \cap B_1 = \{\phi(S) | \phi \in \Phi\} \cap B_1 = B_2 \cap B_1.$$

Consider a bijection $\pi : \mathbb{N} \times \mathbb{N} \rightarrow \mathbb{N}$. Let $U_{-1} = V_{-1} = \emptyset$. Suppose that for each t ($t < k$, $k \geq 0$), finite subsets U_t and V_t of S_0 have already been defined inductively, such that $U_i \subseteq U_{i+1}$ and $V_i \subseteq V_{i+1}$ for all i ($0 \leq i < k-1$). The finiteness of both U_{k-1} and V_{k-1} implies the existence of an element u_k in $S_0 - (U_{k-1} \cup V_{k-1})$. Let i and j be the unique nonnegative integers satisfying $\pi(i,j) = k$. Let $U_k = U_{k-1} \cup \{u_k\}$ whenever $\phi_i(S_0 - V_{k-1}) \neq \emptyset$ and $\phi_i(S_0 - V_{k-1}) \subseteq S_j$. Suppose $\phi_i(S_0 - V_{k-1})$ is empty or not a subset of S_j. Since ϕ_i is compact, there exists a finite subset W_k of $S_0 - V_{k-1}$, such that $\phi_i(W_k)$ is either empty or not a subset of S_j. Define $U_k = U_{k-1} \cup \{u_k\} \cup W_k$. Since Q_k is infinite, in either case there exists an element v_k in $Q_k - U_k$. Finally, let $V_k = V_{k-1} \cup \{v_k\}$, which extends the induction. Define $S = \bigcup\{U_n | n \geq 0\}$. Then S is a countably infinite subset of S_0, and by Definition 4.2(0), S is in A, i.e., $\Gamma(B_2 \cup \{S\})$ is a subalgebra of A that properly includes B_2.

The verification of (ii) and (iii) – for which we need the compactness and the monotonicity of each ϕ in Φ – is similar as in [21] Lemma 1.1, (cf. [17] pp. 208-211. The objects are languages in [17,21] and $\{a\}^*$ is the central object; see also §5.4 below.) and it is therefore not repeated here. □

Following Theorem 4.3 results on incomparable elements in A1-lattices (Section 3) are now applicable to $Sa(A)$ for each A1-algebra (A, Ω) satisfying $A = O$.

If X is a subalgebra of A, then the *exterior* $E(X)$ of X in $Sa(A)$ [21] is defined by $E(X) = \bigcup\{C | C \in Sa(A); C \text{ is finitely generated}; X \not\subseteq C\}$. It is straightforward to show that for each strongly compact subalgebra X of an A1-algebra A, we have $E(X) = X \blacklozenge X_l$ provided $X \neq 0$; cf. Theorem 3.5 and the proof of Theorem 3.7. Note that in the algebraic lattice $Sa(A)$, we also have for the set of non-generators $N(X)$ of a subalgebra X that the equality $N(X) = E(X) \cap X$ holds.

We leave the formulation of Theorems 3.3, 3.5 and 3.7 in terms of A1-algebras and their exteriors to the reader.

5. Applications to Formal Language Theory

In this section the reader is assumed to be familiar with basic concepts from formal language theory (e.g. [33] Chapters I-III) and with the rudiments of L system theory [32] and of full Abstract Families of Languages ([16] or [33] Chapter IV). However, we will recall the main notions in order to explain our notational conventions; cf. §5.1-5.3. Then in §5.4-5.5 we will apply results from previous sections to language-theoretic algebraic structures.

5.1. Preliminaries – Languages and Families of Languages

Let Σ_ω be a countably infinite set of symbols; each finite subset Σ of Σ_ω is called an *alphabet*. A *word* over an alphabet Σ is a finite sequence of symbols: $a_1 a_2 \dots a_n$ $(n \geq 0)$ with $a_i \in \Sigma$ for each i $(1 \leq i \leq n)$; the word $a_1 a_2 \dots a_n$ has *length* n. The empty word will be denoted by λ.

A *language* L (over Σ) is a subset of Σ^* for some alphabet Σ. So each language is a countable set. A *family* (of languages over Σ_ω) is a collection of languages of which each language L satisfies $L \subseteq \Sigma_L^*$ for some finite subset (or alphabet) Σ_L from Σ_ω. Apart from well-known language families like FIN (the family of finite languages), REG (the family of regular languages), CF (the family of context-free languages), and ETOL (the family of extended tabled context-independent Lindenmayer languages) we will mainly deal with the family ALL, i.e., the family of all languages over finite subsets of Σ_ω. A family is called *nontrivial* if it contains a nontrivial language L, i.e., $L \neq \varnothing$ and $L \neq \{\lambda\}$. Obviously, FIN, REG, CF, ETOL and ALL are nontrivial families.

In view of Definition 4.2 and the condition in Theorem 4.3 we take O equal to ALL in the remaining part of this paper, i.e., henceforth our objects are languages with $\{a\}^*$ as central object (a is an arbitrary but fixed symbol from Σ_ω). Note that each (countably infinite) subset of $\{a\}^*$ is a language and hence it belongs to ALL; cf. Definition 4.2(0).

In the sequel the family ALL will serve as the carrier set of several (types of) algebras. In other words we will provide ALL with different sets of operations Ω and apply our results from previous sections to Sa(ALL). But first we discuss some operations on languages from which we constitute different sets Ω.

5.2. Operations on Languages

First, we consider the regular operations, viz. union $L_1 \cup L_2$, concatenation $L_1 \cdot L_2$, and Kleene $*$ or iteration L_1^*. So these operations have arity 2, 2 and 1, respectively.

Let K be a family. A *K-substitution* τ is a function on an alphabet such that for each a in Σ, $\tau(a)$ is a language in K. The function τ is extended to words over Σ by $\tau(\lambda) = \{\lambda\}$, $\tau(a_1 a_2 \dots a_n) = \tau(a_1) \cdot \tau(a_2) \cdot \dots \cdot \tau(a_n)$ for each $a_1 a_2 \dots a_n$ in Σ^* $(n \geq 1;\ a_i \in \Sigma$ for $1 \leq i \leq n)$, and to languages by $\tau(L) = \bigcup\{\tau(w) \mid w \in L\}$ for each L with $L \subseteq \Sigma^*$. Note that $\tau(\varnothing) = \varnothing$. A *$K$-substitution over* Σ is a K-substitution τ with $\tau(a) \subseteq \Sigma^*$ for each a in Σ. If K equals FIN or REG, then a K-substitution is usually called *finite* or *regular substitution*, respectively. If Σ consists of n symbols, then each K-substitution $\tau : \Sigma \to K$ may be considered as an $(n+1)$-ary operation since τ maps the languages $L, \tau(a_1), \dots, \tau(a_n)$ onto the language $\tau(L)$.

For each m and n $(m,n \geq 1)$, we introduce in a similar way a few $(mn+1)$-ary operations. Let $\Sigma = \{a_1, \dots, a_n\}$, L be a language over Σ, and $\tau_1, \dots, \tau_m$ be K-substitutions over Σ. A (multiple) *iterated K-substitution* μ is an operation transforming the $mn+1$ languages L,

$\tau_1(a_1), \tau_1(a_2), \ldots, \tau_m(a_n)$ onto the language

$$\bigcup\{f_1 \ldots f_k(L) \mid k \geq 0;\ f_p \in \{\tau_1, \ldots, \tau_m\} \text{ for } 1 \leq p \leq k\}$$

over Σ. If $m = 1$, then μ is called a *single iterated K-substitution.* A *nested* [single] iterated K-substitution is a [single] iterated K-substitution satisfying $a \in \tau_i(a)$ for each a in Σ and each i $(1 \leq i \leq m)$. In the names of these substitution operations we will omit the prefix "K-" whenever the family K is clear from the context.

Finally, we need two 0-ary or constant operations ; viz. $c_1 = \{a\}$ and $c_0 = \{a\}^*$ for some arbitrary but fixed symbol a from Σ_ω.

5.3. Algebraic Structures in Formal Language Theory

Clearly, the family ALL is closed under all the operations mentioned in §5.2. Now we form several sets of operations Ω with which we will provide the family ALL. This results in different types of well-known language-theoretic algebraic structures.

• Let $\Omega_I = \varnothing$. Then Sa (ALL) – or better Sa ((ALL,$\varnothing$)) which we will denote by **ALL** – is equal to the power set of ALL and to the class of all language families over Σ_ω including the trivial families $\varnothing$, $\{\varnothing\}$, $\{\{\lambda\}\}$ and $\{\varnothing,\{\lambda\}\}$. The existence of largest incomparable language families is now trivial, as **ALL** is a Boolean algebra with $0 = \varnothing$ and $1 =$ ALL.

• Define Ω_Π by

$$\Omega_\Pi = \{c_1\} \cup \{\delta_R \mid \delta_R(L) = L \cap R;\ R \in \text{REG}\} \cup \{\tau \mid \tau \text{ is a finite substitution}\}.$$

The pseudo-unary type of algebra corresponding to this set of operations is called *prequasoid* [1,2]; it is a slightly more primitive structure than the *quasoid* [35] which is defined by means of $\Omega_{\Pi_0} = \Omega_\Pi \cup \{c_0\}$. So each quasoid is a prequasoid, but the converse implication does not hold [1,2]. FIN [REG, respectively] is the smallest prequasoid [quasoid] and FIN is the only prequasoid that is not a quasoid. Note that c_1 ensures that each prequasoid K is a nontrivial family, and that $K \supseteq$ FIN. Similarly, c_0 causes each quasoid K to contain an infinite language and that $K \supseteq$ REG; cf. [1,2]. We will also denote Sa ((ALL,Ω_Π)) and Sa ((ALL,Ω_{Π_0})) by **PQ** and **Q**, respectively.

• Next we turn to a few algebraic structures related to full Abstract Families of Languages or full AFL's [16,33]. As starting point we consider the set of operations $\Omega_{\hat{\mathcal{M}}}$, defined by

$$\Omega_{\hat{\mathcal{M}}} = \Omega_{\Pi_0} \cup \{\tau \mid \tau \text{ is a regular substitution}\},$$

which determines the notion of *full trio* [16] or *full M-AFL* [2], i.e., a family (of languages) is a full trio if it is closed under operations from $\Omega_{\hat{\mathcal{M}}}$. Similarly, we obtain the concepts of *full semi-AFL* [25,16] or *full S-AFL* [2], of *full pseudo-AFL* or *full P-AFL* [2], of *full Kleene-AFL* or *full C-AFL* [2], and of *full AFL* [19,16] or *full F-AFL* [2] by augmenting $\Omega_{\hat{\mathcal{M}}}$ with some regular operations, viz.

$$\Omega_{\hat{\mathcal{S}}} = \Omega_{\hat{\mathcal{M}}} \cup \{\cup\},$$

$$\Omega_{\hat{\mathcal{P}}} = \Omega_{\hat{\mathcal{M}}} \cup \{\cup, \cdot\},$$

$$\Omega_{\hat{\mathcal{C}}} = \Omega_{\hat{\mathcal{M}}} \cup \{^*\},$$

$$\Omega_{\hat{\mathcal{F}}} = \Omega_{\hat{\mathcal{M}}} \cup \{\cup, \cdot, ^*\},$$

respectively. In [19,16] it has been shown that closure under the operations from $\Omega_{\hat{\mathcal{M}}} \cup \{\cdot\}$ implies closure under $\cup$, and that a family closed under the operations from $\Omega_{\hat{C}}$ augmented with either $\cup$ or $\cdot$ is a full AFL. So we may restrict ourselves to the above-mentioned combinations of regular operations.

- Finally, we extend $\Omega_{\hat{\mathcal{F}}}$ with substitutions and their generalizations.

$$\Omega_{\hat{\mathcal{R}}} = \Omega_{\hat{\mathcal{F}}} \cup \{\tau | \tau \text{ is a substitution}\},$$

$$\Omega_{\hat{\mathcal{A}}} = \Omega_{\hat{\mathcal{F}}} \cup \{\mu | \mu \text{ is a nested iterated substitution}\},$$

$$\Omega_{\hat{\mathcal{H}}_1} = \Omega_{\hat{\mathcal{F}}} \cup \{\mu | \mu \text{ is a single iterated substitution}\},$$

$$\Omega_{\hat{\mathcal{H}}} = \Omega_{\hat{\mathcal{F}}} \cup \{\mu | \mu \text{ is a multiple iterated substitution}\}.$$

The corresponding type of algebras are called *full substitution-closed AFL* [20,16] or *full R-AFL* [2,5], *full super-AFL* [26] or *full A-AFL* [2,5], *full hyper(1)-AFL* [35,2,13] or *full H_1-AFL* [2,13], and *full hyper-AFL* [35,36,1,2] or *full H-AFL* [2,5], respectively.

Under minor conditions on a family K, each nested multiple iterated K-substitution can be replaced by an equivalent nested single iterated K-substitution; cf. [2,4,5] for details. And since each substitution-closed full M-AFL is a full (substitution-closed) AFL by Proposition 3.3.1 from [16], we restrict ourselves to the above-mentioned sets of operations and the corresponding algebraic structures.

For each set of operations Ω_Z we denote the algebraic closure operator associated with Ω_Z by Z. So Z may be equal to I, Π, Π_0, $\hat{\mathcal{M}}$, $\hat{\mathcal{S}}$, etc. Note that I is the identity mapping on **ALL**. For **X** equal to **M, S, P, C, F, R, A, H$_1$**, or **H**, let **X-AFL** denote the lattice of all full sub-X-AFL's of the full X-AFL ALL. So **X-AFL** $= Sa((\text{ALL}, \Omega_{\hat{X}}))$ and it equals the collection of all full X-AFL's over Σ_ω.

5.4. The Lattice of Subalgebras

First we determine the structure of the lattice of subalgebras $Sa((\text{ALL}, \Omega))$ for each set of operations Ω discussed in §5.3.

Theorem 5.4.1.

(1) **PQ** *is a strongly algebraic lattice but neither an A1-lattice nor an A2-lattice. The zero element in this lattice equals* FIN.

(2) **Q**, **M-AFL** *and* **C-AFL** *are A2-lattices. In these lattices we have* $\bigwedge\{K_i | i \in I\} = \bigcap\{K_i | i \in I\}$, $\bigvee\{K_i | i \in I\} = \bigcup\{K_i | i \in I\}$, *and* $0 = \text{REG}$.

(3) *The closure operators* Π, Π_0, $\hat{\mathcal{M}}$ *and* $\hat{C}$ *are completely additive.*

(4) *For* **X** *equal to* **S, P, F, R, A, H$_1$**, *or* **H**, **X-AFL** *is an A1-lattice. The zero in these lattices is equal to* $\hat{X}$ (FIN). *In particular,* $\hat{X}$ (FIN) *equals*

- REG *if* **X** *equals* **S, P, F** *or* **R**,
- CF *if* **X** *equals* **A**,
- ETOL *if* **X** *equals* **H**.

(5) *In all these lattices the unit element equals* ALL.

Proof: First, note that each $\Omega_{\hat{X}}$ involved is countable and only contains operations that are both compact and monotone. Apart from the case $\hat{X} = \Pi$, in all cases mentioned in 5.4.1, $(\text{ALL}, \Omega_{\hat{X}})$ is an A1-algebra with languages as objects and $c_0 = \{a\}^*$ as central object.

Following Theorem 4.3, the corresponding lattice of subalgebras is an A1-lattice. Since Ω_{Π}, Ω_{Π_0}, $\Omega_{\hat{\mathcal{M}}}$ and $\Omega_{\hat{\mathcal{C}}}$ only contain unary and 0-ary operations **PQ**, **Q**, **M-AFL** and **C-AFL** are strongly algebraic lattices and the corresponding closure operators are completely additive; cf. Theorem 2.8. Consequently, the latter three lattices are even A2-lattices.

PQ possesses an element, viz. REG, that has no incomparable element and that lies properly between the zero FIN and the unit ALL [1,2]. By Theorem 3.3(2), **PQ** is neither an A1- nor an A2-lattice.

The zero elements in these lattices have been determined for **PQ** in [1], for **Q** in [35], for **X-AFL** with **X** equal to **M**, **C**, **P**, **S**, **F** and **R** in [19,20,2], for **A-AFL** in [26], for **$\mathbf{H_1}$-AFL** in [13], and for **H-AFL** in [2,36].

The remaining statements directly follow from Theorem 4.3. □

A characterization of $\hat{\mathcal{H}}_1$(FIN) in terms of a well-known language family is still an open problem; cf. [13].

We show that an improvement of Theorem 5.4.1(4) to A2-lattices in the cases X equals S, P, or F is impossible.

Theorem 5.4.2. *Let* **X** *be equal to* **S**, **P** *or* **F**.

(1) **X-AFL** *is a non-distributive A1-lattice.*

(2) **X-AFL** *is not an A2-lattice.*

(3) *The corresponding closure operator* $\hat{X}$ *is not (completely) additive.*

Proof: (1) Ginsburg and Spanier [21] established the existence of a full F-AFL which is not coprime and does not split. Their argument also applies in the cases X equals S or P. According to Proposition 2.5, **X-AFL** is not distributive. Now (2) follows from (1), Theorem 2.8(2) and Definition 4.2, whereas (3) is a consequence of (1) and Theorem 2.7(2). □

The problem whether the lattice **X-AFL** is (completely) distributive is still open in the cases **X** equals **R**, **A**, $\mathbf{H_1}$, **H**.

5.5. Incomparable Language-Theoretic Algebras

By Theorem 5.4.1 results on incomparable elements in A1- and A2-lattices become applicable to quasoids and to full X-AFL's. The following results are direct consequences from Theorems 3.3, 3.5, 3.7, 4.3 and 5.4.1; cf. [21] Theorems 1.1, 2.1 and 2.2, respectively.

Theorem 5.5.1.

(1) *Let X be equal to M, C, S, P, F, R, A, H_1 or H. Then each full X-AFL K which is unequal to* $\hat{X}$ (FIN) *and to* ALL, *possesses a countable full X-AFL incomparable with K.*

(2) *Each quasoid K unequal to* REG *and to* ALL *possesses a countable quasoid incomparable with K.* □

Theorem 5.5.2. *Let X be equal to M, C, S, P, F, R, A, H_1 or H, and let K be a full X-AFL. Then the following propositions are equivalent.*

(1) *K is finitely generated and coprime.*

(2) *There exists a largest full X-AFL K' in* **X-AFL** *such that K and K' are incomparable.*

(3) *There exists a largest full X-AFL K_l in* **X-AFL** *such that $K_l \subset K$, whereas $K \blacklozenge K_l$ exists and $K \blacklozenge K_l$ is (the largest element) incomparable with K.* □

Theorem 5.5.3. *Let X be equal to M, C, S, P, F, R, A, H_1 or H, and let K and K' be full X-AFL's. Then the following statements are equivalent.*

(1) *K' is the largest full X-AFL incomparable with K.*

(2) *K is the smallest full X-AFL incomparable with K'.* □

Statements similar to Theorems 5.5.2 and 5.5.3 also hold for **Q**; the precise formulations are left to the reader.

According to Theorems 3.5, 4.3 and 5.4.1 strongly compact full X-AFL's play an essential rôle in establishing the existence of largest incomparable full X-AFL's. So we now focus our attention to finitely generated and coprime full X-AFL's.

From Proposition 4.1 it follows that the weakly compact elements in **X-AFL** [**PQ**,**Q**] are precisely the countable full X-AFL's [quasoids, prequasoids, respectively]. For the compact elements, i.e., the finitely generated algebras we have a result similar to the cases X equal to either S or F established in [18,16]. Note that a cyclic full F-AFL is usually called full principal.

Theorem 5.5.4. [18,16]. *Let X be equal to S, P, F, R, A, H_1 or H. Then a full X-AFL is finitely generated if and only if it is cyclic.* □

Remark that Theorem 5.5.4 does not hold for full C-AFL's, full M-AFL's and (pre)quasoids, i.e., there exist finitely generated full C-AFL's, full M-AFL's and (pre)quasoids that are not cyclic.

In the remaining part of this section we compare the power of different sets of operations. So in order to avoid confusion we use prefixed phrases like X-cyclic, X-coprime, etc. Note that F-coprime is called fully prime in [21]. We first consider the question whether an X-cyclic full X-AFL is also X'-cyclic. The proof of the following result is straightforward.

Proposition 5.5.5. *Let* $\Omega_{\hat{X}} \subseteq \Omega_{\hat{X}'}$.

(1) *If K is an X-cyclic full X-AFL, then* $\hat{X}'(K)$ *is X'-cyclic.*

(2) *The following statements are equivalent.*

  (2.1) *Each X'-cyclic full X'-AFL is X-cyclic.*

  (2.2) *If K is an X-cyclic full X-AFL, then* $\hat{X}'(K)$ *is X-cyclic.* □

Note that we defined the sets of operations in a way such that $\Omega_{\hat{X}} \subseteq \Omega_{\hat{X}'}$ holds if and only if **X-AFL** $\supseteq$ **X'-AFL**. Thus pairs $(\hat{X}, \hat{X}')$ which satisfy Proposition 5.5.5(1) can be determined from the inclusion diagram in Figure 2(a). All inclusions in Figure 2(a) are proper; cf. [2].

We write $\hat{X} \rightarrow \hat{X}'$ if the pair $(\hat{X}, \hat{X}')$ satisfies Proposition 5.5.5(2.1) or 5.5.5(2.2), and $\hat{X} \not\rightarrow \hat{X}'$ for the negation of $\hat{X} \rightarrow \hat{X}'$. Remark that $\rightarrow$ is a reflexive and transitive, but not a symmetric relation which satisfies: if $\hat{X} \rightarrow \hat{X}'$ and $\hat{X}' \not\rightarrow \hat{X}''$, then $\hat{X} \not\rightarrow \hat{X}''$.

Theorem 5.5.6. (1) $\hat{\mathcal{M}} \rightarrow \hat{\mathcal{S}}$, (2) $\hat{\mathcal{S}} \rightarrow \hat{\mathcal{F}}$, (3) $\hat{\mathcal{M}} \rightarrow \hat{\mathcal{C}}$, (4) $\hat{\mathcal{C}} \rightarrow \hat{\mathcal{F}}$, (5) $\hat{\mathcal{F}} \rightarrow \hat{\mathcal{A}}$, (6) $\hat{\mathcal{A}} \rightarrow \hat{\mathcal{H}}$, (7) $\hat{\mathcal{F}} \not\rightarrow \hat{\mathcal{R}}$, (8) $\hat{\mathcal{R}} \rightarrow \hat{\mathcal{A}}$, (9) $\hat{\mathcal{A}} \not\rightarrow \hat{\mathcal{H}}_1$, (10) $\hat{\mathcal{H}}_1 \rightarrow \hat{\mathcal{H}}$, (11) $\hat{\mathcal{P}} \rightarrow \hat{\mathcal{F}}$.

Proof: The statements follow from Proposition 5.5.5 and, respectively, (1) [16] Proposition 5.1.1; (2-4) [18] Theorem 3.1 or [16] Theorem 5.4.1, and [2] Chapter 5 Theorem 4.4; (5) [26] Theorem 4.7; (6) [14]; (7) [25] Theorem 3.2, [16] p. 239, [27]; (8) from (5); (9) [13]; (10) from (6); and (11) from (2). □

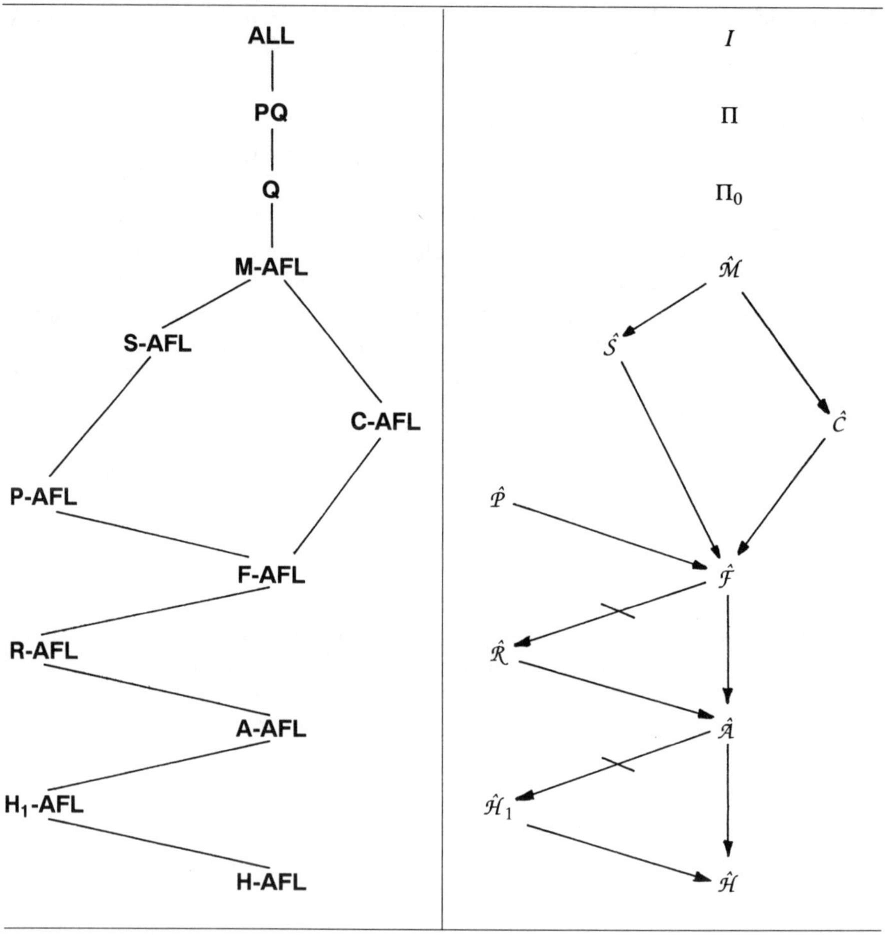

(a) **(b)**

Figure 2.

These results are summarized in Figure 2(b). The question whether $\hat{\mathcal{S}} \rightarrow \hat{\mathcal{P}}$ holds, is still open.

Finally, we turn to coprime language-theoretic algebraic structures. Using one of the main results from [21] we are able to establish the following statement.

Theorem 5.5.7. (1) *Each full F-AFL closed under* $\hat{\mathcal{R}}$ *is F-coprime.*

(2) *Let K be an F-cyclic full F-AFL closed under* $\hat{\mathcal{R}}$ *and unequal to* REG.

(i) *There exists a largest full F-AFL K' that is incomparable with K. Moreover, K' is closed under* $\hat{\mathcal{R}}$.

(ii) *There exists a largest proper full sub-F-AFL* K_l *of K. Furthermore,* K_l *is closed under* $\hat{\mathcal{R}}$ *and* K_l *equals the set of non-F-generators of K.*

Proof: As already mentioned, the statements follow from a result of Ginsburg and Spanier; cf. Theorem 2.3 in [21]. However, before we can apply this result an observation is due. Viz. the definition of non-generator used in this part of formal language theory differs from the one usually encounters in algebra. The latter one has been recalled in §2; the former reads: a language L is a *non-X-generator* of a family K, if $\hat{X}(\{L\}) \neq K$. Clearly, this is a natural definition in view of Theorem 5.5.4. For a comparison of these two definitions we refer to [6]. Fortunately, both definitions are equivalent for the algebraic structures in which we are interested here. □

Remark that Theorem 5.5.7(2-ii) is a special case of a general result on so-called "syntactic operators" due to Greibach; cf. [27] Theorem 3.3. It is very likely that Theorem 5.5.7(2-i) also allows a similar extension in terms of these syntactic operators.

By Theorem 5.5.7 each F-cyclic full H_1-AFL [A-AFL, H-AFL] K unequal to $\hat{\mathcal{H}}_1$(FIN) [unequal to CF, unequal to ETOL, respectively] possesses a largest incomparable full F-AFL K' which is closed under $\hat{\mathcal{R}}$. The question remains, however, whether K' is also closed under $\hat{\mathcal{H}}_1$ [$\hat{\mathcal{A}}$, $\hat{\mathcal{H}}$, respectively].

6. Concluding Remarks

In the previous sections we generalized theorems on the characterization of incomparable full F-AFL's due to Ginsburg and Spanier [21] to lattice theory and universal algebra. Whereas some straightforward results can already be established for complete lattices (Theorems 2.1, 2.2 and Corollary 2.3) or for algebraic lattices (Theorem 2.4) and universal algebras (Theorem 2.9 and 2.10), the main results of this paper have only been proved for A1-lattices (Section 3) and A1-algebras (Section 4). Then we applied our results from Sections 3 and 4 to other algebraic structures in formal language theory, viz. to full X-AFL's with $X \neq F$ and to related structures like quasoids and prequasoids. So it suffices to prove certain propositions only once at the level of lattice theory or of universal algebra, rather than to modify the original arguments from [21] for each type of full X-AFL. Apart from the operations considered in this paper, there are more operations on languages studied in formal language theory – cf. e.g. [16,26,3,4,5] – each of which leads to a corresponding type of full X-AFL. Whether results similar to those in §5.4 and §5.5 also holds for these other types of full X-AFL remains to be investigated.

The most striking point of this extension from a part of formal language theory to algebra is – in our opinion – that it becomes clear that one actually needs only a few AFL-properties (abstracted in the notion of A1-algebra) to establish Ginsburg and Spanier-like results. It also shows for instance that in Theorem 2.2(2) – cf. Theorem 2.10 – the "non-splitting" condition (introduced in [27]) is rather obvious. According to Theorem 2.2(1) there exists a maximal element c_b with $b \leq c_b < c$ and the additional requirement that c does not split only forbids c_b to possess an incomparable element in the set $\{x \mid 0 \leq x \leq c\}$.

This observation leads to another interesting problem. Apart from the obvious question whether the definitions of A1-lattice and of A1-algebra could be weaken while Theorems 3.3, 3.5, 3.7 and 4.3 still remain valid, we consider a further restriction. Let a *relative A1-lattice L* be an algebraic lattice which satisfies 3.1(1) and

(2′) For each element t in L that is not weakly compact, and for each pair of weakly compact elements a and b with $a < t$ and $b < t$, there exists a weakly compact element c in L such that $a < c < t$ and $c \wedge b = a \wedge b$.

It follows almost directly that for each element t of a relative A1-lattice, the fact that t is not weakly compact implies that the set $\{x | 0 \leq x \leq t\}$ is also a relative A1-lattice. The question arises, of course, which additional conditions one should require on an A1-algebra A in order that $Sa(A)$ becomes a relative A1-lattice. Results in this direction would certainly have applications in AFL-theory; cf. [21] Section 4.

On the other hand particular types of language-theoretic algebraic structures can serve as (counter)example in lattice theory. As an example we show the correctness of the diagram in Figure 1. Firstly, we note that the right-upper part – consisting of the square AL, CL, CDCL and CDAL – is well known [9,10,31]. By Theorem 2.6 we have SAL = CDAL, and Theorem 3.3(1) implies that the inclusions A1-L$\subseteq$AL and A2-L$\subseteq$SAL are both proper. From Theorems 5.4.1(1) and 5.4.2, it follows that SAL$\not\subseteq$A1-L, A1-L$\not\subseteq$SAL, and that the inclusion A2-L$\subseteq$A1-L is also proper.

Acknowledgments. I am grateful to Joost Engelfriet and Leo Verbeek for many discussions and critical remarks. I am also indebted to Joel Berman and F. Loonstra for their comment on an earlier version of this paper.

References

1. Asveld, P.R.J.: Controlled iteration grammars and full hyper-AFL's. *Inform. and Contr.* **34**, 248-269 (1977)
2. Asveld, P.R.J.: *Iterated Context-Independent Rewriting – An Algebraic Approach to Families of Languages.* Doctoral Dissertation, Department of Applied Mathematics, Twente University of Technology, Enschede, The Netherlands 1978
3. Asveld, P.R.J.: Complexity aspects of iterated rewriting – a survey. In: *Essays on Concepts, Formalisms, and Tools* (P.R.J. Asveld & A. Nijholt, eds.). CWI Tract, Vol. **42**, pp. 89-105. Amsterdam: Centre for Mathematics and Computer Science 1987
4. Asveld, P.R.J.: Abstract grammars based on transductions. *Theoret. Comput. Sci.* **81** (1991) (To appear).
5. Asveld, P.R.J.: Generalizations of context-free grammars and the non-self-embedding property. (In preparation).
6. Asveld, P.R.J. & Engelfriet, J.: A note on non-generators of full AFL's. *Internat. J. Comput. Math.* **12**, 13-17 (1982)
7. Balachandran, V.K.: A characterization of ΣΔ-rings of subsets. *Fund. Math.* **41**, 38-41 (1954)
8. Balachandran, V.K.: On complete lattices and a problem of Birkhoff and Frink. *Proc. Amer. Math. Soc.* **6**, 548-553 (1955)
9. Balbes, R. & Dwinger, Ph.: *Distributive Lattices.* Columbia. MO: University of Missouri Press 1974
10. Birkhoff, G.: *Lattice Theory.* 3rd ed. Providence. RI: Amer. Math. Soc. 1967
11. Cohn, P.M.: *Universal Algebra.* New York: Harper & Row 1965; 2nd ed. Dordrecht. The Netherlands: Reidel 1981
12. Crawley, P. & Dilworth, R.P.: *Algebraic Theory of Lattices.* Englewood Cliffs. NJ: Prentice-Hall 1973
13. Engelfriet, J.: Iterating iterated substitution. *Theoret. Comput. Sci.* **5**, 85-100 (1977)
14. Engelfriet, J. & Slutzki, G.: Extended macro grammars and stack controlled machines. *J. Comput. Systems Sci.* **29**, 366-408 (1984)
15. Geissinger, L. & Graves, W.: The category of complete algebraic lattices. *J. Combin. Theor. (A)* **13**, 332-338 (1972)
16. Ginsburg, S.: *Algebraic and Automata-Theoretic Properties of Formal Languages.* Amsterdam: North-Holland 1975
17. Ginsburg, S. & Goldstine, J.: Intersection-closed full AFL and the recursively enumerable languages. *Inform. and Contr.* **22**, 210-231 (1973)
18. Ginsburg, S. & Greibach, S.A.: Principal AFL. *J. Comput. Systems Sci.* **4**, 308-338 (1970)
19. Ginsburg, S., Greibach, S.A. & Hopcroft, J.E.: Studies in Abstract Families of Languages. *Mem. Amer. Math. Soc.* **87** (1969)

20. Ginsburg, S. & Spanier, E.H.: Substitution in families of languages. *Inform. Sci.* **2**, 83-110 (1970)
21. Ginsburg, S. & Spanier, E.H.: On incomparable Abstract Families of Languages (AFL). *J. Comput. Systems Sci.* **9**, 88-108 (1974)
22. Grätzer, G.: On the family of certain subalgebras of a universal algebra. *Indag. Math.* **27**, 790-802 (1965); also *Proc. Nederl. Akad. Wetensch. (A)* **68**, 790-802 (1965)
23. Grätzer, G.: *Universal Algebra.* Princeton. NJ. Van Nostrand 1968; 2nd ed. Berlin-Heidelberg-New York: Springer 1979
24. Grätzer, G.: *Lattice Theory – First Concepts and Distributive Lattices.* San Francisco. CA: Freeman 1971
25. Greibach, S.A.: Chains of full AFL's. *Math. Systems Theory* **4**, 231-242 (1970)
26. Greibach, S.A.: Full AFL's and nested iterated substitution. *Inform. and Contr.* **16**, 7-35 (1970)
27. Greibach, S.A.: Syntactic operators on full semi-AFL's. *J. Comput. Systems Sci.* **6**, 30-76 (1972)
28. Halmos, P.R.: *Naive Set Theory.* Berlin-Heidelberg-New York: Springer 1960
29. Jónsson, B.: *Topics in Universal Algebra.* Lecture Notes in Mathematics, Vol. **250**. Berlin-Heidelberg-New York: Springer 1972
30. Mal'cev, A.I.: *Algebraic Systems.* Berlin-Heidelberg-New York: Springer 1973
31. Raney, G.N.: Completely distributive complete lattices. *Proc. Amer. Math. Soc.* **3**, 677-680 (1952)
32. Rozenberg, G. & Salomaa, A.: *The Mathematical Theory of L Systems.* New York: Academic Press 1980
33. Salomaa, A.: *Formal Languages.* New York: Academic Press 1973
34. Schmidt, E.T.: *Kongruenzrelationen algebraischen Strukturen.* Berlin: VEB 1969
35. Van Leeuwen, J.: *F*-iteration languages. Memorandum. Berkeley. CA: University of California 1973
36. Van Leeuwen, J.: A study of complexity in hyper-algebraic families. In: *Automata, Languages, Development* (A. Lindenmayer & G. Rozenberg, eds.). pp. 323-333. Amsterdam: North-Holland 1976

On the Inverse Neighborhoods of Reversible Cellular Automata

Jarkko Kari

Mathematics Department, University of Turku, SF-20500 Turku, Finland

Abstract: This work investigates the possible neighborhood sizes of the inverse automata of reversible cellular automata. It is known that in case of two- and higher dimensiaonal CA the neighborhood of the inverse automaton can be very large — there is no upper bound on its size computable from the number of states. In one-dimensional case there exists, on the other hand, a polynomial upper bound. We improve the known upper bound in some special cases. We show, for instance, that if the original automaton has p states, p prime, and its neighborhood has two elements, then the inverse automaton uses a neighborhood of size at most $p-1$. We also construct examples of cellular automata with large inverse neighborhoods, and use them to prove that our upper bounds are best possible.

Keywords: Cellular automata, reversibility.

1 Introduction

Cellular automata, like L-systems, are parallel processes in which symbols in strings are rewritten on discrete time steps. The following two properties of one-dimensional cellular automata essentially distinguish them from L-systems : A symbol is always replaced by a symbol, never with a longer word or with the empty word λ, and usually no finite strings are considered. The strings, or the configurations of the automata, are supposed to extend to infinity in both directions.

One-dimensional cellular automata (CA for short) are discrete dynamical systems. They consist of an infinite string of cells. The cells contain always an element from a finite state set (or alphabet) of the automaton. The cells change their states synchronously on discrete time steps. To determine their new states cells are able to see the old states of some nearby cells, so called neighbors. The local transition rule of the cellular automaton defines the new state of each cell as a function of the old states of its neighbors. The automaton is homogeneous which means that all cells are using the same local rule.

More generally, for any $d \geq 1$ a d-dimensional CA consists of an infinite collection of cells arranged in the shape of an infinite d-dimensional checker board. The cells change their states in the same manner as in the one-dimensional case above.

Cellular automata are widely used as simple models of dynamical systems encountered in various contexts. In many cases systems we meet in physics and other natural sciences can be described as large collections of simple basic objects that interact with each other only locally. The local interactions can be very simple even though the system as whole behaves in a complicated way. In situations like this it is possible that the system can be modelled with simple cellular automata.

Closed physical systems are reversible. This means that no information is lost during the evolution of the system. Cellular automata can also conserve information. In reversible cellular automata the previous configurations of the system are uniquely determined by the present one. In fact, there always exists another CA, so called inverse automaton, that simulates the inverse process. This means that to trace back one time step of the computation each cell needs to study only the states of the cells in some finite surrounding of the cell. In the present work we study the problem how far the cells might need to look to determine their state one time step earlier.

The article is organized as follows. First, formal definitions of cellular automata and reversibility are presented together with some known basic results. Next reversible one-dimensional CA whose neighborhood contain only two elements are studied. We introduce important notions of left and right predecessor sets, and prove some upper bounds for the size of the neighborhood of the inverse automaton with respect to the number of states. In section 5 the results are extended for 1D automata with neighborhoods containing more than two elements. In the next section we present examples of reversible automata whose inverse automata need a large neighborhood. These examples show that the upper bounds of sections 4 and 5 are optimal or close to optimal. Finally, in section 7 we briefly discuss the higher dimensional cases.

2 Definitions and notations

Formally, a d-dimensional *cellular automaton* (CA) is a quadruple

$$\mathcal{A} = (d, S, N, f),$$

where d is a positive integer indicating the *dimension* of $\mathcal{A}$, S is a finite *state set*, N is a *neighborhood vector*

$$N = (\bar{x}_1, \bar{x}_2, \ldots, \bar{x}_n)$$

of n different elements of $\mathbf{Z}^d$ and f is the *local rule* of the CA presented as a function from S^n into S. The cells are laid on an infinite d-dimensional array and their positions are indexed by $\mathbf{Z}^d$, the set of d-tuples of integers. The *neighbors* of a cell situated in $\bar{x} \in \mathbf{Z}^d$ are the cells in positions

$$\bar{x} + \bar{x}_i \qquad \text{for } i = 1, 2, \ldots, n.$$

The local rule f gives the new state of a cell from the old states of its neighbors.

A *configuration* of a CA $\mathcal{A} = (d, S, N, f)$ is a function

$$c : \mathbf{Z}^d \to S$$

that assigns states to all cells. Let $\mathcal{C}$ denote the set of all configurations. The local rule f determines the *global transition function*

$$G_f : \mathcal{C} \to \mathcal{C}$$

that describes the dynamics of the CA. At each time step a configuration c is transformed into a new configuration $G_f(c)$ where

$$G_f(c)(\bar{x}) = f(c(\bar{x} + \bar{x}_1), c(\bar{x} + \bar{x}_2), \ldots, c(\bar{x} + \bar{x}_n))$$

for all $\bar{x}$ in $\mathbf{Z}^d$. If there is no danger of confusion we write simply G instead of G_f.

Often we are not interested in the states of all cells in the infinite array, but only in some of them. In this case we can use a so called *partial configuration.* A partial configuration c_D on the set $D \subseteq \mathbf{Z}^d$ is a function

$$c_D : D \to S$$

that assigns states only to the cells in D. The states of cells outside D are not specified.

Sometimes a *quiescent state* q in S is distinguished. The quiescent state must have the property

$$f(q, q, \ldots, q) = q.$$

A configuration is called *finite* if it has only finitely many cells in non-quiescent states. Let $\mathcal{C}_F$ denote the set of finite configurations. It follows from the special property of the quiescent state that a finite configuration remains finite in the evolution of the CA. Let G_f^F denote the restriction of the global function G_f to the set of finite configurations.

Note that the order in which the neighbors are listed in the neighborhood vector is not important. We use a neighborhood vector instead of a neighborhood set only to specify in which order the arguments of the local transition rule are given. In many cases, however, it is more convenient to talk about the *neighborhood set* : The neighborhood set of a CA $\mathcal{A}$ is simply the set containing all elements of its neighborhood vector.

In the neighborhood set of a CA $\mathcal{A}$ there can be elements that are not actually needed. The local rule does not depend on these elements. If we remove all such unnecessary elements we get the minimal neighborhood of $\mathcal{A}$.

Definition. Let $\mathcal{A} = (d, S, N, f)$ be a CA. The *minimal neighborhood* of $\mathcal{A}$ (or the minimal neighborhood of G_f) is

$$\{i \in \mathbf{Z}^d \mid \exists c_1, c_2 \in \mathcal{C} : c_1(i) \neq c_2(i) \text{ but } c_1(j) = c_2(j) \text{ for all } j \neq i, \text{ and } G_f(c_1)(\bar{0}) \neq G_f(c_2)(\bar{0}) \}.$$

(Here $\bar{0}$ denotes $(0, 0, \ldots, 0) \in \mathbf{Z}^d$.)

The following lemma is an easy observation.

Lemma 1 *Let G be the global transition function of a CA. There exists a CA with the neighborhood set N whose global function equals G if and only if the minimal neighborhood of G is a subset of N.*

3 Reversible Cellular automata

In reversible systems no information is lost during the process. A cellular automaton $\mathcal{A} = (d, S, N, f)$ is called *injective* if its global function G_f is one-to-one, that is, there do not exist any two different configurations $c_1 \neq c_2$ such that $G_f(c_1) = G_f(c_2)$. The class of injective cellular automata is the class of information conserving cellular automata.

A cellular automaton $\mathcal{A} = (d, S, N, f)$ is called *reversible* if there exists a cellular automaton $\mathcal{B} = (d, S, N', f')$ such that $G_{f'} = (G_f)^{-1}$. This means that the CA $\mathcal{B}$ is simulating the inverse process. In this case we call $\mathcal{B}$ the *inverse automaton* of $\mathcal{A}$ and denote $\mathcal{B} = \mathcal{A}^{-1}$.

Theorem 1 (Richardson [6]) *A CA is reversible if and only if it is injective.*

According to theorem 1 the inverse of the global function of an injective CA is always computed by another CA. It is enough for every cell to look into the current states in some finite neighborhood to uniquely determine its state in the previous configuration. The purpose of this article is to study how large this neighborhood can be.

A property of CA, closely related to injectivity, is surjectivity. A cellular automaton is called *surjective* if its global transition function G_f is surjective, that is, for every configuration c there exists a configuration c' such that $G_f(c') = c$. The following theorem was proved in [4] ("only if") and in [5] ("if").

Theorem 2 (Garden-of-Eden Theorem) *A CA is surjective if and only if G_f^F, the restriction of the global function to finite configurations, is one-to-one.*

Corollary *All injective cellular automata are surjective.*

A naturally arising problem is how one can decide if a given CA is reversible or not. Amoroso and Patt gave an algorithm for deciding this in case of one-dimensional cellular automata.

Theorem 3 (Amoroso, Patt [1]) *It is decidable if a given one-dimensional cellular automaton is reversible.*

In higher dimensional cases the situation is completely different, as proved in [3]. Even if we restrict ourselves to the class of two-dimensional CA that use the *von Neumann* neighborhood

$$N_{vN} = ((0,0),(1,0),(0,1),(-1,0),(0,-1)),$$

(see figure 1) the reversibility problem is undecidable.

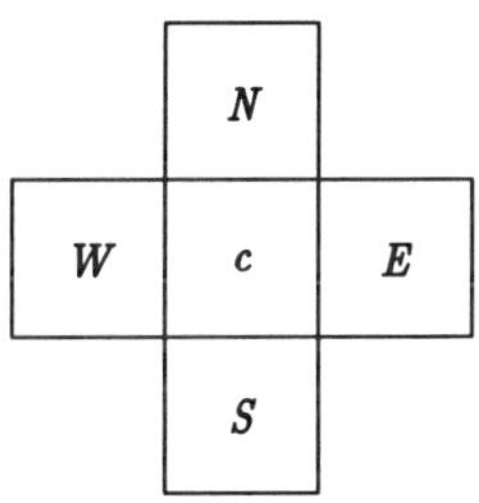

Figure 1: The von Neumann neighborhood of cell c.

Theorem 4 (Kari [3]) *It is undecidable if a given two-dimensional CA using the von Neumann neighborhood N_{vN} is reversible.*

From the proof of theorem 3 in [1] one can compute an upper bound on the size of the neighborhood of the inverse automaton in case of one-dimensional cellular automata. We can assume that the neighborhood set of the reversible one-dimensional CA is *contiguous*, that is, it is of the form $\{k, k+1, \ldots, k+m-1\}$ for some k and m. The size of the neighborhood is m. Certainly every neighborhood set is included in a contiguous one.

Proposition 1 *Let $\mathcal{A} = (1, S, (k, k+1, \ldots, k+m-1), f)$ be a reversible CA, and let $n = |S|$ denote the size of the state set. The minimal neighborhood of the inverse automaton $\mathcal{A}^{-1}$ is included in $\{-\frac{1}{2}n^m(n^{m-1}-1) - k - m + 1, \ldots, \frac{1}{2}n^m(n^{m-1}-1) - k\}$. So $n^m(n^{m-1}-1) + m - 1$ is an upper bound for the size of the neighborhood of $\mathcal{A}^{-1}$.*

In the following two sections the upper bound of the previous proposition is improved in some special cases.

4 The case of size 2 neighborhoods

In this section we will concentrate on those one-dimensional CA whose neighborhood vector contains just two elements. In this case the local rule $f : S^2 \rightarrow S$ can be given as matrix, the row indicating the first argument and the column the second argument (see table 1). What two-element neighborhood we use is not important, because all CA with the same matrix are essentially simulating the same process. In the following we are using the neighborhood vector $N = (0, 1)$. The CA with the same matrix but with the neighborhood vector (a, b) would simulate $|a - b|$ copies of the original CA simultaneously, possibly with reversed configurations (if $b - a < 0$) and shifting the configurations to the right or to the left on each time step.

	$\cdots$	b	$\cdots$
$\vdots$		$\vdots$	
a	$\cdots$	$f(a, b)$	$\cdots$
$\vdots$		$\vdots$	

Table 1: The matrix of the local transition rule f.

From now on let $\mathcal{A} = (1, S, (0, 1), f)$ be a reversible one-dimensional CA with the neighborhood vector $(0, 1)$ and global transition function G. In our considerations the right (and left) predecessor sets of $\mathcal{A}$ play a central role.

Definition. Let $\mathbf{Z}_-$ and $\mathbf{N}$ denote the sets $\{\ldots, -2, -1\}$ and $\{0, 1, \ldots\}$, respectively. For all partial configurations

$$c' : \mathbf{Z}_- \rightarrow S$$
$$c'' : \mathbf{N} \rightarrow S$$

the sets

$$R(c') = \{s \in S \mid \exists c \in C : c(0) = s \text{ and } G(c)_{|\mathbf{Z}_-} = c'\}$$
$$L(c'') = \{s \in S \mid \exists c \in C : c(0) = s \text{ and } G(c)_{|\mathbf{N}} = c''\}$$

denote the right and left predecessor sets of c' and c'', respectively.

By definition, the right predecessor set of c' contains the states of cell 0 in the predecessors of c' (=the predecessors of all extensions of c' to $\mathbf{Z}$). Similarily, the left predecessor set of c'' includes the states appearing in cell 0 in the predecessors of c''. Intuitively, the right (left) predecessor sets tell how much information is moving to the right (to the left) in the inverse automaton of $\mathcal{A}$. The smaller the right predecessor set is, the more information the states on the left side of a cell contain about the state of the cell one time step earlier. Note that the cell 0 is the only cell that directly, on one time step, affects a cell in both $\mathbf{Z}_-$ and $\mathbf{N}$.

Proposition 2 *For all partial configurations*

$$c' : \mathbf{Z}_- \to S$$
$$c'' : \mathbf{N} \to S$$

we have

$$|R(c')| \cdot |L(c'')| = |S| .$$

Proof. We show that for each $s \in S$ there are unique $a \in R(c')$ and $b \in L(c'')$ such that $f(a, b) = s$, which proves the proposition.

Let $s \in S$ be given. Construct the configuration

$$c_s : \mathbf{Z} \to S$$

by combining c', s and c'' as follows :

$$c_s(i) = \begin{cases} c'(i), & \text{if } i < 0, \\ s, & \text{if } i = 0, \\ c''(i-1), & \text{if } i > 0. \end{cases}$$

Note that c'' was shifted to the right to make space for s.

Because the CA $\mathcal{A}$ is reversible there exists a configuration c such that $G(c) = c_s$. According to the definitions of right and left predecessor sets, $c(0) \in R(c')$ and $c(1) \in L(c'')$. Moreover, $f(c(0), c(1)) = c_s(0) = s$, so that $c(0)$ and $c(1)$ can be chosen as a and b.

To show that the choise of a and b is unique, assume that there exist $a_1, a_2 \in R(c')$ and $b_1, b_2 \in L(c'')$ such that

$$f(a_1, b_1) = f(a_2, b_2) = s, \text{ and}$$
$$a_1 \neq a_2 \text{ or } b_1 \neq b_2.$$

According to the definition of predecessor sets there exist configurations c_1', c_1'', c_2' and c_2'' such that

$$\begin{aligned} c_1'(0) &= a_1 \text{ and } G(c_1')_{|\mathbf{Z}_-} &= c', \\ c_2'(0) &= a_2 \text{ and } G(c_2')_{|\mathbf{Z}_-} &= c', \\ c_1''(0) &= b_1 \text{ and } G(c_1'')_{|\mathbf{N}} &= c'', \text{ and} \\ c_2''(0) &= b_2 \text{ and } G(c_2'')_{|\mathbf{N}} &= c''. \end{aligned}$$

Let us combine c_1' and c_1'' into a new configuration c_1, and c_2' and c_2'' into a configuration c_2 as follows :

$$c_1(i) = \begin{cases} c_1'(i), & \text{if } i \leq 0, \\ c_1''(i-1), & \text{if } i \geq 1. \end{cases}$$

$$c_2(i) = \begin{cases} c_2'(i), & \text{if } i \leq 0, \\ c_2''(i-1), & \text{if } i \geq 1. \end{cases}$$

Then $G(c_1) = G(c_2) = c_s$ but $c_1 \neq c_2$. This contradicts the fact that in a reversible CA each configuration has a unique predecessor. □

Corollary $|c_1'| = |c_2'|$ *for all partial configurations* $c_1', c_2' : \mathbf{Z}_- \to S$ *and* $|c_1''| = |c_2''|$ *for all* $c_1'', c_2'' : \mathbf{N} \to S$.

According to the previous corollary all right (left) predecessor sets have the same number of elements. Let us denote these cardinalities n_R (n_L, respectively). The proposition 2 can be formulated in the following form : $n_R \cdot n_L = |S|$. Let us also denote the families of all right (left) predecessor sets of $\mathcal{A}$ by $\mathcal{R}$ ($\mathcal{L}$, respectively), that is,

$$\begin{aligned} \mathcal{R} &= \{R(c') \mid c' : \mathbf{Z}_- \to S\}, \text{ and} \\ \mathcal{L} &= \{L(c'') \mid c'' : \mathbf{N} \to S\}. \end{aligned}$$

Proposition 3 $|L \cap R| = 1$ *for all* $R \in \mathcal{R}$ *and* $L \in \mathcal{L}$.

Proof. Let $c' : \mathbf{Z}_- \to S$ and $c'' : \mathbf{N} \to S$ be partial configurations and denote $R = R(c')$ and $L = L(c'')$. Combine c' and c'' into a total configuration c''' :

$$c'''(i) = \begin{cases} c'(i), & \text{if } i < 0, \\ c''(i), & \text{if } i \geq 0. \end{cases}$$

There is a unique configuration $c \in \mathcal{C}$ such that $G(c) = c'''$. Because $c(0) \in R$ and $c(0) \in L$ we conclude that $|R \cap L| \geq 1$. If $|R \cap L| > 1$ then one can construct two different predecessors of c''', which contradicts the fact that $\mathcal{A}$ is reversible. □

The rest of this section concentrates on the special case when $n_L = 1$ and $n_R = |S|$ (or, symmetrically, $n_R = 1$ and $n_L = |S|$). This is the case with a reversible CA if and only if the columns (rows, symmetrically) of the matrix of its local rule are permutations of S :

Lemma 2 $n_R = |S|$ *iff* $\forall b, s \in S\ \exists a \in S : f(a,b) = s$.

Proof. Let us prove the two implications separately :

- $\Rightarrow$ Let $n_R = |S|$. Assume, contrary to the lemma, that, for some $b, s \in S$, there does not exist $a \in S$ such that $f(a,b) = s$. This means that a partial configuration $c' : \mathbf{Z}_- \to S$, in which $c'(-1) = s$, can not have b in its right predecessor set. This means that $|R(c')| < |S|$, a contradiction.
- $\Leftarrow$ Let $c' : \mathbf{Z}_- \to S$ be a partial configuration. It follows easily from the assumption $\forall b, s \in S\ \exists a \in S : f(a,b) = s$ that there is, for every $b \in S$, a configuration c such that $c(0) = b$ and $G(c)(i) = c'(i)$ for all $i < 0$. □

Note that necessarily $n_L = 1$ or $n_L = |S|$ if the number of states is a prime number.

In the following we show that the minimal neighborhood of the inverse automaton is contained in $\{0, 1, \ldots, n-2\}$ if $n_L = 1$, $n_R = |S|$ (in {-n+1,-n+2,...,-1}, symmetrically, if $n_R = 1, n_L = |S|$).

Proposition 4 *Let* $\mathcal{A} = (1, S, (0,1), f)$ *be a reversible one-dimensional CA. Denote* $n = |S|$. *If* $n_L = 1$ *and* $n_R = n$ *then the minimal neighborhood of the inverse CA* $\mathcal{A}^{-1}$ *is included in* $\{0, 1, \ldots, n-2\}$.

Proof. Note first that the minimal neighborhood of $\mathcal{A}^{-1}$ is certainly included in $\mathbf{N} = \{0, 1, 2, \ldots\}$. Assume, on the contrary, that $i < 0$ belongs to the minimal neighborhood. This means that there are configurations c_1 and c_2 such that $c_1(0) \neq c_2(0)$ and $G(c_1)(i) \neq G(c_2)(i)$ but $G(c_1)(j) = G(c_2)(j)$ for all $j \neq i$. Especially, $G(c_1)_{|\mathbf{N}} = G(c_2)_{|\mathbf{N}}$, so that $c_1(0)$ and $c_2(0)$ belong to the same left predecessor set. This contradicts the fact that $n_L = 1$.

According to lemma 2 for each b and s in S there exists exactly one $a \in S$ satisfying the property $f(a,b) = s$. This allows us to define a function $\varphi_s : S \to S$ for each $s \in S$ as follows :

$$\varphi_s(b) = a, \text{ if } f(a,b) = s.$$

This definition is then extended for arbitrary words $w \in S^*$ as follows :

$$\begin{aligned}\varphi_\lambda(b) &= b, \text{ for all } b \in S,\\ \varphi_{sw}(b) &= \varphi_s(\varphi_w(b)), \text{ for all } w \in S^* \text{ and } b, s \in S.\end{aligned}$$

Let us use the functions φ_w to define a hierarchy of relations $\rho_i \subseteq S \times S$, $i \geq 0$, as follows :

$$(a, b) \in \rho_i \text{ if and only if } \varphi_w(a) = \varphi_w(b) \text{ for all } w \in S^i.$$

We show in the following that

(i) ρ_i are equivalence relations,

(ii) ρ_0 is the identity relation id_S of S,

(iii) $\rho_i \subseteq \rho_{i+1}$ for every $i \geq 0$,

(iv) ρ_k is the full relation $S \times S$ for some $k \geq 0$, and

(v) $\rho_i \neq \rho_{i+1}$ if $\rho_i \neq S \times S$.

(i) and (ii) are obvious — they follow immediately from the definition of the relations ρ_i.

Let us show (iii). If $(a, b) \in \rho_i$ then $\varphi_w(a) = \varphi_w(b)$ for each $w \in S^i$. So, for every $s \in S$ and $w \in S^i$,

$$\varphi_{sw}(a) = \varphi_s(\varphi_w(a)) = \varphi_s(\varphi_w(b)) = \varphi_{sw}(b),$$

which proves that $(a, b) \in \rho_{i+1}$.

In order to prove (iv) the following interpretation of the functions φ_w is used. Let $a_{-k}, a_{-k+1}, \ldots, a_{-1} \in S$ be any states and let $c' : \mathbf{Z}_- \to S$ be a partial configuration such that $c'(j) = a_j$ for all $j \in \{-k, -k+1, \ldots, -1\}$. Because $n_R = n$ there exists, for each $b \in S$, a configuration c such that $c(0) = b$ and $G(c)_{|\mathbf{Z}_-} = c'$. Moreover, necessarily

$$c(-k) = \varphi_w(b),$$

where $w = a_{-k}a_{-k+1} \ldots a_{-1}$. This follows directly from the way the functions φ_w were defined.

To see (iv) take such a k that the minimal neighborhood of $\mathcal{A}^{-1}$ is included in $\{0, 1, \ldots, k-1\}$. For all choices of $a_{-k}, a_{-k+1}, \ldots, a_{-1}$ the state $c(-k)$ in the previous interpretation does not depend on the choice of b, because it is uniquely determined by the inverse automaton from the states $c'(-k), c'(-k+1), \ldots, c'(-1)$. This means that the value of $\varphi_w(b)$ does not depend on b for any $w \in S^k$. So $\rho_k = S \times S$.

Let us show finally (v). Note that

$$(a,b) \in \rho_{i+1} \iff (\varphi_s(a), \varphi_s(b)) \in \rho_i \text{ for all } s \in S.$$

If $\rho_{i+1} = \rho_i$ then

$$\begin{aligned}(a,b) \in \rho_{i+1} &\iff (\varphi_s(a), \varphi_s(b)) \in \rho_i \text{ for all } s \in S \\ &\iff (\varphi_s(a), \varphi_s(b)) \in \rho_{i+1} \text{ for all } s \in S \\ &\iff (a,b) \in \rho_{i+2}.\end{aligned}$$

So $\rho_{i+2} = \rho_{i+1}$. This reasoning can be repeated arbitrarily many times, so that $\rho_k = \rho_i$ for all $k \geq i$. If $\rho_i \neq S \times S$ this contradicts (iv).

Consider the number of the equivalence classes of the relations ρ_i. Let e_i denote this number. From (ii) it follows that $e_0 = n$. From (iii) and (v) we conclude that $e_{i+1} \leq e_i$ for all $i \geq 0$, and $e_{i+1} < e_i$ if $e_i > 1$. This means that $e_{n-1} = 1$, that is, $\rho_{n-1} = S \times S$. This implies further (see the interpretation given to the functions φ_w earlier) that the state $c(-n+1)$ is always uniquely determined by the states $c'(-n+1), c'(-n+2), \ldots, c'(-1)$ of its successor $c' = G(c)$, which proves the proposition. □

Corollary *If the number n of the states of a reversible CA $\mathcal{A} = (1, S, (0,1), f)$ is a prime then the minimal neighborhood of the inverse CA $\mathcal{A}^{-1}$ is included either in $\{0, 1, \ldots, n-2\}$ or in $\{-1, -2, \ldots, -n+1\}$.*

The section 6 contains examples of cellular automata that reach the minimal neighborhood of proposition 4. This shows that the upper bound proved in the proposition is the best possible.

5 Wider neighborhoods

Let us turn our attention to one-dimensional reversible cellular automata whose neighborhood contains more than two elements. Such a neighborhood is always included in a contiguous neighborhood $\{k, k+1, \ldots, k+m-1\}$ for some k and m. We can restrict ourselves to cases where $k = 0$, as the effect of k is only to shift the configurations to the left (if $k < 0$) or to the right (if $k > 0$) at each time step. For any given automaton $\mathcal{A} = (1, S, (0, 1, \ldots, m-1), f)$ one can construct an automaton $\mathcal{A}' = (1, S^{m-1}, (0,1), f')$ that uses a two element neighborhood and simulates the original automaton $\mathcal{A}$. The cells of $\mathcal{A}'$ simply represent the blocks of $m-1$ consecutive cells of $\mathcal{A}$. The local rule f' is defined in the natural way :

$$f'\left((a_1, \ldots, a_{m-1}), (a_m, \ldots, a_{2m-2})\right) = \left(f(a_1, \ldots, a_m), \ldots, f(a_{m-1}, \ldots, a_{2m-2})\right).$$

Using this simulation we can generalize the notions of right and left predecessor sets to the 1D CA with arbitrary contiguous neighborhoods :

Definition. Let $\mathcal{A} = (1, S, (0, 1, \ldots, m-1), f)$ be a cellular automaton and let

$$\begin{aligned} c' &: \mathbf{Z}_- \to S \\ c'' &: \mathbf{N} \to S \end{aligned}$$

be partial configurations. The subsets

$$\begin{aligned} R(c') &= \{(c(0), c(1), \ldots, c(m-2)) \mid c \in \mathcal{C} : G(c)_{|\mathbf{Z}_-} = c'\} \\ L(c'') &= \{(c(0), c(1), \ldots, c(m-2)) \mid c \in \mathcal{C} : G(c)_{|\mathbf{N}} = c''\} \end{aligned}$$

of S^{m-1} denote the right and left predecessor sets of c' and c'', respectively.

The predecessor sets in the definition are simply the predecessor sets in the sence of the previous section of the CA $\mathcal{A}'$ which is simulating $\mathcal{A}$. The propositions proved in the previous section can be applied to $\mathcal{A}'$ to obtain similar results for $\mathcal{A}$.

Proposition 2' *For all partial configurations*

$$\begin{aligned} c' &: \mathbf{Z}_- \to S \\ c'' &: \mathbf{N} \to S \end{aligned}$$

we have

$$|R(c')| \cdot |L(c'')| = |S|^{m-1}.$$

Corollary $|c'_1| = |c'_2|$ *for all partial configurations* $c'_1, c'_2 : \mathbf{Z}_- \to S$ *and* $|c''_1| = |c''_2|$ *for all* $c''_1, c''_2 : \mathbf{N} \to S$.

Let us denote, as before, the (unique) cardinalities of the right (left) predecessor sets n_R (n_L, respectively). The proposition 2' can be formulated in the following form : $n_R \cdot n_L = |S|^{m-1}$. Let $\mathcal{R}$ ($\mathcal{L}$) denote the families of all right (left) predecessor sets of $\mathcal{A}$.

Proposition 3' $|L \cap R| = 1$ *for all* $R \in \mathcal{R}$ *and* $L \in \mathcal{L}$.

Finally, the theorem concerning the size of the inverse neighborhood of $\mathcal{A}$ if $n_L = 1$ (or $n_R = 1$) can be stated as follows:

Proposition 4' *Let* $\mathcal{A} = (1, S, (0, 1, \ldots, m), f)$ *be a reversible one-dimensional CA. Denote* $n = |S|$. *If* $n_L = 1$ *and* $n_R = n^{m-1}$ *then the minimal neighborhood of the inverse CA* $\mathcal{A}^{-1}$ *is included in* $\{0, 1, \ldots, n^{m-1} - 2\}$. *Symmetrically, if* $n_R = 1$ *and* $n_L = n^{m-1}$ *then the minimal neighborhood of* $\mathcal{A}^{-1}$ *is included in* $\{-n^{m-1} + 1, -n^{m-1} + 2, \ldots, -1\}$.

6 Examples of reversible CA with large inverse neighborhoods

In this section we present examples demonstrating that

(i) the upper bound in proposition 4 is tight, and

(ii) the upper bound in proposition 4' is almost tight.

First we consentrate on CA with two element neighborhoods:

(i) Let $S = \{1, 2, \dots, n\}$ be a state set containing n elements. We will construct a reversible one-dimensional CA $\mathcal{A}_n$ (shortly $\mathcal{A}$ when there is no risk of confusion about n) with the neighborhood vector $(0, 1)$ and the state set S such that

- $n_R = n$ and $n_L = 1$, and
- the minimal neighborhood of the inverse CA is $\{0, 1, \dots, n-2\}$.

The local transition rule f of $\mathcal{A}$ is defined as follows :

$$f(a,b) = \begin{cases} a, & \text{if } b \le a, \\ 1, & \text{if } b = a+1, \\ a+1, & \text{if } b > a+1. \end{cases}$$

The local rule is shown also in the table 2.

Let us first show that $\mathcal{A}$ is injective (and thus reversible). Assume, on the contrary, that there exist two configurations c_1 and c_2 such that $G_f(c_1) = G_f(c_2)$ but $c_1 \ne c_2$. Let $i \in \mathbf{Z}$ be such that

$$c_1(i) \ne c_2(i), \text{ and} \tag{1}$$

$$c_1(j) \ne c_2(j) \implies \max\{c_1(j), c_2(j)\} \le \max\{c_1(i), c_2(i)\}, \tag{2}$$

that is, in the position i the configurations c_1 and c_2 differ and, moreover, there cannot be a position j where the configurations differ and where the state of either c_1 or c_2 is bigger that both $c_1(i)$ and $c_2(i)$ (the natural ordering $1 < 2 < \dots < n$ of the state set is used).

Let us denote $a = c_1(i)$, $a' = c_2(i)$, $b = c_1(i+1)$ and $b' = c_2(i+1)$. We can assume without loss of generality that $a > a'$. We have three possibbilities :

- $b > a+1$. In this case $b' = b$ (because of (2)), so that $b' > a'+1$. This means that

  $$a + 1 = G_f(c_1)(i) = G_f(c_2)(i) = a' + 1,$$

  which contradicts $a \ne a'$.

	1	2	3	4	...	$n-1$	n
1	1	1	2	2	...	2	2
2	2	2	1	3	...	3	3
3	3	3	3	1	...	4	4
$\vdots$	$\vdots$	$\vdots$	$\vdots$	$\vdots$		$\vdots$	$\vdots$
$n-1$	$n-1$	$n-1$	$n-1$	$n-1$	...	$n-1$	1
n	n	n	n	n	...	n	n

Table 2: The local transition rule of $\mathcal{A}_n$

- $b = a + 1$. Again $b' = b = a + 1 > a' + 1$, so that

$$1 = G_f(c_1)(i) = G_f(c_2)(i) = a' + 1,$$

  a contradiction.

- $b \leq a$. In this case $G_f(c_2)(i) = G_f(c_1)(i) = a > a'$, so that we must have $b' > a'+1$ and $a = a' + 1$. This means that $b' > a$ which contradicts (2).

In all three cases we ran into a contradiction, which means that there cannot exist configurations c_1 and c_2, and the cellular automata $\mathcal{A}_n$ are injective.

It is easy to see that for every $b, s \in S$ there exists $a \in S$ such that $f(a, b) = s$. This implies that $n_R = n$ (see lemma 2). According to proposition 2 $n_L = 1$.

Let us next investigate the neighborhood vector of the inverse automaton. To prove that $i \in \mathbf{Z}$ is in the minimal neighborhood it is enough to construct two configurations c_1 and c_2 with the properties

a) $c_1(-i) \neq c_2(-i)$, and

b) $G_f(c_1)(0) \neq G_f(c_2)(0)$, but $G_f(c_1)(j) = G_f(c_2)(j)$ for all $j \neq 0$.

Let c_1 and c_2 be defined as follows:

$$c_1(j) = 1, \text{ for all } j \in \mathbf{Z}, \text{ and}$$

$$c_2(j) = \begin{cases} 1, & \text{if } j \leq -n, \\ n+j, & \text{if } -n < j \leq 0, \\ 1, & \text{if } j > 0. \end{cases}$$

Then

$$G_f(c_1)(j) = 1, \text{ for all } j \in \mathbf{Z}, \text{ and}$$

$$G_f(c_2)(j) = \begin{cases} 1, & \text{if } j \neq 0, \\ n, & \text{if } j = 0. \end{cases}$$

We conclude that all elements in the set

$$\{-j \mid c_1(j) \neq c_2(j)\} = \{0, 1, \ldots, n-2\}$$

must belong to the minimal neighborhood of $\mathcal{A}^{-1}$. According to proposition 4 this neighborhood cannot contain any other elements.

(ii) Let us then turn our attention to one-dimensional CA having a contiguous neighborhood containing more than two elements. Let $m > 2$ be fixed and let us consider the neighborhood $N_m = (0, 1, \ldots, m-1)$ of size m. In the following we present for each $n \geq 1$ an example of a CA $\mathcal{B}_{m,n}$ (or $\mathcal{B}$ for short), which has $2n$ states, uses the neighborhood N_m and satisfies the conditions

- $n_R = (2n)^{m-1}$ and $n_L = 1$, and
- the minimal neighborhood of the inverse CA $\mathcal{B}^{-1}$ is $\{0, 1, \ldots, n^{m-1} - 1\}$. This does not reach the upper bound of proposition 4', because the automaton has $2n$ states instead of n states.

In the construction we need the following two lemmas.

Lemma 3 *For all alphabets $\Sigma = \{a_1, a_2, \ldots, a_n\}$ and natural numbers $k \geq 2$ there exists a word $w \in \Sigma^*$ that contains each word $\alpha \in \Sigma^k$ of length k exactly once as a subword.*

Proof. Construct the digraph

$$D = \left(\Sigma^{k-1}, A\right)$$

whose vertices are the words of length $k-1$ over Σ, and whose arc-family A contains an arc

$$u_1 \longrightarrow u_2$$

if and only if

$$\begin{aligned} u_1 &= a_i x, \\ u_2 &= x a_j, \end{aligned}$$

for some $a_i, a_j \in \Sigma$ and $x \in \Sigma^{k-2}$. Let us label such an arc $u_1 \longrightarrow u_2$ with $a_i x a_j$. It is easy to see that in the digraph D there is exactly one arc labelled with α for each $\alpha \in \Sigma^k$.

The digraph D is Eulerian (that is, there exists a path that uses each arc exactly once), because it is connected and the numbers of incoming and outgoing arcs at each vertice are equal ($= n$). Let

$$\alpha_1, \alpha_2, \ldots, \alpha_{n^k}$$

be the labels of the consecutive arcs on an Eulerian path. Because of the way the arcs are defined there is a word $w \in \Sigma^{n^k+k-1}$ such that

$$\forall i, 1 \leq i \leq n^m, \exists x_i, y_i \in \Sigma^*, |x_i| = i-1 : w = x_i \alpha_i y_i,$$

that is, the subword of w starting at position i and having length k is α_i. Because the words α_i are different from each other, w satisfies the requirements of the lemma. □

Lemma 4 *Let $w \in \Sigma^*$ be a word satisfying the conditions of lemma 3. Every word w' which is not a subword of w contains a subword $u \in \Sigma^{k+1}$ of length $k+1$ that is not a subword of w. Especially, all words longer than w contain such a subword u.*

Proof. Necessarily $|w'| > k$ because w contains all words of length k as its subword. Let us prove the lemma using induction on the length of w'.

1° $|w'| = k+1$. We can choose $u = w'$.

2° $|w'| = k' > k+1$, and the lemma has been proved for words shorter than w'. Let $a_i, a_j \in \Sigma$ and $x \in \Sigma^{k'-2}$ be such that

$$w' = a_i x a_j.$$

If one of the words $a_i x$ and $x a_j$ is not a subword of w then, according to the inductive hypothesis, a subword u can be found inside this shorter word. Assume then that both $a_i x$ and $x a_j$ are subwords of w. Because $|x| \geq k$ the word x can appear only in one position in w, which means that $a_i x a_j$ is also a subword of w, a contradiction. □

We are now ready to give the definition of the CA $\mathcal{B}_{m,n}$ ($= \mathcal{B}$, for short). Let $\Sigma = \{a_1, a_2, \ldots, a_n\}$ and let us use the lemma 3 with $k = m-1$ to obtain a word w that contains each word of Σ^{m-1} exactly once as a subword. The state set of the CA $\mathcal{B}$ is

$$S = \{0,1\} \times \Sigma.$$

Let $\pi_1 : S \to \{0,1\}$ and $\pi_2 : S \to \Sigma$ be the two projections of the state set, and let $\oplus : \{0,1\}^2 \to \{0,1\}$ denote the exclusive-or (=XOR, modulo 2 addition) operator. The local transition rule g is defined as follows (The symbol $\preceq$ denotes the relation "is a subword of".) :

$$g\left((b_1, a_{i_1}), (b_2, a_{i_2}), \ldots, (b_m, a_{i_m})\right) = \begin{cases} (b_1, a_{i_1}), & \text{if } a_{i_1} a_{i_2} \ldots a_{i_m} \npreceq w, \\ (b_1 \oplus b_2, a_{i_1}), & \text{if } a_{i_1} a_{i_2} \ldots a_{i_m} \preceq w, \end{cases}$$

that is, the state of a cell is not changed if the Σ-components of the states in the neighborhood form a word that is not a subword of w, otherwise the bit-component is changed by performing the XOR-operation with the bit-component of the right neighbor.

Let us show that the CA $\mathcal{B}$ is injective. Assume, on the contrary, that there are configurations c_1 and c_2 such that $c_1(0) \neq c_2(0)$, but $G_g(c_1) = G_g(c_2)$. Since the Σ-components of the states are not changed by the local rule g, $\pi_2(c_1(i)) = \pi_2(c_2(i))$ for all $i \in \mathbf{Z}$. Because $\pi_1(c_1(0)) \neq \pi_1(c_2(0))$ we must have

$$\pi_2(c_1(0))\ \pi_2(c_1(1))\ \ldots\ \pi_2(c_1(m-1))\ \preceq\ w, \text{ and}$$
$$\pi_1(c_1(1)) \neq \pi_1(c_2(1)).$$

The reasoning can be repeated for the states $c_1(1)$ and $c_2(1)$ and, by continuing further, for all states $c_1(i)$ and $c_2(i)$, $i \geq 0$. This means that

$$\pi_2(c_1(i))\ \pi_2(c_1(i+1))\ \ldots\ \pi_2(c_1(i+m-1))\ \preceq\ w,$$

for all $i \geq 0$. This contradicts lemma 4.

We have shown that the CA $\mathcal{B}$ is injective. It is easy to see that $n_R = (2n)^{m-1}$ and $n_L = 1$. This follows immediately from the fact that for all states $(b_2, a_{i_1}), \ldots, (b_m, a_{i_m})$ and (b'_1, a_{i_1}) there exists $b_1 \in \{0,1\}$ such that

$$g\left((b_1, a_{i_2}), (b_2, a_{i_2}), \ldots, (b_m, a_{i_m})\right) = (b'_1, a_{i_1}).$$

Let us finally investigate the minimal neighborhood of $\mathcal{B}^{-1}$. In order to show that $i \in \mathbf{Z}$ is in this neighborhood it is enough to construct two configurations c_1 and c_2 satisfying the previously given properties a) and b). Let c_1 be defined as follows (note that the length of w is $n^{m-1} + m - 2$) :

$$\pi_2(c_1(-n^{m-1}+1))\ \pi_2(c_1(-n^{m-1}+2))\ \ldots\ \pi_2(c_1(m-2))\ =\ w,$$
$$\pi_2(c_1(i)) = a_1, \text{ for } i \leq -n^{m-1} \text{ and } i \geq m-1,$$
$$\pi_1(c_1(i)) = 0, \text{ for all } i \in \mathbf{Z},$$

and let c_2 be defined in the following way :

$$\pi_2(c_2(i)) = \pi_2(c_1(i)), \text{ for all } i \in \mathbf{Z},$$

$$\pi_1(c_2(i)) = \begin{cases} 1, \text{ if } -n^{m-1}+1 \leq i \leq 0, \text{ and} \\ 0, \text{ if } i \leq -n^{m-1} \text{ or } i \geq 1. \end{cases}$$

i		$\ldots$	$-n^{m-1}$	$-n^{m-1}+1\ldots 0$	$1\ldots m-2$	$m-1$	$\ldots$
$c_1(i)$	π_1	$\ldots$ 0	0	$00\ldots 0$	$00\ldots 0$	0	0 $\ldots$
	π_2	$\ldots$ a_1	a_1	w		a_1	a_1 $\ldots$
$c_2(i)$	π_1	$\ldots$ 0	0	$11\ldots 1$	$00\ldots 0$	0	0 $\ldots$
	π_2	$\ldots$ a_1	a_1	w		a_1	a_1 $\ldots$

Table 3: The configurations c_1 and c_2

The configurations c_1 and c_2 are summarized in table 3. The Σ-components of c_1 and c_2 are the same and they form the word w surrounded by a_1's. The bit-components are different inside w (except in the last $m-2$ symbols of w) and the same otherwise.

It is easy to see that

$$\begin{aligned} \pi_1(G_g(c_1)(i)) &= 0, \text{ for all } i \in \mathbf{Z}, \text{ and,} \\ \pi_1(G_g(c_2)(i)) &= \begin{cases} 0, \text{ if } i \neq 0, \\ 1, \text{ if } i = 0. \end{cases} \end{aligned}$$

(Of course $\pi_2(G_g(c_1)(i)) = \pi_2(G_g(c_2)(i))$, for all $i \in \mathbf{Z}$.)

This means that all numbers in $\{0, 1, \ldots, n^{m-1}-1\}$ are included in the minimal neighborhood of $\mathcal{B}^{-1}$. We skip the proof of the fact that no other elements are included in the neighborhood. (Intuitively the reason is that there cannot be more than n^{m-1} consecutive cells such that the Σ-components of the states of the following m cells form subwords of w.)

7 Higher dimensional cases

The theory of two- and higher dimensional reversible cellular automata is totally different from the one-dimensional case. According to theorem 4 it is undecidable if a given two- or higher dimensional CA is reversible. This is true even if only 2D automata with von Neumann neighborhood N_{vN} are used. It follows from this fact that the neighborhood of the inverse automaton can be extremely large.

Proposition 5 *There does not exist a computable function* $f : \mathbf{N} \to \mathbf{N}$ *such that the minimal neighborhood of the inverse CA* $\mathcal{A}^{-1}$ *of every reversible CA* $\mathcal{A} = (2, S, N_{vN}, g)$ *is contained in*

$$M_{f(n)} = \{(a,b) \mid -f(n) \leq a \leq f(n) \text{ and } -f(n) \leq b \leq f(n)\},$$

where $n = |S|$.

Proof. Assume, on the contrary, that such a function f exists. Let an arbitrary CA $\mathcal{A} = (2, S, N_{vN}, g)$ be given. There exist only finitely many 2D cellular automata with the state set S whose neighborhood is contained in $M_{f(|S|)}$, and all such automata can be effectively constructed. Let $\mathcal{H}$ be the set of these CA. Necessarily $\mathcal{A}^{-1} \in \mathcal{H}$ if $\mathcal{A}$ is reversible. One can effectively list all elements $\mathcal{B}$ of $\mathcal{H}$ and check one after the other if $G_{\mathcal{A}} \circ G_{\mathcal{B}} = id$. Note that this can be effectively checked. $\mathcal{A}$ is reversible if and only if such a CA $\mathcal{B}$ is found. This algorithm violates the theorem 4, so that our initial assumption about the existence of f is false. □

The situation is different if the size of the neighborhood of a two- or higher dimensional cellular automaton is two. In this case the automaton is actually simulating infinitely many copies of a one-dimensional CA, and the results of section 4 can be applied.

References

[1] Amoroso, S. and Patt, Y.: Decision Procedures for Surjectivity and Injectivity of Parallel Maps for Tessellation Structures. J. Comput. System Sci. 6, 448-464 (1972)

[2] Culik II, K.: On Invertible Cellular Automata. Complex Systems 1, 1035-1044 (1987)

[3] Kari, J.: Decision Problems concerning Cellular Automata. University of Turku, Ph.D. Thesis (1990)

[4] Moore, E.F.: Machine Models of Self-reproduction. Proc. Sympos. Appl. Math. 14, 17-33 (1962)

[5] Myhill, J.: The Converse to Moore's Garden-of-Eden Theorem. Proc. Amer. Math. Soc. 14, 685-686 (1963)

[6] Richardson, D.: Tessellations with Local Transformations. J. Comput. System Sci. 6, 373-388 (1972)

Templates, Fragments and Skins

Brian Mayoh

Computer Science Department, Aarhus University, DK-8000 Aarhus C, Denmark

Abstract: Scientists build elegant theories in an attempt to explain phenomena in the world of nature. In this chapter changes in the salinity of the Baltic Sea are formulated in five popular formalisms: strings, graphs, cellular automata, pictures and processes. Not only do we give a precise definition of the usual notions: stages, productions, templates and steps, but two novelties are introduced: fragments and skins. Not only are the novelties securely founded on indexed categories, they are also useful in formalising realistic models.

Keywords: Knowledge representation, modelling, grammars, modularity.

#0 Introduction

Scientists build elegant and intricate theories in an attempt to explain phenomena in the world of nature. This is only possible if they ignore much detail when formulating the formalism and concepts of their theories. In this chapter we will discuss five popular formalisms: strings, graphs, cellular automata, pictures and processes. All these formalisms capture the concurrent activity, so characteristic of natural phenomena. For each of the formalisms we will give a precise definition of their

- *stages*, formal descriptions of static structures in nature;
- *productions*, formal descriptions of possible structure changing actions;
- *templates*, formal description of allowed combinations of productions;
- *steps*, precise definition of the new stage given by applying a template to a stage;
- *fragments*, formal description of how stages can be subdivided;
- *skins*, formal description of "boundaries" of stages and fragments.

We will also discuss various translations and interrelations between the formalisms. The last section of the paper describes the connection to category theory (stages are objects, fragmented stages are products, interrelations are functors).

Our running example shows that the "fragment" and "skin" concepts are as natural as they are novel. There is considerable evidence (e.g. raised shorelines) for the theory that the Baltic sea has twice been a freshwater lake in the last 10000 years. The accepted theory is:

- 8000BC Lake because global glaciers have lowered the sea level 100 meters
- 7700BC Sea because global melting has raised the sea level and the sea breaks through Denmark
- 6500BC Lake because, relieved of its load of glaciers, Scandinavia rises and shuts out the North Sea
- 5000BC Sea because further global melting raises the sealevel so that the North Sea once more breaks through.

Although lack of space forces our five formalisations of this theory to be crude oversimplifications, each of our formalisations can mimic these "catastrophic" changes in the salinity of the Baltic Sea when they are allowed to use fragmentation.

#1 String world: $N \Rightarrow L$

The string world is the home of many formalisms, both from the formal language industry and the Lindenmayer system tradition. These formalisms agree that *stages* should be words on an alphabet L (formally $W : n \to L$ where n is an initial segment of the natural numbers). A *production* is a rule for rewriting a letter in a word. A *grammar* consists of

- a set of productions
- context and synchronicity requirements for the application of productions.

Suppose W is a stage. A *template* from W is an assignment of productions to the letters of W, that pays due respect to the context and synchronicity requirements. The grammar may be such that there are zero, one or many templates for a given W. A *step* is the application of a template for a stage W and its result is the stage W' given by substituting the right sides of the productions concurrently.

Example: To model the Baltic Sea in the string world, we choose

- a line of latitude
- an alphabet BS={d,s,f,l,h} whose letters abbreviate "deep sea", "saltwater", "freshwater", "lowland", and "highland" respectively
- a way of displaying stages
- to use the L-system conventions (simultaneous rewriting of all letters)
- to represent the 8000BC stage by the word

  d d d d d d d d d s s s l l l f f f f f f f f f f f f f l l l h h h h h h h h h

The top display in figure 1a shows the visual representation of the 8000BC stage. The geological processes are represented by two templates: the Searising template with productions

deepening	$s \to d$	in context dd
erosion	$l \to s$	in context ss
collapse	$h \to l$	in context sl
saltening	$f \to s$	in context s

and the Landrising template with productions

shallowing	$d \to s$	in context s
dilation	$s \to l$	in context ds
eruption	$l \to h$	in contexts l..l,h..l,l..h
freshening	$s \to f$	in contexts f..f,f..s,hl

The second (third, fourth) display of figure 1a shows the result of repeated use of the Searising (Landrising, Searising) template. We see that this model does not capture the second breakthrough of the sea (and our model presentations here do not capture the drama of their NeXT-animated versions).

Usually string formalisers need *context-sensitive* productions, so they have an undefined context problem at the beginning and end of words. One solution to this problem is to introduce skins and fragments. Stages may be fragmented. Suppose the alphabet L has a *separator* #, which cannot be rewritten by a production. Suppose # is the first and last letter of every stage (remember the end-markers in formal language theory). Now # separates every stage into one or more *fragments*. As there is no reason why fragments should be rewritten by the same template, a step can have different templates for each fragment. From any fragmented string we can get an unfragmented string by *hiding*; one just removes all separators.

Example: A simple way of fragmenting the string model of the Baltic Sea is to

- put a separator in the alphabet BS={d,s,f,l,h}
- represent the separator in displays by a kink
- keep the unfragmented productions and templates
- to represent the 8000BC stage by the word

d d d d d d # d d s s s l l l f f f f f f f f f f f f f l l l h h h # h h h h h h

Figure 1b shows that the fragmented string model *does* capture the second breakthrough of the sea.

If one thinks of separators as 'markers for undefined context', it is natural to define

- the *skin* of a stage is its first and its last letter;
- the *skin* of a fragment is its first and its last letter.

In our example the 8000BC stage skin is (d,h) and the fragment strings are (d,d), (d,h) and (h,h).

#2 Graph world: $V \Rightarrow L$

The graph world is the home of those who chafe at the structures of the one dimensional string world. In the graph world all agree that *stages* should be labelled graphs, but few agree on the definition of graph grammars and productions. However most will accept that a *production* is a rule for rewriting subgraphs and a *grammar* consists of

- a set of productions
- context, glueing and synchronicity requirements for the application of productions.

Suppose G is a stage. A *template* for G is an assignment of productions to subgraphs of G, that pays due respect to context, *glueing* and synchronicity requirements. Often graph formalisers need 'context-sensitive' productions, so they have an undefined context problem around the 'edges' of graphs. A *step* is the application of a template for a stage G and its result is the stage G' given by glueing the right sides of the productions

Example: To model the Baltic Sea in the graph world, we choose

- an alphabet BS={d,s,f,l,h} as above
- a way of displaying stages
- to use the L-system conventions (simultaneous rewriting of all letters)
- to represent the 8000BC stage by the graph

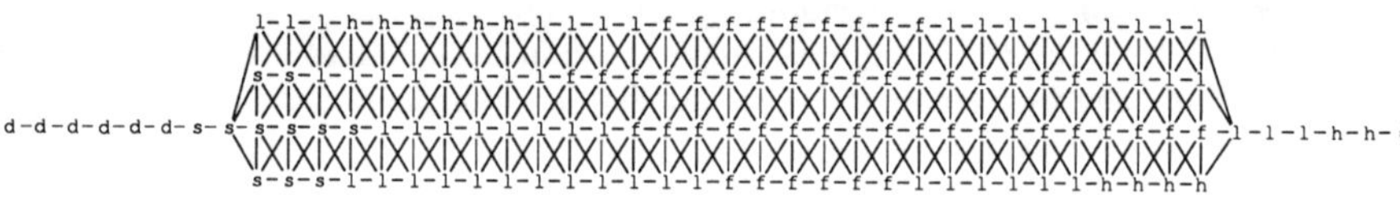

The top display in figure 2a shows the visual representation of the 8000BC stage. Again the geological processes are represented by two templates: the Searising template with productions

deepening	$s \rightarrow d$	if s-vertex has d-neighbour
erosion	$l \rightarrow s$	if l-vertex has l-neighbour but no h-neighbour
collapse	$h \rightarrow l$	if h-vertex has no h-neighbour
saltening	$f \rightarrow s$	if f-vertex has s-neighbour

and the Landrising template with productions

shallowing	$d \rightarrow s$	if d-vertex has an s-neighbour
dilation	$s \rightarrow l$	if s-vertex has an l-neighbour
eruption	$l \rightarrow h$	if l-vertex has an h-neighbour
freshening	$s \rightarrow f$	if s-vertex has an f-neighbour but no l-neighbour

To make model comparison easier, we have chosen graph productions similar to the string productions in the last section- none of them create or destroy vertices, they just change the labels on vertices. Figure 2a shows the historical development of the Baltic Sea in this model. We see that it does not capture the second breakthrough of the sea.

Graph stages may be fragmented. Suppose the label set L has a separator #, which cannot be rewritten by a production. The separator partitions every stage G into *fragments* $G_1, G_2 \ldots G_m$ such that:

v is in the same fragment as w $\equiv$ there is a #-free path from v to w.

There is no objection to different templates for different fragments of a graph stage. From any fragmented stage we can get an unfragmented stage by *hiding*; one just removes all separator vertices and their incident edges.

Example: A simple way of fragmenting the graph model of the Baltic Sea is to

- put a separator in the alphabet BS={d,s,f,l,h}
- represent the separator in displays by a kink
- keep the unfragmented productions and templates
- to represent the 8000BC stage by the graph

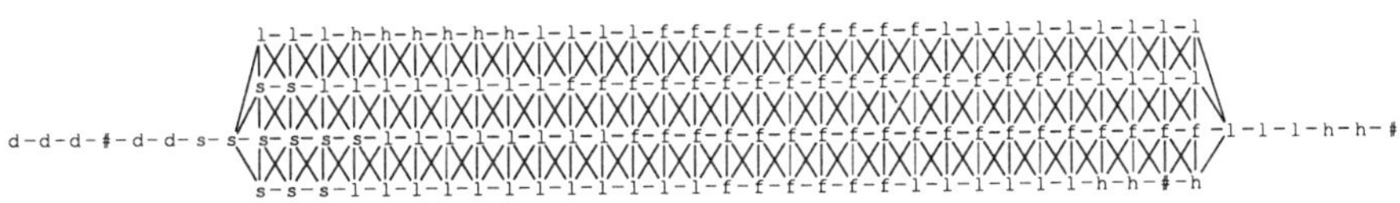

Figure 2b shows that the fragmented graph model *does* capture the second breakthrough of the sea.

If one thinks of separators as 'markers for undefined context', it is natural to define

- the *skin* of a graph stage G is the set of labelled vertices attached to #-vertices that do NOT separate fragments
- the *skin* of a fragment F is the set of its labelled vertices attached to #-vertices that separate F from the other fragments.

The 8000BC stage skin consists of an f- and two h-vertices. The skin of the northsea fragment is a d-vertex; the skin of the other fragment consists of a d- and an h-vertex.

#3 Cellular World: $C \Rightarrow L$

The cellular world is the home of those who like grids – usually integral points of Euclidean space. In the cellular world all agree that *stages* should be labellings of grids C. Moreover all agree that productions and grammars use *transforms*. Informally a transform is a map from a cell to its neighbours; formally it is a function $tr : C \rightarrow C^*$. Usually cells have a fixed numbers of neighbours but there are problems with boundary cells. Because of these problems we define a *production* as a label-combinator and a *template* as an assignment of a production and a transform to each grid point.

A *step* is the application of a template to a stage S and its result is the stage S' given by relabelling the nodes as proscribed by the template

$$S'(y) = \Gamma\{'S(tr(y))\}$$

where Γ is the label combinator for the template and $'S$ is the extension of S to lists.

Example: On moving from the string world to the cellular world, our example becomes two dimensional. The top display in figure 3a shows the visual representation of the 8000BC stage. Again the geological processes are represented by two templates: the Searising template with productions

deepening	$s \to d$	if s-cell has at least 6 d -neighbours
erosion	$l \to s$	if l-cell has at least 3 s-neighbours
collapse	$h \to l$	if h-cell has at least 4 l-neighbours
saltening	$f \to s$	if f-cell has an s-neighbour

and the Landrising template with productions

shallowing	$d \to s$	if d-cell has at least 4 s-neighbours
dilation	$s \to l$	if s-cell has at least 4 l-neighbours
eruption	$l \to h$	if l-cell has at least 3 h-neighbours
freshening	$s \to f$	if s-cell has an f-neighbour.

Rather complicated rules determine choices between dilation and freshening productions.

Both templates use the transform tr8 that takes each point into its 8 nearest neighbours. Figure 3a shows the historical development of the Baltic Sea in this model. We see that it does capture the second breakthrough of the sea.

Cellular stages may be fragmented. Suppose the label set L has a separator #, which cannot be rewritten by a template. The separator partitions every stage S into fragments $S_1, S_2 \cdots S_m$ such that:

v is in the same fragment as w $\equiv$ there is a #-free path from v to w.

Cellular fragments are somewhat unusual in that subsequent development cannot change their size (= number of internal nodes). Even so there is no objection to different templates for different fragments of a cellular stage. The *skin* of a cellular stage or fragment is the boundary of its "support" (non-quiescent nodes). Cellular skins are somewhat unusual in that they may not be connected. From any fragmented cell stage we can get an unfragmented cell stage by *hiding*; one just replaces all separators by quiescent cells.

Example: The top display in figure 3b shows the fragmented cellular 8000BC stage. In all our figures for fragmented stages separators are represented by a kink, so a line of separators is represented by a gash. Thus the skin of the 8000BC stage consists of the edge cells; the skin of the northsea (fareast, middle) fragment is a line of d-cells (a line of h-cells, a line of d-cells and a line of h-cells) and some edge cells. Figure 3b shows that the fragmented graph model *does* capture the second breakthrough of the sea.

#4 Picture world: $E \Rightarrow L$

The picture world is the home of those who are impressed both by the 3-dimensional world of nature and the crucial importance of natural phenomena on 2-dimensional interfaces. In the picture world most agree that *stages* should be labellings of Euclidean subspaces, $P : E \Rightarrow L$. In these days of satellite and tomography images one must grudgingly accept the pressures for standardisation. In the image algebra standard [RWD] functions $t : Y \Rightarrow L^X$ are called 'templates'. For us such functions are transforms, *productions* are label-combinators, and *templates* are assignments of a production and a transform to each picture point. A *step* is the application of the template to a picture stage P and its result is the stage P' by convoluting the template with P –

$$P'(y) = \Gamma\{P(x) \circ t_y(x)\}$$

where Γ is the label combinator for the template and $\circ$ is a binary operation on labels. It is easy to show that

- steps in the cellular world are special cases of steps in the picture world
- picture transforms are more powerful than cellular transforms.

Example: The top display in figure 4a shows the visual representation of the 8000BC stage. Again the geological processes are represented by two templates: the Searising template with productions

deepening	$s \rightarrow d$	if s-pixel has at least 13 d-neighbours
erosion	$l \rightarrow s$	if l-pixel has at least 7 s-neighbours
collapse	$h \rightarrow l$	if h-pixel has at least 13 l-neighbours
saltening	$f \rightarrow s$	if f-pixel has an immediate s-neighbour

and the Landrising template with productions

shallowing	$d \rightarrow s$	if d-pixel has at least 8 s-neighbours
dilation	$s \rightarrow l$	if s-pixel has at least 6 l-neighbours
eruption	$l \rightarrow h$	if l-pixel has at least 8 h-neighbours
freshening	$s \rightarrow f$	if s-pixel has an immediate f-neighbour

Rather complicated rules determine choices between dilation and freshening productions. Each of the Searising (Landrising) productions is a variant of the usual erosion (dilation) operators for black/white pictures [Se]. One of the productions is

eruption $l \rightarrow h$ if the l-pixel has at least 7 h-neighbours

Both templates use the transform t8 that gives the 8 nearest neighbours of a pixel one L-colour, and the 16 next-nearest neighbours another L-colour. Figure 4a shows the historical development of the Baltic Sea in this model. We see that it does not capture the second breakthrough of the sea. Picture stages may be fragmented. A fragmented picture stage is a set of unfragmented stages $\{P_i, E_i, L_i\}$ and a set of separator subspaces $\{S_j\}$. As our definition of a production $t : Y \Rightarrow L^X$ allows space changes and separators can control space overlapping there is no objection to fragments overlapping within separators.

The *skin* of a picture stage (fragment) is the restriction of the E_i to the union of the separators (separators meeting the fragment). From any fragmented picture we can get an unfragmented picture by *hiding*; one just takes the union of the subspaces E_i as E and for skin points defines $P(x)$ as

$$\Delta\{P_i(x)\} \parallel x \in E_i\}$$

where Δ is a label combinator.

Example: The top display in figure 4b shows the fragmented picture 8000BC stage. The fragments of this 8000BC stage are given by the fragments of the 8000BC cellular stage. The two separators are given by the gashes. The 8000BC picture stage and the 8000BC cellular stage have the same stage skins, fragments, and fragment skins. Figure 4b shows that the fragmented picture model *does* capture the second breakthrough of the sea.

#5 Process world: $V \Rightarrow \mathcal{P}$

The process world is the home of the evergrowing throng of Heracliteans – "All is flux". In the process world all agree that *stages* are "nets of parallel distributed processes" – more precisely, graphs whose nodes are labelled by processes. There is some confusion about what process productions are (see [Ma] for some attempts), but most would not protest against our choice of CCS agents [Mi] as processes. In a recent simulation of an ant colony [To] individual ants were modelled as CCS agents. Process *stages* are graphs whose arcs and nodes are labelled by 'port names' and CCS agents respectively. There is a *production* for each 'capability' of the CCS agents. A *template* is a collection of productions, and a *step* is the application of a template.

Example: To model the Baltic Sea in the process world, we choose to represent the 8000BC stage by the graph

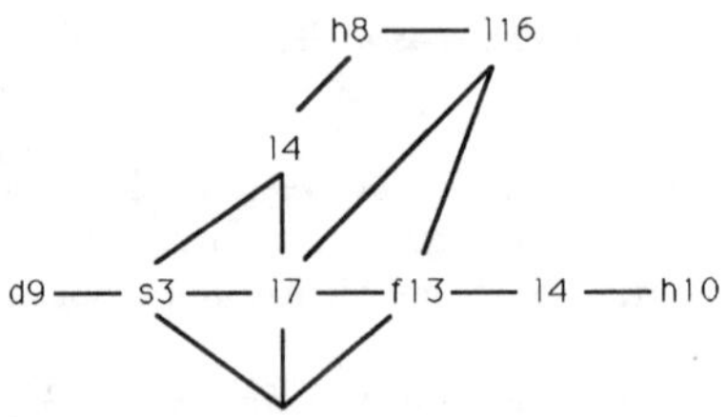

This graph is much simpler than the graph in section 2 because the vertex processes can be rather subtle. The top display in figure 5a shows the visual representation of the 8000BC stage. The definition of these nondeterministic processes incorporates deepening, erosion, collapse, saltening, shallowing, dilation, eruption and freshening productions. The Searising and Landrising templates are heuristics determining the choices of productions. The

Searising template has productions that are slight modifications of

deepening	s	decreases and d increases when connected
erosion	l	decreases and s increases when connected
collapse	h	decreases and l increases when connected
saltening	f	changes to s when connected to s

The Landrising template has productions that are slight modifications of

shallowing	d	decreases and s increases when connected
dilation	s	decreases and l increases when connected
eruption	l	decreases and h increases when connected
freshening	s	changes to f when connected to f

Figure 5a shows the historical development of the Baltic Sea in this model. We see that it does capture the second breakthrough of the sea.

Process stages may be fragmented. As stages are graphs, one can take their connected components as *fragments* if one allows 'observers' as separators. An *observer* is just a passive set of nodes.

- The *skin* of a process stage G is the set of labelled vertices attached to observers that do NOT separate fragments
- the *skin* of a fragment F is the set of its labelled vertices attached to observers that separate F from the other fragments.

From any fragmented stage we can get an unfragmented stage by *hiding* ; one just removes all observer processes and connects any process nodes that were joined by observers.

Example: A simple way of fragmenting the process model of the Baltic Sea is to split the leftmost and rightmost processes.

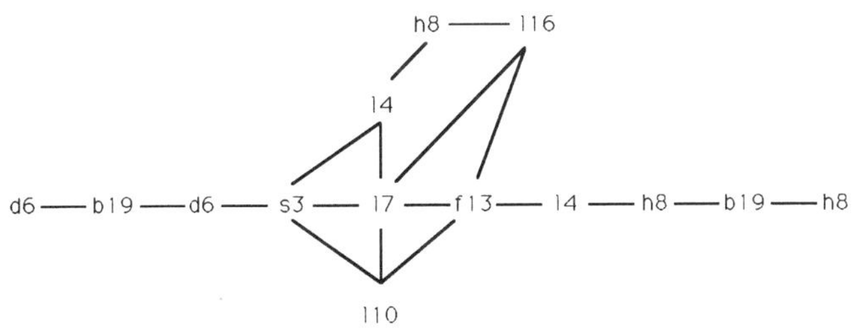

This splitting gives northsea and fareast observers. The skins of the 8000BC fragments are$\{d6\}, \{d6, h8\}$, and $\{h8\}$. The skin of the 8000BC stage is empty because both observers separate fragments. Figure 5b shows that the fragmented graph model *does* capture the second breakthrough of the sea.

#6 Fragmentation $\Leftrightarrow$ Shape

In all our worlds we have the possibility of fragmentation. The increase of expressive power given by fragmentation is both a help and a hindrance to formalisers; fragmentation should only be used when there is some evidence that it corresponds to reality, that its consequences can be observed.

One consequence of fragmentation is that stages have extra skins, one for each fragment. In the cellular world there are many algorithms that interactively shrink skins down to *skeletons* (collections of lines) and some of them mark each node of the stage or fragment with the point on the skeleton it shrinks to. If the stage skeleton is close to the sum of the fragment skeletons, then it is likely that the fragmentation corresponds to reality in some way. If the skeletonisation algorithm marks each stage or fragment node, then the partition of skeletons into lines gives an embedding $\mathcal{E} : C \to V$ which maps the cellular stage into a graph stage (edges between vertices for adjacent skeleton lines).

In the picture world there are also skeletonisation algorithms. One of the most popular is the grassfire algorithm in which an (imaginary) fire is started simultaneously at all *skin* points, and the skeleton consists of the points where firefronts meet [Bl]. If the stage grassfire skeleton is close to the sum of the grassfire skeletons of its fragments, then it is likely that the fragmentation is not unreal. Figure 6a shows the skeletons for the pictures in figure 4a, figure 6b shows the skeletons for an unrealistic fragmentation of the pictures; if we had taken the actual fragmentation used in figure 4b, then figure 6b would have been much closer to figure 6a.

One can modify the grassfire algorithm so that it marks each pixel with a particular point on the grassfire skeleton – the point reached by the fire front from the pixel. The partition of skeletons into lines gives an embedding $\mathcal{E} : E \to V$ which maps the picture stage into a graph stage.

In the elegant series of papers [Le] Leyton has described how the shape of an object or organism can suggest its formation processes, its development. Instead of grassfire skeletons he uses "symmetry axes" defined by moving a circle along the outline of a shape so that it is always tangential to the curve at *two* points (the axis is the trace of the midpoint of the are between these tangent points). Two of the Leyton rules for recovering process history from smooth natural shapes are [HL]:

> *Symmetry – Curvature Duality*: Any section of a smooth curve with one and only one curvature extremum has just one symmetry axis, which terminates at the extremum itself.
>
> *Interaction Principle*: the symmetry axes of a perceptual organisation are interpreted as the directions along which processes are most likely to act or have acted.

The definition of symmetry axes gives a new kind of skeleton to stages and fragments in the picture world. As with the grassfire algorithm this gives a validity check for fragmentation, but the Leyton approach claims and gives more. One also has the heuristic: Size–is–Time: "In the absence of information to the contrary, assume size corresponds to time – the larger a boundary feature is, the longer it has taken to develop". Both this heuristic and the interaction principle claim that skeletonisation can give information about the *templates* driving (*steps* in) development, not just the fragmentation in a particular *stage*.

#7 World Neutrality

One should not be too fixed in one of our five worlds. From the viewpoint of category theory

- stages & skins form *objects*
- productions & templates & steps are just ways of proscribing *morphisms* (cf. [Me] where the viewpoint is described for a formalism we have not discussed: the term-rewriting world)
- fragmentation is given by *products*
- maps between worlds are given by *functors*.

Perhaps we should elaborate on the somewhat surprising "fragmentation is given by *products*". Fragments and skins were defined in terms of separators, and separators are no more than representation devices to hide the indexing in an indexed category. In the string world an index is just a number, a three fragment string is just an ordered list of three strings, and hiding is concatenation. Thus an object is no more than a stage and its separators. Fragmented stages are "collections of fragments" and these are product objects because we have incorporated separators into the definition of objects. When we say "product objects", we mean that the individual fragments are independent of one another. In other words: stages with more than one skin are products because the extra 'fragment skins' give a unique decomposition of the stage into fragments.

We should certainly elaborate on "maps between worlds are functors". We will describe *tessellation* functors, *skeletonization* functors, and *skinning* functors. Real pictures – think of the satellite pictures of mountain regions, microscope pictures of plants, medical pictures and the like – are often discrete, because they consist of pixels. The pattern recognition and scene analysis community have devised sophisticated methods for identifying significant regions in pictures and labelling them appropriately. In essence these methods map the picture world into the graph world, individual pictures $\mathcal{P} : E \Rightarrow L$ are mapped into particular graphs $\mathcal{G} : V \Rightarrow L$ by some embedding function $\mathcal{E} : E \to V$. Vertex v gets region $\mathcal{E}^{-1}[v]$. Thus we have a tessellation functor from the picture world to the graph world.

Maps from the graph world to the picture world are also common. One can convert a graph $\mathcal{G} : V \Rightarrow L$ into a picture $\mathcal{P} : E \Rightarrow L$, if one has Euclidean coordinates for each vertex in V; define $\mathcal{E}(x)$ as the nearest vertex to x, then define $\mathcal{P}(x)$ be $\mathcal{G}(\mathcal{E}(x))$. That was the Dirichlet tesselation functor between the graph world and the picture world but there are many others. We used ad hoc tessellation functors to get the pictures in figure 3 (5) from the underlying graph (process) stages. As stages in the cellular world are "Grid graph" stages, any tessellation functor from graphs to pictures also gives a tessellation functor from the cellular world to the picture world. In the other direction representing pictures by pixels gives a tessellation functor to the cellular world.

One can easily make skeletonization functors from the picture-to-graph and cell-to-graph maps in the last section. In each of our five worlds we have defined the skins of stages and fragments. In essence skinning is an endofunctor, a map from a world to itself. Often one can modify this endofunctor to get a useful skinning functor into another world. Thus skins in the graph world are subsets and if we order separators and vertices, we get a skinning functor from the graph world to the string world.

Note that the graph world is more observable than the process world, but the only natural map between the two worlds (the forgetful functor) goes from the process to the graph world. This is frustratingly typical. Maps usually go from the less observable world to the more observable world, but we are particularly interested in the other direction – finding laws in an unobservable world that explain behaviour in an observable world. There may be zero, one or many explanations for a behaviour – and none of them may be correct in the sense that they explain the actual behaviour if one makes more and more detailed observations.

References

[Bl] H. Blum *Biological shape and visual science (part 1).* J. Theor. Biol. 38 (1973) 205-287.

[HL] P. Hayes, M. Leyton *Processes at discontinuities.* Proc. IJCAI89, pp. 1267-1272, M. Kaufmann, 1989.

[Le] M. Leyton *A process grammar for shape.* Artificial Intelligence 34 (1988) 213-247.

[Ma] B. Mayoh *A uniform model for the growth of biological organisms: coöperating sequential processes.* In eds. G. Rozenberg, A. Salomaa *The book of L.* Springer, 1985, pp. 291-302.

[Me] J. Meseguer *Rewriting as a unified model of concurrency.* To appear in Theoretical Computer Science or *A logical theory of concurrent objects.* OOPSLA ECOOP90 proceedings, Sigplan Notices 25 (1990) no. 10, pp. 101-115.

[Mi] R. Milner *Communication and concurrency.* Prentice Hall, 1989.

[Na] P. Naur *Programming as theory building.* Microprocessing and microprogramming 15 (1985) 253-261.

[RWD] G. X. Ritter, J. N. Wilson, J. L. Davidson *Image algebra: An overview.* Computer Vision, Graphics and Image Processing 49 (1990) 297-331.

[Se] J. Serra *Image analysis and mathematical morphology.* Academic Press, 1982.

[To] C. Tofts *The autosynchronisation of Leptothorax Aceruotum (Fabricius) described in WCCS.* Report ECS-LFCS-90-128, Edinburgh Univ.

Postscript Since the discovery of the "greenhouse" effect, there have been dire predictions that global warming will raise the sea level 50cm in the next 25 years. Recently a rival theory has become popular – the sealevel will fall because the global warming will change the climate so that more water will be locked into glaciers, not less. This shows the effect of ignoring climate in a geological theory; the ignoring of tilting, folding and continental drift in our formalisations of the Baltic Sea theory is just as bad.

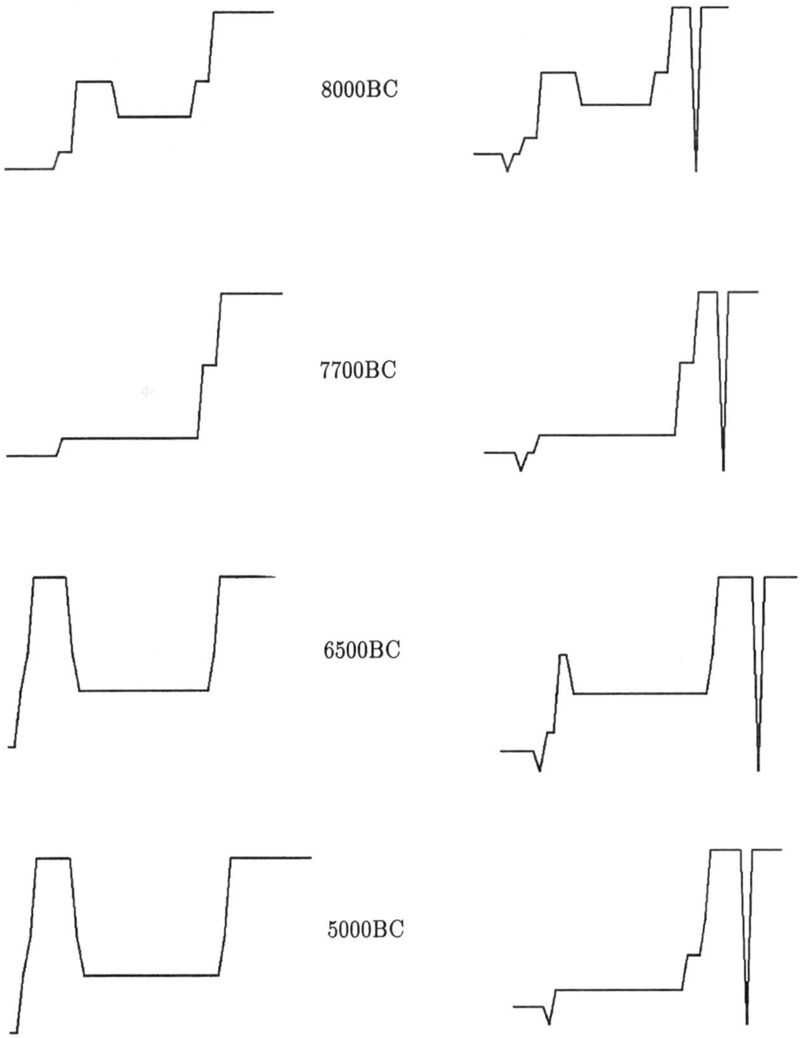

Fig. 1 (a) Unfragmented Strings (b) Fragmented Strings

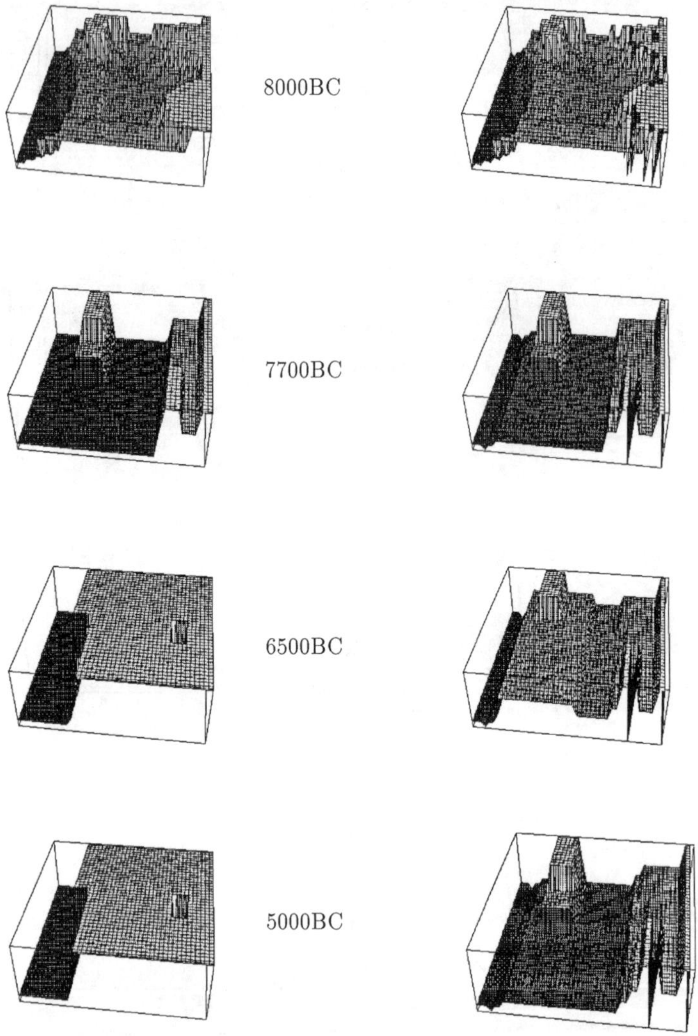

Fig. 2 (a) Unfragmented graphs (b) Fragmented graphs

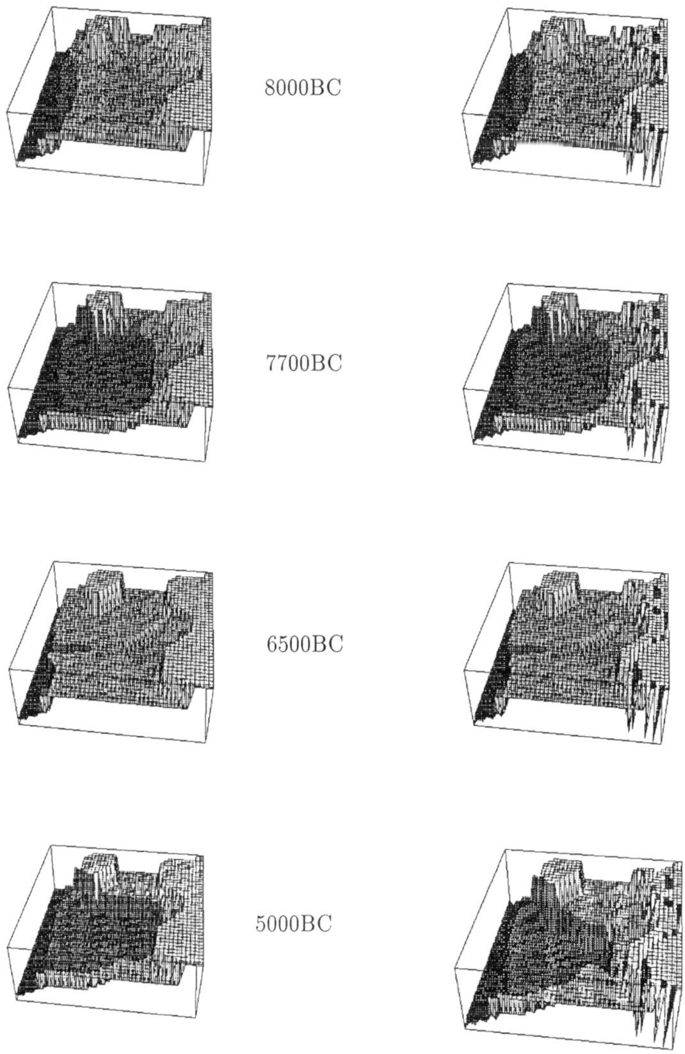

Fig. 3 (a) Unfragmented cells (b) Fragmented cells

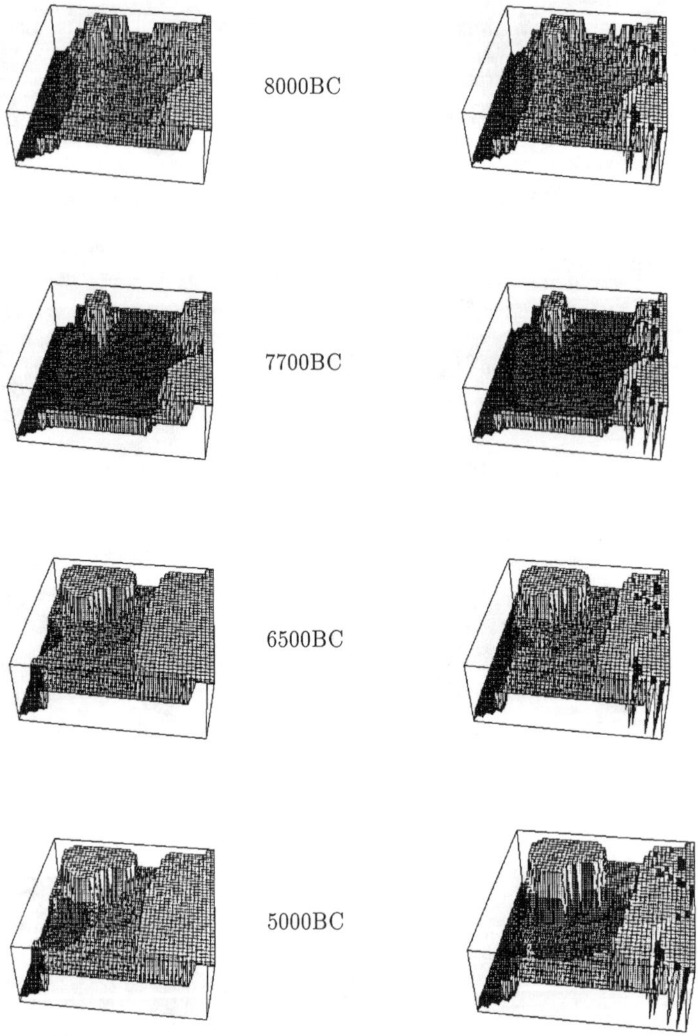

Fig. 4 (a) Unfragmented pictures (b) Fragmented pictures

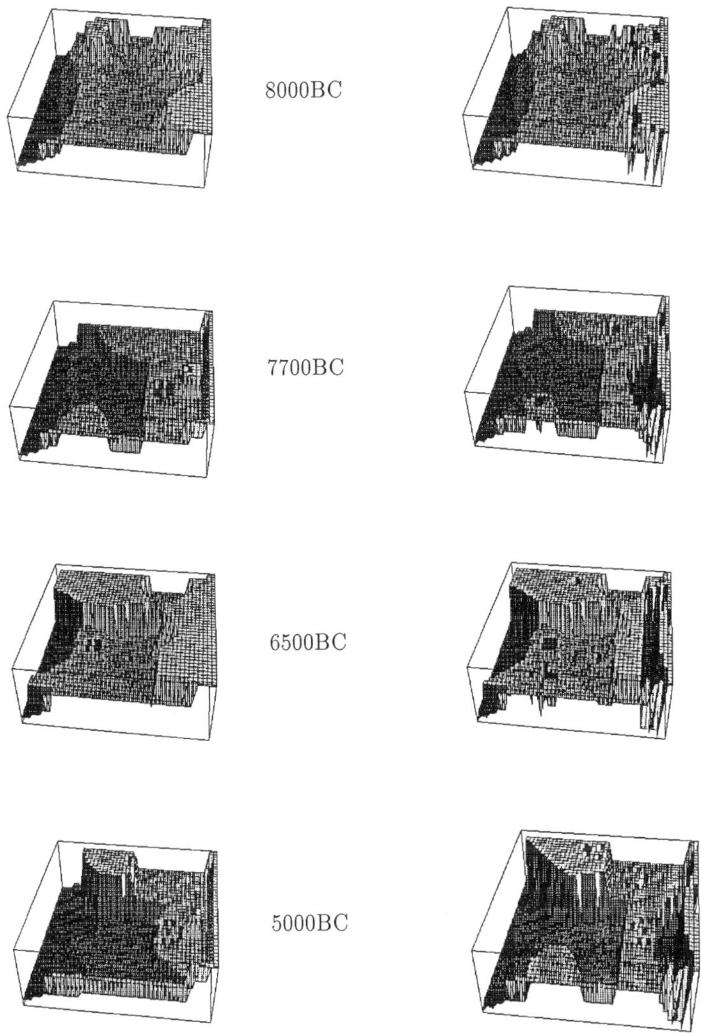

Fig. 5 (a) Unfragmented processes (b) Fragmented processes

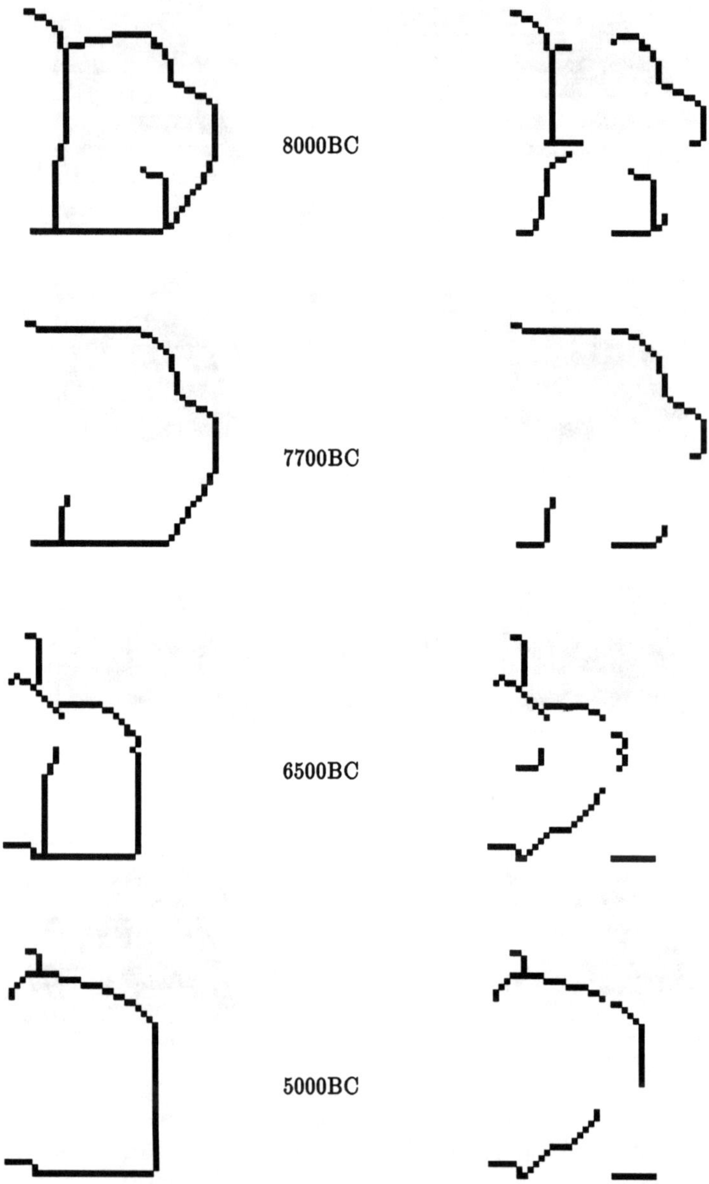

Fig. 6 (a) Skeleton of an unfragmented picture (b) Skeleton of an unwisely fragmented picture